RADIATIVE TRANSFER

BY

S. CHANDRASEKHAR

Morton D. Hull Distinguished Service Professor

University of Chicago

DOVER PUBLICATIONS, INC.
NEW YORK

Published in Canada by General Publishing Com-
pany, Ltd., 30 Lesmill Road, Don Mills, Toronto,
Ontario.
Published in the United Kingdom by Constable
and Company, Ltd., 10 Orange Street, London
WC 2.

This Dover edition, first published in 1960, is an
unabridged and slightly revised version of the work
originally published in 1950. It is published through
special arrangements with the Oxford University
Press.

International Standard Book Number: 0-486-60590-6

Library of Congress Catalog Card Number: 58-162

Manufactured in the United States of America
Dover Publications, Inc.
180 Varick Street
New York, N. Y. 10014

PREFACE

THE problem of specifying the radiation field in an atmosphere which scatters light in accordance with well-defined physical laws originated in Lord Rayleigh's investigations in 1871 on the illumination and polarization of the sunlit sky. But the fundamental equations governing Rayleigh's particular problem had to wait seventy-five years for their formulation and solution. However, the subject was given a fresh start under more tractable conditions, when Arthur Schuster formulated in 1905 a problem in Radiative Transfer in an attempt to explain the appearance of absorption and emission lines in stellar spectra, and Karl Schwarzschild introduced in 1906 the concept of radiative equilibrium in stellar atmospheres. Since that time the subject of Radiative Transfer has been investigated principally by astrophysicists, though in recent years the subject has attracted the attention of physicists also, since essentially the same problems arise in the theory of the diffusion of neutrons.

In this book I have attempted to present the subject of Radiative Transfer in plane-parallel atmospheres as a branch of mathematical physics with its own characteristic methods and techniques. On the physical side the novelty of the methods used consists in the employment of certain general principles of invariance (Chapters IV and VII) which on the mathematical side leads to the systematic use of nonlinear integral equations and the development of the theory of a special class of such equations (Chapters V and VIII). On these accounts the subject would seem to have an interest which is beyond that of the specialist alone: at any rate, that has been my justification for writing this book. However, my own partiality has led me to include two chapters (Chapters XI and XII) which are probably of interest only to the astrophysicist.

I am grateful to Dr. G. Münch, who drew all the diagrams which illustrate this volume, to Miss Donna D. Elbert, who assisted with the proofs and the preparation of the index, and most of all to the Clarendon Press for bringing to this book that excellence of craftsmanship and typography which is characteristic of all their work.

S. C.

YERKES OBSERVATORY
1 *August* 1949

CONTENTS

I

THE EQUATION OF TRANSFER

1. Introduction

IN this chapter we shall define the fundamental quantities which the subject of Radiative Transfer deals with and derive the basic equation —the equation of transfer—which governs the radiation field in a medium which absorbs, emits, and scatters radiation. In formulating the various concepts and equations we shall not aim at the maximum generality possible but limit ourselves, rather, by the situations which the problems considered in this book actually require.

The chapter also includes a classification and discussion of the various types of problems which will be treated in this book.

2. Definitions

2.1. *The specific intensity*

The analysis of a radiation field often requires us to consider the amount of radiant energy, dE_ν, in a specified frequency interval $(\nu, \nu+d\nu)$ which is transported across an element of area $d\sigma$ and in directions confined to an element of solid angle $d\omega$, during a time dt (see Fig. 1). This energy, dE_ν, is expressed in terms of the *specific intensity* (or, more simply, *the intensity*), I_ν, by

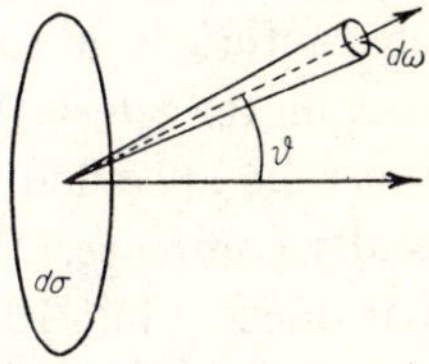

$$dE_\nu = I_\nu \cos\vartheta \, d\nu d\sigma d\omega dt, \qquad (1)$$

FIG. 1.

where ϑ is the angle which the direction considered makes with the outward normal to $d\sigma$. The construction we have used here defines also a *pencil of radiation*.

It follows from the definition of intensity that in a medium which absorbs, emits, and scatters radiation, I_ν may be expected to vary from point to point and also with direction through every point. Thus, for a general radiation field, we may write

$$I_\nu \equiv I_\nu(x, y, z; l, m, n; t), \qquad (2)$$

where (x, y, z) and the direction cosines (l, m, n) define the point and the direction to which I_ν refers.

A radiation field is said to be *isotropic* at a point, if the intensity is independent of direction at that point. And if the intensity is the same at all points and in all directions, the radiation field is said to be *homogeneous* and isotropic.

The case of greatest interest in astrophysical (and terrestrial) contexts is that of an atmosphere stratified in parallel planes in which all the physical properties are invariant over a plane. In this case we can write

$$I_\nu \equiv I_\nu(z, \vartheta, \varphi; t), \tag{3}$$

where z denotes the height measured normal to the plane of stratification and ϑ and φ are the polar and azimuthal angles, respectively. If I_ν should be further independent of φ we have a field which has *axial symmetry* about the z-axis.

Another case of interest which also arises in practice is that of *spherical symmetry* when

$$I_\nu \equiv I_\nu(r, \vartheta; t), \tag{4}$$

where r is the distance from the centre of symmetry and ϑ is the inclination of the direction considered to the radius vector.

The intensity I_ν integrated over all the frequencies is denoted by I and is called the *integrated intensity*; thus

$$I = \int_0^\infty I_\nu \, d\nu. \tag{5}$$

While for most purposes the intensity $I_\nu(x, y, z; l, m, n)$ sufficiently characterizes a radiation field, it is important to note that further parameters describing the state of polarization of the radiation field must be specified before we can regard the description of the field as really complete. We shall consider the characterization of these further parameters in § 15.

2.2. *The net flux*

Equation (1) gives the energy in the frequency interval $(\nu, \nu + d\nu)$ which flows across an element area of $d\sigma$ in a direction which is inclined at an angle ϑ to its outward normal and confined to an element of solid angle $d\omega$. The net flow in all directions is therefore given by

$$d\nu d\sigma dt \int I_\nu \cos\vartheta \, d\omega, \tag{6}$$

where the integration is to be effected over all solid angles. The quantity

$$\pi F_\nu = \int I_\nu \cos\vartheta \, d\omega \tag{7}$$

which occurs in the expression (6) is called the *net flux* and defines the rate of flow of radiant energy across $d\sigma$ per unit area and per unit frequency interval.

For a system of polar coordinates with the z-axis in the direction of the outward normal to $d\sigma$

$$d\omega = \sin\vartheta\, d\vartheta d\varphi, \qquad (8)$$

and the expression for the net flux can be written in the form

$$\pi F_\nu = \int_0^\pi \int_0^{2\pi} I_\nu(\vartheta, \varphi)\sin\vartheta \cos\vartheta\, d\vartheta d\varphi. \qquad (9)$$

As F_ν has been defined, it depends on the direction of the outward normal to the elementary surface across which the flow of radiant energy has been considered. However, this dependence of the flux on direction is simple and is of the nature of a vector. For, considering the flux across a surface the direction cosines of whose normal are l, m, and n, we have

$$\pi F_{\nu;l,m,n} = \int I_\nu(l', m', n')\cos\Theta\, d\omega, \qquad (10)$$

where Θ is the angle between the directions (l', m', n') and (l, m, n). Hence

$$\pi F_{\nu;l,m,n} = \int I_\nu(l', m', n')(ll' + mm' + nn')\, d\omega, \qquad (11)$$

or

$$F_{\nu;l,m,n} = lF_{\nu;x} + mF_{\nu;y} + nF_{\nu;z}, \qquad (12)$$

where $F_{\nu;x}$, $F_{\nu;y}$, and $F_{\nu;z}$ define the fluxes across surfaces normal to the x, y, and z directions, respectively.

For a radiation field which has an axial symmetry the expression for F_ν along the axis of symmetry is

$$F_\nu = 2\int_0^\pi I_\nu(\vartheta)\sin\vartheta \cos\vartheta\, d\vartheta. \qquad (13)$$

2.3. *The density of radiation*

The energy density $u_\nu\, d\nu$ of the radiation in the frequency interval $(\nu, \nu+d\nu)$ at any given point is the amount of radiant energy per unit volume, in the stated frequency interval, which is in course of transit in the immediate neighbourhood of the point considered.

To find the expression for the energy density at a point P we construct around P an infinitesimal volume v with a convex bounding surface σ. We next surround v by another convex surface Σ such that the linear dimensions of Σ are large compared with those of σ; nevertheless, we arrange that the volume element enclosed by Σ is still so small that we can regard the intensity in any given direction as the same for all points inside Σ.

Now all the radiation traversing the volume v must have crossed some element of the surface Σ. Let $d\Sigma$ be such an element; further let

Θ and ϑ denote the angles which the normals to $d\Sigma$ and an element $d\sigma$ of σ make with the line joining the two elements. The energy flowing across $d\Sigma$ which also flows across $d\sigma$ is

$$I_\nu \cos\Theta \, d\Sigma d\omega' d\nu = I_\nu \, d\nu \frac{\cos\vartheta \cos\Theta \, d\sigma d\Sigma}{r^2} \tag{14}$$

since the solid angle $d\omega'$ subtended by $d\sigma$ at $d\Sigma$ is

$$d\sigma \cos\vartheta / r^2,$$

where r is the distance between $d\sigma$ and $d\Sigma$. If l is the length traversed by the pencil of radiation considered through the volume element v, the radiation (14) incident on $d\sigma$ per unit time will have traversed the element in a time l/c, where c denotes the velocity of light. The contribution to the total amount of radiant energy in course of transit through v by the pencil of radiation considered is

$$I_\nu \, d\nu \frac{\cos\vartheta \cos\Theta \, d\sigma d\Sigma}{r^2} \frac{l}{c} = \frac{1}{c} I_\nu \, dv d\nu d\omega, \tag{15}$$

where

$$d\omega = d\Sigma \cos\Theta / r^2$$

is the solid angle subtended by $d\Sigma$ at P and

$$dv = l \, d\sigma \cos\vartheta$$

is the volume intercepted in v by the pencil of radiation. Therefore the total energy in the frequency interval $(\nu, \nu+d\nu)$ in course of transit through v, due to the radiation coming from all directions, is obtained by integrating (15) over all v and ω: thus,

$$\frac{d\nu}{c} \int d v \int d\omega \, I_\nu = \frac{v}{c} \, d\nu \int I_\nu \, d\omega. \tag{16}$$

Hence

$$u_\nu = \frac{1}{c} \int I_\nu \, d\omega. \tag{17}$$

The *integrated energy density*, u, is similarly given in terms of the integrated intensity I; thus,

$$u = \int_0^\infty u_\nu \, d\nu = \frac{1}{c} \int I \, d\omega. \tag{18}$$

It is often convenient to introduce the *average intensity*

$$J_\nu = \frac{1}{4\pi} \int I_\nu \, d\omega, \tag{19}$$

which is related to the energy density by

$$u_\nu = \frac{4\pi}{c} J_\nu. \tag{20}$$

For an axially symmetric radiation field (cf. eq. [13])

$$J_\nu = \tfrac{1}{2} \int_0^\pi I_\nu \sin\vartheta \, d\vartheta. \tag{21}$$

3. Absorption coefficient. True absorption and scattering. Phase function

A pencil of radiation traversing a medium will be weakened by its interaction with matter. If the specific intensity I_ν therefore becomes $I_\nu + dI_\nu$ after traversing a thickness ds in the direction of its propagation, we write

$$dI_\nu = -\kappa_\nu \rho I_\nu \, ds, \tag{22}$$

where ρ is the density of the material. The quantity κ_ν introduced in this manner defines the *mass absorption coefficient* for radiation of frequency ν. Now it should not be assumed that this reduction in intensity, which a pencil of radiation in passing through matter experiences, is necessarily lost to the radiation field. For it can very well happen that the energy lost from the incident pencil may all reappear in other directions as *scattered radiation.* In general we may, however, expect that only a part of the energy lost from an incident pencil will reappear as scattered radiation in other directions and that the remaining part will have been 'truly' absorbed in the sense that it represents the transformation of radiation into other forms of energy (or even of radiation of other frequencies). We shall therefore have to distinguish between *true absorption* and *scattering.*

Considering first the case of scattering, we say that a material is characterized by a *mass scattering coefficient* κ_ν if from a pencil of radiation incident on an element of mass of cross-section $d\sigma$ and height ds, energy is scattered from it at the rate

$$\kappa_\nu \rho \, ds \times I_\nu \cos\vartheta \, d\nu d\sigma d\omega \tag{23}$$

in all directions. Since the mass of the element is

$$dm = \rho \cos\vartheta \, d\sigma ds, \tag{24}$$

we can also write $\kappa_\nu I_\nu \, dm d\nu d\omega. \tag{25}$

It is now evident that to formulate quantitatively the concept of scattering we must specify in addition the angular distribution of the scattered radiation (25). We shall therefore introduce a *phase function* $p(\cos\Theta)$ such that

$$\kappa_\nu I_\nu p(\cos\Theta) \frac{d\omega'}{4\pi} \, dm d\nu d\omega \tag{26}$$

gives the rate at which energy is being scattered into an element of solid angle $d\omega'$ and in a direction inclined at an angle Θ to the direction of incidence of a pencil of radiation on an element of mass dm. According to (26) the rate of loss of energy from the incident pencil due to scattering in all directions is

$$\kappa_\nu I_\nu \, dm d\nu d\omega \int p(\cos\Theta)\frac{d\omega'}{4\pi}; \tag{27}$$

this agrees with (25) if

$$\int p(\cos\Theta)\frac{d\omega'}{4\pi} = 1, \tag{28}$$

i.e. if the phase function is *normalized to unity*.

Returning to the general case when both scattering and true absorption are present, we shall still write for the scattered energy the *same expression* (26). But in this case (in contrast to the case of scattering only) the total loss of energy from the incident pencil must be less than (25); accordingly

$$\int p(\cos\Theta)\frac{d\omega'}{4\pi} = \varpi_0 \leqslant 1. \tag{29}$$

Thus the general case differs from the case of pure scattering only by the fact that the phase function is not normalized to unity.

It is evident from our definitions that ϖ_0 represents the fraction of the light lost from an incident pencil due to scattering, while $(1-\varpi_0)$ represents the remaining fraction which has been transformed into other forms of energy (or of radiation of other wave-lengths).

We shall refer to ϖ_0 as the *albedo for single scattering*. Moreover, when $\varpi_0 = 1$ we shall say that we have a *conservative case* of *perfect scattering*: perfect scattering is, in our present context, the analogue of the concept of conservatism in dynamics.

The simplest example of a phase function is

$$p(\cos\Theta) = \text{constant} = \varpi_0. \tag{30}$$

In this case the radiation scattered by each element of mass is *isotropic*. Next to this isotropic case greatest interest is attached to *Rayleigh's phase function* (cf. § 16)

$$p(\cos\Theta) = \tfrac{3}{4}(1+\cos^2\Theta). \tag{31}$$

This phase function is normalized to unity so that this is an example of a conservative case of perfect scattering. Another phase function which is of particular interest in problems relating to planetary illumination is

$$p(\cos\Theta) = \varpi_0(1+x\cos\Theta) \quad (-1 \leqslant x \leqslant +1). \tag{32}$$

In general we may suppose that the phase function can be expanded as a series in Legendre polynomials of the form

$$p(\cos\Theta) = \sum_{l=0}^{\infty} \varpi_l P_l(\cos\Theta), \tag{33}$$

where the ϖ_l's are constants. In practice the series on the right-hand side is a terminating one with only a finite number of terms.

4. The emission coefficient

The *emission coefficient* j_ν is defined in such a way that an element of mass dm emits in directions confined to an element of solid angle $d\omega$, in the frequency interval $(\nu, \nu+d\nu)$ and in time dt, an amount of radiant energy given by

$$j_\nu \, dm d\omega d\nu dt. \tag{34}$$

In the case of a medium which scatters radiation (not necessarily with an albedo $\varpi_0 = 1$) there will be a contribution to the emission coefficient from the scattering of radiation from all other directions into the pencil of directions considered. Thus it follows from (26) that the scattering of a pencil of radiation from a direction (ϑ', φ') (say) contributes to a pencil in the direction (ϑ, φ), energy at the rate

$$\kappa_\nu \, dm d\nu d\omega \, p(\vartheta, \varphi; \vartheta', \varphi') I_\nu(\vartheta', \varphi') \frac{\sin\vartheta' \, d\vartheta' d\varphi'}{4\pi}, \tag{35}$$

where we have written $p(\vartheta, \varphi; \vartheta', \varphi')$ to denote the phase function for the angle between the directions specified by (ϑ, φ) and (ϑ', φ'). Hence the contribution, $j_\nu^{(s)}$, to the emission coefficient by scattering alone is

$$j_\nu^{(s)}(\vartheta, \varphi) = \kappa_\nu \frac{1}{4\pi} \int_0^\pi \int_0^{2\pi} p(\vartheta, \varphi; \vartheta', \varphi') I_\nu(\vartheta', \varphi') \sin\vartheta' \, d\vartheta' d\varphi'. \tag{36}$$

We may expect that in general there will be contributions to the emission coefficient from causes other than scattering. When this is *not* the case we shall say that we have a *scattering atmosphere*. In other words, for a scattering atmosphere

$$j_\nu \equiv j_\nu^{(s)}. \tag{37}$$

(Note that a scattering atmosphere does *not* imply that we have a case of perfect scattering.)

A case which is in some sense the opposite of a scattering atmosphere is that of an atmosphere in *local thermodynamic equilibrium*. In this latter case, it is assumed that the circumstances are such that we can define at each point in the atmosphere a *local temperature* T such that the emission coefficient at that point is given in terms of the absorption coefficient by Kirchhoff's law: i.e. at each point we have the relation

$$j_\nu = \kappa_\nu B_\nu(T), \tag{38}$$

where

$$B_\nu(T) = \frac{2h\nu^3}{c^2} \frac{1}{e^{h\nu/kT}-1} \tag{39}$$

is the Planck function (k and h are the Boltzmann and Planck constants, respectively).

5. The source function

The ratio of the emission to the absorption coefficient plays an important role in the subsequent developments of the theory. It is called the *source function*. We shall denote it by $\mathfrak{J}_\nu$. Thus

$$\mathfrak{J}_\nu = \frac{j_\nu}{\kappa_\nu}. \tag{40}$$

According to equations (36) and (37), for a scattering atmosphere

$$\mathfrak{J}_\nu(\vartheta, \varphi) = \frac{1}{4\pi} \int_0^\pi \int_0^{2\pi} p(\vartheta, \varphi; \vartheta', \varphi') I_\nu(\vartheta', \varphi') \sin\vartheta' \, d\vartheta' d\varphi', \tag{41}$$

while for an atmosphere in local thermodynamic equilibrium

$$\mathfrak{J}_\nu = B_\nu(T). \tag{42}$$

6. The equation of transfer

We shall now derive the fundamental equation which governs the variation of intensity in a medium characterized by an absorption coefficient κ_ν and an emission coefficient j_ν. (It should be noted that the emission coefficient can itself depend on the radiation field as will be the case, for example, in a scattering atmosphere.) For this purpose consider a small cylindrical element of cross-section $d\sigma$ and height ds in the medium. From the definition of intensity, it now follows that the difference in the radiant energy in the frequency interval $(\nu, \nu+d\nu)$ crossing the two faces normally, in a time dt and confined to an element of solid angle, is given by

$$\frac{dI_\nu}{ds} ds\,d\nu\,d\sigma\,d\omega\,dt. \tag{43}$$

This difference in energy must arise from the excess of emission over absorption in the frequency interval and element of solid angle considered. Now the amount absorbed is (cf. eq. [23])

$$\kappa_\nu \rho \, ds \times I_\nu \, d\nu\,d\sigma\,d\omega\,dt, \tag{44}$$

while the amount emitted is

$$j_\nu \, \rho \, d\sigma ds dv d\omega dt. \tag{45}$$

Counting up the gains and losses in the pencil of radiation during its traversal of the cylinder, we have

$$\frac{dI_\nu}{ds} = -\kappa_\nu \, \rho \, I_\nu + j_\nu \, \rho. \tag{46}$$

In terms of the source function $\mathfrak{J}_\nu$ (eq. [40]) we can rewrite this equation in the form

$$-\frac{dI_\nu}{\kappa_\nu \, \rho \, ds} = I_\nu - \mathfrak{J}_\nu. \tag{47}$$

This is *the equation of transfer*.

For a scattering atmosphere and an atmosphere in local thermodynamic equilibrium the source functions are given by equations (41) and (42).

In a Cartesian system of coordinates the equation of transfer can be written in the form

$$-\frac{1}{\kappa_\nu \, \rho}\left(l\frac{\partial}{\partial x} + m\frac{\partial}{\partial y} + n\frac{\partial}{\partial z}\right)I_\nu(x,y,z;l,m,n)$$
$$= I_\nu(x,y,z;l,m,n) - \mathfrak{J}_\nu(x,y,z;l,m,n). \tag{48}$$

Since the source function is functionally dependent on the intensity at a point, the equation of transfer is generally an *integro-differential equation*. We shall presently have examples of such integro-differential equations.

7. The formal solution of the equation of transfer

In our further discussion in this and the following chapters it is convenient to suppress the suffixes ν to the various quantities: no ambiguity is likely to arise from this. Thus we shall write the equation of transfer (47) in the form

$$-\frac{dI}{\kappa\rho \, ds} = I - \mathfrak{J}. \tag{49}$$

The formal solution of equation (49) is readily written down. We have (see Fig. 2)

$$I(s) = I(0)e^{-\tau(s,0)} + \int_0^s \mathfrak{J}(s')e^{-\tau(s,s')}\kappa\rho \, ds', \tag{50}$$

where $\tau(s,s')$ is the *optical thickness* of the material between the points s and s'; thus

$$\tau(s,s') = \int_{s'}^s \kappa\rho \, ds. \tag{51}$$

The physical meaning of the solution (50) is clear: It expresses the fact that the intensity at any point and in a given direction results from the emission at all anterior points, s', reduced by the factor $e^{-\tau(s,s')}$ to allow for the absorption by the intervening matter.

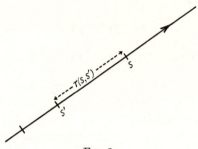

Fig. 2.

Sometimes it is convenient not to stop the range of integration over the source function at some definite point as we have done in equation (50) but write instead

$$I(s) = \int_{-\infty}^{s} \mathfrak{J}(s')e^{-\tau(s,s')}\kappa\rho \, ds'. \tag{52}$$

If the medium extends to $-\infty$ in the direction $\mathbf{s}$, no ambiguity arises by expressing the intensity $I(s)$ in the manner of equation (52). But if for decreasing s' we should encounter a 'radiating surface', for example, then we should stop the integration over s' at this point and add an extra term (as in eq. [50]) to take into account the intensity radiated by the surface. We shall take equation (52) to mean this.

It is of course clear that equations (50) and (52) do not in any real sense 'solve' the equation of transfer. But it is clear that if the source function should depend on the intensity in some specified way, then we can convert the formal solution (52) into an *integral equation* for the source function. We shall encounter examples of such integral equations in § 11.

8. The equation of transfer for a scattering atmosphere. The flux integral for conservative cases

For a scattering atmosphere the source function can be written in the form (cf. eq. [41])

$$\mathfrak{J}(\mathbf{r}, \mathbf{s}) = \frac{1}{4\pi} \int p(\mathbf{s}, \mathbf{s}')I(\mathbf{r}, \mathbf{s}') \, d\omega_{\mathbf{s}'}, \tag{53}$$

where $\mathbf{s}$ is a unit vector specifying some direction through a point $\mathbf{r}$.

The equation of transfer (48) can accordingly be written, in this case, in the form

$$-\frac{1}{\kappa\rho}(\mathbf{s}\cdot\mathrm{grad})I(\mathbf{r},\mathbf{s}) = I(\mathbf{r},\mathbf{s}) - \frac{1}{4\pi}\int p(\mathbf{s},\mathbf{s}')I(\mathbf{r},\mathbf{s}')\,d\omega_{\mathbf{s}'}. \qquad (54)$$

Integrating this equation over all directions s we have

$$-\frac{1}{\kappa\rho}\int(\mathbf{s}\cdot\mathrm{grad})I(\mathbf{r},\mathbf{s})\,d\omega_{\mathbf{s}}$$

$$= \int I(\mathbf{r},\mathbf{s})\,d\omega_{\mathbf{s}} - \frac{1}{4\pi}\int\int p(\mathbf{s},\mathbf{s}')I(\mathbf{r},\mathbf{s}')\,d\omega_{\mathbf{s}}\,d\omega_{\mathbf{s}'}. \qquad (55)$$

It is evident that the quantity on the left-hand side is the divergence of the vector $\pi\mathbf{F}$ whose components are the fluxes parallel to the x-, y-, and z-axes. And the first term on the right-hand side is clearly $4\pi J$ (cf. eq. [19]); the second term is also expressible in terms of J by evaluating the integral of $p(\mathbf{s},\mathbf{s}')$ over the directions $\mathbf{s}$ first (cf. eq. [29]). We therefore have

$$-\frac{1}{4\kappa\rho}\,\mathrm{div}\,\mathbf{F} = (1-\varpi_0)J, \qquad (56)$$

where ϖ_0 is the albedo for single scattering.

In cases of perfect scattering $\varpi_0 = 1$ and

$$\mathrm{div}\,\mathbf{F} = 0. \qquad (57)$$

This represents the *flux integral* for conservative problems.

For a plane-parallel atmosphere equation (57) reduces to

$$\frac{dF_z}{dz} = 0 \quad \text{or} \quad F_z = \text{constant}; \qquad (58)$$

the net flux normal to the plane of stratification is therefore constant through the atmosphere.

For radiation fields having spherical symmetry the flux integral reduces to

$$F_r = \frac{F_0}{r^2}, \qquad (59)$$

where F_0 is a constant; similarly for fields having cylindrical symmetry

$$F_r = \frac{F_0}{r}. \qquad (60)$$

9. The equation of transfer for plane-parallel problems

In problems of radiative transfer in plane-parallel atmospheres it is convenient to measure linear distances normal to the plane of stratification.

If z is this distance, the equation of transfer becomes

$$-\cos\vartheta \, \frac{dI(z,\vartheta,\varphi)}{\kappa\rho \, dz} = I(z,\vartheta,\varphi) - \Im(z,\vartheta,\varphi), \tag{61}$$

where ϑ denotes the inclination to the outward normal and φ the azimuth referred to a suitably chosen x-axis.

Introducing the *normal optical* thickness

$$\tau = \int_z^\infty \kappa\rho \, dz, \tag{62}$$

measured from the boundary *inward* we have

$$\mu \, \frac{dI(\tau,\mu,\varphi)}{d\tau} = I(\tau,\mu,\varphi) - \Im(\tau,\mu,\varphi). \tag{63}$$

In equation (63) we have further let $\mu = \cos\vartheta$.

Equation (63) is the standard form of the equation of transfer for plane-parallel atmospheres.

In considering transfer problems in plane-parallel atmospheres we shall distinguish two cases: (i) *the semi-infinite atmosphere* which is bounded on one side ($\tau = 0$) and extends to infinity in the direction $\tau \to \infty$; and (ii) *the finite atmosphere* which is bounded on two sides at $\tau = 0$ and at $\tau = \tau_1$ (say).

In the case of an atmosphere with a finite optical thickness the formal solution (50) reduces to

$$I(\tau, +\mu, \varphi) = I(\tau_1, \mu, \varphi)e^{-(\tau_1-\tau)/\mu} + \int_\tau^{\tau_1} \Im(t, \mu, \varphi)e^{-(t-\tau)/\mu} \, \frac{dt}{\mu} \quad (1 \geqslant \mu > 0), \tag{64}$$

and

$$I(\tau, -\mu, \varphi) = I(0, -\mu, \varphi)e^{-\tau/\mu} + \int_0^\tau \Im(t, -\mu, \varphi)e^{-(\tau-t)/\mu} \, \frac{dt}{\mu} \quad (1 \geqslant \mu > 0), \tag{65}$$

giving respectively the *outward* and the *inward* intensities at each level. In particular for the *emergent intensities* we have

$$I(0, +\mu, \varphi) = I(\tau_1, \mu, \varphi)e^{-\tau_1/\mu} + \int_0^{\tau_1} e^{-t/\mu}\Im(t, +\mu, \varphi) \, \frac{dt}{\mu}, \tag{66}$$

and $\quad I(\tau_1, -\mu, \varphi) = I(0, -\mu, \varphi)e^{-\tau_1/\mu} + \int_0^{\tau_1} e^{-(\tau_1-t)/\mu}\Im(t, -\mu, \varphi) \, \frac{dt}{\mu}. \tag{67}$

In the case of a semi-infinite atmosphere the foregoing equations reduce to

$$I(\tau, +\mu, \varphi) = \int_{\tau}^{\infty} \mathfrak{J}(t, +\mu, \varphi) e^{-(t-\tau)/\mu} \frac{dt}{\mu}, \tag{68}$$

$$I(\tau, -\mu, \varphi) = I(0, -\mu, \varphi) e^{-\tau/\mu} + \int_{0}^{\tau} \mathfrak{J}(t, -\mu, \varphi) e^{-(\tau-t)/\mu} \frac{dt}{\mu}, \tag{69}$$

and $$I(0, +\mu, \varphi) = \int_{0}^{\infty} \mathfrak{J}(t, +\mu, \varphi) e^{-t/\mu} \frac{dt}{\mu}. \tag{70}$$

10. Plane-parallel scattering atmospheres. The K-integral

For a plane-parallel scattering atmosphere the equation of transfer can be written in the form (cf. eq. [41])

$$\mu \frac{dI(\tau, \mu, \varphi)}{d\tau} = I(\tau, \mu, \varphi) - \frac{1}{4\pi} \int_{-1}^{+1} \int_{0}^{2\pi} p(\mu, \varphi; \mu', \varphi') I(\tau, \mu', \varphi') \, d\mu' d\varphi'. \tag{71}$$

In conservative cases ($\varpi_0 = 1$) equation (71) admits the flux integral (eq. [58])

$$F = \text{constant}, \tag{72}$$

where πF represents the flux of radiation normal to the plane of stratification.

There is another integral of importance which conservative problems generally admit. This is the so-called K-integral and can be obtained in the following manner: Multiplying equation (71) by μ and integrating over all solid angles we have

$$\frac{d}{d\tau} \int_{-1}^{+1} \int_{0}^{2\pi} I(\tau, \mu, \varphi) \mu^2 \, d\mu d\varphi$$

$$= \pi F - \frac{1}{4\pi} \int_{-1}^{+1} \int_{0}^{2\pi} d\mu' d\varphi' I(\tau, \mu', \varphi') \int_{-1}^{+1} \int_{0}^{2\pi} d\mu d\varphi \, \mu p(\mu, \varphi; \mu', \varphi'). \tag{73}$$

Now we shall suppose that $p(\cos \Theta)$ can be expanded, as in equation (33), in a series in Legendre polynomials. Then

$$p(\mu, \varphi; \mu', \varphi') = \sum \varpi_l P_l[\mu\mu' + (1-\mu^2)^{\frac{1}{2}}(1-\mu'^2)^{\frac{1}{2}} \cos(\varphi-\varphi')], \tag{74}$$

where it may be recalled that $\varpi_0 = 1$. Expanding P_l in equation (74) by the addition theorem for spherical harmonics we readily find that

$$\frac{1}{4\pi} \int_{-1}^{+1} \int_{0}^{2\pi} p(\mu, \varphi; \mu', \varphi') \mu \, d\mu d\varphi = \tfrac{1}{2} \varpi_1 \mu' \int_{-1}^{+1} \mu^2 \, d\mu = \tfrac{1}{3} \varpi_1 \mu'. \tag{75}$$

Hence equation (73) reduces to

$$\frac{1}{4\pi}\frac{d}{d\tau}\int_{-1}^{+1}\int_{0}^{2\pi} I(\tau,\mu,\varphi)\mu^2\,d\mu d\varphi = \tfrac{1}{4}(1-\tfrac{1}{3}\varpi_1)F. \qquad (76)$$

Now writing
$$K(\tau) = \frac{1}{4\pi}\int_{-1}^{+1}\int_{0}^{2\pi} I(\tau,\mu,\varphi)\mu^2\,d\mu d\varphi, \qquad (77)$$

we have
$$\frac{dK}{d\tau} = \tfrac{1}{4}(1-\tfrac{1}{3}\varpi_1)F, \qquad (78)$$

or, since F is a constant,

$$K = \tfrac{1}{4}F[(1-\tfrac{1}{3}\varpi_1)\tau+Q], \qquad (79)$$

where Q is a constant. This is the K-integral.

It is of interest also to notice that in conservative cases the equation of transfer admits a solution of the form

$$I(\tau,\mu) = \text{constant}\left(\tau+\frac{\mu}{1-\tfrac{1}{3}\varpi_1}\right). \qquad (80)$$

For, inserting this form for $I(\tau,\mu)$ in equation (71) and remembering that $\varpi_0 = 1$, we readily verify that the equation is in fact satisfied. Normalizing the solution (80) to yield a net flux πF we have

$$I(\tau,\mu) = \tfrac{3}{4}F[(1-\tfrac{1}{3}\varpi_1)\tau+\mu]. \qquad (81)$$

It should be emphasized again that equation (81) does not represent the solution of any physical problem we have formulated. But we shall see (Chap. III, § 25) that there are physical problems for which the solutions tend to (81) as $\tau \to \infty$. The solution (81) is also useful in certain other contexts (Chap. IV, § 29.3).

11. Problems in semi-infinite plane-parallel atmospheres with a constant net flux

We have seen that in conservative cases of perfect scattering the equation of transfer admits the flux integral (72). Because of this constancy of net flux, a type of problem which arises in these contexts is that of a semi-infinite plane-parallel atmosphere with no incident radiation and with a constant net flux, πF, of radiation flowing through the atmosphere normal to the plane of stratification. The particular importance of this type of problem for astrophysics arises from the fact that in *stellar atmospheres* (idealized as plane-parallel atmospheres) the constant net flux is provided by the radiation coming from the 'deep interior'.

It is evident that for the type of problem we have formulated, solutions of the equation of transfer must be sought which exhibit axial symmetry about the z-axis. The intensity and the source function must therefore be azimuth independent, and the equation of transfer (71) becomes

$$\mu \frac{dI(\tau,\mu)}{d\tau} = I(\tau,\mu) - \tfrac{1}{2} \int\limits_{-1}^{+1} p^{(0)}(\mu,\mu') I(\tau,\mu')\, d\mu', \tag{82}$$

where

$$p^{(0)}(\mu,\mu') = \frac{1}{2\pi} \int\limits_{0}^{2\pi} p(\mu,\varphi;\mu',\varphi')\, d\varphi'. \tag{83}$$

And we require solutions of equation (82) which satisfy the boundary condition

$$I(0,-\mu) = 0 \quad (0 < \mu \leqslant 1); \tag{84}$$

also, that as $\tau \to \infty$ all integrals of the type which occur in the formal solution (eqs. [68]–[70]) converge. Quite generally, the convergence of these integrals over the source function require that

$$\mathfrak{J}(\tau,\mu) \prec e^{\tau} \quad (\tau \to \infty), \tag{85}$$

or, explicitly,

$$\mathfrak{J}(\tau,\mu) e^{-\tau} \to 0 \quad \text{as} \quad \tau \to \infty. \tag{85'}$$

In view of the axial symmetry of the radiation field, the flux and the K-integrals which equation (82) admits can be expressed in the forms

$$F = 2 \int\limits_{-1}^{+1} I(\tau,\mu)\mu\, d\mu = \text{constant},$$

and

$$K = \tfrac{1}{2} \int\limits_{-1}^{+1} I(\tau,\mu)\mu^2\, d\mu = \tfrac{1}{4}F[(1-\tfrac{1}{3}\varpi_1)\tau + Q]. \tag{86}$$

Finally, it may be stated that in the problem with a constant net flux in semi-infinite atmospheres the greatest interest is attached to the angular distribution, or, as it is often called, *the law of darkening*, of the emergent radiation:

$$I(0,\mu) \quad \text{for} \quad \tau = 0 \quad \text{and} \quad 0 \leqslant \mu \leqslant 1. \tag{87}$$

We shall now consider two special cases of the problem we have formulated.

11.1. *Isotropic case*

For isotropic scattering $p^{(0)}(\mu;\mu') \equiv 1$ and equation (82) becomes

$$\mu \frac{dI(\tau,\mu)}{d\tau} = I(\tau,\mu) - J(\tau), \tag{88}$$

where

$$J(\tau) = \mathfrak{J}(\tau) = \tfrac{1}{2} \int\limits_{-1}^{+1} I(\tau,\mu)\, d\mu. \tag{89}$$

This is the simplest case of an equation of transfer and has been studied most extensively in the literature.

Since there is no radiation incident on the surface $\tau = 0$, the formal solution of this problem is given by (cf. eqs. [68] and [69]),

$$I(\tau, +\mu) = \int_\tau^\infty e^{-(t-\tau)/\mu} J(t) \frac{dt}{\mu} \quad (0 < \mu \leqslant 1),$$

and

$$I(\tau, -\mu) = \int_0^\tau e^{-(\tau-t)/\mu} J(t) \frac{dt}{\mu} \quad (0 < \mu \leqslant 1). \tag{90}$$

With the foregoing solution the integral

$$I_n(\tau) = \int_{-1}^{+1} I(\tau, \mu) \mu^n \, d\mu \tag{91}$$

can be expressed directly as one over $J(\tau)$. Thus, substituting for $I(\tau, \mu)$ according to equation (90) we have

$$I_n(\tau) = \int_\tau^\infty \int_0^1 e^{-(t-\tau)/\mu} J(t) \mu^{n-1} \, dt d\mu + (-1)^n \int_0^\tau \int_0^1 e^{-(\tau-t)/\mu} J(t) \mu^{n-1} \, dt d\mu \tag{92}$$

or, inverting the order of the integrations, we have

$$I_n(\tau) = \int_\tau^\infty dt \, J(t) \int_0^1 d\mu \, \mu^{n-1} e^{-(t-\tau)/\mu} + (-1)^n \int_0^\tau dt \, J(t) \int_0^1 d\mu \, \mu^{n-1} e^{-(\tau-t)/\mu}. \tag{93}$$

With the substitution $\mu = 1/x$ equation (93) becomes

$$I_n(\tau) = \int_\tau^\infty dt \, J(t) \int_1^\infty \frac{dx}{x^{n+1}} e^{-x(t-\tau)} + (-1)^n \int_0^\tau dt \, J(t) \int_1^\infty \frac{dx}{x^{n+1}} e^{-x(\tau-t)}, \tag{94}$$

or, in terms of the *exponential integral*,

$$E_n(y) = \int_1^\infty \frac{dx}{x^n} e^{-xy}, \tag{95}\dagger$$

we have

$$I_n(\tau) = \int_\tau^\infty J(t) E_{n+1}(t-\tau) \, dt + (-1)^n \int_0^\tau J(t) E_{n+1}(\tau-t) \, dt. \tag{96}$$

Remembering the definitions of J and F, we have, in particular from equation (96), that

$$J(\tau) = \tfrac{1}{2} \int_0^\infty J(t) E_1(|t-\tau|) \, dt, \tag{97}$$

and

$$F = 2 \int_\tau^\infty J(t) E_2(t-\tau) \, dt - 2 \int_0^\tau J(t) E_2(\tau-t) \, dt. \tag{98}$$

† For a discussion of the properties of these and other related integrals see Appendix I.

We observe that equation (97) is an integral equation for J. This is the *Schwarzschild–Milne integral equation*; and the solution of this equation is clearly equivalent to solving the equation of transfer (88).

11.2. *The case of Rayleigh's phase function*

As a second example we shall consider the case of scattering in accordance with Rayleigh's phase function (31). In this case

$$p(\mu,\varphi;\mu',\varphi') = \tfrac{3}{4}[1+\mu^2\mu'^2+(1-\mu^2)(1-\mu'^2)\cos^2(\varphi-\varphi')+$$
$$+2\mu\mu'(1-\mu^2)^{\frac{1}{2}}(1-\mu'^2)^{\frac{1}{2}}\cos(\varphi-\varphi')]. \quad (99)$$

Hence
$$p^{(0)}(\mu,\mu') = \tfrac{3}{4}[1+\mu^2\mu'^2+\tfrac{1}{2}(1-\mu^2)(1-\mu'^2)],$$

or
$$p^{(0)}(\mu,\mu') = \tfrac{3}{8}[3-\mu^2+(3\mu^2-1)\mu'^2]. \quad (100)$$

The equation of transfer (82) is therefore

$$\mu\,\frac{dI(\tau,\mu)}{d\tau}$$
$$= I(\tau,\mu)-\frac{3}{16}\Bigg[(3-\mu^2)\int_{-1}^{+1} I(\tau,\mu')\,d\mu'+(3\mu^2-1)\int_{-1}^{+1} I(\tau,\mu')\mu'^2\,d\mu'\Bigg],$$
$$(101)$$

or, in terms of J and K defined as usual,

$$\mu\,\frac{dI(\tau,\mu)}{d\tau} = I(\tau,\mu)-\tfrac{3}{8}[(3-\mu^2)J(\tau)+(3\mu^2-1)K(\tau)]. \quad (102)$$

The source function for this problem is therefore

$$\mathfrak{J}(\tau,\mu) = \tfrac{3}{8}[(3-\mu^2)J(\tau)+(3\mu^2-1)K(\tau)]. \quad (103)$$

Now quite generally (cf. eq. [92])

$$\int_{-1}^{+1} I(\tau,\mu)\mu^n\,d\mu = \int_{\tau}^{\infty}\int_0^1 \mathfrak{J}(t,\mu)e^{-(t-\tau)/\mu}\mu^{n-1}\,dt\,d\mu+$$

$$+(-1)^n\int_0^{\tau}\int_0^1 \mathfrak{J}(t,-\mu)e^{-(\tau-t)/\mu}\,\mu^{n-1}\,dt\,d\mu. \quad (104)$$

With $\mathfrak{J}(\tau,\mu)$ given by equation (103), the various integrals occurring on the right-hand side can be reduced in the manner of equations (93)–(96). We thus find

$$\int_{-1}^{+1} I(\tau,\mu)\mu^n\,d\mu = \frac{3}{8}\Bigg[\int_{\tau}^{\infty}(3E_{n+1}-E_{n+3})_{(t-\tau)}J(t)\,dt+$$

$$+\int_{\tau}^{\infty}(3E_{n+3}-E_{n+1})_{(t-\tau)}K(t)\,dt+(-1)^n\int_0^{\tau}(3E_{n+1}-E_{n+3})_{(\tau-t)}J(t)\,dt+$$

$$+(-1)^n\int_0^{\tau}(3E_{n+3}-E_{n+1})_{(\tau-t)}K(t)\,dt\Bigg]. \quad (105)$$

Considering the cases $n = 0$ and $n = 2$ of equation (105) we therefore have

$$J(\tau) = \frac{3}{16}\left[\int_0^\infty (3E_1 - E_3)_{|t-\tau|} J(t)\, dt + \int_0^\infty (3E_3 - E_1)_{|t-\tau|} K(t)\, dt \right], \quad (106)$$

and

$$K(\tau) = \frac{3}{16}\left[\int_0^\infty (3E_3 - E_5)_{|t-\tau|} J(t)\, dt + \int_0^\infty (3E_5 - E_3)_{|t-\tau|} K(t)\, dt \right].$$
$$(107)$$

Equations (106) and (107) represent a pair of integral equations for J and K and the solution of these equations is equivalent to solving the equation of transfer (102).

It is apparent from the two examples we have considered that the linear integral equations which generally replace the equation of transfer become increasingly of higher order as the phase function considered includes more and more terms in the expansion (33).

12. Axially symmetric problems in semi-infinite atmospheres and in non-conservative cases

In the preceding section we have formulated the transfer problem in semi-infinite atmospheres with a constant net flux. Similar axially symmetric problems can be formulated also in non-conservative cases. To illustrate the nature of these latter problems, we shall consider the case of isotropic scattering with an albedo $\varpi_0 < 1$. The equation of transfer appropriate to the circumstances is

$$\mu \frac{dI(\tau, \mu)}{d\tau} = I(\tau, \mu) - \tfrac{1}{2}\varpi_0 \int_{-1}^{+1} I(\tau, \mu')\, d\mu'. \quad (108)$$

We first observe that equation (108) admits a solution of the form

$$I(\tau, \mu) = e^{k\tau} g(\mu), \quad (109)$$

where k is a constant (unspecified for the present) and $g(\mu)$ is a function of μ only. Thus inserting this form for $I(\tau, \mu)$ in equation (108) we have

$$(1 - k\mu) g(\mu) = \tfrac{1}{2}\varpi_0 \int_{-1}^{+1} g(\mu)\, d\mu. \quad (110)$$

Hence $g(\mu)$ must be of the form

$$g(\mu) = \frac{\text{constant}}{1 - k\mu}. \quad (111)$$

Substituting this expression for $g(\mu)$ back into equation (110) we are left with

$$1 = \tfrac{1}{2}\varpi_0 \int_{-1}^{+1} \frac{d\mu'}{1-k\mu'} = \frac{\varpi_0}{2k} \log\left(\frac{1+k}{1-k}\right); \tag{112}$$

k must therefore be a root of the *characteristic equation*

$$\varpi_0 = \frac{2k}{\log\{(1+k)/(1-k)\}}. \tag{113}$$

From this equation it follows that if k is a root then so is $-k$. And it can also be shown that for a given $\varpi_0 < 1$, there is a unique value of $k^2 < 1$ which satisfies equation (113). This is apparent, for example, from Table I, where the characteristic roots for various values of ϖ_0 are listed.

TABLE I

The Characteristic Roots k

ϖ_0	k	ϖ_0	k
0	1·00000	0·8	0·71041
0·2	0·99991	0·9	0·52543
0·3	0·99741	0·925	0·45993
0·4	0·98562	0·950	0·37948
0·5	0·95750	0·975	0·27111
0·6	0·90733	1·000	0
0·7	0·82864		

We have thus shown that the equation of transfer (108) admits a solution of the form

$$I(\tau, \mu) = \text{constant} \, \frac{e^{\pm k\tau}}{1 \mp k\mu}, \tag{114}$$

where $0 < k < 1$ is the positive real root of equation (113) for $0 < \varpi_0 < 1$. The source function corresponding to the solution (114) is

$$\mathfrak{J}(\tau) = \tfrac{1}{2}\varpi_0 \int_{-1}^{+1} I(\tau, \mu) \, d\mu = \text{constant} \, e^{\pm k\tau}. \tag{115}$$

[It may be noticed here that $e^{\pm k\tau}$ is also a solution of the integral equation

$$J(\tau) = \tfrac{1}{2}\varpi_0 \int_{-\infty}^{+\infty} J(t) E_1(|t-\tau|) \, dt, \tag{116}$$

which is appropriate in the present context for a plane-parallel atmosphere extending to infinity on *both* sides (cf. eq. [97]); for it may be readily verified that

$$\tfrac{1}{2}\varpi_0 \int_{-\infty}^{+\infty} e^{k\tau} E_1(|t-\tau|) \, dt = \left[\frac{\varpi_0}{2k} \log\left(\frac{1+k}{1-k}\right)\right] e^{k\tau}, \tag{117}$$

which shows that for k which is a root of the characteristic equation (113), $e^{k\tau}$ is a solution of equation (116).]

The source function $\mathfrak{J}(\tau) = \text{constant } e^{k\tau}$ $(k < 1)$ clearly satisfies the condition (85) at infinity. Accordingly, solutions of equation (108) may be expected to exist which behave at ∞ as

$$I(\tau, \mu) \to \frac{\text{constant}}{1 - k\mu} e^{+k\tau} \quad (\tau \to \infty). \tag{118}$$

In particular, we can ask for solutions having this behaviour at infinity and which also satisfy the boundary condition

$$I(0, -\mu) = 0 \quad (0 < \mu \leqslant 1), \tag{119}$$

at $\tau = 0$. This problem is therefore analogous to the one considered in § 11.1. And it is apparent from our discussion of this isotropic case that similar problems can be formulated for equations more general than (108).

13. Diffuse reflection and transmission

In some ways the most fundamental problem in the theory of radiative transfer in plane-parallel atmospheres is the *diffuse reflection* and *transmission* of a parallel beam of radiation; for it will appear that the solutions of all other problems can be reduced to this one. In this section we shall formulate the basic problem and make only some general comments; the detailed discussion of the various aspects of this problem will be taken up in subsequent chapters.

The problem of diffuse reflection and transmission by a plane-parallel atmosphere is the following (cf. Fig. 3).

A parallel beam of radiation of net flux πF per unit area normal itself is incident on a plane-parallel atmosphere of optical thickness τ_1 in some specified direction $-\mu_0, \varphi_0$. It is required to find the angular distributions of the intensities diffusely reflected from the surface $\tau = 0$ and diffusely transmitted below the surface $\tau = \tau_1$.

We shall find it convenient to express the resulting laws of diffuse reflection and transmission in terms of a *scattering function*

$$S(\tau_1; \mu, \varphi; \mu_0, \varphi_0)$$

and a *transmission function*

$$T(\tau_1; \mu, \varphi; \mu_0, \varphi_0)$$

such that the reflected and the transmitted intensities are given by

$$I(0, +\mu, \varphi) = \frac{F}{4\mu} S(\tau_1; \mu, \varphi; \mu_0, \varphi_0)$$

and

$$I(\tau_1, -\mu, \varphi) = \frac{F}{4\mu} T(\tau_1; \mu, \varphi; \mu_0, \varphi_0) \quad (0 \leqslant \mu \leqslant 1). \tag{120}$$

It is to be specially noted that the reflected and the transmitted intensities refer only to the light which has suffered one or more scattering processes; $I(\tau_1, -\mu, \varphi)$ does not include, for example, the directly transmitted flux $\frac{1}{4}Fe^{-\tau_1/\mu_0}$ in the direction $(-\mu_0, \varphi_0)$.

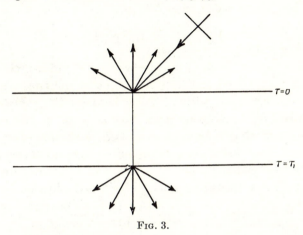

FIG. 3.

The reason for introducing the factor $1/\mu$ in the expressions defining $I(0, +\mu, \varphi)$ and $I(\tau_1, -\mu, \varphi)$ is for securing the symmetry of S and T in the pair of variables (μ, φ) and (μ_0, φ_0) as required by Helmholtz's principle of reciprocity (Chap. VII, § 52):

$$S(\tau_1; \mu, \varphi; \mu_0, \varphi_0) = S(\tau_1; \mu_0, \varphi_0; \mu, \varphi),$$

and
$$T(\tau_1; \mu, \varphi; \mu_0, \varphi_0) = T(\tau_1; \mu_0, \varphi_0; \mu, \varphi). \tag{121}$$

Though the problem of diffuse reflection and transmission has been formulated for an incident parallel beam of light, it is apparent that the solution for an arbitrary incident field of radiation (with the same angular distribution at all points on the surface $\tau = 0$) can be expressed in terms of S and T. Thus if $I_{\text{inc.}}(\mu', \varphi')$ represents the intensity incident on $\tau = 0$ in the direction $(-\mu', \varphi')$, the angular distributions of the reflected and the transmitted light will be given by

$$I_{\text{ref}}(0, +\mu, \varphi) = \frac{1}{4\pi\mu} \int_0^1 \int_0^{2\pi} S(\tau_1; \mu, \varphi; \mu', \varphi') I_{\text{inc}}(\mu', \varphi') \, d\mu' d\varphi' \tag{122}$$

and

$$I_{\text{trans}}(\tau_1, -\mu, \varphi) = \frac{1}{4\pi\mu} \int_0^1 \int_0^{2\pi} T(\tau_1; \mu, \varphi; \mu', \varphi') I_{\text{inc}}(\mu', \varphi') \, d\mu' d\varphi'. \tag{123}$$

The agreement of these expressions with the definitions (121) is apparent when it is observed that a parallel beam of radiation, incident in the direction $(-\mu_0, \varphi_0)$, has an angular distribution expressible in terms of Dirac's δ-functions in the form

$$I_{\mathrm{inc}}(\mu', \varphi') = \pi F\, \delta(\mu'-\mu_0)\delta(\varphi'-\varphi_0). \tag{124}$$

For a semi-infinite atmosphere we are naturally interested only in the law of reflection. We shall then write

$$I(0, \mu, \varphi) = \frac{F}{4\mu}\, S(\mu, \varphi; \mu_0, \varphi_0). \tag{125}$$

One general remark relating to the problem of diffuse reflection and transmission may be made here. It is that in the treatment of this problem it is convenient to distinguish between the *reduced incident radiation* $\pi F e^{-\tau/\mu_0}$ which penetrates to the level τ without having suffered any scattering or absorption process, and the *diffuse radiation field* which has arisen in consequence of one or more scattering processes. It is this diffuse radiation field which we shall characterize by the intensity $I(\tau, \mu, \varphi)$ in these problems. With this distinction between the two fields of radiation, we can write the equation of transfer appropriate for the problem of diffuse reflection and transmission in the form

$$\mu\, \frac{dI(\tau, \mu, \varphi)}{d\tau} = I(\tau, \mu, \varphi) - \frac{1}{4\pi} \int_{-1}^{+1} \int_0^{2\pi} p(\mu, \varphi; \mu', \varphi') I(\tau, \mu', \varphi')\, d\mu' d\varphi' - $$
$$- \tfrac{1}{4} F e^{-\tau/\mu_0} p(\mu, \varphi; -\mu_0, \varphi_0). \tag{126}$$

Solutions of this equation are required which satisfy the boundary conditions

$$I(0, -\mu, \varphi) = 0 \quad (0 < \mu \leqslant 1), \tag{127}$$

and

$$I(\tau_1, +\mu, \varphi) = 0 \quad (0 < \mu \leqslant 1), \tag{128}$$

at $\tau = 0$ and $\tau = \tau_1$. It will be noticed that in writing the boundary condition (128) we have assumed that at $\tau = \tau_1$ we have a perfect absorber (or, equivalently, a vacuum). However, we shall show later that the solution for the case when different 'ground conditions' are imposed at $\tau = \tau_1$ can be reduced to the 'standard problem' we have formulated.

For the case of diffuse reflection by a semi-infinite atmosphere we have to impose the condition (127) and the *boundedness* of the solution for $\tau \to \infty$.

For isotropic scattering with an albedo ϖ_0, the radiation field will have axial symmetry, also for the problem of diffuse reflection and transmission, and the appropriate equation of transfer is

$$\mu\, \frac{dI(\tau, \mu)}{d\tau} = I(\tau, \mu) - \tfrac{1}{2}\varpi_0 \int_{-1}^{+1} I(\tau, \mu')\, d\mu' - \tfrac{1}{4}\varpi_0 F e^{-\tau/\mu_0}. \tag{129}$$

For more general phase functions the radiation field will show dependence on azimuth as well. However, for phase functions which can be expanded as a series in Legendre polynomials, the intensity $I(\tau, \mu, \varphi)$ can be expressed in the form

$$I(\tau, \mu, \varphi) = \sum I^{(m)}(\tau, \mu) \cos m(\varphi - \varphi_0). \tag{130}$$

14. Problems with spherical symmetry

When the medium in which we are considering the transfer of radiation has spherical symmetry, the mass absorption coefficient, κ, and the density, ρ, will be functions of the distance r from the centre of symmetry only. And when, further, no radiation from the outside is incident, the intensity and the source function will be functions only of the distance r and the inclination ϑ to the radius vector.

Now if ds denotes the element of length in the direction ϑ at r,

$$dr = \cos \vartheta \, ds \quad \text{and} \quad r \, d\vartheta = -\sin \vartheta \, ds, \tag{131}$$

and the equation of transfer (49) becomes

$$\cos \vartheta \, \frac{\partial I}{\partial r} - \frac{\sin \vartheta}{r} \frac{\partial I}{\partial \vartheta} = -\kappa \rho [I(r, \vartheta) - \mathfrak{J}(r, \vartheta)], \tag{132}$$

or, writing μ for $\cos \vartheta$, we have

$$\mu \frac{\partial I}{\partial r} + \frac{1 - \mu^2}{r} \frac{\partial I}{\partial \mu} = -\kappa \rho [I(r, \mu) - \mathfrak{J}(r, \mu)]. \tag{133}$$

For a scattering atmosphere the source function is (cf. eqs. [82] and [83])

$$\mathfrak{J}(r, \mu) = \tfrac{1}{2} \int_{-1}^{+1} p^{(0)}(\mu, \mu') I(r, \mu') \, d\mu', \tag{134}$$

while for an atmosphere in local thermodynamic equilibrium,

$$\mathfrak{J}(r, \mu) = B_\nu(T_r), \tag{135}$$

where $B_\nu(T_r)$ denotes the Planck function for the temperature prevailing at r.

For isotropic scattering the equation of transfer (133) reduces to

$$\mu \frac{\partial I}{\partial r} + \frac{1 - \mu^2}{r} \frac{\partial I}{\partial \mu} = -\kappa \rho \left[I(r, \mu) - \tfrac{1}{2} \varpi_0 \int_{-1}^{+1} I(r, \mu') \, d\mu' \right]. \tag{136}$$

In investigations of transfer problems with spherical symmetry attention has mostly been restricted to equation (136), though it is not difficult to generalize the results to include, for example, the case of Rayleigh's phase function. Also, it may be stated that in astrophysical

contexts the case of greatest interest is when the medium extends to
infinity and $\kappa\rho$ varies as some inverse power of r (greater than 1);
in physical contexts, on the other hand, it appears that interest is
principally centred on 'diffusion' in homogeneous spheres and spherical
shells (with additional sources distributed through the medium).

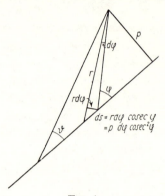

Fɪɢ. 4.

Finally, we may note that the formal solution of § 7 adapted to the
present case is (cf. Fig. 4)

$$I(r, \vartheta) = p \int_{\vartheta}^{\pi} \Im(p \operatorname{cosec} \varphi; \varphi) \exp\left\{ -p \int_{\vartheta}^{\varphi} (\kappa\rho)_{p \operatorname{cosec} \varphi} \operatorname{cosec}^2\varphi \, d\varphi \right\} \times$$

$$\times (\kappa\rho)_{p \operatorname{cosec} \varphi} \operatorname{cosec}^2\varphi \, d\varphi. \quad (137)$$

In the case of isotropic scattering and for homogeneous spheres this
solution can be used to obtain an integral equation for J as in the
analogous problem in plane-parallel atmospheres (§ 11).

15. The representation of polarized light

So far we have not included the state of polarization in our specifica-
tion of the radiation field. But it is clear that we must allow for this
in any exact treatment of scattering problems since, on scattering, light
in general gets polarized. For example, on Rayleigh's classical laws
(cf. § 16), an initially unpolarized beam, when scattered in a direction
making an angle Θ to the direction of incidence, becomes partially
plane-polarized with a ratio of intensities $1 : \cos^2\Theta$ in directions perpendicular and parallel to the *plane of scattering*. (This is the plane which
contains the directions of the incident and the scattered light.) The
diffuse radiation field in a scattering atmosphere must therefore be
partially polarized, and the question arises as to how best we can
characterize the radiation field under these circumstances, in order that

the relevant equations of transfer may be most conveniently formulated. In many ways this is a fundamental question: on its answer will depend the solution of a variety of problems, including the important one of the illumination and polarization of the sunlit sky.

Now it is evident that to describe a general radiation field, four parameters should be specified which will give the intensity, the degree of polarization, the plane of polarization, and the ellipticity of the radiation at each point and in any given direction. But it is apparent that it would be impossible to include such diverse quantities as an intensity, a ratio, an angle, and a pure number in any symmetrical way in formulating the equations of transfer. A proper parametric representation of polarized light is therefore a matter of some importance.

It appears that for the purposes of formulating the equations of transfer in a gaseous medium the most convenient representation of polarized light is by a set of four parameters, introduced by Sir George Stokes in 1852. With slight modifications, Stokes's representation will be used in this book.

In view of the general inaccessibility of Stokes's considerations it may be useful to have them presented here in a form suitable for our purposes.

15.1. *An elliptically polarized beam*

As is well known, in an elliptically polarized beam the vibrations of the electric (and the magnetic) vector in the plane transverse to the direction of propagation are such that the ratio of the amplitudes and the difference in phases of the components in any two directions at right angles to each other are absolute constants. A regular vibration of this character can be represented by

$$\xi_l = \xi_l^{(0)} \sin(\omega t - \epsilon_l) \quad \text{and} \quad \xi_r = \xi_r^{(0)} \sin(\omega t - \epsilon_r), \tag{138}$$

where ξ_l and ξ_r are the components of the vibration along two directions l and r at right angles to each other (see Fig. 5), ω the circular frequency of the vibration, and $\xi_l^{(0)}$, $\xi_r^{(0)}$, ϵ_l, and ϵ_r are constants.

If the principal axes of the ellipse described by (ξ_l, ξ_r) are in directions making angles χ and $\chi + \tfrac{1}{2}\pi$ to the direction l, the equations representing the vibration take the simplified forms

$$\xi_\chi = \xi^{(0)} \cos\beta \sin\omega t \quad \text{and} \quad \xi_{\chi+\frac{1}{2}\pi} = \xi^{(0)} \sin\beta \cos\omega t, \tag{139}$$

where β denotes an angle whose tangent is the ratio of the axes of the ellipse traced by the end point of the electric vector. We shall suppose that the numerical value of β lies between 0 and $\tfrac{1}{2}\pi$ and that the sign of β is positive or negative according as the polarization is right-handed or left-handed. Further, in (139) $\xi^{(0)}$ denotes a quantity proportional to the mean amplitude of the electric vector and whose square is equal to the intensity of the beam:

$$I = [\xi^{(0)}]^2 = [\xi_l^{(0)}]^2 + [\xi_r^{(0)}]^2 = I_l + I_r. \tag{140}$$

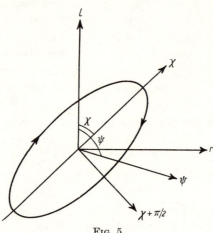

Fig. 5.

The formulae connecting the representations (138) and (139) are important and can be obtained in the following manner.

Starting from the representation (139) we obtain for the vibrations in the directions l and r, the expressions

$$\xi_l = \xi^{(0)}(\cos\beta\cos\chi\sin\omega t - \sin\beta\sin\chi\cos\omega t),$$

and $\quad\quad \xi_r = \xi^{(0)}(\cos\beta\sin\chi\sin\omega t + \sin\beta\cos\chi\cos\omega t). \tag{141}$

These equations can be reduced to the form (138) by letting

$$\xi_l^{(0)} = \xi^{(0)}(\cos^2\beta\cos^2\chi + \sin^2\beta\sin^2\chi)^{\frac{1}{2}},$$

$$\xi_r^{(0)} = \xi^{(0)}(\cos^2\beta\sin^2\chi + \sin^2\beta\cos^2\chi)^{\frac{1}{2}}, \tag{142}$$

$$\tan\epsilon_l = \tan\beta\tan\chi \quad \text{and} \quad \tan\epsilon_r = -\tan\beta\cot\chi. \tag{143}$$

The intensities I_l and I_r in the directions l and r are therefore given by

$$I_l = [\xi_l^{(0)}]^2 = I(\cos^2\beta\cos^2\chi + \sin^2\beta\sin^2\chi),$$

and $\quad\quad I_r = [\xi_r^{(0)}]^2 = I(\cos^2\beta\sin^2\chi + \sin^2\beta\cos^2\chi). \tag{144}$

Further, according to equations (142) and (143) we may readily verify that

$$2\xi_l^{(0)}\xi_r^{(0)}\cos(\epsilon_l - \epsilon_r) = 2[\xi^{(0)}]^2(\cos^2\beta - \sin^2\beta)\cos\chi\sin\chi$$

$$= I\cos 2\beta\sin 2\chi. \tag{145}$$

Similarly, $\quad\quad 2\xi_l^{(0)}\xi_r^{(0)}\sin(\epsilon_l - \epsilon_r) = I\sin 2\beta. \tag{146}$

From the foregoing equations it follows that whenever the regular vibrations representing an elliptically polarized beam can be expressed in the form (138) we can at once write the relations

$$I = [\xi_l^{(0)}]^2 + [\xi_r^{(0)}]^2 = I_l + I_r,$$

$$Q = [\xi_l^{(0)}]^2 - [\xi_r^{(0)}]^2 = I \cos 2\beta \cos 2\chi = I_l - I_r,$$

$$U = 2\xi_l^{(0)}\xi_r^{(0)} \cos(\epsilon_l - \epsilon_r) = I \cos 2\beta \sin 2\chi = (I_l - I_r)\tan 2\chi,$$

and $\quad V = 2\xi_l^{(0)}\xi_r^{(0)} \sin(\epsilon_l - \epsilon_r) = I \sin 2\beta = (I_l - I_r)\tan 2\beta \sec 2\chi. \quad (147)$

These are the *Stokes parameters* representing an elliptically polarized beam.

We observe that among the quantities I, Q, U, and V defined as in equations (147) there exists the relation

$$I^2 = Q^2 + U^2 + V^2. \tag{148}$$

Further, the plane of polarization and the ellipticity follow from the equations

$$\tan 2\chi = \frac{U}{Q} \quad \text{and} \quad \sin 2\beta = \frac{V}{\sqrt{(Q^2 + U^2 + V^2)}}. \tag{149}$$

In representing the vibration by (138) we have considered the amplitudes and the phases to be constants. But in practice this is not attainable; for even in the simplest case of approximately monochromatic light, the amplitudes and the phases must be regarded as liable to incessant variations, though they may remain constant, or sensibly constant, for a great number of vibrations. The known high frequency of the electromagnetic oscillations representing light allows us to suppose that the phases and the amplitudes may be constant for millions of vibrations and yet change irregularly millions of times a second. However, in an elliptically polarized beam these irregular variations must be such that the *ratio* of the amplitudes, $(\xi_l^{(0)} : \xi_r^{(0)})$ and the *difference* of phases, $\delta = \epsilon_l - \epsilon_r$, should be *absolute constants*. Consequently, all we will be able to appreciate is the (apparent) mean intensity in any direction in the transverse plane. Thus, the apparent intensities I_l and I_r in the directions l and r will be given by the mean values

$$I_l = \overline{[\xi_l^{(0)}]^2} \quad \text{and} \quad I_r = \overline{[\xi_r^{(0)}]^2}. \tag{150}$$

Also, if we now let

$$Q = I_l - I_r = \overline{[\xi_l^{(0)}]^2} - \overline{[\xi_r^{(0)}]^2},$$

$$U = 2\overline{[\xi_l^{(0)}\xi_r^{(0)}]}\cos \delta,$$

and $\quad V = 2\overline{[\xi_l^{(0)}\xi_r^{(0)}]}\sin \delta, \tag{151}$

it follows from equations (142) and (143) that

$$Q = \overline{[\xi^{(0)}]^2} \cos 2\beta \cos 2\chi = I \cos 2\beta \cos 2\chi,$$

$$U = \overline{[\xi^{(0)}]^2} \cos 2\beta \sin 2\chi = I \cos 2\beta \sin 2\chi,$$

and $$V = \overline{[\xi^{(0)}]^2} \sin 2\beta = I \sin 2\beta, \tag{152}$$

since the shape and the orientation of the ellipse remain constant through all variations.

It should be noticed that with the definitions (150) and (151) equation (148) continues to be valid (cf. § 15.3 below).

15.2. *The Stokes parameters for an arbitrarily polarized light*

An arbitrarily polarized beam of light can be completely analysed by the following procedure: We introduce a known amount of retardation in the phase of vibrations in one direction relative to the phase of vibrations in a direction at right angles to it, and then measure the intensity in all directions in the transverse plane.

Let $$\xi_l = \xi_l^{(0)} \sin(\omega t - \epsilon_l) \quad \text{and} \quad \xi_r = \xi_r^{(0)} \sin(\omega t - \epsilon_r) \tag{153}$$

represent the instantaneous vibration in the beam. As we have already explained, the amplitudes and the phases are subject to irregular variations. But certain correlations among them will persist during all the variations and it is these correlations which give to the light the character of partial or complete polarization as the case may be; thus it is the constancy of the ratio of the amplitudes and the difference in phases of any two components at right angles to each other that distinguishes an elliptically polarized beam from one which is not.

Since we are at present concerned only with phase differences, we may rewrite (153) in the form

$$\xi_l = \xi_l^{(0)} \sin \omega t \quad \text{and} \quad \xi_r = \xi_r^{(0)} \sin(\omega t - \delta). \tag{154}$$

Let the second component be subject to a constant retardation ϵ so that

$$\xi_l = \xi_l^{(0)} \sin \omega t \quad \text{and} \quad \xi_r = \xi_r^{(0)} \sin(\omega t - \delta - \epsilon). \tag{155}$$

Now resolving the vibration (155) in a direction making an angle ψ with the l direction we have

$$\xi_l^{(0)} \sin \omega t \cos \psi + \xi_r^{(0)} \sin(\omega t - \delta - \epsilon) \sin \psi$$
$$= [\xi_l^{(0)} \cos \psi + \xi_r^{(0)} \cos(\delta + \epsilon) \sin \psi] \sin \omega t -$$
$$- \xi_r^{(0)} \sin(\delta + \epsilon) \sin \psi \cos \omega t. \tag{156}$$

The momentary intensity is therefore given by

$$\xi^2(\psi; \epsilon) = [\xi_l^{(0)}]^2 \cos^2\psi + [\xi_r^{(0)}]^2 \sin^2\psi +$$
$$+ 2\xi_l^{(0)}\xi_r^{(0)}(\cos \delta \cos \epsilon - \sin \delta \sin \epsilon)\sin \psi \cos \psi. \tag{157}$$

To get the apparent intensity in the direction ψ we must take the mean of this expression, keeping ψ and ϵ constant. Thus

$$I(\psi; \epsilon) = \overline{[\xi_l^{(0)}]^2}\cos^2\psi + \overline{[\xi_r^{(0)}]^2}\sin^2\psi +$$
$$+ \{2\overline{[\xi_l^{(0)}\xi_r^{(0)}\cos\delta]}\cos\epsilon - 2\overline{[\xi_l^{(0)}\xi_r^{(0)}\sin\delta]}\sin\epsilon\}\sin\psi\cos\psi. \qquad (158)$$

From this equation it is clear that the intensities in the directions l and r are independent of ϵ and are given by

$$I_l = \overline{[\xi_l^{(0)}]^2} \quad \text{and} \quad I_r = \overline{[\xi_r^{(0)}]^2}. \qquad (159)$$

Now let

$$U = 2\overline{[\xi_l^{(0)}\xi_r^{(0)}\cos\delta]} \quad \text{and} \quad V = 2\overline{[\xi_l^{(0)}\xi_r^{(0)}\sin\delta]}. \qquad (160)$$

It will be observed that these expressions for U and V are in agreement with the earlier definitions (151) for an elliptically polarized beam since in the latter case the phase difference is a constant.

According to equations (159) and (160) we can rewrite equation (158) in the form

$$I(\psi; \epsilon) = I_l\cos^2\psi + I_r\sin^2\psi + \tfrac{1}{2}(U\cos\epsilon - V\sin\epsilon)\sin 2\psi, \qquad (161)$$

or, defining

$$I = I_l + I_r = \overline{[\xi_l^{(0)}]^2} + \overline{[\xi_r^{(0)}]^2}$$

and

$$Q = I_l - I_r = \overline{[\xi_l^{(0)}]^2} - \overline{[\xi_r^{(0)}]^2}, \qquad (162)$$

we can also write

$$I(\psi; \epsilon) = \tfrac{1}{2}[I + Q\cos 2\psi + (U\cos\epsilon - V\sin\epsilon)\sin 2\psi]. \qquad (163)$$

From equation (163) it follows that the character of an arbitrarily polarized light, in so far as experimental tests can reveal, is completely determined by the intensities in two directions at right angles to each other (or, equivalently, the total intensity I and $Q = I_l - I_r$) and the parameters U and V. The intensities I, Q, U, and V are the general Stokes parameters representing light.

Two beams characterized by the same set of Stokes's parameters are said to be *equivalent* since two such beams cannot be distinguished by any optical analysis such as could be effected by transmission through doubly refracting crystals, reflection, etc.

Again, from equation (163) it follows that *when several independent streams of light are combined, the Stokes parameters for the mixture is the sum of the respective Stokes parameters of the separate streams.* It is to be particularly emphasized that *this additivity of the Stokes parameters holds only so long as the component streams forming a mixture have no permanent phase relations between themselves.* This is what is understood by the streams being *independent.*

For a beam resulting from a mixture of several independent streams of elliptically polarized light the Stokes parameters are therefore given by (cf. eqs. [147])

$$I = \sum I^{(n)}; \qquad I_l = \sum I_l^{(n)}; \qquad I_r = \sum I_r^{(n)},$$

$$Q = \sum Q^{(n)} = \sum I^{(n)} \cos 2\beta_n \cos 2\chi_n,$$

$$U = \sum U^{(n)} = \sum I^{(n)} \cos 2\beta_n \sin 2\chi_n,$$

and

$$V = \sum V^{(n)} = \sum I^{(n)} \sin 2\beta_n, \tag{164}$$

where $I^{(n)}$, χ_n, and β_n define the intensity, the plane of polarization, and the ellipticity of the component streams.

15.3. *Natural light as a mixture of two independent oppositely polarized streams of equal intensity*

The experimental definition of natural light is that when resolved in any direction in the transverse plane the apparent intensity is the same, and this is further unaffected by any previous retardation of one of the rectangular components relative to the other into which it may have been resolved. In other words, for natural light we must require

$$I(\psi; \epsilon) = \tfrac{1}{2}I, \quad \text{independently of } \psi \text{ and } \epsilon. \tag{165}$$

Hence *the necessary and sufficient condition that light be natural* is

$$Q = U = V = 0. \tag{166}$$

This is the analytical representation of natural light.

We shall now examine the circumstances under which two independent streams of polarized light when mixed will result in natural light.

Let (χ_1, β_1) refer to the first and (χ_2, β_2) refer to the second stream, and let the intensities of the two streams be in the ratio $1:q$. The resulting mixture will be equivalent to natural light, if and only if (cf. eqs. [164])

$$\sin 2\beta_1 + q \sin 2\beta_2 = 0,$$

$$\cos 2\chi_1 \cos 2\beta_1 + q \cos 2\chi_2 \cos 2\beta_2 = 0,$$

and

$$\sin 2\chi_1 \cos 2\beta_1 + q \sin 2\chi_2 \cos 2\beta_2 = 0. \tag{167}$$

Transposing, squaring, and adding, we find $q^2 = 1$ and since q is positive, $q = 1$. Since β_1 and β_2 are supposed to lie between $-\tfrac{1}{2}\pi$ and $+\tfrac{1}{2}\pi$ we next conclude from the first of the equations (167) that

$$-\beta_2 = \beta_1 \quad \text{or} \quad -\beta_2 = \pm\tfrac{1}{2}\pi - \beta_1 \tag{168}$$

where $+$ or $-$ sign in the second alternative should be taken depending on whether β_1 is positive or negative.

Now it is apparent that the same physical situation can be analytically expressed in two ways: Thus (β_1, χ_1) and $(\tfrac{1}{2}\pi - \beta_1, \chi_1 + \tfrac{1}{2}\pi)$ represent the same physical situation for right-handed polarization and similarly (β_1, χ_1) and $(-\tfrac{1}{2}\pi - \beta_1, \chi_1 + \tfrac{1}{2}\pi)$ represent the same physical situation for left-handed polarization. We may therefore reject the second alternative in (168) as expressing the same solution as the first in a different way. The second and third equations then give

$$\cos 2\chi_1 = -\cos 2\chi_2 \quad \text{and} \quad \sin 2\chi_1 = -\sin 2\chi_2.$$

Hence χ_1 and χ_2 differ by $90°$:

$$|\chi_1 - \chi_2| = \tfrac{1}{2}\pi. \tag{169}$$

The equations (167) are also satisfied by $\beta_1 = -\beta_2 = \pm 45°$; but this solution is only a particular case of (168).

We have thus proved that two independent streams of elliptically polarized light of equal intensity are together equivalent to natural light, if and only if the ellipses described in the case of the two streams are similar, their major axes perpendicular to each other, and the sense of rotation in one stream contrary to that in the other (see Fig. 6).

Two streams related in the manner

$$(\beta, \chi) \quad \text{and} \quad (-\beta, \chi + \tfrac{1}{2}\pi) \tag{170}$$

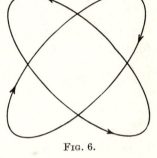

FIG. 6.

are said to be *oppositely polarized*. We may now restate the result enunciated in the preceding paragraph as follows: *Natural light is equivalent to any two independent oppositely polarized streams of half the intensity; and no two independent polarized streams can together be equivalent to natural light unless they be oppositely polarized and of equal intensity.*

The concept of opposite polarization (due to Stokes) to which we have been led is of importance also in another connexion. For it can be shown that the total intensity of a mixture of oppositely polarized beams (independent or not) is unaffected by any retardation of phase of one stream relative to another. Two oppositely polarized streams can therefore never interfere.

15.4. *The representation of an arbitrarily polarized light as a mixture of two independent oppositely polarized streams*

We shall first prove a theorem due to Stokes that *the most general mixture of light can be regarded as a mixture of an elliptically polarized stream and an independent stream of natural light.*

To prove Stokes's theorem, it is first necessary to observe that among the quantities I, Q, U, and V there always exists the inequality

$$I^2 \geqslant Q^2 + U^2 + V^2, \tag{171}$$

or, according to the definitions of these quantities (cf. eqs. [160]–[163])

$$\overline{[\xi_l^{(0)}]^2} \times \overline{[\xi_r^{(0)}]^2} \geqslant \overline{[\xi_l^{(0)}\xi_r^{(0)}\cos\delta]}^2 + \overline{[\xi_l^{(0)}\xi_r^{(0)}\sin\delta]}^2. \tag{172}$$

This last inequality can be established in the following manner.

Let the amplitudes and the phase difference remaining constant (or, sensibly constant) during a time interval proportional to t_1 have the values $\xi_l^{(0,1)}$, $\xi_r^{(0,1)}$, and δ_1, respectively. Similarly, let the values distinguished by the indices 2, 3,..., occur during times proportional to t_2, t_3,.... Then

$$\overline{[\xi_l^{(0)}]^2} \times \overline{[\xi_r^{(0)}]^2} = \left(\sum_n t_n [\xi_l^{(0,n)}]^2\right)\left(\sum_m t_m [\xi_r^{(0,m)}]^2\right)$$
$$= \sum_n t_n^2 [\xi_l^{(0,n)}\xi_r^{(0,n)}]^2 + \sum_{(n,m)} t_n t_m \{[\xi_l^{(0,n)}\xi_r^{(0,m)}]^2 + [\xi_l^{(0,m)}\xi_r^{(0,n)}]^2\}, \tag{173}$$

where (n, m) indicates that the summation is to be extended over all distinct pairs of intervals n and m. Similarly

$$\overline{[\xi_l^{(0)}\xi_r^{(0)}\cos\delta]}^2 + \overline{[\xi_l^{(0)}\xi_r^{(0)}\sin\delta]}^2 = \sum_n t_n^2 [\xi_l^{(0,n)}\xi_r^{(0,n)}]^2 +$$
$$+ 2\sum_{(n,m)} t_n t_m \xi_l^{(0,n)}\xi_r^{(0,n)}\xi_l^{(0,m)}\xi_r^{(0,m)}[\cos\delta_n \cos\delta_m + \sin\delta_n \sin\delta_m]. \tag{174}$$

Hence

$$I^2 - Q^2 - U^2 - V^2 = \sum_{(n,m)} t_n t_m \{[\xi_l^{(0,n)}\xi_r^{(0,m)}]^2 + [\xi_l^{(0,m)}\xi_r^{(0,n)}]^2 -$$
$$- 2\xi_l^{(0,n)}\xi_r^{(0,m)}\xi_l^{(0,m)}\xi_r^{(0,n)}\cos(\delta_n - \delta_m)\}. \tag{175}$$

Each of the summands on the right-hand side is essentially positive. And the only case in which the entire sum can vanish is when

$$\delta_n = \delta_m \quad \text{and} \quad \frac{\xi_l^{(0,n)}}{\xi_r^{(0,n)}} = \frac{\xi_l^{(0,m)}}{\xi_r^{(0,m)}}, \tag{176}$$

for all pairs (n, m), i.e. when the ratio of the amplitudes and the difference in phase remain constant through all fluctuations: these are the conditions for the light to be elliptically polarized.

We have thus proved that

$$I^2 \geqslant Q^2 + U^2 + V^2$$

and that the equality can occur, when and only when the light is elliptically polarized.

Since we have already shown that for an elliptically polarized beam $I^2 = Q^2 + U^2 + V^2$ (eq. [148]), it follows that the existence of this equality is a necessary and sufficient condition for a beam to be elliptically polarized. However, in general

$$I > (Q^2 + U^2 + V^2)^{\frac{1}{2}}, \tag{177}$$

and the original light can be resolved into two independent groups characterized by the Stokes parameters

$$\{I - (Q^2 + U^2 + V^2)^{\frac{1}{2}}, \ 0, \ 0, \ 0\} \tag{178}$$

and

$$\{(Q^2 + U^2 + V^2)^{\frac{1}{2}}, \ Q, \ U, \ V\}. \tag{179}$$

The former represents natural light and the latter, as we have seen, must represent an elliptically polarized beam, the plane of polarization and the ellipticity of which are given by

$$\tan 2\chi = \frac{U}{Q} \quad \text{and} \quad \sin 2\beta = \frac{V}{(Q^2 + U^2 + V^2)^{\frac{1}{2}}}. \tag{180}$$

These formulae can always be satisfied. It is therefore always possible to represent a general mixture of light by a stream of natural light and a stream of elliptically polarized light independent of the former. Moreover, there is only one way in which this resolution can be accomplished. For though the second of the equations (180) gives two values of β complementary to each other, these values, as we have already explained, represent only two different ways of expressing the same result. If we choose that value of β which is numerically the smaller, then among the different values of χ differing by 90° which satisfy the first of the equations (180) we must choose that which makes $\cos 2\chi$ the same sign as Q.

An alternative way of resolving a partially polarized beam defined by the Stokes parameters I, Q, U, and V is to express it as the resultant of two independent streams of elliptically polarized light in the states of opposite polarization (β, χ) and $(-\beta, \chi + \frac{1}{2}\pi)$ where β and χ are given by equations (180). The intensities in the two states can be readily written down when it is remembered that natural light is equivalent to a mixture of any two independent streams of oppositely polarized light of equal intensity. The component (178) of natural light is therefore equivalent to two independent polarized beams, each of intensity

$$\tfrac{1}{2}[I - (Q^2 + U^2 + V^2)^{\frac{1}{2}}], \tag{181}$$

in the states of polarization (β, χ) and $(-\beta, \chi + \frac{1}{2}\pi)$. Combining these components of the natural light (178) with the component (179) already in the state of polarization (β, χ), we conclude that *a beam* *characterized by the parameters I, Q, U, and V is equivalent to two independent streams of elliptically polarized light of intensities*

$$I^{(+)} = \tfrac{1}{2}[I + (Q^2 + U^2 + V^2)^{\frac{1}{2}}]$$

and

$$I^{(-)} = \tfrac{1}{2}[I - (Q^2 + U^2 + V^2)^{\frac{1}{2}}], \tag{182}$$

in the states of opposite polarization

$$(\beta, \chi) \quad and \quad (-\beta, \chi + \tfrac{1}{2}\pi), \tag{183}$$

where

$$\tan 2\chi = \frac{U}{Q} \quad and \quad \sin 2\beta = \frac{V}{(Q^2 + U^2 + V^2)^{\frac{1}{2}}}. \tag{184}$$

15.5. *The law of transformation of the Stokes parameters for a rotation of the axes*

In our discussion of the Stokes parameters we have so far referred them to some chosen fixed rectangular axes. We shall now obtain the law of transformation of these parameters for a rotation of the axes.

Since any general mixture of light is equivalent to two independent streams of oppositely polarized light, it is clear that the required law of transformation can be obtained by considering the case of an elliptically polarized beam for which the Stokes parameters are given by equations (151) and (152). From these equations it is at once apparent that *the total intensity I and the parameter V are invariant for a rotation of the axes*. But Q and U change with the axes, and if Q' and U' are the values of the parameters when the axes are rotated through an angle ϕ in the *clockwise* direction, then clearly

$$Q' = I \cos 2\beta \cos 2(\chi - \phi) \quad and \quad U' = I \cos 2\beta \sin 2(\chi - \phi), \tag{185}$$

or $\quad Q' = Q \cos 2\phi + U \sin 2\phi \quad and \quad U' = -Q \sin 2\phi + U \cos 2\phi. \tag{186}$

In our subsequent work we shall find it more convenient to use the intensities (I_l and I_r) in two directions at right angles to each other and the parameters U and V than the original set I, Q, U, and V used by Stokes.

The law of transformation of I_l, I_r, U, and V for a rotation of the axes can be readily written down from equations (186) and the invariance of I and V. Thus

$$I_\phi + I_{\phi + \frac{1}{2}\pi} = I_l + I_r; \qquad V' = V,$$
$$I_\phi - I_{\phi + \frac{1}{2}\pi} = (I_l - I_r)\cos 2\phi + U \sin 2\phi,$$

and

$$U' = -(I_l - I_r)\sin 2\phi + U \cos 2\phi, \tag{187}$$

or

$$I_\phi = I_l \cos^2\phi + I_r \sin^2\phi + \tfrac{1}{2}U \sin 2\phi,$$
$$I_{\phi + \frac{1}{2}\pi} = I_l \sin^2\phi + I_r \cos^2\phi - \tfrac{1}{2}U \sin 2\phi,$$
$$U' = -I_l \sin 2\phi + I_r \sin 2\phi + U \cos 2\phi,$$

and

$$V' = V. \tag{188}$$

Hence *if* $$I = (I_l, I_r, U, V) \qquad (189)$$

denotes a vector whose components are the four parameters I_l, I_r, U, *and* V *which characterize an arbitrarily (partially) polarized light, then the effect of a rotation of the axes through an angle* ϕ *in the clockwise direction is to subject* I *to the linear transformation*

$$L(\phi) = \begin{pmatrix} \cos^2\phi & \sin^2\phi & \tfrac{1}{2}\sin 2\phi & 0 \\ \sin^2\phi & \cos^2\phi & -\tfrac{1}{2}\sin 2\phi & 0 \\ -\sin 2\phi & \sin 2\phi & \cos 2\phi & 0 \\ 0 & 0 & 0 & 1 \end{pmatrix}. \qquad (190)$$

It is to be noticed that $L(\phi)$ is reducible with respect to V.

It is evident that $L(\phi)$ must satisfy the group relations

$$L(\phi_1)L(\phi_2) = L(\phi_1 + \phi_2) \quad \text{and} \quad L^{-1}(\phi) = L(-\phi), \qquad (191)$$

relations which can be directly verified by using the representation (190).

16. Rayleigh scattering

The simplest and in some ways the most important example of a physical law of light scattering which has found wide application is that discovered by Lord Rayleigh in 1871 in the context of his accounting for the blue of the sky. Though Rayleigh's earliest considerations were related to the scattering by a dielectric sphere of radius small compared with the wave-length of light, it was soon recognized by Maxwell and later by Rayleigh himself that the law is of much wider scope and generality. For example, we know now that the angular distribution and the state of polarization of the scattered light according to Rayleigh applies also to the Thomson scattering by free electrons.

Rayleigh's law as commonly formulated states that when a pencil of natural light of wave-length λ, intensity I, and solid angle $d\omega$, is incident on a particle of polarizability α, energy at the rate

$$\frac{128\pi^5}{3\lambda^4}\, \alpha^2 I\, d\omega \times \tfrac{3}{4}(1+\cos^2\Theta)\frac{d\omega'}{4\pi} \qquad (192)$$

is scattered in a direction making an angle Θ with the direction of incidence and in a solid angle $d\omega'$; that the scattered light is partially plane-polarized; that the plane of polarization is at right angles to the *plane of scattering*†; and finally, that the intensities of the scattered light in directions (in the transverse plane containing the electric and the magnetic vectors) parallel and perpendicular respectively to the plane of scattering are in the ratio $\cos^2\Theta : 1$.

---

† This is the plane which contains the directions of the incident and the scattered light.

According to (192) the scattering coefficient, σ, per particle is

$$\sigma = \frac{128\pi^5}{3\lambda^4}\alpha^2. \tag{193}$$

For electrons, the *Thomson scattering coefficient*

$$\sigma_e = \frac{8\pi e^4}{3m_e^2 c^4} \tag{194}$$

can be obtained by setting

$$\alpha = \left(\frac{\lambda}{c}\right)^2 \frac{e^2}{4\pi^2 m_e} \tag{195}$$

in (193). (In eqs. [194] and [195] c denotes the velocity of light, e the charge on the electron, and m_e the mass of the electron.)

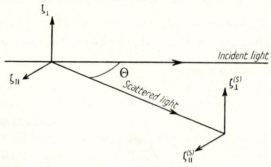

FIG. 7. The coordinate system for referring to the state of polarization of the incident and the scattered light in single scattering: the scattering plane is the plane containing the directions of the incident and the scattered light; and ∥ and ⊥ refer to directions in the transverse planes (of the incident and the scattered light) parallel and perpendicular respectively to the plane of scattering.

The usual statement of Rayleigh's law which we have given is inadequate for the purposes of formulating the relevant equations of transfer, for it does not tell us how a partially polarized beam will be scattered. It is clear that what we require is a statement of the law which will enable us to relate the Stokes parameters of the scattered light with those of the incident light. Consider, then, the incidence of an arbitrarily polarized beam on a particle. Let the momentary vibrations representing the incident beam resolved along directions parallel and perpendicular respectively to the plane of scattering be (see Fig. 7)

$$\xi_{\parallel} = \xi_{\parallel}^{(0)}\sin(\omega t - \epsilon_1) \quad \text{and} \quad \xi_{\perp} = \xi_{\perp}^{(0)}\sin(\omega t - \epsilon_2). \tag{196}$$

The complete statement of Rayleigh's law is that the vibrations representing the light scattered in a direction making an angle Θ with the direction of incidence is

$$\xi_{\parallel}^{(s)} = (\tfrac{3}{2}\sigma)^{\frac{1}{2}}\xi_{\parallel}^{(0)}\cos\Theta\sin(\omega t - \epsilon_1)$$

and
$$\xi_{\perp}^{(s)} = (\tfrac{3}{2}\sigma)^{\frac{1}{2}}\xi_{\perp}^{(0)}\sin(\omega t - \epsilon_2), \tag{197}$$

where *the phase, (ϵ_1, ϵ_2), and the amplitude, $(\xi_\parallel^{(0)}, \xi_\perp^{(0)})$, relations in the incident beam are maintained, unaltered, in the scattered beam.* Accordingly, the parameters representing the scattered light are proportional to

$$\tfrac{3}{2}\sigma\overline{[\xi_\parallel^{(0)}]^2}\cos^2\Theta = \tfrac{3}{2}\sigma I_\parallel\cos^2\Theta,$$

$$\tfrac{3}{2}\sigma\overline{[\xi_\perp^{(0)}]^2} = \tfrac{3}{2}\sigma I_\perp,$$

$$\tfrac{3}{2}\sigma\overline{[2\xi_\parallel^{(0)}\xi_\perp^{(0)}\cos(\epsilon_1-\epsilon_2)]}\cos\Theta = \tfrac{3}{2}\sigma U\cos\Theta,$$

and $\qquad \tfrac{3}{2}\sigma\overline{[2\xi_\parallel^{(0)}\xi_\perp^{(0)}\sin(\epsilon_1-\epsilon_2)]}\cos\Theta = \tfrac{3}{2}\sigma V\cos\Theta. \qquad (198)$

Therefore, denoting the incident light by the vector

$$\boldsymbol{I} = (I_\parallel, I_\perp, U, V), \qquad (199)$$

we can express the scattered intensity in the direction Θ by

$$\left(\sigma\frac{d\omega'}{4\pi}\right)\boldsymbol{R}\boldsymbol{I}\,d\omega, \qquad (200)$$

where $\qquad \boldsymbol{R} = \dfrac{3}{2}\begin{pmatrix} \cos^2\Theta & 0 & 0 & 0 \\ 0 & 1 & 0 & 0 \\ 0 & 0 & \cos\Theta & 0 \\ 0 & 0 & 0 & \cos\Theta \end{pmatrix}. \qquad (201)$

For natural light $I_\parallel = I_\perp = \tfrac{1}{2}I$ and $U = V = 0$ and we verify that Rayleigh's law as expressed by (200) and (201) reduces to the earlier, more usual, statement of the law.

We may call $\boldsymbol{R}$ the *phase matrix* for Rayleigh scattering (cf. [26] and [200]).

17. The equation of transfer for an atmosphere scattering radiation in accordance with Rayleigh's law

When we consider the radiation field in an atmosphere in which each particle (atom, molecule, or electron as the case may be) scatters according to a definite physical law (such as Rayleigh's) we must first inquire how an element of volume containing a large number of particles, as distinct from a single particle, will scatter light. The question at issue here is whether the light scattered by the different particles in a small element of volume can be considered independent or not. While a rigorous examination of this question requires some careful consideration, it is, nevertheless, fairly evident that if the scattering centres are distributed in a perfectly random fashion (as in a gas obeying Maxwell's law) there would be no permanent correlations in the phases of the light scattered by the different particles. And if this is the case, the light scattered by the different particles will indeed be independent and we can add the Stokes parameters. In our future discussions we shall assume that these circumstances always prevail. A further consequence of this assumption is that the light in the different directions through the same point may also be considered independent.

According to our remarks in the preceding paragraph, we can introduce a mass scattering coefficient κ given by

$$\kappa = \frac{\sigma}{\rho} N, \tag{202}$$

where N is the number of particles per unit volume and ρ is the density; further, we may write for the scattering by an element of mass dm the expression (cf. [200])

$$\left(\kappa \, dm \, \frac{d\omega'}{4\pi}\right) \boldsymbol{RI} \, d\omega, \tag{203}$$

where it should be particularly noted that $\boldsymbol{I}$ is defined in a rectangular system of coordinates in which the directions parallel and perpendicular to the plane of scattering define the axes.

For electrons we have already given Thomson's expression for κ (eq. [194]). For the case of molecular scattering, Rayleigh's formula for κ can be obtained from (193) by setting

$$\alpha = \frac{n^2 - 1}{4\pi N}, \tag{204}$$

where n denotes the refractive index of the medium; thus

$$\kappa = \frac{8\pi^3}{3} \frac{(n^2 - 1)^2}{\lambda^4 N \rho}. \tag{205}$$

17.1 The equation of transfer for $\boldsymbol{I}(\vartheta, \varphi)$

Proceeding now to the formulation of the equation of transfer, we shall characterize the radiation field at each point by the four intensities $I_l(\vartheta, \varphi)$, $I_r(\vartheta, \varphi)$, $U(\vartheta, \varphi)$, and $V(\vartheta, \varphi)$, where ϑ and φ are the polar angles referred to an appropriately chosen coordinate system through the point under consideration (see Fig. 8) and l and r refer to the directions in the meridian plane and at right angles to it, respectively.

Letting $\quad \boldsymbol{I}(\vartheta, \varphi) = [I_l(\vartheta, \varphi), I_r(\vartheta, \varphi), U(\vartheta, \varphi), V(\vartheta, \varphi)], \tag{206}$

we can formally write the equation of transfer in the vector form

$$-\frac{d\boldsymbol{I}(\vartheta, \varphi)}{\kappa \rho \, ds} = \boldsymbol{I}(\vartheta, \varphi) - \mathfrak{J}(\vartheta, \varphi), \tag{207}$$

where $\mathfrak{J}(\vartheta, \varphi)$ denotes the vector source function for $\boldsymbol{I}(\vartheta, \varphi)$.

To evaluate $\mathfrak{J}(\vartheta, \varphi)$, consider the contribution $d\mathfrak{J}(\vartheta, \varphi; \vartheta', \varphi')$, to the source function arising from the scattering of a pencil of radiation of solid angle $d\omega'$ in the direction (ϑ', φ'). This will be given by

$$RI\frac{d\omega'}{4\pi} \tag{208}$$

if $I(\vartheta', \varphi')$ is referred to the directions parallel and perpendicular to the plane of scattering. But as we have introduced it, $I(\vartheta', \varphi')$ is referred

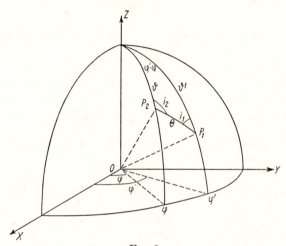

Fig. 8.

to directions along the meridian plane $OP_1 Z$ and at right angles to it. However, according to § 15 (§ 15.5, eq. [190]) we can transform $I(\vartheta', \varphi')$ to the directions required for using (208) by applying to $I(\vartheta', \varphi')$ the linear transformation $L(-i_1)$, where i_1 denotes the angle between the meridian plane $OP_1 Z$ through $P_1 \ (= (\vartheta', \varphi'))$ and the plane of scattering $OP_1 P_2$. Consequently, the contribution to the source function from the scattering of the pencil in the direction (ϑ', φ') is

$$R(\cos\Theta)L(-i_1)I(\vartheta', \varphi')\frac{d\omega'}{4\pi}. \tag{209}$$

But (209) refers the Stokes parameters to directions at P_2, parallel and perpendicular to the plane of scattering. To transform (209) to the chosen coordinate axes at P_2 (namely, the directions along the great circle arc $P_2 Z$ and the direction at right angles to this) we must apply to (209) the linear transformation $L(\pi - i_2)$, where i_2 denotes the angle between the planes $OP_2 Z$ and $OP_1 P_2$. Thus, we finally obtain for

$d\mathfrak{J}(\vartheta,\varphi;\vartheta',\varphi')$ the expression

$$d\mathfrak{J}(\vartheta,\varphi;\vartheta',\varphi') = \boldsymbol{L}(\pi-i_2)\boldsymbol{R}(\cos\Theta)\boldsymbol{L}(-i_1)\boldsymbol{I}(\vartheta',\varphi')\frac{d\omega'}{4\pi}. \qquad (210)$$

The source function $\mathfrak{J}(\vartheta,\varphi)$ is now obtained by integrating (210) over all directions (ϑ',φ'). Thus

$$\mathfrak{J}(\vartheta,\varphi) = \frac{1}{4\pi}\int\limits_0^\pi\int\limits_0^{2\pi} \boldsymbol{L}(\pi-i_2)\boldsymbol{R}(\cos\Theta)\boldsymbol{L}(-i_1)\boldsymbol{I}(\vartheta',\varphi')\sin\vartheta'\,d\vartheta'd\varphi'. \qquad (211)$$

We can now write the equation of transfer (207) in the form

$$-\frac{d\boldsymbol{I}(\vartheta,\varphi)}{\kappa\rho\,ds} = \boldsymbol{I}(\vartheta,\varphi)-\frac{1}{4\pi}\int\limits_0^\pi\int\limits_0^{2\pi} \boldsymbol{P}(\vartheta,\varphi;\vartheta',\varphi')\boldsymbol{I}(\vartheta',\varphi')\sin\vartheta'\,d\vartheta'd\varphi', \qquad (212)$$

where the *phase-matrix* $\boldsymbol{P}(\vartheta,\varphi;\vartheta',\varphi')$ is given by

$$\boldsymbol{P}(\vartheta,\varphi;\vartheta',\varphi') = \boldsymbol{L}(\pi-i_2)\boldsymbol{R}(\cos\Theta)\boldsymbol{L}(-i_1). \qquad (213)$$

17.2. *The explicit form of the phase-matrix*

Substituting for $\boldsymbol{L}$ and $\boldsymbol{R}$ according to equations (190) and (201) we have

$$\boldsymbol{P}(\vartheta,\varphi;\vartheta',\varphi') = \frac{3}{2}\begin{pmatrix} \cos^2 i_2 & \sin^2 i_2 & -\frac{1}{2}\sin 2i_2 & 0 \\ \sin^2 i_2 & \cos^2 i_2 & +\frac{1}{2}\sin 2i_2 & 0 \\ \sin 2i_2 & -\sin 2i_2 & \cos 2i_2 & 0 \\ 0 & 0 & 0 & 1 \end{pmatrix} \times$$

$$\times\begin{pmatrix} \cos^2\Theta & 0 & 0 & 0 \\ 0 & 1 & 0 & 0 \\ 0 & 0 & \cos\Theta & 0 \\ 0 & 0 & 0 & \cos\Theta \end{pmatrix}\begin{pmatrix} \cos^2 i_1 & \sin^2 i_1 & -\frac{1}{2}\sin 2i_1 & 0 \\ \sin^2 i_1 & \cos^2 i_1 & +\frac{1}{2}\sin 2i_1 & 0 \\ \sin 2i_1 & -\sin 2i_1 & \cos 2i_1 & 0 \\ 0 & 0 & 0 & 1 \end{pmatrix}, \qquad (214)$$

or,

$$\boldsymbol{P}(\vartheta,\varphi;\vartheta',\varphi') = \frac{3}{2}\begin{pmatrix} \cos^2 i_2 & \sin^2 i_2 & -\frac{1}{2}\sin 2i_2 & 0 \\ \sin^2 i_2 & \cos^2 i_2 & +\frac{1}{2}\sin 2i_2 & 0 \\ \sin 2i_2 & -\sin 2i_2 & \cos 2i_2 & 0 \\ 0 & 0 & 0 & 1 \end{pmatrix} \times$$

$$\times\begin{pmatrix} \cos^2\Theta\cos^2 i_1 & \cos^2\Theta\sin^2 i_1 & -\frac{1}{2}\cos^2\Theta\sin 2i_1 & 0 \\ \sin^2 i_1 & \cos^2 i_1 & \frac{1}{2}\sin 2i_1 & 0 \\ \cos\Theta\sin 2i_1 & -\cos\Theta\sin 2i_1 & \cos\Theta\cos 2i_1 & 0 \\ 0 & 0 & 0 & \cos\Theta \end{pmatrix}. \qquad (215)$$

Introducing the abbreviations

$$(l, l) = \sin i_1 \sin i_2 - \cos i_1 \cos i_2 \cos \Theta,$$

$$(r, r) = \sin i_1 \sin i_2 \cos \Theta - \cos i_1 \cos i_2,$$

$$(r, l) = \sin i_1 \cos i_2 \cos \Theta + \cos i_1 \sin i_2$$

and

$$(l, r) = -(\sin i_2 \cos i_1 \cos \Theta + \cos i_2 \sin i_1), \tag{216}$$

we readily verify that the matrix product on the right-hand side of equation (215) is

$$\boldsymbol{P}(\vartheta, \varphi; \vartheta', \varphi')$$

$$= \frac{3}{2} \begin{pmatrix} (l, l)^2 & (r, l)^2 & (l, l)(r, l) & 0 \\ (l, r)^2 & (r, r)^2 & (l, r)(r, r) & 0 \\ 2(l, l)(l, r) & 2(r, r)(r, l) & (l, l)(r, r) + (r, l)(l, r) & 0 \\ 0 & 0 & 0 & (l, l)(r, r) - (r, l)(l, r) \end{pmatrix}.$$

$$\tag{217}$$

On the other hand, from the spherical triangle ZP_1P_2 we have the relations

$$(l, l) = \sin \vartheta \sin \vartheta' + \cos \vartheta \cos \vartheta' \cos(\varphi' - \varphi),$$

$$(r, l) = +\cos \vartheta \sin(\varphi' - \varphi),$$

$$(l, r) = -\cos \vartheta' \sin(\varphi' - \varphi),$$

and

$$(r, r) = \cos(\varphi' - \varphi). \tag{218}$$

Using these expressions for (l, l) etc., we find that the various matrix elements of $\boldsymbol{P}$ are

$$(l, l)^2 = \tfrac{1}{2}[2(1 - \mu^2)(1 - \mu'^2) + \mu^2 \mu'^2] +$$
$$+ 2\mu\mu'(1 - \mu^2)^{\frac{1}{2}}(1 - \mu'^2)^{\frac{1}{2}} \cos(\varphi' - \varphi) + \tfrac{1}{2}\mu^2\mu'^2 \cos 2(\varphi' - \varphi),$$

$$(r, l)^2 = \tfrac{1}{2}\mu^2[1 - \cos 2(\varphi' - \varphi)]; \quad (l, r)^2 = \tfrac{1}{2}\mu'^2[1 - \cos 2(\varphi' - \varphi)],$$

$$(r, r)^2 = \tfrac{1}{2}[1 + \cos 2(\varphi' - \varphi)],$$

$$(l, l)(r, l) = \mu(1 - \mu^2)^{\frac{1}{2}}(1 - \mu'^2)^{\frac{1}{2}} \sin(\varphi' - \varphi) + \tfrac{1}{2}\mu^2\mu' \sin 2(\varphi' - \varphi),$$

$$(l, l)(l, r) = -\mu'(1 - \mu^2)^{\frac{1}{2}}(1 - \mu'^2)^{\frac{1}{2}} \sin(\varphi' - \varphi) - \tfrac{1}{2}\mu\mu'^2 \sin 2(\varphi' - \varphi),$$

$$(l, r)(r, r) = -\tfrac{1}{2}\mu' \sin 2(\varphi' - \varphi); \quad (r, l)(r, r) = \tfrac{1}{2}\mu \sin 2(\varphi' - \varphi),$$

$$(l, l)(r, r) + (r, l)(l, r) = (1 - \mu^2)^{\frac{1}{2}}(1 - \mu'^2)^{\frac{1}{2}} \cos(\varphi' - \varphi) + \mu\mu' \cos 2(\varphi' - \varphi),$$

and

$$(l, l)(r, r) - (r, l)(l, r) = \mu\mu' + (1 - \mu^2)^{\frac{1}{2}}(1 - \mu'^2)^{\frac{1}{2}} \cos(\varphi' - \varphi), \tag{219}$$

where we have written μ and μ' for $\cos \vartheta$ and $\cos \vartheta'$, respectively.

According to equations (219) we may express the phase-matrix $P(\mu,\varphi;\mu',\varphi')$ in the form

$$P(\mu,\varphi;\mu',\varphi')$$
$$= Q[P^{(0)}(\mu,\mu')+(1-\mu^2)^{\frac{1}{2}}(1-\mu'^2)^{\frac{1}{2}}P^{(1)}(\mu,\varphi;\mu',\varphi')+P^{(2)}(\mu,\varphi;\mu',\varphi')]$$
$$(220)$$

where

$$P^{(0)}(\mu,\mu') = \frac{3}{4}\begin{pmatrix} 2(1-\mu^2)(1-\mu'^2)+\mu^2\mu'^2 & \mu^2 & 0 & 0 \\ \mu'^2 & 1 & 0 & 0 \\ 0 & 0 & 0 & 0 \\ 0 & 0 & 0 & \mu\mu' \end{pmatrix},$$
$$(221)$$

$$P^{(1)}(\mu,\varphi;\mu',\varphi') = \frac{3}{4}\begin{pmatrix} 4\mu\mu'\cos(\varphi'-\varphi) & 0 & 2\mu\sin(\varphi'-\varphi) & 0 \\ 0 & 0 & 0 & 0 \\ -2\mu'\sin(\varphi'-\varphi) & 0 & \cos(\varphi'-\varphi) & 0 \\ 0 & 0 & 0 & \cos(\varphi'-\varphi) \end{pmatrix},$$
$$(222)$$

$$P^{(2)}(\mu,\varphi;\mu',\varphi')$$
$$= \frac{3}{4}\begin{pmatrix} \mu^2\mu'^2\cos 2(\varphi'-\varphi) & -\mu^2\cos 2(\varphi'-\varphi) & \mu^2\mu'\sin 2(\varphi'-\varphi) & 0 \\ -\mu'^2\cos 2(\varphi'-\varphi) & \cos 2(\varphi'-\varphi) & -\mu'\sin 2(\varphi'-\varphi) & 0 \\ -\mu\mu'^2\sin 2(\varphi'-\varphi) & \mu\sin 2(\varphi'-\varphi) & \mu\mu'\cos 2(\varphi'-\varphi) & 0 \\ 0 & 0 & 0 & 0 \end{pmatrix},$$
$$(223)$$

and

$$Q = \begin{pmatrix} 1 & 0 & 0 & 0 \\ 0 & 1 & 0 & 0 \\ 0 & 0 & 2 & 0 \\ 0 & 0 & 0 & 2 \end{pmatrix}.$$
$$(224)$$

It will be observed that

$$\tilde{P}^{(i)}(\mu,\varphi;\mu',\varphi') = P^{(i)}(\mu,\varphi;\mu',\varphi') \quad (i=0,1,2), \tag{225}$$

where $\tilde{P}^{(i)}$ stands for the matrix $P^{(i)}$ after its rows and columns have been interchanged and the variables (μ,φ) and (μ',φ') have also been interchanged. This symmetry of the phase-matrix for transposition is, as we shall see (Chap. VII, § 52), simply the mathematical expression of Helmholtz's principle of reciprocity for single scattering when due allowance is made for the polarization of the scattered light.

It will also be observed that *the matrix P is reducible with respect to the parameter V*; V therefore satisfies an equation of transfer which is independent of the other three parameters.

Finally, we may note that for a plane-parallel atmosphere the equation of transfer can be written in the form

$$\mu \frac{d\mathbf{I}(\tau,\mu,\varphi)}{d\tau} = \mathbf{I}(\tau,\mu,\varphi) - \frac{1}{4\pi} \int_{-1}^{+1} \int_{0}^{2\pi} \mathbf{P}(\mu,\varphi;\mu',\varphi')\mathbf{I}(\tau,\mu',\varphi')\,d\mu'd\varphi',$$

(226)

where τ denotes, as usual, the normal optical thickness, measured, in this instance, in terms of the scattering coefficient κ.

17.3. *The equations of radiative transfer for an electron scattering atmosphere*

Rayleigh scattering is clearly a case of conservative scattering. Consequently, the axially symmetric problem in semi-infinite plane-parallel atmospheres with a constant net flux in the total intensity (I_l+I_r) is one which is physically significant. Indeed, the problem is one of particular astrophysical interest in view of the fairly definite indications we now have that the transfer of radiation in the atmospheres of early type stars with surface temperatures exceeding $15{,}000°$ K. is predominantly controlled by the scattering by free electrons. And, as we have already remarked, Thomson scattering by free electrons agrees with Rayleigh's in the prediction of the angular distribution and the state of polarization of the scattered radiation.

Now in a plane-parallel atmosphere with no incident radiation, the axial symmetry of the radiation field clearly requires that the plane of polarization be along the meridian plane (or, at right angles to it). Consequently, under the circumstances, $U = V = 0$ and the two intensities $I_l(\tau,\mu)$ and $I_r(\tau,\mu)$ suffice to characterize the radiation field. And the equation of transfer governing the intensities I_l and I_r can be written in the form (cf. eqs. [220], [221], and [226]),

$$\mu \frac{d}{d\tau}\begin{pmatrix} I_l(\tau,\mu) \\ I_r(\tau,\mu) \end{pmatrix}$$
$$= \begin{pmatrix} I_l(\tau,\mu) \\ I_r(\tau,\mu) \end{pmatrix} - \frac{3}{8} \int_{-1}^{+1} \begin{pmatrix} 2(1-\mu^2)(1-\mu'^2)+\mu^2\mu'^2 & \mu^2 \\ \mu'^2 & 1 \end{pmatrix}\begin{pmatrix} I_l(\tau,\mu') \\ I_r(\tau,\mu') \end{pmatrix} d\mu'.$$

(227)

Solutions of this equation are required which satisfy the boundary conditions

$$I_l(0,-\mu) = I_r(0,-\mu) \equiv 0 \quad (\tau = 0 \text{ and } 0 < \mu \leqslant 1)$$

and $\quad I_l(\tau,\mu) \prec e^\tau \quad$ and $\quad I_r(\tau,\mu) \prec e^\tau \quad$ for $\quad \tau \to \infty;$ (228)

and, as in similar transfer problems in the theory of stellar atmospheres, greatest interest is again attached to the angular distribution of the emergent radiation.

17.4. *The basic problem in the theory of the illumination of the sky*

The problem of diffuse reflection and transmission by a plane-parallel atmosphere scattering radiation in accordance with Rayleigh's law is clearly the basic one in the theory of the illumination and polarization of the sky. The problem is also of interest for the illumination of the other planets by the sun, particularly Venus and Jupiter.

Though in practice we are mostly interested in the case of incidence of a natural beam of light, from a theoretical point of view it is advantageous to consider the following somewhat more general problem:

A parallel beam of radiation of net flux

$$\pi \boldsymbol{F} = \pi(F_l, F_r, F_U, F_V) \tag{229}$$

per unit area normal to itself in the four Stokes parameters is incident on a plane-parallel atmosphere of optical thickness τ_1 in some specified direction $(-\mu_0, \varphi_0)$. It is required to find the angular distribution and the state of polarization of the light diffusely reflected by the surface $\tau = 0$ and diffusely transmitted below the surface $\tau = \tau_1$.

We shall find it convenient to express the resulting laws of diffuse reflection and transmission in terms of a *scattering matrix* $\boldsymbol{S}(\tau_1; \mu, \varphi; \mu_0, \varphi_0)$ and a *transmission matrix* $\boldsymbol{T}(\tau_1; \mu, \varphi; \mu_0, \varphi_0)$ such that the reflected and the transmitted intensities are given by

$$\boldsymbol{I}(0; +\mu, \varphi) = \frac{1}{4\mu} \boldsymbol{S}(\tau_1; \mu, \varphi; \mu_0, \varphi_0) \boldsymbol{F}$$

and

$$\boldsymbol{I}(\tau_1, -\mu, \varphi) = \frac{1}{4\mu} \boldsymbol{T}(\tau_1; \mu, \varphi; \mu_0, \varphi_0) \boldsymbol{F}. \tag{230}$$

In the definitions (230) we have introduced the factor $1/\mu$ to secure for $\boldsymbol{S}$ and $\boldsymbol{T}$ the symmetry for transposition similar to what we have already noticed for the phase-matrix (eq. [225]).

Distinguishing, as in § 15, between the reduced incident flux prevailing at the various depths and the diffuse radiation field which has arisen in consequence of multiple scattering, we may write the equation of transfer appropriate for the problem of diffuse reflection and transmission in the form

$$\mu \frac{d\boldsymbol{I}(\tau, \mu, \varphi)}{d\tau} = \boldsymbol{I}(\tau, \mu, \varphi) - \frac{1}{4\pi} \int_{-1}^{+1} \int_{0}^{2\pi} \boldsymbol{P}(\mu, \varphi; \mu', \varphi') \boldsymbol{I}(\tau, \mu', \varphi') \, d\mu' d\varphi' -$$
$$- \tfrac{1}{4} e^{-\tau/\mu_0} \boldsymbol{P}(\mu, \varphi; -\mu_0, \varphi_0) \boldsymbol{F}. \tag{231}$$

The *standard problem* in the context of this equation is to solve it with the boundary conditions

$$I(0, -\mu, \varphi) \equiv 0 \quad (0 < \mu \leqslant 1,\ 0 < \varphi \leqslant 2\pi)$$

and $\quad I(\tau_1, +\mu, \varphi) \equiv 0 \quad (0 < \mu \leqslant 1,\ 0 < \varphi \leqslant 2\pi).$ \hfill (232)

We shall later show (Chap. X, § 72) how the solution of the case when there is a 'ground' at $\tau = \tau_1$ can be reduced to this standard problem.

18. Scattering by anisotropic particles

On Rayleigh's law, when light is scattered at right angles to the direction of incidence it is plane-polarized, and this is true independently of the state of polarization of the incident light. (This follows directly from eqs. [200] and [201].) On the other hand, experiments on scattering by gases and liquids have shown that in practice this situation is never fully realized and that the light scattered at right angles to the direction of incidence is always weakly admixtured with a certain proportion of natural light. This imperfection in the polarization of the light transversely scattered has been interpreted by Rayleigh, Cabannes, and L. V. King as due to the anisotropy of the scattering particles. While it is not within the scope of this book to go in any detail into the purely physical aspects of these and similar theories, it is nevertheless important for our purposes to generalize the current treatments of the problem to include the case of scattering of an arbitrarily polarized light characterized by a set of Stokes parameters. This generalization will be required in an eventual complete treatment of the illumination of the sky (cf. Chap. X, § 74).

The basic idea underlying Rayleigh and King's classical theory of scattering by anisotropic particles (molecules) is that a particle is characterized by three planes of symmetry such that if an electric (light) vector is incident along any of the three principal axes defined by the planes of symmetry, it induces dipole moments which are proportional, respectively, to three constants A, B, and C. Thus, if OX, OY, and OZ define the principal axes of a particle, an incident light vector

$$\boldsymbol{\xi} = (\xi_x, \xi_y, \xi_z)$$

induces dipole moments of amounts

$$\alpha_x = A\xi_x, \qquad \alpha_y = B\xi_y, \quad \text{and} \quad \alpha_z = C\xi_z, \hfill (233)$$

along the three axes. In this respect an anisotropic particle differs from a spherical particle for which $A = B = C$ and the induced dipole moment is always parallel to the instantaneous light vector.

Consider now the incidence of an electric vector $\boldsymbol{\xi} = (\xi_1, \xi_2, \xi_3)$ on a particle whose principal axes are along directions specified by the direction cosines (l_1, m_1, n_1), (l_2, m_2, n_2), and (l_3, m_3, n_3). To find the dipole moment induced by ξ in the directions of the chosen axes, $(1, 2, 3)$, we must first resolve ξ along the principal axes of the particle, then use equations (233) to find the induced moment α in the coordinate system (X, Y, Z), and finally apply to α the transformation required to refer it to the coordinate system, $(1, 2, 3)$. Thus

$$\alpha = \begin{pmatrix} \alpha_1 \\ \alpha_2 \\ \alpha_3 \end{pmatrix} = \begin{pmatrix} l_1 & l_2 & l_3 \\ m_1 & m_2 & m_3 \\ n_1 & n_2 & n_3 \end{pmatrix} \begin{pmatrix} A[l_1\xi_1 + m_1\xi_2 + n_1\xi_3] \\ B[l_2\xi_1 + m_2\xi_2 + n_2\xi_3] \\ C[l_3\xi_1 + m_3\xi_2 + n_3\xi_3] \end{pmatrix}.$$

(234)

Introducing the symmetric *polarizability tensor*, (p_{ij}), with the components

$$p_{11} = Al_1^2 + Bl_2^2 + Cl_3^2; \qquad p_{12} = p_{21} = Al_1 m_1 + Bl_2 m_2 + Cl_3 m_3,$$
$$p_{22} = Am_1^2 + Bm_2^2 + Cm_3^2; \quad p_{23} = p_{32} = Am_1 n_1 + Bm_2 n_2 + Cm_3 n_3,$$
$$p_{33} = An_1^2 + Bn_2^2 + Cn_3^2; \quad p_{31} = p_{13} = An_1 l_1 + Bn_2 l_2 + Cn_3 l_3, \qquad (235)$$

we can write

$$\alpha_i = \sum_{j=1}^{n} p_{ij}\xi_j \quad (i = 1, 2, 3). \tag{236}$$

Let the instantaneous vibration of the electric vector representing the incident light be (see Fig. 7)

$$\xi_1 = \xi_\| = \xi_\|^{(0)} \sin(\omega t - \epsilon_1); \quad \xi_2 = 0; \quad \xi_3 = \xi_\perp = \xi_\perp^{(0)} \sin(\omega t - \epsilon_2). \tag{237}$$

According to equation (236), the components of the induced dipole moment are

$$\alpha_1 = p_{11}\xi_\| + p_{13}\xi_\perp; \quad \alpha_2 = p_{21}\xi_\| + p_{23}\xi_\perp; \quad \alpha_3 = p_{31}\xi_\| + p_{33}\xi_\perp. \tag{238}$$

Now on the classical theory *the vibration representing the scattered light is proportional to the projection of the induced dipole moment on the plane at right angles to the direction of scattering.* For the light scattered in a direction Θ, we can therefore write

$$\xi_\|^{(s)} = \alpha_1 \cos\Theta - \alpha_2 \sin\Theta \quad \text{and} \quad \xi_\perp^{(s)} = \alpha_3, \tag{239}$$

where for the sake of convenience we have omitted writing a constant of proportionality.

From equations (237)–(239) we now have

$$\xi_\|^{(s)} = (p_{11}\cos\Theta - p_{21}\sin\Theta)\xi_\|^{(0)}\sin(\omega t - \epsilon_1) +$$
$$+ (p_{13}\cos\Theta - p_{23}\sin\Theta)\xi_\perp^{(0)}\sin(\omega t - \epsilon_2)$$

and

$$\xi_\perp^{(s)} = p_{31}\xi_\|^{(0)}\sin(\omega t - \epsilon_1) + p_{33}\xi_\perp^{(0)}\sin(\omega t - \epsilon_2). \tag{240}$$

It is important to note that in these expressions representing the scattered light, the phase, (ϵ_1, ϵ_2), and the amplitude, $(\xi_\parallel^{(0)}, \xi_\perp^{(0)})$, relations in the incident beam are maintained, unaltered.

Expanding the sines and cosines in equations (240) we have

$$\xi_\parallel^{(s)} =$$

$$[(p_{11}\cos\Theta - p_{21}\sin\Theta)\xi_\parallel^{(0)}\cos\epsilon_1 + (p_{13}\cos\Theta - p_{23}\sin\Theta)\xi_\perp^{(0)}\cos\epsilon_2]\sin\omega t -$$

$$-[(p_{11}\cos\Theta - p_{21}\sin\Theta)\xi_\parallel^{(0)}\sin\epsilon_1 + (p_{13}\cos\Theta - p_{23}\sin\Theta)\xi_\perp^{(0)}\sin\epsilon_2]\cos\omega t$$

$$= a_1\sin\omega t - b_1\cos\omega t \quad \text{(say)}, \tag{241}$$

and

$$\xi_\perp^{(s)} = [p_{31}\xi_\parallel^{(0)}\cos\epsilon_1 + p_{33}\xi_\perp^{(0)}\cos\epsilon_2]\sin\omega t -$$

$$-[p_{31}\xi_\parallel^{(0)}\sin\epsilon_1 + p_{33}\xi_\perp^{(0)}\sin\epsilon_2]\cos\omega t$$

$$= a_2\sin\omega t - b_2\cos\omega t \quad \text{(say)}. \tag{242}$$

The Stokes parameters for the scattered light must now be found according to equations (159) and (160), where it must be noted that in finding the various mean values we must now average not only over the fluctuating amplitudes and phases in the incident beam but also over the different orientations of the particle (i.e. also over all the directions l_1, m_1, n_1, etc.). However, the two averagings we have just mentioned can be performed independently of each other.†

According to equations (241) and (242), the required Stokes parameters are proportional to

$$\overline{a_1^2 + b_1^2}, \quad \overline{a_2^2 + b_2^2}, \quad 2\overline{(a_1 a_2 + b_1 b_2)}, \quad \text{and} \quad 2\overline{(a_2 b_1 - a_1 b_2)}, \ddagger \tag{243}$$

respectively. The evaluation of these various mean values is considerably simplified when use is made of the fact that when averaging over the different orientations of the particle, the only non-vanishing contributions come from terms of the type

$$\overline{p_{ii}^2}, \quad \overline{p_{ij}^2}, \quad \text{and} \quad \overline{p_{ii}p_{jj}}. \tag{244}$$

† In the quantum theory, at this stage, difficulties can arise (cf. § 19).

‡ These expressions for the Stokes parameters become apparent when we write the equations (241) and (242) in the standard form

$$\xi_\parallel^{(s)} = \xi_\parallel^{(0,s)}\sin(\omega t - \delta_1) \quad \text{and} \quad \xi_\perp^{(s)} = \xi_\perp^{(0,s)}\sin(\omega t - \delta_2)$$

by making the substitutions

$$\xi_\parallel^{(0,s)} = (a_1^2 + b_1^2)^{\frac{1}{2}}, \qquad \xi_\perp^{(0,s)} = (a_2^2 + b_2^2)^{\frac{1}{2}}$$

$$\tan\delta_1 = \frac{b_1}{a_1} \quad \text{and} \quad \tan\delta_2 = \frac{b_2}{a_2}.$$

From these equations it readily follows that

$$[\xi_\parallel^{(0,s)}]^2 = \overline{a_1^2 + b_1^2}; \qquad [\xi_\perp^{(0,s)}]^2 = \overline{a_2^2 + b_2^2},$$

$$2[\xi_\parallel^{(0,s)}\xi_\perp^{(0,s)}\cos(\delta_1 - \delta_2)] = 2\overline{(a_1 a_2 + b_1 b_2)},$$

and

$$2[\xi_\parallel^{(0,s)}\xi_\perp^{(0,s)}\sin(\delta_1 - \delta_2)] = 2\overline{(a_2 b_1 - a_1 b_2)}.$$

These averages have the values

$$\overline{p_{11}^2} = \overline{p_{22}^2} = \overline{p_{33}^2} = \overline{m_1^4} \sum A^2 + 2\overline{m_1^2 m_2^2} \sum BC$$

$$= \tfrac{1}{15}[3 \sum A^2 + 2 \sum BC] = \overline{K^2} \quad \text{(say)},$$

$$\overline{p_{12}^2} = \overline{p_{23}^2} = \overline{p_{31}^2} = \overline{m_1^2 n_1^2} \sum A^2 + 2\overline{m_1 n_1 m_2 n_2} \sum BC$$

$$= \tfrac{1}{15}[\sum A^2 - \sum BC] = \overline{H^2} \quad \text{(say)},$$

and

$$\overline{p_{11} p_{33}} = \overline{p_{33} p_{22}} = \overline{p_{22} p_{11}} = \overline{m_1^2 n_1^2} \sum A^2 + 2\overline{m_2^2 n_3^2} \sum BC$$

$$= \tfrac{1}{15}[\sum A^2 + 4 \sum BC], \tag{245}$$

where for the sake of brevity we have written

$$\sum A^2 = A^2 + B^2 + C^2 \quad \text{and} \quad \sum BC = AB + BC + CA. \tag{246}$$

We thus find from equations (241) and (242) that the Stokes parameters, (243), representing the scattered light are, respectively,

$$[\overline{p_{11}^2} \cos^2\Theta + \overline{p_{21}^2} \sin^2\Theta][\overline{\xi_{\parallel}^{(0)}}]^2 + [\overline{p_{13}^2} \cos^2\Theta + \overline{p_{23}^2} \sin^2\Theta][\overline{\xi_{\perp}^{(0)}}]^2,$$

$$\overline{p_{31}^2}[\overline{\xi_{\parallel}^{(0)}}]^2 + \overline{p_{33}^2}[\overline{\xi_{\perp}^{(0)}}]^2,$$

$$(\overline{p_{33} p_{11}} + \overline{p_{13}^2})2[\overline{\xi_{\parallel}^{(0)} \xi_{\perp}^{(0)} \cos(\epsilon_1 - \epsilon_2)}]\cos\Theta,$$

and

$$(\overline{p_{33} p_{11}} - \overline{p_{13}^2})2[\overline{\xi_{\parallel}^{(0)} \xi_{\perp}^{(0)} \sin(\epsilon_1 - \epsilon_2)}]\cos\Theta. \tag{247}$$

The various averages over the amplitudes and phases which occur in the foregoing expressions are simply the Stokes parameters of the incident light.

Using for the various means of the squares and products of the tensor components in (247) their values given by equations (245), we have

$$(\overline{K^2} \cos^2\Theta + \overline{H^2} \sin^2\Theta)I_{\parallel} + \overline{H^2}I_{\perp},$$

$$\overline{H^2}I_{\parallel} + \overline{K^2}I_{\perp},$$

$$\tfrac{1}{15}[2 \sum A^2 + 3 \sum BC]U \cos\Theta = (\overline{K^2} - \overline{H^2})U \cos\Theta,$$

and

$$[\tfrac{1}{3} \sum BC]V \cos\Theta = (\overline{K^2} - 3\overline{H^2})V \cos\Theta, \tag{248}$$

respectively. Therefore, denoting the incident light by the vector $I = (I_{\parallel}, I_{\perp}, U, V)$, we can write for the intensity scattered in a solid angle $d\omega'$ and in a direction making an angle Θ with the direction of an incident pencil having a solid angle $d\omega$, the expression

$$\left(\sigma \frac{d\omega'}{4\pi}\right)RI \, d\omega, \tag{249}$$

where σ is the scattering coefficient per particle,

$$R = \frac{3}{2(1+2\gamma)} \begin{pmatrix} \cos^2\Theta + \gamma\sin^2\Theta & \gamma & 0 & 0 \\ \gamma & 1 & 0 & 0 \\ 0 & 0 & (1-\gamma)\cos\Theta & 0 \\ 0 & 0 & 0 & (1-3\gamma)\cos\Theta \end{pmatrix},$$

(250)

and

$$\gamma = (\overline{H^2}/\overline{K^2}).$$ (251)†

For an incident beam of natural light the scattered intensities in directions parallel and perpendicular to the plane of scattering are proportional to

$$\frac{3}{4(1+2\gamma)}[2\gamma + (1-\gamma)\cos^2\Theta]I \quad (\text{$\parallel$-component}),$$

and

$$\frac{3}{4(1+2\gamma)}(1+\gamma)I \quad (\text{$\perp$-component}).$$ (252)

From these expressions it follows that for light scattered at right angles to the direction of incidence

$$\left(\frac{I_\parallel^{(s)}}{I_\perp^{(s)}}\right)_{\Theta=90°} = \frac{2\gamma}{1+\gamma}.$$ (253)

This is the so-called *depolarization factor*, generally denoted by ρ_n. Hence

$$\rho_n = \frac{2\gamma}{1+\gamma}.$$ (254)

(The inequality $\gamma \leqslant \tfrac{1}{3}$ now implies that $\rho_n \leqslant \tfrac{1}{2}$.)

Again, according to the expressions (252) the angular distribution of the scattered intensity, for an incident natural beam, is governed by the phase function

$$[p(\cos\Theta)]_{\text{natural light}} = \frac{3}{4(1+2\gamma)}[(1+3\gamma)+(1-\gamma)\cos^2\Theta].$$ (255)

It will be observed that this phase function is normalized to unity. (This is indeed the reason for introducing the factor $3/2(1+2\gamma)$ in [250].)

For scattering by anisotropic molecules the scattering coefficient is given by

$$\sigma = \frac{8\pi^3}{3} \frac{(n^2-1)^2}{\lambda^4 N^2} \frac{3(2+\rho_n)}{6-7\rho_n},$$ (256)

a formula due to Cabannes.

† From the equations defining $\overline{H^2}$ and $\overline{K^2}$ it follows that $\gamma \leqslant \tfrac{1}{3}$; that the case $\gamma = \tfrac{1}{3}$ can occur when, and only when, two of the three constants A, B, and C vanish; that in all other cases the inequality $\gamma < \tfrac{1}{3}$ is a strict one.

For air, experiments indicate that the depolarization factor, ρ_n, is 0·04.

The formulation of the equation of transfer for an atmosphere scattering according to the law expressed by (249) and (250) raises no difficulties: it can indeed be written down directly from the equations we have already derived in § 17. For, the matrix R being reducible with respect to V, this parameter is scattered independently of others and in fact, according to the phase function, $3(1-3\gamma)\cos\Theta/2(1+2\gamma)$. Considering next the part of the matrix R referring to I_l, I_r, and U we can write it as the sum of the two matrices

$$\frac{3(1-\gamma)}{2(1+2\gamma)}\begin{pmatrix}\cos^2\Theta & 0 & 0\\ 0 & 1 & 0\\ 0 & 0 & \cos\Theta\end{pmatrix} \text{ and } \frac{3\gamma}{2(1+2\gamma)}\begin{pmatrix}1 & 1 & 0\\ 1 & 1 & 0\\ 0 & 0 & 0\end{pmatrix}. \quad (257)$$

The first of these two matrices is, apart from a constant of proportionality, the same as the matrix which appeared in the theory of Rayleigh scattering (cf. eq. [201]), and the second represents isotropic scattering of I_l and I_r independently of the state of polarization of the incident light. Consequently, the phase matrix which must be used in an equation of transfer of the form (226) for I_l, I_r, and U is

$$\frac{1-\gamma}{1+2\gamma}\,P(\mu,\varphi;\mu',\varphi')+\frac{3\gamma}{2(1+2\gamma)}\begin{pmatrix}1 & 1 & 0\\ 1 & 1 & 0\\ 0 & 0 & 0\end{pmatrix}, \quad (258)$$

where P denotes the same matrix as defined in equations (220)–(223) with the fourth row and column omitted.

We may summarize the results of this section by the statement that the scattering of partially plane-polarized light by an anisotropic particle may be regarded as equivalent to a superposition of Rayleigh scattering with a weight $(1-\gamma)/(1+2\gamma)$ and isotropic scattering of each of the components I_l and I_r with a weight $3\gamma/(1+2\gamma)$.

19. Resonance line scattering

The quantum theory of resonance fluorescence leads to a law of scattering which is formally of the same type as the classical law of scattering by anisotropic particles which we have described in the preceding section.

In the context of resonance line scattering we are concerned with transitions from an initial ground state to an excited intermediate state and back to the ground state. However, in discussing these transitions we must distinguish between the different *substates* of each level as specified by the magnetic quantum number m (which are the eigen-values of the z-component of the total angular momentum in units of $\hbar$). Let the relevant states of the radiating atom be designated by A_k, B_n, and A_p, where A and B refer to the ground (initial or final) and the intermediate (excited) states, respectively, and the subscripts refer to the m-values of the different substates in question.

Now the probability of a transition $A_k \rightarrow B_n$ between a single pair of m-states is easily calculable for any given stream of incident radiation; similarly, the angular distribution and the state of polarization of the quantum emitted in a transition $B_n \rightarrow A_p$ between a single pair of m-states are also known. But attention must be paid in investigating resonance fluorescence to the fact that the different sequences of transitions which are possible starting from the same state A_k are not uncorrelated; for, when transitions from a given state A_k to different substates B_n are possible, the wave functions belonging to these substates have phases which are related in a definite manner to the phase of the wave function belonging to A_k; the resulting transitions $B_n \rightarrow A_p$ from the different substates B_n cannot therefore be regarded as independent of each other. A theory allowing for these correlations has recently been worked out by D. R. Hamilton, whose results we shall quote.

In resonance fluorescence, the Stokes parameters I_l, I_r, and U are scattered in accordance with a phase-matrix of the form (cf. eq. [257])

$$\tfrac{3}{2}E_1 \begin{pmatrix} \cos^2\Theta & 0 & 0 \\ 0 & 1 & 0 \\ 0 & 0 & \cos\Theta \end{pmatrix} + \tfrac{1}{2}E_2 \begin{pmatrix} 1 & 1 & 0 \\ 1 & 1 & 0 \\ 0 & 0 & 0 \end{pmatrix}, \tag{259}$$

where E_1 and E_2 are certain constants depending on the initial j-value and the Δj ($= \pm 1$ or 0) involved in the transition. Further, as in Rayleigh scattering and in scattering by anisotropic particles, the parameter V is scattered independently of the rest and according to a phase function of the form

$$\tfrac{3}{2}E_3 \cos\Theta, \tag{260}$$

where E_3 is another constant depending also on j and Δj.

In Table II we list the values of E_1, E_2, and E_3 as given by Hamilton. It will be observed from this table that

$$E_1 + E_2 = 1, \tag{261}$$

a condition which is clearly necessary for conservative scattering.

It will also be noticed that for $j = 0$ and $\Delta j = 1$,

$$E_1 = 1, \quad E_2 = 0, \quad \text{and} \quad E_3 = 1; \tag{262}$$

this case is therefore the same as 'Rayleigh scattering'.

TABLE II

The Constants E_1, E_2, and E_3

Δj	E_1	E_2	E_3
1	$\dfrac{(2j+5)(j+2)}{10(j+1)(2j+1)}$	$\dfrac{3j(6j+7)}{10(j+1)(2j+1)}$	$\dfrac{j+2}{2(j+1)}$
0	$\dfrac{(2j-1)(2j+3)}{10j(j+1)}$	$\dfrac{3(2j^2+2j+1)}{10j(j+1)}$	$\dfrac{1}{2j(j+1)}$
−1	$\dfrac{(2j-3)(j-1)}{10j(2j+1)}$	$\dfrac{3(6j^2+5j-1)}{10j(2j+1)}$	$\dfrac{j-1}{2j}$

Since the phase-matrix (259) is of the same form as the one considered in § 18, no additional remarks are necessary as to how the appropriate equations of transfer are to be formulated.

BIBLIOGRAPHICAL NOTES

The following general references may be noted.

1. M. PLANCK, *Wärmestrahlung*, 5th ed., Leipzig, 1923; English translation (of the 2nd German ed.) by M. MASIUS, *Planck's Heat Radiation*, Blackiston, Philadelphia, 1914.
2. E. A. MILNE, 'Thermodynamics of the Stars', *Handbuch der Astrophysik*, Band iii/1, pp. 65–173, Springer, Berlin, 1930.
3. S. CHANDRASEKHAR, *An Introduction to the Study of Stellar Structure*, chap. v, University of Chicago Press, Chicago, 1939.
4. E. HOPF, *Mathematical Problems of Radiative Equilibrium*, Cambridge Mathematical Tracts, No. 31, Cambridge, 1934.
5. S. CHANDRASEKHAR, 'The Transfer of Radiation in Stellar Atmospheres' (Josiah Willard Gibbs Lecture), *Bulletin of the American Mathematical Society*, **53**, 641 (1947).

§ 10. In the context of conservative isotropic scattering the K-integral first occurs in—

6. A. S. EDDINGTON, *The Internal Constitution of the Stars*, p. 321, Cambridge, England, 1926.

§ 11.1. The derivation of the 'Schwarzschild–Milne' integral equation follows—

7. E. A. MILNE, *Monthly Notices Roy. Astron. Soc. London*, **81**, 361 (1921).

The classical papers in this context are those of—

8. K. SCHWARZSCHILD, *Göttinger Nachrichten*, p. 41 (1906) and *Sitzungs-Berichte d. Preuss. Akad. Berlin*, p. 1183 (1914).
9. A. SCHUSTER, *Astrophys. J.* **21**, 1 (1905).

§ 15. The discussion and the representation of polarized light in this section follows—

10. G. G. STOKES, *Trans. Camb. Philos. Soc.* **9**, 399 (1852); also *Mathematical and Physical Papers of Sir George Stokes* (Cambridge, England, 1901), **3**, 233–51.

Accounts of Stokes's representation from slightly different points of view will be found in—

11. LORD RAYLEIGH, *Scientific Papers* (Cambridge, England, 1902), iii, p. 48; see particularly § 20, pp. 140–7.

12. S. CHANDRASEKHAR, *Astrophys. J.* **105**, 424 (1946).

§ 16. The classical papers in this subject are those of—

13. LORD RAYLEIGH, *Scientific Papers* (Cambridge, England, 1899), i, pp. 87, 104, 518.

§ 17. The equations of transfer allowing properly for the polarization of the scattered radiation were first formulated by—

14. S. CHANDRASEKHAR, *Astrophys. J.* **103**, 351 (1946); **104**, 110 (1946); **105**, 424 (1947).

However, the treatment in the text is much more concise and makes use of the linear transformation of Stokes's parameters given in § 16.

§ 18. For general accounts of the theory of light scattering see—

15. C. V. RAMAN, *Molecular Diffraction of Light*, Calcutta, 1922.

16. J. CABANNES, *La Diffusion moléculaire de la lumière*, Paris, 1929.

17. S. BHAGAVANTAM, *Scattering of Light and Raman Effect*, Chemical Publishing Company, Brooklyn, N.Y., 1942.

But the most completely satisfactory account is that of—

18. G. PLACZEK, *Rayleigh Streuung und Raman Effekt*, Marx, *Handbuch d. Radiologie*, vi–2, p. 209, 1934.

§ 19. The results quoted in this section are those of—

19. D. R. HAMILTON, *Astrophys. J.* **106**, 457 (1947).
See also PLACZEK (ref. 18).

QUADRATURE FORMULAE

20. The method of replacing the equations of transfer by a system of linear equations

IN Chapter I we have seen how the various problems in the theory of Radiative Transfer lead to integrodifferential equations of varying degrees of complexity. It is not to be expected that exact and complete solutions of all the problems will be feasible. In a study of the equations of transfer we must therefore have two objectives: First, the development of approximate methods of solution which will have sufficient flexibility for adaptation to any practical situation which may arise; and second, the development of methods of sufficient power and generality which will enable us to discover the various integral relations which may exist and also to obtain exact solutions for at least those aspects of the problem (such as the angular distribution of the emergent radiations) which are of particular interest. The two objectives we have mentioned are not as unrelated as they may appear at first sight; for in many problems the exact relationships which exist are not at all obvious ones and, indeed, in many cases it would be extremely difficult even to guess at their origin or nature if we did not already have some reason for suspecting their existence; and the grounds for 'suspicion' must often lie in the form of the solutions obtained by the approximate methods. Therefore, in developing approximate methods of solution we can take one of two points of view: *Either* develop methods exclusively in the context of a particular problem (such as isotropic scattering in a homogeneous medium) aiming at the highest accuracy possible, *or* develop methods which, though in particular cases may not be as accurate as the special methods, will yet have sufficient generality for disclosing all the invariant relationships among the problems. Roughly speaking, the former appears to be the point of view of the physicists in their investigations on neutron diffusion; the latter will be the point of view adopted in this book.

Now in trying to develop a systematic method for solving equations of transfer it would appear most natural to start with the method first used by Schuster (1905) and Schwarzschild (1906)† in the context of the

---

† Actually Schuster considered the more general case of isotropic scattering with an albedo (Chap. I, § 12, eq. [108]); but Schwarzschild considered only the conservative case.

simplest equation of transfer (Chap. I, § 11.1, eqs. [88] and [89])

$$\mu \frac{dI(\tau,\mu)}{d\tau} = I(\tau,\mu) - \tfrac{1}{2} \int_{-1}^{+1} I(\tau,\mu')\, d\mu'. \tag{1}$$

Dividing the radiation field into an outward and an inward stream of intensities I_+ and I_-, respectively, Schuster and Schwarzschild replaced equation (1) by the pair of equations

$$+\frac{1}{2}\frac{dI_+}{d\tau} = I_+ - \tfrac{1}{2}(I_+ + I_-)$$

and

$$-\frac{1}{2}\frac{dI_-}{d\tau} = I_- - \tfrac{1}{2}(I_+ + I_-), \tag{2}$$

where the factor $\pm\tfrac{1}{2}$ on the left-hand sides of these equations is to allow for the mean obliquity of the rays to the outward (inward) direction. This division of the radiation field into two streams is reminiscent of the method devised by Joule (1851) in the kinetic theory of gases, in which the molecules in a rectangular box are divided into three equal pairs of streams moving parallel to the length, breadth, and depth of the box, the streams belonging to a given pair moving in opposite directions; the method of Schuster and Schwarzschild has played a similar useful role in the early developments of the subject.

The solution of equations (2) subject to the usual boundary conditions ($I_- = 0$ at $\tau = 0$, for example) is, of course, immediate, and provides in fact a 'first' approximation to the solution of the problem formulated in Chapter I, § 11.1. It is, however, evident that the division of the radiation field into two streams is far too crude to include any of the effects of anisotropic scattering or polarization in which we are primarily interested. Nevertheless, the directness of the conception underlying Schuster and Schwarzschild's method suggests that we retain its basic idea but refine it by dividing the radiation field into more than just two streams. We can then preserve the principal advantage of the method, namely, its visualization of the radiation field in terms of a number of discrete streams. The possibility of this visualization is particularly valuable in the formulation of boundary conditions (such as the vanishing of the radiation field in a hemisphere) with a directness and a freedom from ambiguity which would be unattainable otherwise. Basically this is the reason why the method derived from the original Schuster and Schwarzschild's method is as successful as we shall find it to be in preserving the fundamental relationships of a problem in all orders of approximation.

Dividing then the radiation field into $2n$ streams in the directions μ_i ($i = \pm 1,..., \pm n$ and $\mu_{-i} = -\mu_i$), we can replace the .equation of transfer (1), for example, by the system of $2n$ linear equations

$$\mu_i \frac{dI(\tau, \mu_i)}{d\tau} = I(\tau, \mu_i) - \tfrac{1}{2} \sum_j a_j I(\tau, \mu_j) \quad (i = \pm 1,..., \pm n), \qquad (3)$$

where the a_j's ($j = 1,..., n$) are the *weights* appropriate for a quadrature formula based on the division (μ_i) of the interval $(-1, +1)$ (see § 21 below). (In eq. [3] the summation over j is to be extended over all values of j, positive and negative; there is, however, no term with $j = 0$.)

Since equation (3) represents a system of linear equations with constant coefficients, the solution of this or any other similar system can hardly introduce any difficulties of principle. Indeed, it might even be argued that by making the division of the interval $(-1, +1)$ finer and finer we can approach the exact solution as a limit. However, the *practical* efficacy of the method will largely depend on how well the *lower* orders of approximation ($n = 3$ or 4) represent the true solution. Consequently, in the choice of the division of the radiation field into streams we must be guided by the accuracy with which the integrals over μ can be replaced by weighted means of the intensities at the points of the division. The problem which we encounter here is basically the same as that considered by Gauss in 1814 in deriving his formula for numerical quadratures. In Gauss's formula the interval $(-1, +1)$ is divided according to the zeros (μ_j) of the Legendre polynomial $P_m(\mu)$ and the integral of a function $f(\mu)$ over the interval $(-1, +1)$ is expressed as a sum in the form

$$\int_{-1}^{+1} f(\mu) \, d\mu \simeq \sum_{j=1}^{m} a_j f(\mu_j), \qquad (4)$$

where the weights a_j are given by

$$a_j = \frac{1}{P'_m(\mu_j)} \int_{-1}^{+1} \frac{P_m(\mu)}{\mu - \mu_j} \, d\mu. \qquad (5)$$

The reason Gauss's formula is superior to other formulae for quadratures in the interval $(-1, +1)$ is that for a given m it evaluates the integral exactly for all polynomials of degree less than $2m$ and not merely those of degree less than m; in other words, Gauss's formula is almost twice as accurate as a formula using only m values of the function in the interval would be expected to be. It would therefore seem that in replacing the equations of transfer by systems of linear equations in the manner of (3), we use the Gaussian division of the interval $(-1, +1)$ and evaluate the various integrals over μ which occur in the equations of transfer by sums according to Gauss's quadrature formula. This is one of the principal methods we shall develop.

In view of the fundamental role which Gauss's quadrature formula plays in the subsequent developments of the theory, it will be useful to set out the derivation of the formula particularly, as it will give us an occasion, also, for establishing other quadrature formulae which are useful in other parts of the subject. For example, according to the formal solution (Chap. I, §§ 7 and 9, eq. [70]) of the equation of transfer, the emergent radiation in a semi-infinite plane-parallel atmosphere is given by

$$I(0, \mu) = \int_0^\infty e^{-\tau/\mu} \Im(\tau, \mu) \, \frac{d\tau}{\mu}. \tag{6}$$

The question as to the best quadrature formula for numerically evaluating integrals of the form

$$\int_0^\infty e^{-x} f(x) \, dx \tag{7}$$

is therefore of considerable practical interest. As we shall see (§ 22), in this case a quadrature formula based on the division of the interval $(0, \infty)$ according to the zeros of the Laguerre polynomials provides an accuracy comparable to Gauss's in the interval $(-1, +1)$.

Again, for computing the mean intensities and the fluxes in a plane-parallel atmosphere according to (cf. Chap. I, eqs. [97] and [98])

$$J(\tau) = \tfrac{1}{2} \int_0^\infty \Im(t) E_1(|t-\tau|) \, dt$$

and

$$F(\tau) = 2 \int_\tau^\infty \Im(t) E_2(t-\tau) \, dt - 2 \int_0^\tau \Im(t) E_2(\tau-t) \, dt, \tag{8}$$

it would be convenient to have quadrature formulae analogous to Gauss's and allowing for the fact that weight functions with singularities (but with finite moments) occur. We shall consider quadrature formulae appropriate to these circumstances also (§ 23).

21. The construction of quadrature formulae

Consider quite generally the integral

$$\int_b^a f(x) w(x) \, dx, \tag{9}$$

between two assigned limits, a and b, of a function $f(x)$ *weighted* by some given function $w(x)$. We shall assume that $f(x)$ is continuous in the interval (b, a) and that the integral exists.

Suppose that we wish to evaluate the integral (9) by a quadrature formula using only m values of $f(x)$ in the interval (b, a). The question which arises concerns the choice of the points which will lead to the 'best' evaluation of the integral. We shall investigate this question in the following manner:

Let
$$(x_j) = (x_1, x_2, ..., x_m)$$

represent a *division* of the interval (b, a). Using Lagrange's formula for interpolation we can construct a polynomial, $\phi(x)$, of degree less than or equal to $m-1$, which will take the same values at the points (x_j) as $f(x)$ does. Thus

$$\phi(x) = \sum_{j=1}^{m} f(x_j) \frac{F(x)}{(x-x_j) F'(x_j)}, \tag{10}$$

where
$$F(x) = \prod_{j=1}^{m} (x-x_j) \tag{11}$$

is a polynomial of degree m whose zeros are the points of the division (x_j) and

$$F'(x_j) = \left(\frac{dF}{dx} \right)_{x=x_j} = \prod_{i \neq j} (x_j - x_i). \tag{12}$$

We may regard $\phi(x)$ as an approximate representation of $f(x)$. And the integral (9) evaluated with $\phi(x)$, instead of $f(x)$, is

$$\int_b^a \phi(x) w(x)\, dx = \sum_{j=1}^{m} a_j f(x_j), \tag{13}$$

where
$$a_j = \frac{1}{F'(x_j)} \int_b^a \frac{F(x) w(x)}{x - x_j}\, dx \quad (j = 1, ..., m). \tag{14}$$

The weights a_j (sometimes called the *Christoffel numbers* associated with $F(x)$) are all independent of $f(x)$ and can be evaluated once for all when the division (x_j) is prescribed. In the method of Newton and Cotes, the interval (b, a) is divided equally into $(m+1)$ segments by the points of the division and it is assumed that an approximate evaluation of the integral is obtained by the formula

$$\int_b^a f(x) w(x)\, dx = \sum_{j=1}^{m} a_j f(x_j). \tag{15}$$

But Gauss showed that it would be advantageous to choose (x_j) differently. And the criterion underlying this alternative choice is the following:

The error made in the value of (9) by taking (13) as its approximate value is

$$\int_b^a f(x)w(x)\,dx - \sum_{j=1}^m a_j f(x_j). \tag{16}$$

This clearly vanishes if $f(x)$ is a polynomial of degree less than or equal to $m-1$. We now ask whether, by an appropriate choice of (x_j), we can make the error vanish for polynomials $f(x)$ of degree higher than $m-1$. And if we can, what is the highest degree of $f(x)$ for which this can be accomplished? And what is the corresponding division (x_j)?

We shall first show that by a proper choice of (x_j) we can arrange that the error (16) can be made to vanish for an arbitrary polynomial $f(x)$ of degree less than or equal to $2m-1$. For, in order that (16) may vanish for an arbitrary polynomial of degree $2m-1$, it is clearly necessary and sufficient that it vanish individually for every power of x *less* than $2m$. Thus, letting

$$\alpha_l = \int_b^a x^l w(x)\,dx \quad (l = 0, 1, ..., 2m-1) \tag{17}$$

denote the moments of the given weight function, we require that

$$\alpha_l = \sum_{j=1}^m a_j x_j^l \quad (l = 0, 1, ..., 2m-1). \tag{18}$$

This provides us with $2m$ equations for the $2m$ 'unknowns' $a_j\,(j = 1, ..., m)$ and $x_j\,(j = 1, ..., m)$. If equations (18) can be solved for (x_j) and (a_j) we shall have constructed a quadrature formula which will be the 'best' in the sense we have described. And, in any case, it is clear that under no circumstances can we make the error (16) vanish for an arbitrary polynomial of degree *higher* than $2m-1$.

The following method of solving equations (18) for (x_j) and (a_j) may be noted:

Considering the m sets of $(m+1)$ equations

$$\alpha_{i+l} = \sum_{j=1}^m a_j x_j^{i+l} \quad (l = 0, 1, ..., m; \quad 0 \leqslant i \leqslant m-1), \tag{19}$$

we form the sums

$$\alpha_{i+m} + \sum_{l=0}^{m-1} c_l \alpha_{i+l} \quad (i = 0, 1, ..., m-1), \tag{20}$$

where the c_l's $(l = 0, 1,..., m-1)$ are m constants unspecified for the present. Using equations (18) we can rewrite (20) in the form

$$\sum_{j=1}^{m} a_j x_j^i \left(x_j^m + \sum_{l=0}^{m-1} c_l x_j^l \right) \quad (i = 0, 1,..., m-1). \tag{21}$$

From (21) it is apparent that if we let the m constants c_l be determined (uniquely) by the m linear equations

$$\alpha_{i+m} + \sum_{l=0}^{m-1} c_l \alpha_{i+l} = 0 \quad (i = 0, 1,..., m-1), \tag{22}$$

then

$$x_j^m + \sum_{l=0}^{m-1} c_l x_j^l = 0 \quad (j = 1,..., m). \tag{23}$$

The x_j's $(j = 1,..., m)$ are therefore the roots of the equation

$$F(x) = x^m + \sum_{l=0}^{m-1} c_l x^l = 0. \tag{24}$$

Once the c_l's and the x_j's have been determined in this fashion, the weights a_j can then be determined from any m of the $2m$ equations included in (18): the remaining m equations will then be satisfied identically in view of the fact that of the $2m$ equations (18) only m are linearly independent (cf. eqs. [22] and [23]).

The foregoing considerations show how we can always construct an m-point quadrature formula which will evaluate exactly integrals of the form (9) for all polynomials $f(x)$ of degree less than or equal to $2m-1$. However, it is often convenient to characterize $F(x)$ (and therefore also (x_j) as the zeros of F) somewhat differently as follows:

Let $f(x)$ be a polynomial of degree $2m-1$. Since $f(x)-\phi(x)$ vanishes for $x = x_j$ $(j = 1,..., m)$, we can clearly write

$$f(x) = \phi(x) + F(x) \sum_{l=0}^{m-1} q_l x^l, \tag{25}$$

where the q_l's $(l = 0, 1,..., m-1)$ are certain constants. If the error (16) in evaluating the integral (9) according to (13) is to vanish for an arbitrary polynomial of degree less than $2m$, it is clearly necessary and sufficient that

$$\int_b^a F(x)w(x)x^l \, dx = 0 \quad (l = 0, 1,..., m-1). \tag{26}$$

If the nature of $F(x)$ can be determined from these conditions, the x_j's can then be defined as the zeros of $F(x)$ and the weights a_j determined according to equation (14).

22. Various special quadrature formulae

22.1. *Gauss's formula*

Considering the integral

$$\int_{-1}^{+1} f(\mu)\, d\mu, \tag{27}$$

in the range $(-1, +1)$, we have the conditions (eq. [26])

$$\int_{-1}^{+1} F(\mu)\mu^l\, d\mu = 0 \quad (l = 0, 1, ..., m-1). \tag{28}$$

(For uniformity with the rest of the book we have reverted to the variable μ in this and the following subsections.)

Now it is known that the Legendre polynomial $P_m(\mu)$ of order m is orthogonal to every power of μ less than m. For, using Rodrigues's formula

$$P_m(\mu) = \frac{1}{2^m m!} \frac{d^m}{d\mu^m} (\mu^2 - 1)^m, \tag{29}$$

and integrating

$$\frac{1}{2^m m!} \int_{-1}^{+1} \mu^l \frac{d^m}{d\mu^m} (\mu^2 - 1)^m\, d\mu \tag{30}$$

l times by parts and remembering that all derivatives of $(\mu^2-1)^m$ of order less than l vanish for $\mu = \pm 1$, we have

$$\frac{(-1)^l l!}{2^m m!} \int_{-1}^{+1} \frac{d^{m-l}}{d\mu^{m-l}} (\mu^2 - 1)^m\, d\mu. \tag{31}$$

But this is zero for all $l \leqslant m-1$. Thus the conditions (28) on $F(\mu)$ are satisfied by $P_m(\mu)$. Conversely it may be shown† that a polynomial $F(\mu)$ of degree m which is orthogonal to every power of μ less than m is necessarily a numerical multiple of $P_m(\mu)$. We have therefore shown that if $f(\mu)$ is an arbitrary polynomial of degree $2m-1$, then

$$\int_{-1}^{+1} f(\mu)\, d\mu = \sum_{j=1}^{m} a_j f(\mu_j), \tag{32}$$

where $\mu_1, ..., \mu_m$ are the zeros of $P_m(\mu)$ and

$$a_j = \frac{1}{P'_m(\mu_j)} \int_{-1}^{+1} \frac{P_m(\mu)}{\mu - \mu_j}\, d\mu. \tag{33}$$

This is Gauss's formula.

† Cf. E. W. Hobson, *Spherical and Ellipsoidal Harmonics* (Cambridge, England, 1931), p. 36.

In our future work we shall find it convenient to restrict ourselves always to divisions of the interval $(-1, +1)$ according to the zeros of the even-order Legendre polynomials $P_{2n}(\mu)$. For such even divisions it is clear that

$$a_j = a_{-j} \quad \text{and} \quad \mu_{-j} = -\mu_j \quad (j = 1,...,n). \tag{34}$$

Also, the fact that (32) evaluates the integrals of μ^l exactly for all $l \leqslant 4n-1$ can now be expressed in the form

$$\sum_{j=\pm 1}^{\pm n} a_j \mu_j^l = \frac{2\delta_{l,e}}{l+1} \quad (l \leqslant 4n-1), \tag{35}$$

where

$$\delta_{l,e} = 1 \quad \text{if } l \text{ is even},$$
$$= 0 \quad \text{if } l \text{ is odd}. \tag{36}$$

When we are using Gauss's formula based on the zeros of $P_{2n}(\mu)$ we shall say that we are working in the nth approximation.

In Table III (which is an extract from a more extensive table by Lowan, Davids, and Levenson) we list the zeros of $P_{2n}(\mu)$ and the corresponding Christoffel numbers for the first four approximations.

TABLE III

The Gaussian Divisions and the Gaussian Weights

$$\left.\begin{array}{l} n = 1 \\ m = 2 \end{array}\right\} \mu_{\pm 1} = \pm 0.5773503 \qquad a_1 = a_{-1} = 1$$

$$\begin{array}{l} n = 2 \\ m = 4 \end{array} \left\{\begin{array}{l} \mu_{\pm 1} = \pm 0.3399810 \\ \mu_{\pm 2} = \pm 0.8611363 \end{array}\right. \qquad \begin{array}{l} a_1 = a_{-1} = 0.6521452 \\ a_2 = a_{-2} = 0.3478548 \end{array}$$

$$\begin{array}{l} n = 3 \\ m = 6 \end{array} \left\{\begin{array}{l} \mu_{\pm 1} = \pm 0.2386192 \\ \mu_{\pm 2} = \pm 0.6612094 \\ \mu_{\pm 3} = \pm 0.9324695 \end{array}\right. \qquad \begin{array}{l} a_1 = a_{-1} = 0.4679139 \\ a_2 = a_{-2} = 0.3607616 \\ a_3 = a_{-3} = 0.1713245 \end{array}$$

$$\begin{array}{l} n = 4 \\ m = 8 \end{array} \left\{\begin{array}{l} \mu_{\pm 1} = \pm 0.1834346 \\ \mu_{\pm 2} = \pm 0.5255324 \\ \mu_{\pm 3} = \pm 0.7966665 \\ \mu_{\pm 4} = \pm 0.9602899 \end{array}\right. \qquad \begin{array}{l} a_1 = a_{-1} = 0.3626838 \\ a_2 = a_{-2} = 0.3137066 \\ a_3 = a_{-3} = 0.2223810 \\ a_4 = a_{-4} = 0.1012285 \end{array}$$

So far we have considered only the evaluation of integrals over polynomials. It is of course clear that when $f(\mu)$ is not a polynomial Gauss's formula will still provide the 'best' approximation to the value for the integral. We shall not concern ourselves here with the errors which are involved in the use of Gauss's formula of a certain order. Reference may, however, be made to the books listed in the Bibliographical Notes.

22.2. *Radau's formula*

A quadrature formula slightly inferior to Gauss's is due to Radau. Radau's formula is obtained with the choice

$$F(\mu) = (1-\mu^2)P'_m(\mu),\qquad(37)$$

for the basic polynomial. The division (μ_j) now consists of $(m+1)$ points, which are

$$\mu = \pm 1 \text{ and the } (m-1) \text{ zeros of } P'_m(\mu).\qquad(38)$$

An alternative form of $F(\mu)$ is

$$F(\mu) = \frac{m(m+1)}{2m+1}[P_{m-1}(\mu)-P_{m+1}(\mu)].\qquad(39)$$

From this equation it is evident that

$$\int_{-1}^{+1} F(\mu)\mu^l = 0 \quad \text{for} \quad l \leqslant m-2.\qquad(40)$$

On the other hand, since $F(\mu)$ is a polynomial of degree $(m+1)$, it follows (cf. eq. [25]) that a quadrature formula based on the zeros of (39) will evaluate integrals of the form (27) exactly for all polynomials of degree less than or equal to $(m+1)+(m-2) = 2m-1$. In other words, Radau's formula achieves with $(m+1)$ points the same accuracy as Gauss's with m points.

For the particular form (37) for $F(\mu)$ the formula for the Christoffel numbers (eq. [14]) can be reduced to the form

$$a_j = \frac{1}{m(m+1)}[P_m(\mu_j)]^{-2},\qquad(41)\dagger$$

and the quadrature formula becomes

$$\int_{-1}^{+1} f(\mu)\,d\mu \simeq \frac{1}{m(m+1)}\sum_{j=1}^{m+1}\frac{f(\mu_j)}{[P_m(\mu_j)]^2}.\qquad(42)$$

This is Radau's formula.

In Table IV (due to Z. Kopal) we have listed the zeros of $(1-\mu^2)P'_m(\mu)$ and the corresponding Christoffel numbers for use of Radau's formula.

---

† Cf. E. T. Whittaker and G. Robinson, *The Calculus of Observations* (Blackie and Sons, London, 1924), p. 161.

TABLE IV

The Zeros of $(1-\mu^2)P'_m(\mu)$ and the Corresponding Christoffel Numbers

$m=1$	$\mu_{1,2} = 1$	$a_{1,2} = 1$
$m=2$	$\begin{cases} \mu_{1,3} = \pm 1 \\ \mu_2 = 0 \end{cases}$	$\begin{array}{l} a_{1,3} = 0\cdot3333333 \\ a_2 = 1\cdot3333333 \end{array}$
$m=3$	$\begin{cases} \mu_{1,4} = \pm 1 \\ \mu_{2,3} = \pm 0\cdot4472136 \end{cases}$	$\begin{array}{l} a_{1,4} = 0\cdot1666667 \\ a_{2,3} = 0\cdot8333333 \end{array}$
$m=4$	$\begin{cases} \mu_{1,5} = \pm 1 \\ \mu_{2,4} = \pm 0\cdot6546537 \\ \mu_3 = 0 \end{cases}$	$\begin{array}{l} a_{1,5} = 0\cdot1000000 \\ a_{2,4} = 0\cdot5444444 \\ a_3 = 0\cdot7111111 \end{array}$
$m=5$	$\begin{cases} \mu_{1,6} = \pm 1 \\ \mu_{2,5} = \pm 0\cdot7650553 \\ \mu_{3,4} = \pm 0\cdot0813570 \end{cases}$	$\begin{array}{l} a_{1,6} = 0\cdot0666667 \\ a_{2,5} = 0\cdot3784750 \\ a_{3,4} = 0\cdot5548584 \end{array}$
$m=6$	$\begin{cases} \mu_{1,7} = \pm 1 \\ \mu_{2,6} = \pm 0\cdot8302239 \\ \mu_{3,5} = \pm 0\cdot4688488 \\ \mu_4 = 0 \end{cases}$	$\begin{array}{l} a_{1,7} = 0\cdot0476190 \\ a_{2,6} = 0\cdot2768260 \\ a_{3,5} = 0\cdot4317454 \\ a_4 = 0\cdot4876190 \end{array}$
$m=7$	$\begin{cases} \mu_{1,8} = \pm 1 \\ \mu_{2,7} = \pm 0\cdot8717401 \\ \mu_{3,6} = \pm 0\cdot5917002 \\ \mu_{4,5} = \pm 0\cdot2092992 \end{cases}$	$\begin{array}{l} a_{1,8} = 0\cdot0357143 \\ a_{2,7} = 0\cdot2107042 \\ a_{3,6} = 0\cdot3411227 \\ a_{4,5} = 0\cdot4124591 \end{array}$
$m=8$	$\begin{cases} \mu_{1,9} = \pm 1 \\ \mu_{2,8} = \pm 0\cdot8997580 \\ \mu_{3,7} = \pm 0\cdot6771863 \\ \mu_{4,6} = \pm 0\cdot3631175 \\ \mu_5 = 0 \end{cases}$	$\begin{array}{l} a_{1,9} = 0\cdot0277778 \\ a_{2,8} = 0\cdot1654953 \\ a_{3,7} = 0\cdot2745387 \\ a_{4,6} = 0\cdot3464284 \\ a_5 = 0\cdot3715193 \end{array}$

22.3. *The quadrature formula based on the zeros of the Laguerre polynomials*

Turning next to integrals of the form (7) we have to consider the conditions (eq. [26])

$$\int_0^\infty e^{-x}F(x)x^l \, dx = 0 \quad (l = 0, 1,..., m-1). \tag{43}$$

From the definition

$$L_m(x) = \frac{e^x}{m!}\frac{d^m}{dx^m}(e^{-x}x^m), \tag{44}$$

of the Laguerre polynomial of order m, it follows after repeated integration by parts that

$$\int_0^\infty e^{-x}\left[\frac{e^x}{m!}\frac{d^m}{dx^m}(e^{-x}x^m)\right]x^l \, dx = (-1)^l\frac{l!}{m!}\int_0^\infty \frac{d^{m-l}}{dx^{m-l}}(e^{-x}x^m) \, dx, \tag{45}$$

and this vanishes for $l \leqslant m-1$. Conversely, it can be shown that the conditions (43) imply that $F(x)$ is a numerical multiple of $L_m(x)$. We have accordingly shown that if $f(x)$ is an arbitrary polynomial of degree $2m-1$, then

$$\int_0^\infty e^{-x} f(x)\, dx = \sum_{j=1}^m a_j f(x_j), \qquad (46)$$

where the x_j's $(j = 1,...,m)$ are the zeros of $L_m(x)$ and

$$a_j = \frac{1}{L_m'(x_j)} \int_0^\infty \frac{e^{-x} L_m(x)}{x - x_j}\, dx = \frac{1}{x_j}[L_m'(x_j)]^{-2}. \qquad (47)$$

In Table V (due to A. Reiz) we list the zeros of the Laguerre polynomials and the corresponding Christoffel numbers.

TABLE V

The Zeros of $L_m(x)$ and the Corresponding Christoffel Numbers

$m = 1$	$x_1 = 1$	$a_1 = 1$
$m = 2$	$x_1 = 0\cdot5857864$ $x_2 = 3\cdot4142136$	$a_1 = 0\cdot8535534$ $a_2 = 0\cdot1464466$
$m = 3$	$x_1 = 0\cdot4157746$ $x_2 = 2\cdot2942803$ $x_3 = 6\cdot2899451$	$a_1 = 0\cdot7110930$ $a_2 = 0\cdot2785177$ $a_3 = 0\cdot0103893$
$m = 4$	$x_1 = 0\cdot3225477$ $x_2 = 1\cdot7457611$ $x_3 = 4\cdot5366203$ $x_4 = 9\cdot3950709$	$a_1 = 0\cdot6031541$ $a_2 = 0\cdot3574187$ $a_3 = 0\cdot0388879$ $a_4 = 0\cdot0005393$
$m = 5$	$x_1 = 0\cdot2635603$ $x_2 = 1\cdot4134030$ $x_3 = 3\cdot5964258$ $x_4 = 7\cdot0858102$ $x_5 = 12\cdot6408007$	$a_1 = 0\cdot5217556$ $a_2 = 0\cdot3986668$ $a_3 = 0\cdot0759424$ $a_4 = 0\cdot0036118$ $a_5 = 0\cdot0000234$

23. Quadrature formulae for evaluating mean intensities and fluxes in a stellar atmosphere

As we have already indicated (p. 57), in certain parts of the theory of radiative equilibrium of stellar atmospheres (Chap. XI, § 81.3) we have to evaluate, numerically, integrals of the form

$$J(\tau) = \tfrac{1}{2} \int_0^\infty \Im(t) E_1(|t-\tau|)\, dt$$

and
$$F(\tau) = 2 \int_\tau^\infty \Im(t) E_2(t-\tau)\, dt - 2 \int_0^\tau \Im(t) E_2(\tau-t)\, dt, \qquad (48)$$

where $\Im(t)$ is some known tabulated function. The evaluation of these integrals by any of the standard formulae is inaccurate (or, tiresome!) because the weight functions E_1 and E_2 have singularities at $t = \tau$. Thus, $E_1(x)$ has a logarithmic singularity at $x = 0$ and $E_2(x)$ has a similar singularity in its derivative. Quadrature formulae constructed according to the principles of § 21 and allowing for the 'mild' singularities in the weight functions would therefore be specially useful in these connexions since all the moments of the weight functions exist.

Rewriting the integrals (48) in the forms

$$J(\tau) = \tfrac{1}{2} \int_0^\infty \Im(\tau+x)E_1(x)\,dx + \tfrac{1}{2}\int_0^\tau \Im(\tau-x)E_1(x)\,dx$$

and

$$F(\tau) = 2\int_0^\infty \Im(\tau+x)E_2(x)\,dx - 2\int_0^\tau \Im(\tau-x)E_2(x)\,dx, \tag{49}$$

we can construct quadrature formulae for evaluating $J(\tau)$ and $F(\tau)$ according to the scheme described in § 21 (eqs. [18]–[24]) in terms of the moments

$$\int_0^\infty x^l E_1(x)\,dx, \quad \int_0^\tau x^l E_1(x)\,dx, \quad \int_0^\infty x^l E_2(x)\,dx, \quad \text{and} \quad \int_0^\tau x^l E_2(x)\,dx. \tag{50}$$

All these moments can be evaluated by the formulae given in Appendix I (§ 92, eq. [6]). In this manner A. Reiz has recently constructed two- and three-point quadrature formulae for evaluating $J(\tau)$ and $F(\tau)$.

Considering first integrals of the form

$$\tfrac{1}{2}\int_0^\infty f(x)E_1(x)\,dx \quad \text{and} \quad 2\int_0^\infty f(x)E_2(x)\,dx, \tag{51}$$

which occur in the expressions for $J(\tau)$ and $F(\tau)$, we give in Table VI the divisions and weights for two- and three-point formulae as computed by A. Reiz.

Next, for evaluating the integrals

$$\tfrac{1}{2}\int_0^\tau f(x)E_1(\tau-x)\,dx \quad \text{and} \quad 2\int_0^\tau f(x)E_2(\tau-x)\,dx,$$

the divisions and the weights must be computed for each value of τ separately and A. Reiz has constructed two-point quadrature formulae for various values of τ. His results are given in Table VII.

TABLE VI

Quadrature Formulae for Evaluating the Integrals

$$\tfrac{1}{2}\int_0^\infty f(\tau)E_1(\tau)\,d\tau \quad and \quad 2\int_0^\infty f(\tau)E_2(\tau)\,d\tau$$

TABLE VI (*cont.*)

I.
$$\tfrac{1}{2}\int\limits_0^\infty f(\tau)E_1(\tau)\,d\tau = \sum a_j f(\tau_j)$$

$$m = 2 \begin{cases} \tau_1 = 0\cdot292 \\ \tau_2 = 2\cdot507 \end{cases} \qquad \begin{aligned} a_1 &= 0\cdot4532 \\ a_2 &= 0\cdot0468 \end{aligned}$$

$$m = 3 \begin{cases} \tau_1 = 0\cdot210 \\ \tau_2 = 1\cdot651 \\ \tau_3 = 5\cdot173 \end{cases} \qquad \begin{aligned} a_1 &= 0\cdot4053 \\ a_2 &= 0\cdot0923 \\ a_3 &= 0\cdot00239 \end{aligned}$$

II.
$$2\int\limits_0^\infty f(\tau)E_2(\tau)\,d\tau = \sum a_j f(\tau_j)$$

$$m = 2 \begin{cases} \tau_1 = 0\cdot397 \\ \tau_2 = 2\cdot723 \end{cases} \qquad \begin{aligned} a_1 &= 0\cdot8839 \\ a_2 &= 0\cdot1161 \end{aligned}$$

$$m = 3 \begin{cases} \tau_1 = 0\cdot287 \\ \tau_2 = 1\cdot814 \\ \tau_3 = 5\cdot385 \end{cases} \qquad \begin{aligned} a_1 &= 0\cdot7669 \\ a_2 &= 0\cdot2265 \\ a_3 &= 0\cdot00659 \end{aligned}$$

TABLE VII

The Divisions and the Weights for Evaluating the Integrals

$$\tfrac{1}{2}\int\limits_0^\tau f(t)E_1(\tau-t)\,dt \quad and \quad 2\int\limits_0^\tau f(t)E_2(\tau-t)\,dt$$

I.
$$\tfrac{1}{2}\int\limits_0^\tau f(t)E_1(\tau-t)\,dt = a_1 f(t_1) + a_2 f(t_2)$$

$$\tau = 0\cdot2 \begin{cases} t_1 = 0\cdot051 \\ t_2 = 0\cdot169 \end{cases} \begin{aligned} a_1 &= 0\cdot0851 \\ a_2 &= 0\cdot1278 \end{aligned} \qquad \tau = 2\cdot6 \begin{cases} t_1 = 1\cdot051 \\ t_2 = 2\cdot385 \end{cases} \begin{aligned} a_1 &= 0\cdot0857 \\ a_2 &= 0\cdot4056 \end{aligned}$$

$$\tau = 0\cdot4 \begin{cases} t_1 = 0\cdot106 \\ t_2 = 0\cdot342 \end{cases} \begin{aligned} a_1 &= 0\cdot1094 \\ a_2 &= 0\cdot1959 \end{aligned} \qquad \tau = 2\cdot8 \begin{cases} t_1 = 1\cdot168 \\ t_2 = 2\cdot577 \end{cases} \begin{aligned} a_1 &= 0\cdot0819 \\ a_2 &= 0\cdot4113 \end{aligned}$$

$$\tau = 0\cdot6 \begin{cases} t_1 = 0\cdot167 \\ t_2 = 0\cdot520 \end{cases} \begin{aligned} a_1 &= 0\cdot1198 \\ a_2 &= 0\cdot2421 \end{aligned} \qquad \tau = 3\cdot0 \begin{cases} t_1 = 1\cdot290 \\ t_2 = 2\cdot770 \end{cases} \begin{aligned} a_1 &= 0\cdot0783 \\ a_2 &= 0\cdot4164 \end{aligned}$$

$$\tau = 0\cdot8 \begin{cases} t_1 = 0\cdot232 \\ t_2 = 0\cdot699 \end{cases} \begin{aligned} a_1 &= 0\cdot1224 \\ a_2 &= 0\cdot2772 \end{aligned} \qquad \tau = 3\cdot2 \begin{cases} t_1 = 1\cdot417 \\ t_2 = 2\cdot964 \end{cases} \begin{aligned} a_1 &= 0\cdot0751 \\ a_2 &= 0\cdot4208 \end{aligned}$$

$$\tau = 1\cdot0 \begin{cases} t_1 = 0\cdot303 \\ t_2 = 0\cdot881 \end{cases} \begin{aligned} a_1 &= 0\cdot1214 \\ a_2 &= 0\cdot3044 \end{aligned} \qquad \tau = 3\cdot4 \begin{cases} t_1 = 1\cdot549 \\ t_2 = 3\cdot158 \end{cases} \begin{aligned} a_1 &= 0\cdot0721 \\ a_2 &= 0\cdot4246 \end{aligned}$$

$$\tau = 1\cdot2 \begin{cases} t_1 = 0\cdot378 \\ t_2 = 1\cdot064 \end{cases} \begin{aligned} a_1 &= 0\cdot1182 \\ a_2 &= 0\cdot3262 \end{aligned} \qquad \tau = 3\cdot6 \begin{cases} t_1 = 1\cdot686 \\ t_2 = 3\cdot353 \end{cases} \begin{aligned} a_1 &= 0\cdot0694 \\ a_2 &= 0\cdot4280 \end{aligned}$$

$$\tau = 1\cdot4 \begin{cases} t_1 = 0\cdot459 \\ t_2 = 1\cdot249 \end{cases} \begin{aligned} a_1 &= 0\cdot1138 \\ a_2 &= 0\cdot3442 \end{aligned} \qquad \tau = 3\cdot8 \begin{cases} t_1 = 1\cdot828 \\ t_2 = 3\cdot548 \end{cases} \begin{aligned} a_1 &= 0\cdot0670 \\ a_2 &= 0\cdot4310 \end{aligned}$$

$$\tau = 1\cdot6 \begin{cases} t_1 = 0\cdot544 \\ t_2 = 1\cdot436 \end{cases} \begin{aligned} a_1 &= 0\cdot1090 \\ a_2 &= 0\cdot3591 \end{aligned} \qquad \tau = 4\cdot0 \begin{cases} t_1 = 1\cdot974 \\ t_2 = 3\cdot744 \end{cases} \begin{aligned} a_1 &= 0\cdot0648 \\ a_2 &= 0\cdot4336 \end{aligned}$$

$$\tau = 1\cdot8 \begin{cases} t_1 = 0\cdot635 \\ t_2 = 1\cdot624 \end{cases} \begin{aligned} a_1 &= 0\cdot1040 \\ a_2 &= 0\cdot3716 \end{aligned} \qquad \tau = 4\cdot2 \begin{cases} t_1 = 2\cdot125 \\ t_2 = 3\cdot940 \end{cases} \begin{aligned} a_1 &= 0\cdot0628 \\ a_2 &= 0\cdot4359 \end{aligned}$$

$$\tau = 2\cdot0 \begin{cases} t_1 = 0\cdot731 \\ t_2 = 1\cdot812 \end{cases} \begin{aligned} a_1 &= 0\cdot0991 \\ a_2 &= 0\cdot3821 \end{aligned} \qquad \tau = 4\cdot4 \begin{cases} t_1 = 2\cdot280 \\ t_2 = 4\cdot137 \end{cases} \begin{aligned} a_1 &= 0\cdot0610 \\ a_2 &= 0\cdot4380 \end{aligned}$$

$$\tau = 2\cdot2 \begin{cases} t_1 = 0\cdot832 \\ t_2 = 2\cdot002 \end{cases} \begin{aligned} a_1 &= 0\cdot0944 \\ a_2 &= 0\cdot3912 \end{aligned} \qquad \tau = 4\cdot6 \begin{cases} t_1 = 2\cdot438 \\ t_2 = 4\cdot334 \end{cases} \begin{aligned} a_1 &= 0\cdot0594 \\ a_2 &= 0\cdot4398 \end{aligned}$$

$$\tau = 2\cdot4 \begin{cases} t_1 = 0\cdot939 \\ t_2 = 2\cdot193 \end{cases} \begin{aligned} a_1 &= 0\cdot0899 \\ a_2 &= 0\cdot3989 \end{aligned} \qquad \tau = 4\cdot8 \begin{cases} t_1 = 2\cdot601 \\ t_2 = 4\cdot531 \end{cases} \begin{aligned} a_1 &= 0\cdot0580 \\ a_2 &= 0\cdot4414 \end{aligned}$$

TABLE VII (cont.)

II. $\qquad 2 \displaystyle\int_0^{\tau} f(t)E_2(\tau-t)\,dt = a_1 f(t_1) + a_2 f(t_2)$

$\tau = 0.1 \begin{cases} t_1 = 0.022 \\ t_2 = 0.080 \end{cases}$	$\begin{aligned} a_1 &= 0.0786 \\ a_2 &= 0.0889 \end{aligned}$	$\tau = 1.1 \begin{cases} t_1 = 0.311 \\ t_2 = 0.935 \end{cases}$	$\begin{aligned} a_1 &= 0.2734 \\ a_2 &= 0.5349 \end{aligned}$
$\tau = 0.2 \begin{cases} t_1 = 0.046 \\ t_2 = 0.162 \end{cases}$	$\begin{aligned} a_1 &= 0.1346 \\ a_2 &= 0.1615 \end{aligned}$	$\tau = 1.2 \begin{cases} t_1 = 0.346 \\ t_2 = 1.024 \end{cases}$	$\begin{aligned} a_1 &= 0.2730 \\ a_2 &= 0.5591 \end{aligned}$
$\tau = 0.3 \begin{cases} t_1 = 0.071 \\ t_2 = 0.244 \end{cases}$	$\begin{aligned} a_1 &= 0.1753 \\ a_2 &= 0.2247 \end{aligned}$	$\tau = 1.3 \begin{cases} t_1 = 0.382 \\ t_2 = 1.114 \end{cases}$	$\begin{aligned} a_1 &= 0.2715 \\ a_2 &= 0.5814 \end{aligned}$
$\tau = 0.4 \begin{cases} t_1 = 0.096 \\ t_2 = 0.328 \end{cases}$	$\begin{aligned} a_1 &= 0.2055 \\ a_2 &= 0.2800 \end{aligned}$	$\tau = 1.4 \begin{cases} t_1 = 0.420 \\ t_2 = 1.204 \end{cases}$	$\begin{aligned} a_1 &= 0.2690 \\ a_2 &= 0.6018 \end{aligned}$
$\tau = 0.5 \begin{cases} t_1 = 0.124 \\ t_2 = 0.412 \end{cases}$	$\begin{aligned} a_1 &= 0.2279 \\ a_2 &= 0.3289 \end{aligned}$	$\tau = 1.5 \begin{cases} t_1 = 0.459 \\ t_2 = 1.294 \end{cases}$	$\begin{aligned} a_1 &= 0.2659 \\ a_2 &= 0.6206 \end{aligned}$
$\tau = 0.6 \begin{cases} t_1 = 0.152 \\ t_2 = 0.498 \end{cases}$	$\begin{aligned} a_1 &= 0.2443 \\ a_2 &= 0.3726 \end{aligned}$	$\tau = 1.6 \begin{cases} t_1 = 0.499 \\ t_2 = 1.386 \end{cases}$	$\begin{aligned} a_1 &= 0.2622 \\ a_2 &= 0.6380 \end{aligned}$
$\tau = 0.7 \begin{cases} t_1 = 0.181 \\ t_2 = 0.584 \end{cases}$	$\begin{aligned} a_1 &= 0.2561 \\ a_2 &= 0.4118 \end{aligned}$	$\tau = 1.7 \begin{cases} t_1 = 0.540 \\ t_2 = 1.477 \end{cases}$	$\begin{aligned} a_1 &= 0.2581 \\ a_2 &= 0.6540 \end{aligned}$
$\tau = 0.8 \begin{cases} t_1 = 0.212 \\ t_2 = 0.671 \end{cases}$	$\begin{aligned} a_1 &= 0.2642 \\ a_2 &= 0.4472 \end{aligned}$	$\tau = 1.8 \begin{cases} t_1 = 0.583 \\ t_2 = 1.569 \end{cases}$	$\begin{aligned} a_1 &= 0.2537 \\ a_2 &= 0.6688 \end{aligned}$
$\tau = 0.9 \begin{cases} t_1 = 0.244 \\ t_2 = 0.758 \end{cases}$	$\begin{aligned} a_1 &= 0.2694 \\ a_2 &= 0.4792 \end{aligned}$	$\tau = 1.9 \begin{cases} t_1 = 0.627 \\ t_2 = 1.661 \end{cases}$	$\begin{aligned} a_1 &= 0.2492 \\ a_2 &= 0.6826 \end{aligned}$
$\tau = 1.0 \begin{cases} t_1 = 0.276 \\ t_2 = 0.846 \end{cases}$	$\begin{aligned} a_1 &= 0.2723 \\ a_2 &= 0.5083 \end{aligned}$	$\tau = 2.0 \begin{cases} t_1 = 0.672 \\ t_2 = 1.753 \end{cases}$	$\begin{aligned} a_1 &= 0.2444 \\ a_2 &= 0.6953 \end{aligned}$

BIBLIOGRAPHICAL NOTES

§ 20. The classical papers on the subject of radiative transfer are those of—

1. A. Schuster, *Astrophys. J.* **21**, 1 (1905).
2. K. Schwarzschild, *Göttinger Nachrichten*, p. 41 (1906).

The generalizations of Schuster and Schwarzschild's method along the lines indicated in the text are due to—

3. G. C. Wick, *Zs. f. Physik*, **120**, 702 (1943).
4. S. Chandrasekhar, *Astrophys. J.* **100**, 76 (1944). (This is the first of a series of several papers published in the same periodical during the years 1944–8.)

§ 21. The following references on the subject of quadrature formulae may be noted here:

5. E. T. Whittaker and G. Robinson, *The Calculus of Observations* (see particularly pp. 152–63), Blackie, London, 1924.
6. E. W. Hobson, *Spherical and Ellipsoidal Harmonics* (pp. 76–81), Cambridge, 1931.
7. G. Szego, *Orthogonal Polynomials*, American Mathematical Society Colloquium Publications, vol. xxiii (chap. iii, pp. 37–48 and chap. xv, pp. 340–6), 1939.

The last of these references is particularly valuable.

§ 22. The numerical data given in this section are from—

    8. A. N. Lowan, N. Davids, and A. Levenson, *Bull. Amer. Math. Soc.* **48,** 739 (1942) (§ 22.1).

    9. Z. Kopal, *Astrophys. J.* **104,** 74 (1946) (§ 22.2).

  10. A. Reiz, *Arkiv. f. Math., Astr. och. Fys.* **29,** part iv (1943) (§ 22.3).

§ 23. As stated in the text, the quadrature formulae given in this section are due to A. Reiz. The author is indebted to Dr. Reiz for making his calculations available in advance of publication; also to Dr. L. Aller and Miss B. Boutelle for the entries for $\tau = 2 \cdot 2 – 4 \cdot 8$ (inclusive) on page 67. It should also be stated in this connexion that the original idea of constructing such quadrature formulae for use in the theory of stellar atmospheres is due to B. Strömgren.

ISOTROPIC SCATTERING

24. Introduction

IN this and in the following three chapters we shall be concerned with various aspects of transfer problems in semi-infinite plane-parallel atmospheres. The principal problems to be considered are (i) the axially symmetric problem with a constant net flux (Chap. I, § 11), and (ii) the problem of diffuse reflection (Chap. I, § 13). In this chapter we shall consider these two basic problems for an isotropically scattering atmosphere and obtain the solutions of the relevant equations (Chap. I, § 11.1, eq. [88], and Chap. I, § 13, eq. [129]) on the method of approximation outlined in the preceding chapter. The analysis of this chapter discloses in the simplest context the characteristic features of the method of solution adopted, namely, the possibility of obtaining the solutions, generally, in the nth approximation; of expressing the angular distributions of the emergent radiations in closed forms; and, finally, of preserving in all orders of approximation the exact relations of a problem. And it may also be mentioned here that a fundamental relationship disclosed to exist between the two basic problems suggest and originate the further developments of the following chapters.

25. The solution of the problem with a constant net flux under conditions of isotropic scattering

25.1. *The solution of the equation of transfer in the nth approximation*

For the problem with a constant net flux in a semi-infinite, plane-parallel, isotropically scattering atmosphere, the equation of transfer is (Chap. I, eq. [88])

$$\mu \frac{dI(\tau,\mu)}{d\tau} = I(\tau,\mu) - \tfrac{1}{2} \int_{-1}^{+1} I(\tau,\mu')\,d\mu'. \tag{1}$$

According to the ideas developed in Chapter II, in the nth approximation we replace this integrodifferential equation by the system of $2n$ linear equations

$$\mu_i \frac{dI_i}{d\tau} = I_i - \tfrac{1}{2}\sum_j a_j I_j \quad (i = \pm 1,...,\pm n), \tag{2}\dagger$$

† When, as in this equation, the summation over j is not qualified, it is always to be understood that j runs through all integers, $(\pm 1,..., \pm n)$, positive and negative. (Note, however, that there is no term with $j = 0$.)

where the μ_i's $(i = \pm 1,..., \pm n$ and $\mu_{-i} = -\mu_i)$ are the zeros of the Legendre polynomial $P_{2n}(\mu)$ and the a_j's $(j = \pm 1,..., \pm n$ and $a_{-j} = a_j)$ are the corresponding Gaussian weights (cf. Chap. II, § 22.1); further, in equation (2) we have written I_i for $I(\tau, \mu_i)$.

In solving the system of equations (2) we shall first obtain the different linearly independent solutions and later, by combining them, obtain the general solution.

First, we seek a solution of the form

$$I_i = g_i e^{-k\tau} \quad (i = \pm 1,..., \pm n), \tag{3}$$

where the g_i's and k are constants, unspecified for the present. Introducing equation (3) in equation (2), we obtain the relation

$$g_i(1+\mu_i k) = \tfrac{1}{2} \sum_j a_j g_j \quad (i = \pm 1,..., \pm n). \tag{4}$$

Hence
$$g_i = \frac{\text{constant}}{1+\mu_i k} \quad (i = \pm 1,..., \pm n), \tag{5}$$

where the 'constant' is independent of i. Substituting the foregoing form for g_i in equation (4) we obtain the *characteristic equation*

$$1 = \frac{1}{2} \sum_j \frac{a_j}{1+\mu_j k}. \tag{6}$$

Remembering that $a_j = a_{-j}$ and $\mu_{-j} = -\mu_j$, we can rewrite the characteristic equation in the form

$$1 = \sum_{j=1}^{n} \frac{a_j}{1-\mu_j^2 k^2}; \tag{7}$$

k^2 therefore satisfies an algebraic equation of order n. However, since (cf. Chap. II, eq. [35])

$$\sum_{j=1}^{n} a_j = 1, \tag{8}$$

$k^2 = 0$ is a root of equation (7). Accordingly, the characteristic equation admits only $2n-2$ distinct non-zero roots, which must occur in pairs as

$$\pm k_\alpha \quad (\alpha = 1,..., n-1). \tag{9}$$

It can be shown that all these roots are numerically greater than 1 (cf. Table VIII, p. 79).

Corresponding to the $(2n-2)$ roots (9) we have $(2n-2)$ independent solutions of equation (2). The solution is completed by observing that equation (2) admits also a solution of the form

$$I_i = b(\tau+q_i) \quad (i = \pm 1,..., \pm n), \tag{10}$$

where b is an arbitrary constant. For, inserting this form for I_i in equation (2) we find that

$$\mu_i = \tau + q_i - \tfrac{1}{2} \sum_j a_j (\tau + q_j)$$

$$= q_i - \tfrac{1}{2} \sum_j a_j q_j; \tag{11}$$

and this can be satisfied by

$$q_i = Q + \mu_i \quad (i = \pm 1, ..., \pm n), \tag{12}$$

where Q is an arbitrary constant. Thus the system (2) allows the solution

$$I_i = b(\tau + Q + \mu_i) \quad (i = \pm 1, ..., \pm n). \tag{13}$$

The general solution of the system of equations represented by (2) can therefore be written in the form

$$I_i = b\left\{ \sum_{\alpha=1}^{n-1} \frac{L_\alpha e^{-k_\alpha \tau}}{1 + \mu_i k_\alpha} + \sum_{\alpha=1}^{n-1} \frac{L_{-\alpha} e^{+k_\alpha \tau}}{1 - \mu_i k_\alpha} + \tau + \mu_i + Q \right\} \quad (i = \pm 1, ..., \pm n), \tag{14}$$

where b, $L_{\pm \alpha}$ ($\alpha = 1, ..., n-1$) and Q are the $2n$ arbitrary constants of integration.

For the problem on hand the boundary conditions are (Chap. I, eqs. [85] and [85']) that none of the I_i's increase more rapidly than e^τ as $\tau \to \infty$ and that there is no radiation incident on $\tau = 0$. The first of these conditions requires that in the general solution (14) we omit all terms in $\exp(+k_\alpha \tau)$, thus leaving

$$I_i = b\left\{ \sum_{\alpha=1}^{n-1} \frac{L_\alpha e^{-k_\alpha \tau}}{1 + \mu_i k_\alpha} + \tau + \mu_i + Q \right\} \quad (i = \pm 1, ..., \pm n). \tag{15}$$

Next, the absence of any radiation in the directions $-1 \leqslant \mu < 0$ at $\tau = 0$ implies that in our present approximation we require

$$I_{-i} = 0 \text{ at } \tau = 0 \text{ and } i = 1, ..., n. \tag{16}$$

Hence, according to equation (15),

$$\sum_{\alpha=1}^{n-1} \frac{L_\alpha}{1 - \mu_i k_\alpha} - \mu_i + Q = 0 \quad (i = 1, ..., n). \tag{17}$$

These are the equations which determine the n constants of integration L_α ($\alpha = 1, ..., n-1$) and Q. The constant b is left arbitrary, and this, as we shall see in § 25.4, is related to the assigned constant net flux of radiation through the atmosphere.

25.2. *Some elementary identities*

The further discussion of the solution obtained in § 25.1 requires certain relations which we shall now establish.

Let

$$D_m(x) = \sum_i \frac{a_i \mu_i^m}{1+\mu_i x} = (-1)^m \sum_i \frac{a_i \mu_i^m}{1-\mu_i x} \qquad (m = 0, 1, ..., 4n). \qquad (18)$$

There is a simple recursion formula which $D_m(x)$ defined in this manner satisfies. We have (cf. Chap. II, eq. [35])

$$D_m(x) = \frac{1}{x} \sum_i a_i \mu_i^{m-1}\left(1 - \frac{1}{1+\mu_i x}\right)$$

$$= \frac{1}{x}\left[\frac{2}{m}\epsilon_{m,\text{odd}} - D_{m-1}(x)\right], \qquad (19)$$

where

$$\epsilon_{m,\text{odd}} = \begin{cases} 1 & \text{if } m \text{ is odd,} \\ 0 & \text{if } m \text{ is even.} \end{cases} \qquad (20)$$

For odd and even values of m respectively, equation (19) takes the forms

$$D_{2j-1}(x) = \frac{1}{x}\left[\frac{2}{2j-1} - D_{2j-2}(x)\right] \qquad (21)$$

and

$$D_{2j}(x) = -\frac{1}{x} D_{2j-1}(x). \qquad (22)$$

Combining these relations, we have

$$D_{2j-1}(x) = \frac{1}{x}\left[\frac{2}{2j-1} + \frac{1}{x} D_{2j-3}(x)\right] = -x D_{2j}(x). \qquad (23)$$

From this formula we readily deduce that

$$D_{2j-1}(x) = \frac{2}{(2j-1)x} + \frac{2}{(2j-3)x^3} + ... + \frac{2}{3x^{2j-3}} + \frac{1}{x^{2j-1}}[2 - D_0(x)]$$

$$(j = 1, ..., 2n), \qquad (24)$$

and

$$D_{2j}(x) = -\frac{2}{(2j-1)x^2} - \frac{2}{(2j-3)x^4} - ... - \frac{2}{3x^{2j-2}} - \frac{1}{x^{2j}}[2 - D_0(x)]$$

$$(j = 1, .., 2n). \qquad (25)$$

If we now let x be a root k of the characteristic equation (cf. eq. [6])

$$D_0(k) = 2, \qquad (26)$$

and we find from equations (24) and (25) that

$$D_1(k) = D_2(k) = 0, \tag{27}$$

$$D_{2j-1}(k) = \frac{2}{(2j-1)k} + \frac{2}{(2j-3)k^3} + \cdots + \frac{2}{3k^{2j-3}} \quad (j = 2,\ldots, 2n), \tag{28}$$

and

$$D_{2j}(k) = -\frac{2}{(2j-1)k^2} - \frac{2}{(2j-3)k^4} - \cdots - \frac{2}{3k^{2j-2}} \quad (j = 2,\ldots, 2n). \tag{29}$$

25.3. *A relation between the characteristic roots and the zeros of the Legendre polynomial*

In terms of the $D_{2j}(k)$'s introduced in § 25.2, we can express the characteristic equation for k in a form which does not explicitly involve the Gaussian weights and divisions: Let p_{2j} be the coefficient of μ^{2j} in the polynomial representation of the Legendre polynomial $P_{2n}(\mu)$, so that

$$P_{2n}(\mu) = \sum_{j=0}^{n} p_{2j} \mu^{2j}. \tag{30}$$

Now consider

$$\sum_{j=0}^{n} p_{2j} D_{2j}(k) = \sum_{i} \frac{a_i}{1+\mu_i k} \left(\sum_{j=0}^{n} p_{2j} \mu_i^{2j} \right). \tag{31}$$

Since the μ_i's are the zeros of $P_{2n}(\mu)$,

$$\sum_{j=0}^{n} p_{2j} \mu_i^{2j} = 0 \quad (i = \pm 1,\ldots, \pm n). \tag{32}$$

Hence

$$\sum_{j=0}^{n} p_{2j} D_{2j}(k) = 0. \tag{33}$$

With $D_{2j}(k)$ given by equation (29), equation (33) is the required form of the characteristic equation.

Substituting in particular for D_{2n} and D_0 in equation (33) we have

$$-\frac{2}{3} \frac{p_{2n}}{k^{2n-2}} - \cdots + 2p_0 = 0. \tag{34}$$

From this equation it follows that

$$\frac{1}{(k_1 \ldots k_{n-1})^2} = (-1)^n \frac{3p_0}{p_{2n}} = 3(\mu_1 \ldots \mu_n)^2, \tag{35}$$

or

$$k_1 \ldots k_{n-1} \mu_1 \ldots \mu_n = \frac{1}{\sqrt{3}}. \tag{36}$$

25.4. *The flux and the K-integral*

Returning to the solution (14) we shall evaluate F and K according to the formulae

$$F = 2 \sum_{i} a_i \mu_i I_i \quad \text{and} \quad K = \tfrac{1}{2} \sum_{i} a_i \mu_i^2 I_i, \tag{37}$$

which are derived from the standard definitions (cf. Chap. I, eq. [86])
by replacing the integrals by the corresponding Gauss sums.

Considering first the sum defining F, we have

$$F = 2b\left\{\sum_{\alpha=1}^{n-1} L_\alpha e^{-k_\alpha \tau} D_1(k_\alpha) + (\tau+Q)\sum_i a_i \mu_i + \sum_i a_i \mu_i^2\right\}. \qquad (38)$$

Using equation (27) and Chapter II, equation (35), we have

$$F = \tfrac{4}{3}b. \qquad (39)$$

In other words, F defined for the discrete streams, I_i, is a constant.

Expressing b in terms of F, we can rewrite the solution (15) in the
form

$$I_i = \tfrac{3}{4}F\left\{\sum_{\alpha=1}^{n-1}\frac{L_\alpha e^{-k_\alpha \tau}}{1+\mu_i k_\alpha} + \tau + \mu_i + Q\right\} \quad (i = \pm 1,...,\pm n). \qquad (40)$$

Considering next the sum defining K, we have

$$K = \tfrac{3}{8}F\left\{\sum_{\alpha=1}^{n-1} L_\alpha e^{-k_\alpha \tau} D_2(k_\alpha) + (\tau+Q)\sum_i a_i \mu_i^2 + \sum_i a_i \mu_i^3\right\}, \qquad (41)$$

or, again, using equation (27) and Chapter II, equation (35), we have

$$K = \tfrac{1}{4}F(\tau+Q); \qquad (42)$$

this, in our present scheme of approximation, is the K-integral
(Chap. I, § 10).

25.5. *The source function. The radiation field. The law of darkening*
The source function for the problem under discussion is

$$J = \tfrac{1}{2}\int_{-1}^{+1} I(\tau,\mu)\,d\mu \simeq \tfrac{1}{2}\sum_i a_i I_i. \qquad (43)$$

Using the solution (40) for the I_i's, we have

$$J = \tfrac{3}{8}F\left\{\sum_{\alpha=1}^{n-1} L_\alpha e^{-k_\alpha \tau} D_0(k_\alpha) + (\tau+Q)\sum_i a_i + \sum_i a_i \mu_i\right\}$$

$$= \tfrac{3}{4}F\left\{\sum_{\alpha=1}^{n-1} L_\alpha e^{-k_\alpha \tau} + \tau + Q\right\}. \qquad (44)$$

We shall write this in the form

$$J(\tau) = \tfrac{3}{4}F\{\tau+q(\tau)\}, \qquad (45)$$

where

$$q(\tau) = Q + \sum_{\alpha=1}^{n-1} L_\alpha e^{-k_\alpha \tau}. \qquad (46)$$

In terms of the source function (44) we can determine the radiation field in accordance with the equations (Chap. I, eq. [90])

$$I(\tau, +\mu) = \int_{\tau}^{\infty} J(t) e^{-(t-\tau)/\mu} \frac{dt}{\mu} \quad (0 < \mu \leqslant 1),$$

and
$$I(\tau, -\mu) = \int_{0}^{\tau} J(t) e^{-(\tau-t)/\mu} \frac{dt}{\mu} \quad (0 < \mu \leqslant 1). \tag{47}$$

We thus find that

$$I(\tau, +\mu) = \tfrac{3}{4} F \left\{ \sum_{\alpha=1}^{n-1} \frac{L_\alpha e^{-k_\alpha \tau}}{1+k_\alpha \mu} + \tau + \mu + Q \right\} \tag{48}$$

and

$$I(\tau, -\mu) = \tfrac{3}{4} F \left\{ \sum_{\alpha=1}^{n-1} \frac{L_\alpha}{1-k_\alpha \mu} (e^{-k_\alpha \tau} - e^{-\tau/\mu}) + \tau + (Q-\mu)(1-e^{-\tau/\mu}) \right\}. \tag{49}$$

It will be observed that equation (48) is in agreement with the solution (40), for the *outward* streams, at the points of the Gaussian division; as is to be expected, this is not the case for the inward streams.

The angular distribution of the emergent radiation is obtained by putting $\tau = 0$ in equation (48). We have

$$I(0, \mu) = \tfrac{3}{4} F \left\{ \sum_{\alpha=1}^{n-1} \frac{L_\alpha}{1+k_\alpha \mu} + \mu + Q \right\}. \tag{50}$$

Comparing equation (50) with equation (17) which determines the constants of integration L_α and Q, we observe that the angular distribution of the emergent radiation (defined for the interval $0 \leqslant \mu \leqslant 1$) is expressed in terms of a function whose zeros are assigned in the complementary interval $(-1 \leqslant \mu < 0)$. Thus, letting

$$S(\mu) = \sum_{\alpha=1}^{n-1} \frac{L_\alpha}{1-k_\alpha \mu} - \mu + Q, \tag{51}$$

the boundary conditions require that

$$S(\mu_i) = 0 \quad (i = 1,...,n), \tag{52}$$

while the angular distribution of the emergent radiation is given by

$$I(0, \mu) = \tfrac{3}{4} F S(-\mu). \tag{53}$$

This *reciprocity* between the boundary conditions determining the constants of integration and the law of darkening expressing the angular distribution of the emergent radiation which we encounter here, is of quite general occurrence in the theory (cf., e.g., § 26).

25.6. *The elimination of the constants and the expression of $I(0,\mu)$ in closed form. The H-function*

We shall now show how an explicit formula for $S(\mu)$ can be found without solving explicitly for the constants L_α and Q.

Multiply $S(\mu)$ by $\qquad R(\mu) = \prod_{\alpha=1}^{n-1} (1-k_\alpha\mu), \qquad (54)$

to clear (51) of 'fractions'. The resulting function $S(\mu)R(\mu)$ is a polynomial of degree n in μ which vanishes for $\mu = \mu_i$, $i = 1,...,n$. Hence $S(\mu)R(\mu)$ cannot differ from the polynomial

$$P(\mu) = \prod_{i=1}^{n} (\mu-\mu_i), \qquad (55)$$

by more than a constant factor. And the constant of proportionality can be found by comparing the coefficients of the highest power of μ (namely, μ^n) in $P(\mu)$ and $S(\mu)R(\mu)$. In the former it is unity, while in the latter it is

$$(-1)^n k_1...k_{n-1}. \qquad (56)$$

Hence $\qquad S(\mu) = (-1)^n k_1...k_{n-1} \dfrac{P(\mu)}{R(\mu)}. \qquad (57)$

This is the required formula.

According to equations (54), (55), and (57)

$$S(-\mu) = k_1...k_{n-1} \frac{\prod_{i=1}^{n} (\mu+\mu_i)}{\prod_{\alpha=1}^{n-1} (1+k_\alpha\mu)}, \qquad (58)$$

or, using the relation (36), we can write

$$S(-\mu) = \frac{1}{\sqrt{3}} H(\mu), \qquad (59)$$

where $\qquad H(\mu) = \dfrac{1}{\mu_1...\mu_n} \dfrac{\prod_{i=1}^{n} (\mu+\mu_i)}{\prod_{\alpha=1}^{n-1} (1+k_\alpha\mu)}. \qquad (60)$

In terms of the function $H(\mu)$ defined in this manner, we can express the angular distribution of the emergent radiation in the form (cf. eq. [53])

$$I(0,\mu) = \frac{\sqrt{3}}{4} F H(\mu). \qquad (61)$$

25.7. *The Hopf–Bronstein relation*

According to equations (44), (50), and (61),

$$I(0,0) = J(0) = \tfrac{3}{4}F\left[\sum_{\alpha=1}^{n-1} L_\alpha + Q\right] = \frac{\sqrt{3}}{4} F H(0). \qquad (62)$$

But (cf. eq. [60]) $$H(0) = 1. \tag{63}$$

Hence $$J(0) = \frac{\sqrt{3}}{4} F, \tag{64}$$

a result which is seen to be true in *all* orders of approximation. We therefore conclude that equation (64) represents an *exact relation*.

The relation (64) between $J(0)$ and F was first discovered by Hopf and Bronstein from considerations of a different type (see Chap. XIII).

25.8. *The constants of integration*

Explicit formulae for the constants of integration can now be found in the following manner:

From equation (51) it is apparent that

$$L_\alpha = \lim_{\mu \to k_\alpha^{-1}} (1 - k_\alpha \mu) S(\mu), \tag{65}$$

or, using equation (57) for $S(\mu)$, we have

$$L_\alpha = (-1)^n k_1 \dots k_{n-1} \frac{P(1/k_\alpha)}{R_\alpha(1/k_\alpha)} \quad (\alpha = 1, \dots, n-1), \tag{66}$$

where $$R_\alpha(x) = \prod_{\beta \neq \alpha} (1 - k_\beta x). \tag{67}$$

Turning next to the determination of Q, we first observe that, according to equations (51) and (59),

$$\sum_{\alpha=1}^{n-1} L_\alpha + Q = S(0) = \frac{1}{\sqrt{3}}. \tag{68}$$

With the L_α's given by equation (66), the foregoing equation becomes

$$Q = \frac{1}{\sqrt{3}} + (-1)^{n+1} k_1 \dots k_{n-1} \sum_{\alpha=1}^{n-1} \frac{P(1/k_\alpha)}{R_\alpha(1/k_\alpha)}. \tag{69}$$

To evaluate the sum occurring on the right-hand side of equation (69) we introduce the function

$$f(x) = \sum_{\alpha=1}^{n-1} \frac{P(1/k_\alpha)}{R_\alpha(1/k_\alpha)} R_\alpha(x), \tag{70}$$

and express Q in terms of it. Thus

$$Q = \frac{1}{\sqrt{3}} + (-1)^{n+1} k_1 \dots k_{n-1} f(0). \tag{71}$$

Now $f(x)$ defined as in equation (70) is a polynomial of degree $(n-2)$ in x which takes the values $P(1/k_\alpha)$ for $x = k_\alpha^{-1}$, $\alpha = 1, \dots, n-1$. There must accordingly exist a relation of the form

$$f(x) = P(x) + R(x)(Ax + B)$$
$$= \prod_{i=1}^{n} (x - \mu_i) + (Ax + B) \prod_{\alpha=1}^{n-1} (1 - k_\alpha x), \tag{72}$$

where A and B are constants. But as $f(x)$ is a polynomial of degree $n-2$, the coefficients of x^n and x^{n-1} on the right-hand side of equation (72) must vanish. And the conditions for the vanishing of these coefficients are

$$1+(-1)^{n-1}k_1...k_{n-1}A = 0 \tag{73}$$

and

$$-\sum_{i=1}^{n}\mu_i+(-1)^{n-2}k_1...k_{n-1}A\sum_{\alpha=1}^{n-1}\frac{1}{k_\alpha}+(-1)^{n-1}k_1...k_{n-1}B = 0. \tag{74}$$

These equations determine A and B. We find

$$A = \frac{(-1)^n}{k_1...k_{n-1}}; \qquad B = \frac{(-1)^{n-1}}{k_1...k_{n-1}}\left[\sum_{i=1}^{n}\mu_i-\sum_{\alpha=1}^{n-1}\frac{1}{k_\alpha}\right]. \tag{75}$$

From equation (72) it now follows that

$$f(0) = (-1)^n\mu_i...\mu_n+\frac{(-1)^{n-1}}{k_1...k_{n-1}}\left[\sum_{i=1}^{n}\mu_i-\sum_{\alpha=1}^{n-1}\frac{1}{k_\alpha}\right]. \tag{76}$$

Substituting this value of $f(0)$ in equation (71) and making use of the relation (36), we are left with

$$Q = \sum_{i=1}^{n}\mu_i-\sum_{\alpha=1}^{n-1}\frac{1}{k_\alpha}. \tag{77}$$

With this we have completed the formal solution of the problem in the general nth approximation.

25.9. *The numerical form of the solutions in the first four approximations*

The characteristic roots and the constants of integration in the first four approximations are listed in Table VIII. The corresponding laws of darkening (eq. [50]) and the functions $q(\tau)$ (eq. [46]) are exhibited in Tables IX and X.

TABLE VIII

The Characteristic Roots and the Constants of Integration

First approximation

$k = 0$; $Q = 1/\sqrt{3}$

Second approximation

$k = 0$; $Q = +0{\cdot}694025$
$k_1 = 1{\cdot}97203$; $L_1 = -0{\cdot}116675$

Third approximation

$k = 0$; $Q = +0{\cdot}703899$
$k_1 = 1{\cdot}225211$; $L_1 = -0{\cdot}101245$
$k_2 = 3{\cdot}202945$; $L_2 = -0{\cdot}02530$

Fourth approximation

$k = 0$; $Q = +0{\cdot}706920$
$k_1 = 1{\cdot}103188$; $L_1 = -0{\cdot}083921$
$k_2 = 1{\cdot}591778$; $L_2 = -0{\cdot}036187$
$k_3 = 4{\cdot}45808$; $L_3 = -0{\cdot}009461$

TABLE IX

The Laws of Darkening given by the Second, Third, and Fourth Approximations

μ	$I(0,\mu)/F$			$H(\mu)$
	Second approximation	Third approximation	Fourth approximation	Fourth approximation
0·0	0·4330	0·4330	0·4330	1·0000
0·1	0·5224	0·5285	0·5319	1·2284
0·2	0·6078	0·6164	0·6205	1·4330
0·3	0·6905	0·7003	0·7046	1·6272
0·4	0·7716	0·7819	0·7861	1·8154
0·5	0·8515	0·8620	0·8660	2·0000
0·6	0·9304	0·9410	0·9449	2·1822
0·7	1·0088	1·0193	1·0231	2·3627
0·8	1·0866	1·0970	1·1007	2·5420
0·9	1·1640	1·1743	1·1779	2·7202
1·0	1·2411	1·2513	1·2548	2·8978

TABLE X

The Function $q(\tau)$ Derived on the Basis of the Second, Third, and Fourth Approximations

τ	$q(\tau)$			τ	$q(\tau)$		
	Second approximation	Third approximation	Fourth approximation		Second approximation	Third approximation	Fourth approximation
0·00	0·5774	0·5774	0·5774	0·90	0·6743	0·6898	0·6933
0·05	0·5883	0·5938	0·5974	1·0	0·6778	0·6924	0·6954
0·10	0·5982	0·6080	0·6139	1·2	0·6831	0·6959	0·6987
0·15	0·6072	0·6202	0·6274	1·4	0·6867	0·6982	0·7008
0·20	0·6154	0·6307	0·6386	1·6	0·6891	0·6997	0·7024
0·25	0·6228	0·6398	0·6479	1·8	0·6907	0·7008	0·7035
0·30	0·6295	0·6477	0·6557	2·0	0·6918	0·7016	0·7044
0·35	0·6355	0·6544	0·6621	2·2	0·6925	0·7021	0·7050
0·40	0·6410	0·6603	0·6676	2·4	0·6930	0·7025	0·7055
0·50	0·6505	0·6698	0·6761	2·6	0·6933	0·7028	0·7058
0·60	0·6583	0·6770	0·6823	2·8	0·6936	0·7031	0·7064
0·70	0·6647	0·6824	0·6870	3·0	0·6937	0·7033	0·7065
0·80	0·6699	0·6866	0·6905	∞	0·6940	0·7039	0·7069

A comparison of the approximate solutions for the law of darkening given in Table IX with the exact solution we shall later obtain (Chap. V) indicates that an accuracy of about 0·5 per cent. may be expected in the third approximation of our method of solution.

26. The problem of diffuse reflection. The case $\varpi_0 < 1$

The equation of transfer appropriate to the problem of diffuse reflection by a plane-parallel atmosphere scattering radiation isotropically with an albedo $\varpi_0 < 1$ for single scattering is (Chap. I, eq. [129])

$$\mu \frac{dI(\tau,\mu)}{d\tau} = I(\tau,\mu) - \tfrac{1}{2}\varpi_0 \int_{-1}^{+1} I(\tau,\mu')\, d\mu' - \tfrac{1}{4}\varpi_0\, Fe^{-\tau/\mu_0}, \qquad (78)$$

where it may be recalled that $-\mu_0$ is the direction cosine of the angle of incidence referred to the outward normal. The equivalent system of linear equations in the nth approximation is

$$\mu_i \frac{dI_i}{d\tau} = I_i - \tfrac{1}{2}\varpi_0 \sum_j a_j I_j - \tfrac{1}{4}\varpi_0\, Fe^{-\tau/\mu_0} \qquad (i = \pm 1, ..., \pm n). \qquad (79)$$

26.1. *The solution of the associated homogeneous system*

Considering first the homogeneous system associated with equation (79), we have

$$\mu_i \frac{dI_i}{d\tau} = I_i - \tfrac{1}{2}\varpi_0 \sum_j a_j I_j \qquad (i = \pm 1, ..., \pm n). \qquad (80)$$

This system of equations admits integrals of the form

$$I_i = g_i e^{-k\tau} \qquad (i = \pm 1, ..., \pm n), \qquad (81)$$

where the g_i's and k are constants: for, inserting this form for I_i in equation (80), we have

$$g_i(1 + \mu_i k) = \tfrac{1}{2}\varpi_0 \sum_j a_j g_j. \qquad (82)$$

Hence
$$g_i = \frac{\text{constant}}{1 + \mu_i k} \qquad (i = \pm 1, ..., \pm n) \qquad (83)$$

and the characteristic equation for k is

$$1 = \tfrac{1}{2}\varpi_0 \sum_j \frac{a_j}{1 + \mu_j k} = \varpi_0 \sum_{j=1}^{n} \frac{a_j}{1 - \mu_j^2 k^2}. \qquad (84)$$

If $\varpi_0 < 1$, the characteristic equation (84) admits $2n$ distinct non-zero roots which occur in pairs as

$$\pm k_\alpha \qquad (\alpha = 1, ..., n). \qquad (85)$$

Equation (80) therefore allows the $2n$ independent integrals

$$I_i = \frac{\text{constant}}{1 \pm \mu_i k_\alpha} e^{\mp k_\alpha \tau} \qquad (i = \pm 1, ..., \pm n \text{ and } \alpha = 1, ..., n). \qquad (86)$$

26.2. *A particular integral*

To complete the solution of equation (79) we require a particular integral. This can be found in the following manner: Setting

$$I_i = \tfrac{1}{4}\varpi_0 Fh_i e^{-\tau/\mu_0} \qquad (i = \pm 1, ..., \pm n), \qquad (87)$$

where the h_i's are constants, unspecified for the present, we verify that we must have

$$h_i\left(1+\frac{\mu_i}{\mu_0}\right) = \tfrac{1}{2}\varpi_0 \sum_j a_j h_j + 1. \tag{88}$$

The constants h_i must therefore be expressible in the form

$$h_i = \frac{\gamma}{1+\mu_i/\mu_0} \quad (i = \pm 1,...,\pm n), \tag{89}$$

where the constant γ has to be determined in accordance with the condition (cf. eq. [88])

$$\gamma = \tfrac{1}{2}\varpi_0\gamma \sum_j \frac{a_j}{1+\mu_j/\mu_0}+1.$$

Hence

$$\gamma = \frac{1}{1-\varpi_0 \sum\limits_{j=1}^{n} a_j/(1-\mu_j^2/\mu_0^2)}. \tag{90}$$

With γ given by this equation the required particular integral is

$$I_i = \tfrac{1}{4}\varpi_0 F\frac{\gamma e^{-\tau/\mu_0}}{1+\mu_i/\mu_0} \quad (i = \pm 1,...,\pm n). \tag{91}$$

26.3. *The solution in the nth approximation*

Adding to the particular integral (91) the general solution of the homogeneous system (80) compatible with our present requirement of the boundedness of the solution, we have (cf. eq. [86])

$$I_i = \tfrac{1}{4}\varpi_0 F\left\{\sum_{\alpha=1}^{n} \frac{L_\alpha e^{-k_\alpha\tau}}{1+\mu_i k_\alpha}+\frac{\gamma e^{-\tau/\mu_0}}{1+\mu_i/\mu_0}\right\} \quad (i = \pm 1,...,\pm n). \tag{92}$$

The constants L_α ($\alpha = 1,...,n$) in the solution (92) are to be found from the boundary conditions (Chap. I, eq. [127])

$$I_{-i} = 0 \quad \text{at} \quad \tau = 0 \quad (i = 1,...,n). \tag{93}$$

The equations which determine the constants of integration are therefore

$$\sum_{\alpha=1}^{n} \frac{L_\alpha}{1-\mu_i k_\alpha}+\frac{\gamma}{1-\mu_i/\mu_0} = 0 \quad (i = 1,...,n). \tag{94}$$

Now the source function for the problem under discussion is (cf. eq. [78])

$$\mathfrak{J}(\tau) = \tfrac{1}{2}\varpi_0 \int_{-1}^{+1} I(\tau,\mu)\,d\mu+\tfrac{1}{4}\varpi_0 Fe^{-\tau/\mu_0}$$

$$\simeq \tfrac{1}{2}\varpi_0 \sum a_i I_i+\tfrac{1}{4}\varpi_0 Fe^{-\tau/\mu_0}. \tag{95}$$

With I_i given by equation (92), equation (95) becomes

$$\mathfrak{J}(\tau) = \tfrac{1}{8}\varpi_0^2 F\left\{\sum_{\alpha=1}^{n} L_\alpha e^{-k_\alpha \tau} D_0(k_\alpha) + \gamma e^{-\tau/\mu_0} \sum_j \frac{a_j}{1+\mu_j/\mu_0}\right\} + \tfrac{1}{4}\varpi_0 F e^{-\tau/\mu_0}.$$
(96)

But, according to equations (84) and (90),

$$\tfrac{1}{2}\varpi_0 D_0(k_\alpha) = 1 \quad \text{and} \quad \tfrac{1}{2}\varpi_0 \gamma \sum_j \frac{a_j}{1+\mu_j/\mu_0} = \gamma - 1.$$
(97)

Making use of these relations we obtain

$$\mathfrak{J}(\tau) = \tfrac{1}{4}\varpi_0 F\left\{\sum_{\alpha=1}^{n} L_\alpha e^{-k_\alpha \tau} + \gamma e^{-\tau/\mu_0}\right\}.$$
(98)

In terms of this source function the radiation field in the atmosphere can be determined in the usual fashion (Chap. I, eq. [90]). We find

$$I(\tau, +\mu) = \tfrac{1}{4}\varpi_0 F\left\{\sum_{\alpha=1}^{n} \frac{L_\alpha e^{-k_\alpha \tau}}{1+\mu k_\alpha} + \frac{\gamma e^{-\tau/\mu_0}}{1+\mu/\mu_0}\right\}$$

and

$$I(\tau, -\mu) = \tfrac{1}{4}\varpi_0 F\left\{\sum_{\alpha=1}^{n} \frac{L_\alpha}{1-\mu k_\alpha}(e^{-k_\alpha \tau} - e^{-\tau/\mu}) + \frac{\gamma}{1-\mu/\mu_0}(e^{-\tau/\mu_0} - e^{-\tau/\mu})\right\}.$$
(99)

In particular, the angular distribution of the reflected radiation is given by

$$I(0, \mu) = \tfrac{1}{4}\varpi_0 F\left\{\sum_{\alpha=1}^{n} \frac{L_\alpha}{1+k_\alpha \mu} + \frac{\gamma}{1+\mu/\mu_0}\right\}.$$
(100)

Comparing equations (94) and (100) we observe that, as in § 25, the angular distribution of the emergent radiation (defined in the interval $0 \leqslant \mu \leqslant 1$) is described in terms of a function whose zeros are assigned in the complementary interval $-1 \leqslant \mu < 0$. This enables (again as in § 25) the elimination of the constants and the expression of the angular distribution of the emergent radiation in closed form. But first we shall establish an important identity needed in the further developments.

26.4. *An identity*

Consider the function

$$T(z) = 1 - \tfrac{1}{2}\varpi_0 z \sum_j \frac{a_j}{z+\mu_j}$$

$$= 1 - \varpi_0 z^2 \sum_{j=1}^{n} \frac{a_j}{z^2 - \mu_j^2}.$$
(101)

Comparing this with the characteristic equation (84), we conclude that

$$T(z) = 0 \quad \text{for} \quad z = \pm k_\alpha^{-1} \quad (\alpha = 1,...,n). \tag{102}$$

Accordingly

$$\prod_{j=1}^{n} (z^2 - \mu_j^2) T(z), \tag{103}$$

cannot differ from

$$\prod_{\alpha=1}^{n} (1 - k_\alpha^2 z^2), \tag{104}$$

by more than a constant factor, since (103) is a polynomial of degree $2n$ in z and has the same zeros as (104). The constant of proportionality can be ascertained by comparing the constant terms in (103) and (104), or, equivalently, by putting $z = 0$ in both expressions. We thus find that

$$T(z) = (-1)^n \mu_1^2 ... \mu_n^2 \frac{\displaystyle\prod_{\alpha=1}^{n} (1 - k_\alpha^2 z^2)}{\displaystyle\prod_{j=1}^{n} (z^2 - \mu_j^2)}. \tag{105}$$

In terms of the function (cf. eq. [60])

$$H(\mu) = \frac{1}{\mu_1 ... \mu_n} \frac{\displaystyle\prod_{j=1}^{n} (\mu + \mu_j)}{\displaystyle\prod_{\alpha=1}^{n} (1 + k_\alpha \mu)}, \tag{106}$$

we can rewrite equation (105) in the form

$$1 - \varpi_0 z^2 \sum_{j=1}^{n} \frac{a_j}{z^2 - \mu_j^2} = \frac{1}{H(z) H(-z)}. \tag{107}$$

This is the required identity.

26.5. *The elimination of the constants and the expression of the law of diffuse reflection in closed form*

According to equations (90) and (107) the formula for γ becomes

$$\gamma = \frac{1}{T(\mu_0)} = H(\mu_0) H(-\mu_0). \tag{108}$$

Inserting this value of γ in equations (94) and (100), we can express the angular distribution of the emergent radiation and the boundary conditions in terms of the function

$$S(\mu) = \sum_{\alpha=1}^{n} \frac{L_\alpha}{1 - k_\alpha \mu} + \frac{H(\mu_0) H(-\mu_0)}{1 - \mu/\mu_0}. \tag{109}$$

Thus

$$S(\mu_i) = 0 \quad (i = 1,...,n) \tag{110}$$

and

$$I(0, \mu) = \tfrac{1}{4} \varpi_0 F S(-\mu). \tag{111}$$

Next we observe that the function

$$\left(1-\frac{\mu}{\mu_0}\right)\prod_{\alpha=1}^{n}(1-k_\alpha\mu)S(\mu)$$

is a polynomial of degree n in μ which vanishes for $\mu=\mu_i$, $i=1,...,n$. There must, accordingly, exist a relation of the form

$$S(\mu) = X\frac{(-1)^n}{\mu_1\cdots\mu_n}\frac{\prod_{i=1}^{n}(\mu-\mu_i)}{\prod_{\alpha=1}^{n}(1-k_\alpha\mu)}\frac{1}{(1-\mu/\mu_0)}, \tag{112}$$

where X is some constant. An alternative form of this relation is (cf. eq. [106])

$$S(\mu) = X\frac{H(-\mu)}{1-\mu/\mu_0}. \tag{113}$$

The constant X appearing in equation (113) can be determined by making use of the fact that, while according to equation (109)

$$\lim_{\mu\to\mu_0}\left(1-\frac{\mu}{\mu_0}\right)S(\mu) = H(\mu_0)H(-\mu_0), \tag{114}$$

according to equation (113)

$$\lim_{\mu\to\mu_0}\left(1-\frac{\mu}{\mu_0}\right)S(\mu) = XH(-\mu_0). \tag{115}$$

Hence

$$X = H(\mu_0), \tag{116}$$

and

$$S(\mu) = \frac{H(\mu_0)H(-\mu)}{1-\mu/\mu_0}. \tag{117}$$

The law of diffuse reflection therefore takes the form

$$I(0,\mu) = \tfrac{1}{4}\varpi_0\,F\frac{\mu_0}{\mu+\mu_0}\,H(\mu)H(\mu_0). \tag{118}$$

If we express the reflected intensity (118) in terms of a scattering function $S(\mu,\mu_0)$ (cf. Chap. I, eq. [125]) in the form

$$I(0,\mu) = \frac{F}{4\mu}\,S(\mu,\mu_0), \tag{119}$$

then,

$$\left(\frac{1}{\mu}+\frac{1}{\mu_0}\right)S(\mu,\mu_0) = \varpi_0\,H(\mu)H(\mu_0). \tag{120}$$

The separation of the variables, μ and μ_0, in equation (120) is particularly noteworthy. The origin of this separation will become apparent in Chapter IV.

We shall not give here numerical solutions for $H(\mu)$ in the various approximations, for, in a later chapter (Chap. V) we shall give tables of the exact functions.

27. The law of diffuse reflection in the conservative case

The solution of the conservative case can be obtained by simply letting $\varpi_0 = 1$ and one of the characteristic roots (say, k_n) equal zero in the formulae of the preceding section. Thus (cf. eqs. [98] and [108]–[111])

$$\mathfrak{J}(\tau) = \tfrac{1}{4}F\left\{\sum_{\alpha=1}^{n-1} L_\alpha e^{-k_\alpha \tau} + L_n + H(\mu_0)H(-\mu_0)e^{-\tau/\mu_0}\right\} \tag{121}$$

and

$$I(0,\mu) = \tfrac{1}{4}F\left\{\sum_{\alpha=1}^{n-1} \frac{L_\alpha}{1+k_\alpha\mu} + L_n + \frac{H(\mu_0)H(-\mu_0)}{1+\mu/\mu_0}\right\}, \tag{122}$$

where $H(\mu)$ has now the same meaning as in § 25 (eq. [60]).

The elimination of the constants proceeds exactly as in § 26 and leads to the law of diffuse reflection (cf. eq. [118])

$$I(0,\mu) = \tfrac{1}{4}F\,\frac{\mu_0}{\mu+\mu_0}\,H(\mu)H(\mu_0). \tag{123}$$

The scattering function continues to be given by equation (120).

A fact of some interest in this context of conservative scattering is that $\mathfrak{J}(\tau)$ tends to a finite limit as $\tau \to \infty$. Thus

$$\mathfrak{J}(\infty) = \tfrac{1}{4}FL_n. \tag{124}$$

We shall now find the ratio of this value for $\tau \to \infty$, to the value

$$\mathfrak{J}(0) = I(0,0) = \tfrac{1}{4}FH(\mu_0) \tag{125}$$

at $\tau = 0$.

Multiplying the equation (cf. eqs. [122] and [123])

$$\sum_{\alpha=1}^{n-1} \frac{L_\alpha}{1+k_\alpha\mu} + L_n + \frac{H(\mu_0)H(-\mu_0)}{1+\mu/\mu_0} = \frac{1}{\mu_1\cdots\mu_n}\frac{\prod\limits_{i=1}^{n}(\mu+\mu_i)}{\prod\limits_{\alpha=1}^{n-1}(1+k_\alpha\mu)}\frac{H(\mu_0)}{1+\mu/\mu_0}, \tag{126}$$

by

$$\left(1+\frac{\mu}{\mu_0}\right)\prod_{\alpha=1}^{n-1}(1+k_\alpha\mu), \tag{127}$$

and comparing the coefficients of μ^n on either side, we find

$$\frac{1}{\mu_0}k_1\ldots k_{n-1}L_n = \frac{H(\mu_0)}{\mu_1\cdots\mu_n}. \tag{128}$$

Hence

$$L_n = \mu_0\frac{H(\mu_0)}{k_1\ldots k_{n-1}\mu_1\cdots\mu_n}, \tag{129}$$

or, making use of the relation (36), we have

$$L_n = \mu_0 H(\mu_0)\sqrt{3}. \tag{130}$$

Combining equations (124), (125), and (130) we obtain

$$\frac{\mathfrak{J}(\infty)}{\mathfrak{J}(0)} = \mu_0\sqrt{3}, \tag{131}$$

a result which is seen to be independent of the order of the approximation. We therefore conclude that equation (131) represents an exact relation.

Finally, attention may be specially drawn to the fact that the solution for the problem of diffuse reflection involves the same H-function as the solution for the law of darkening in the problem with the constant net flux (cf. § 25, eq. [61]). The origin of this remarkable relationship between the two problems will become apparent in Chapter IV.

BIBLIOGRAPHICAL NOTES

The analysis in this chapter, except for rearrangement of details, follows:

1. S. CHANDRASEKHAR, *Astrophys. J.* **100**, 76 (1944) and **101**, 328 (1945). (§ 25.)
2. S. CHANDRASEKHAR, ibid. **101**, 348 (1945) and **103**, 165 (1946). (§§ 26 and 27.)

See also:

3. S. CHANDRASEKHAR, *Bull. Amer. Math. Soc.* (1946 Gibbs Lecture), **53**, 641 (1947).

The H-functions first appear explicitly in ref. 2.

The following references to earlier investigations on the problems considered in this chapter may be noted:

§ 25. The equation of transfer considered in this section first arose in K. Schwarzschild's investigation on the radiative equilibrium of a grey stellar atmosphere under conditions of local thermodynamic equilibrium:

4. K. SCHWARZSCHILD, *Göttinger Nachrichten, Math.-Phys. Klasse*, p. 41 (1906).

Later investigations are those of:

5. J. H. JEANS, *Monthly Notices Roy. Astron. Soc. London*, **78**, 28 (1917).
6. E. A. MILNE, ibid. **81**, 361 (1921) and *Philos. Mag.* **44**, 872 (1922).
7. B. LINDBLAD, *Acta. Reg. Soc. Sci. Uppsala*, iv, **6**, 1 (1923) and *Uppsala Univ. Årsskrift*, **1** (1920).
8. E. A. MILNE, *Philos. Trans. Roy. Soc. (London)* A, **223**, 201 (1922).

A summary of these and other investigations will be found in:

9. A. S. EDDINGTON, *The Internal Constitution of the Stars*, chap. xii, Cambridge, 1926.
10. E. A. MILNE, *Handbuch der Astrophysik*, vol. iii, part i, 'Thermodynamics of the Stars', pp. 70–172, Springer, Berlin, 1930.
11. A. UNSÖLD, *Physik der Sternatmosphären*, pp. 89–104, Springer, Berlin, 1938.

The article by Milne (ref. 10) is a particularly valuable reference.

The 'Hopf–Bronstein relation' (§ 25.7) first occurs in:

12. M. BRONSTEIN, *Zs. f. Physik*, **58**, 696 (1929); see also *Zs. f. Physik*, **59**, 144 (1929).

13. E. HOPF, *Monthly Notices Roy. Astron. Soc. London*, **90**, 287 (1930).

An important reference in these connexions is:

14. E. HOPF, *Mathematical Problems of Radiative Equilibrium* (Cambridge Mathematical Tract, No. 31), Cambridge, 1934.

§§ 26 and 27. The problem considered in these sections (more particularly, § 27) first arose in discussions relating to the so-called 'reflection effect' in eclipsing binaries; and in this context was first considered by:

15. A. S. EDDINGTON, *Monthly Notices Roy. Astron. Soc. London*, **86**, 320 (1926).

More detailed discussions of this problem are due to:

16. E. A. MILNE, ibid. **87**, 43 (1926); also ref. 9, pp. 134–41.

17. E. A. HOPF, ref. 14.

The result expressed by equation (131) is established in ref. 17 by a different chain of arguments.

PRINCIPLES OF INVARIANCE

28. Principles of invariance

THE solutions for the problems of constant net flux and diffuse reflection obtained in Chapter III under conditions of isotropic scattering have revealed that a relationship between the two problems must exist. Thus the law of darkening in the problem with a constant net flux and the law of diffuse reflection are both expressed in terms of the same H-function in the form (Chap. III, eqs. [61] and [123])

$$I(0,\mu) = \frac{\sqrt{3}}{4} F H(\mu) \tag{1}$$

and $$I(0,\mu;\mu_0) = \tfrac{1}{4} F \frac{\mu_0}{\mu+\mu_0} H(\mu) H(\mu_0). \tag{2}$$

In the nth approximation

$$H(\mu) = \frac{1}{\mu_1 \cdots \mu_n} \frac{\prod\limits_{i=1}^{n}(\mu+\mu_i)}{\prod\limits_{\alpha=1}^{n-1}(1+k_\alpha\mu)}, \tag{3}$$

where the μ_i's are the zeros of the Legendre polynomial $P_{2n}(\mu)$ and the k_α's are the positive non-zero roots of the characteristic equation

$$1 = \sum_{j=1}^{n} \frac{a_j}{1-\mu_j^2 k^2}. \tag{4}$$

Though the forms of the solutions (1) and (2) have been established only in a particular scheme of approximation, it is, nevertheless, clear that the relationship exhibited must be an exact one, since it is present in all orders of approximation and must, consequently, also be present in the limit of infinite approximation when the solutions will become exact ones of the problems. The question now arises as to the origin and meaning of this relationship. In this instance of isotropic scattering it is possible to go back to the original equations of transfer (Chap. III, eqs. [1] and [78], the latter with $\varpi_0 = 1$) and derive the relationship we have noticed as an integral of the equations. But when more general laws of scattering are considered, the relationships which emerge are so involved (cf. Chap. VI) that the attempt to establish them as integrals of the relevant equations, even if successful, would hardly disclose their physical meaning. It is therefore of interest to observe that the real origin of the relationship between the two problems has to be traced to the *invariance of the emergent radiation from a semi-infinite plane-parallel atmosphere to the addition* (or *subtraction*) *of layers of arbitrary optical thickness to* (or *from*) *the atmosphere.*

Principles of invariance similar to the one we have just stated can be formulated also in the context of other problems. Thus, in the problem of diffuse reflection, *the law of diffuse reflection by a semi-infinite plane-parallel atmosphere must be invariant to the addition* (or *subtraction*) *of layers of arbitrary optical thickness to* (or *from*) *the atmosphere.* This invariance governing the law of diffuse reflection by a semi-infinite atmosphere was first formulated by Ambarzumian.

The principles of invariance we have stated in the preceding paragraphs are only two of a large number that can be formulated. These other principles are more appropriately discussed in the context of transfer problems in atmospheres of finite optical thicknesses. Their consideration is accordingly postponed to Chapter VII. But it may be stated here that these principles of invariance and the method of solution of equations of transfer described in Chapter III together provide a powerful means for treating transfer problems in plane-parallel atmospheres.

29. The mathematical formulation of the principles of invariance

We shall now give mathematical expression to the principles of invariance formulated in § 28. For this purpose we shall consider an atmosphere scattering radiation according to a phase function $p(\cos \Theta)$. (The necessary generalizations to include the case of scattering according to a phase-matrix will be given in § 36.)

29.1. *The invariance of the law of diffuse reflection*

Let a parallel beam of radiation of net flux πF per unit area normal to itself be incident on a semi-infinite plane-parallel atmosphere in the direction $(-\mu_0, \varphi_0)$, and let the intensity $I(0, \mu, \varphi)$ diffusely reflected in the direction (μ, φ) be expressed in terms of a scattering function $S(\mu, \varphi; \mu_0, \varphi_0)$ in the form (cf. Chap. I, eq. [125])

$$I(0, \mu, \varphi) = \frac{F}{4\mu} S(\mu, \varphi; \mu_0, \varphi_0). \tag{5}$$

Now, considering the radiation field in such an atmosphere, we can distinguish, at any depth τ, between the reduced incident flux of amount

$$\pi F e^{-\tau/\mu_0} \tag{6}$$

in the direction $(-\mu_0, \varphi_0)$ and a diffuse radiation field characterized by the intensity $I(\tau, \mu, \varphi)$. To distinguish, further, between the outward $(0 \leqslant \mu \leqslant 1)$ and the inward $(0 > \mu \geqslant -1)$ directions, we shall write

$$I(\tau, +\mu, \varphi) \quad (0 \leqslant \mu \leqslant 1) \tag{7}$$

and $\qquad\qquad I(\tau, -\mu, \varphi) \quad (0 < \mu \leqslant 1). \tag{8}$

It is now apparent that the atmosphere below τ will reflect the radiation (6) and (8) according to the same law of diffuse reflection by a semi-infinite atmosphere and will contribute an outward intensity in the direction $(+\mu, \varphi)$ which must equal $I(\tau, +\mu, \varphi)$. In other words,

$$I(\tau, +\mu, \varphi) = \frac{F}{4\mu} e^{-\tau/\mu_0} S(\mu, \varphi; \mu_0, \varphi_0) +$$

$$+ \frac{1}{4\pi\mu} \int_0^1 \int_0^{2\pi} S(\mu, \varphi; \mu', \varphi') I(\tau, -\mu', \varphi') \, d\mu' d\varphi'. \tag{9}$$

This is the statement of the invariance of $S(\mu, \varphi; \mu_0, \varphi_0)$ to the addition or subtraction of layers.

29.2. *The invariance of the law of darkening*

Consider next the axially symmetric radiation field in a semi-infinite atmosphere with a constant net flux. In this case the invariance of the emergent radiation $I(0, \mu)$ to the addition (or subtraction) of layers of arbitrary optical thickness to (or from) the atmosphere is clearly equivalent to the statement that the outward radiation $I(\tau, +\mu)$ $(0 \leqslant \mu \leqslant 1)$, at any level τ, can differ from the emergent radiation, $I(0, \mu)$, only on account of the fact that at τ there is an inward directed radiation field which will be reflected by the atmosphere below τ by the law of diffuse reflection by a semi-infinite atmosphere. For, the removal of the layers above τ must restore $I(\tau, +\mu)$ to $I(0, \mu)$. We must therefore have

$$I(\tau, +\mu) = I(0, \mu) + \frac{1}{4\pi\mu} \int_0^1 \int_0^{2\pi} S(\mu, \varphi; \mu', \varphi') I(\tau, -\mu') \, d\mu' d\varphi', \tag{10}$$

or, in view of the axial symmetry of $I(\tau, -\mu)$,

$$I(\tau, +\mu) = I(0, \mu) + \frac{1}{2\mu} \int_0^1 S^{(0)}(\mu, \mu') I(\tau, -\mu') \, d\mu', \tag{11}$$

where $\qquad\qquad S^{(0)}(\mu, \mu') = \frac{1}{2\pi} \int_0^{2\pi} S(\mu, \varphi; \mu', \varphi') \, d\varphi', \tag{12}$

is the azimuth-independent term in the Fourier expansion of $S(\mu, \varphi; \mu', \varphi')$ in $(\varphi' - \varphi)$.

29.3. *An invariance arising from the asymptotic solution at infinity*

An invariance of a different sort arises from the following considerations:

The equation of transfer appropriate to the problem with a constant net flux admits a solution of the form (Chap. I eq. [81])

$$I(\tau,\mu) = \tfrac{3}{4}F[(1-\tfrac{1}{3}\varpi_1)\tau+\mu], \tag{13}$$

where ϖ_1 is the coefficient of the first Legendre polynomial in the expansion of the phase function in spherical harmonics (Chap. I, eqs. [33] and [74]). The solution (13) does not, of course, satisfy the boundary conditions at $\tau = 0$. We shall therefore let

$$I(\tau,\mu) = \tfrac{3}{4}F[(1-\tfrac{1}{3}\varpi_1)\tau+\mu]+I^*(\tau,\mu) \tag{14}$$

represent the solution for the problem with a constant net flux. With this expression of the solution for a semi-infinite atmosphere as the sum of two terms, a term representing the solution for an infinite unbounded atmosphere and a term allowing for the departure from the asymptotic solution (13) as we approach $\tau = 0$, it is evident that at any level τ the intensity, $I^*(\tau, +\mu)$, in the outward directions must result from the reflection of the inward directed radiation, $I^*(\tau, -\mu')$. Hence

$$I(\tau,+\mu) = \tfrac{3}{4}F[(1-\tfrac{1}{3}\varpi_1)\tau+\mu]+$$
$$+\frac{1}{4\pi\mu}\int_0^1\int_0^{2\pi} S(\mu,\varphi;\mu',\varphi')I^*(\tau,-\mu')\,d\mu'd\varphi'$$
$$= \tfrac{3}{4}F[(1-\tfrac{1}{3}\varpi_1)\tau+\mu]+\frac{1}{2\mu}\int_0^1 S^{(0)}(\mu,\mu')I^*(\tau,-\mu')\,d\mu'. \tag{15}$$

The application of this equation to the boundary at $\tau = 0$ leads to a specially noteworthy result:

Remembering that at $\tau = 0$, $I(\tau,-\mu) \equiv 0$ $(0 < \mu \leqslant 1)$, we have from equation (14)

$$I^*(0,-\mu') = \tfrac{3}{4}F\mu'. \tag{16}$$

With this value of $I^*(0,-\mu')$, equation (15) becomes

$$I(0,\mu) = \tfrac{3}{4}F\left[\mu+\frac{1}{2\mu}\int_0^1 S^{(0)}(\mu,\mu')\mu'\,d\mu'\right]. \tag{17}$$

Equation (17) is an integral equation relating the law of darkening for the problem with a constant net flux and the law of diffuse reflection.

30. The integral equation for the scattering function

The importance of the principles formulated in §§ 29.1 and 29.2 arises from the fact that they can be used to derive integral equations for the functions governing the angular distributions of the emergent radiations in the various problems. As we shall see, these integral equations are generally non-linear; nevertheless, they are easier to treat than the linear integral equations of the Schwarzschild–Milne type (Chap. I, § 11). This is not surprising when it is remembered that the non-linear integral equations are the mathematical expressions of the deeper invariances of the physical problem.

The general principle underlying the derivation of the non-linear integral equations we have referred to is that we differentiate the equations representing the invariances and then pass to the limit $\tau = 0$. Thus, in considering the problem of diffuse reflection we differentiate equation (9) with respect to τ and then set $\tau = 0$. In this manner we obtain

$$\left[\frac{dI(\tau, +\mu, \varphi)}{d\tau}\right]_{\tau=0} = -\frac{F}{4\mu\mu_0} S(\mu, \varphi; \mu_0, \varphi_0) +$$

$$+ \frac{1}{4\pi\mu} \int_0^1 \int_0^{2\pi} S(\mu, \varphi; \mu', \varphi') \left[\frac{dI(\tau, -\mu', \varphi')}{d\tau}\right]_{\tau=0} d\mu' d\varphi'. \quad (18)$$

The derivatives which occur in equation (18) can be found from the equation of transfer (Chap. I, eq. [126])

$$\mu \frac{dI(\tau, \mu, \varphi)}{d\tau} = I(\tau, \mu, \varphi) - \Im(\tau, \mu, \varphi), \quad (19)$$

where

$$\Im(\tau, \mu, \varphi) = \tfrac{1}{4} e^{-\tau/\mu_0} p(\mu, \varphi; -\mu_0, \varphi_0) F +$$

$$+ \frac{1}{4\pi} \int_{-1}^{+1} \int_0^{2\pi} p(\mu, \varphi; \mu'', \varphi'') I(\tau, \mu'', \varphi'') \, d\mu'' d\varphi''. \quad (20)$$

In equation (20)

$$p(\mu, \varphi; \mu', \varphi') = p(\cos\Theta)$$
$$= p[\mu\mu' + (1-\mu^2)^{\frac{1}{2}}(1-\mu'^2)^{\frac{1}{2}} \cos(\varphi'-\varphi)] \quad (21)$$

is the phase function. From equation (19) we obtain

$$\left[\frac{dI(\tau, +\mu, \varphi)}{d\tau}\right]_{\tau=0} = \frac{1}{\mu}[I(0, +\mu, \varphi) - \Im(0, +\mu, \varphi)] \quad (22)$$

and $$\left[\frac{dI(\tau, -\mu', \varphi')}{d\tau}\right]_{\tau=0} = \frac{1}{\mu'} \Im(0, -\mu', \varphi'). \quad (23)$$

In writing the second of these equations we have made use of the boundary condition

$$I(0, -\mu, \varphi) \equiv 0 \quad (0 < \mu \leqslant 1). \quad (24)$$

Inserting (22) and (23) in equation (18) we have

$$I(0, +\mu, \varphi) - \mathfrak{J}(0, +\mu, \varphi) = -\frac{F}{4\mu_0} S(\mu, \varphi; \mu_0, \varphi_0) +$$

$$+\frac{1}{4\pi} \int_0^1 \int_0^{2\pi} S(\mu, \varphi; \mu', \varphi') \mathfrak{J}(0, -\mu', \varphi') \frac{d\mu'}{\mu'} d\varphi', \quad (25)$$

or, using equation (5), we have

$$\tfrac{1}{4} F\left(\frac{1}{\mu} + \frac{1}{\mu_0}\right) S(\mu, \varphi; \mu_0, \varphi_0) = \mathfrak{J}(0, +\mu, \varphi) +$$

$$+\frac{1}{4\pi} \int_0^1 \int_0^{2\pi} S(\mu, \varphi; \mu', \varphi') \mathfrak{J}(0, -\mu', \varphi') \frac{d\mu'}{\mu'} d\varphi'. \quad (26)$$

On the other hand, according to equations (5), (20), and (24)

$$\mathfrak{J}(0, \mu, \varphi) = \tfrac{1}{4} F p(\mu, \varphi; -\mu_0, \varphi_0) +$$

$$+\frac{F}{16\pi} \int_0^1 \int_0^{2\pi} p(\mu, \varphi; \mu'', \varphi'') S(\mu'', \varphi''; \mu_0, \varphi_0) \frac{d\mu''}{\mu''} d\varphi'' \quad (-1 \leqslant \mu \leqslant 1). \quad (27)$$

With this expression for $\mathfrak{J}(0, \mu, \varphi)$ equation (26) becomes

$$\left(\frac{1}{\mu} + \frac{1}{\mu_0}\right) S(\mu, \varphi; \mu_0, \varphi_0) = p(\mu, \varphi; -\mu_0, \varphi_0) +$$

$$+\frac{1}{4\pi} \int_0^1 \int_0^{2\pi} p(\mu, \varphi; \mu'', \varphi'') S(\mu'', \varphi''; \mu_0, \varphi_0) \frac{d\mu''}{\mu''} d\varphi'' +$$

$$+\frac{1}{4\pi} \int_0^1 \int_0^{2\pi} S(\mu, \varphi; \mu', \varphi') p(-\mu', \varphi'; -\mu_0, \varphi_0) \frac{d\mu'}{\mu'} d\varphi' +$$

$$+\frac{1}{16\pi^2} \int_0^1 \int_0^{2\pi} \int_0^1 \int_0^{2\pi} S(\mu, \varphi; \mu', \varphi') p(-\mu', \varphi'; \mu'', \varphi'') \times$$

$$\times S(\mu'', \varphi''; \mu_0, \varphi_0) \frac{d\mu'}{\mu'} d\varphi' \frac{d\mu''}{\mu''} d\varphi''. \quad (28)$$

This is the required integral equation for S.

31. The principle of reciprocity

An important property of the scattering function $S(\mu, \varphi; \mu_0, \varphi_0)$ to which we have made reference before (Chap. I, § 13) is its symmetry in the pair of variables (μ, φ) and (μ_0, φ_0). This is required by the general principle of reciprocity considered in Chapter VII, § 52. In the meantime, it is of interest to see to what extent the principle can be inferred from the integral equation (28) for S.

First, we may observe that for single scattering the principle of reciprocity arises from the fact that the phase function depends only on the cosine of the angle between the directions (μ, φ) and (μ', φ'). Thus (cf. eq. [21])

$$p(\mu, \varphi; \mu', \varphi') = p(\mu', \varphi'; \mu, \varphi),$$

$$p(-\mu, \varphi; -\mu', \varphi') = p(\mu, \varphi; \mu', \varphi'),$$

$$p(\mu, \varphi; -\mu', \varphi') = p(-\mu, \varphi; \mu', \varphi') = p(\mu', \varphi'; -\mu, \varphi). \tag{29}$$

For expressing relations like the foregoing it is convenient to denote by $\tilde{f}(\mu, \varphi; \mu', \varphi')$ the function obtained from $f(\mu, \varphi; \mu', \varphi')$ by *transposing* the variables (μ, φ) and (μ', φ'):

$$\tilde{f}(\mu, \varphi; \mu', \varphi') = f(\mu', \varphi'; \mu, \varphi). \tag{30}$$

In this notation the symmetry properties of $p(\mu, \varphi; \mu', \varphi')$ can be expressed by

$$\tilde{p} = p \quad \text{and} \quad \tilde{p}(\mu, \varphi; -\mu', \varphi') = p(\mu', \varphi'; -\mu, \varphi). \tag{31}$$

Now, transposing the variables (μ, φ) and (μ_0, φ_0) in equation (28) and making use of the relations (29), we observe that $\tilde{S}$ satisfies the same equation as S. Hence, *if S is a solution of equation* (28), *then so is $\tilde{S}$*. But it does not follow from this that S is necessarily symmetrical in (μ, φ) and (μ_0, φ_0). The complete proof of the symmetry of S requires somewhat more elaborate considerations and is given in Chapter VII, § 52. In the meantime we shall assume that

$$S \equiv \tilde{S}, \tag{32}$$

which is certainly compatible with the integral equation for S.

32. An integral equation between $I(0, \mu)$ and $S^{(0)}(\mu, \mu')$

Differentiating equation (11) and letting $\tau = 0$, we obtain

$$\left[\frac{dI(\tau, +\mu)}{d\tau}\right]_{\tau=0} = \frac{1}{2\mu} \int_0^1 S^{(0)}(\mu, \mu') \left[\frac{dI(\tau, -\mu')}{d\tau}\right]_{\tau=0} d\mu'. \tag{33}$$

On the other hand, from the equation of transfer we conclude as in § 30 (cf. eqs. [22] and [23]) that

$$\left[\frac{dI(\tau, +\mu)}{d\tau}\right]_{\tau=0} = \frac{1}{\mu}[I(0, \mu) - \mathfrak{J}(0, +\mu)]$$

and

$$\left[\frac{dI(\tau, -\mu')}{d\tau}\right]_{\tau=0} = \frac{1}{\mu'} \mathfrak{J}(0, -\mu'). \tag{34}$$

where, now (Chap. I, eqs. [82] and [83]),

$$\mathfrak{J}(0,\mu) = \tfrac{1}{2}\int_0^1 p^{(0)}(\mu,\mu'')I(0,\mu'')\,d\mu'' \quad (-1 \leqslant \mu \leqslant +1). \tag{35}$$

Substituting from these equations in equation (33) we obtain

$$I(0,\mu) = \mathfrak{J}(0,+\mu) + \frac{1}{2}\int_0^1 S^{(0)}(\mu,\mu')\mathfrak{J}(0,-\mu')\frac{d\mu'}{\mu'},$$

or

$$I(0,\mu) = \frac{1}{2}\int_0^1 p^{(0)}(\mu,\mu'')I(0,\mu'')\,d\mu'' +$$

$$+ \frac{1}{4}\int_0^1\int_0^1 S^{(0)}(\mu,\mu')p^{(0)}(-\mu',\mu'')I(0,\mu'')\frac{d\mu'}{\mu'}d\mu'', \tag{36}$$

which is the required integral equation between $I(0,\mu)$ and $S^{(0)}(\mu,\mu')$.

33. The explicit forms of the integral equations in the case of isotropic scattering

We shall now illustrate the use of the integral equations derived in the preceding sections by considering the case of isotropic scattering when

$$p = \varpi_0 = \text{constant} \quad (\leqslant 1); \tag{37}$$

ϖ_0 is the albedo for single scattering.

33.1. *The integral equation for* $S(\mu,\mu_0)$

In view of the axial symmetry of the radiation field in this case, equation (28) becomes

$$\left(\frac{1}{\mu}+\frac{1}{\mu_0}\right)S(\mu,\mu_0) = \varpi_0\left\{1 + \frac{1}{2}\int_0^1 S(\mu,\mu')\frac{d\mu'}{\mu'} + \right.$$

$$\left. + \frac{1}{2}\int_0^1 S(\mu'',\mu_0)\frac{d\mu''}{\mu''} + \frac{1}{4}\int_0^1\int_0^1 S(\mu,\mu')S(\mu'',\mu_0)\frac{d\mu'}{\mu'}\frac{d\mu''}{\mu''}\right\}. \tag{38}$$

We now observe that the quantity on the right-hand side is separable in the variables μ and μ_0; thus

$$\left(\frac{1}{\mu}+\frac{1}{\mu_0}\right)S(\mu,\mu_0) = \varpi_0\left\{1 + \frac{1}{2}\int_0^1 S(\mu,\mu')\frac{d\mu'}{\mu'}\right\}\left\{1 + \frac{1}{2}\int_0^1 S(\mu'',\mu_0)\frac{d\mu''}{\mu''}\right\}. \tag{39}$$

Since $S(\mu, \mu')$ is symmetrical in μ and μ', the two factors on the right-hand side of equation (39) must be the values for μ and μ_0, of the *same* function. Therefore, letting

$$H(\mu) = 1 + \frac{1}{2} \int_0^1 S(\mu, \mu') \frac{d\mu'}{\mu'}$$

$$= 1 + \frac{1}{2} \int_0^1 S(\mu', \mu) \frac{d\mu'}{\mu'}, \tag{40}$$

we can express the scattering function in the form

$$\left(\frac{1}{\mu} + \frac{1}{\mu_0}\right) S(\mu, \mu_0) = \varpi_0 H(\mu) H(\mu_0). \tag{41}$$

Substituting for S according to this equation back into equation (40), we obtain the following non-linear integral equation for $H(\mu)$:

$$H(\mu) = 1 + \tfrac{1}{2} \varpi_0 \mu H(\mu) \int_0^1 \frac{H(\mu')}{\mu + \mu'} d\mu'. \tag{42}$$

Comparing equation (41) with the solution obtained in Chapter III (eqs. [119] and [120]), we conclude that the *form* of the solution obtained there is indeed the exact one. Moreover, it would now appear that the H-function as defined in Chapter III (eq. [106]) becomes in the limit of infinite approximation the solution of the integral equation (42). We shall see in Chapter V (§§ 39 and 40) that this expectation is justified.

The conservative case ($\varpi_0 = 1$) requires no special consideration: the relevant equations can be obtained by simply putting $\varpi_0 = 1$ in equations (41) and (42).

33.2. *The law of darkening in the problem with a constant net flux*

For conservative isotropic scattering, ($\varpi_0 = 1$), equation (36) becomes

$$I(0, \mu) = \frac{1}{2} \int_0^1 I(0, \mu'') \, d\mu'' + \frac{1}{4} \int_0^1 \int_0^1 S(\mu, \mu') I(0, \mu'') \frac{d\mu'}{\mu'} d\mu'',$$

or,

$$I(0, \mu) = J(0) \left[1 + \frac{1}{2} \int_0^1 S(\mu, \mu') \frac{d\mu'}{\mu'} \right], \tag{43}$$

where

$$J(0) = \tfrac{1}{2} \int_0^1 I(0, \mu) \, d\mu. \tag{44}$$

But (cf. eq. [41]) $S(\mu, \mu') = \dfrac{\mu\mu'}{\mu+\mu'} H(\mu)H(\mu').$ (45)

Equation (43) therefore becomes

$$I(0,\mu) = J(0)\left[1 + \tfrac{1}{2}\mu H(\mu) \int\limits_0^1 \frac{H(\mu')}{\mu+\mu'}\, d\mu'\right].$$ (46)

Now using the integral equation satisfied by $H(\mu)$ (i.e. eq. [42] with $\varpi_0 = 1$) we obtain

$$I(0,\mu) = J(0)H(\mu).$$ (47)

This agrees with the form for $I(0,\mu)$ found in Chapter III (eq. [61]); however, to get complete agreement we must establish the validity of the Hopf–Bronstein relation. It will be shown in Chapter V (§ 38, eqs. [25] and [26]) that the Hopf–Bronstein relation is indeed a direct consequence of certain elementary integral properties of the H-functions. But it is of interest to establish the relation from the principles of invariance alone. We do this in the following subsection.

33.3. *A derivation of the Hopf–Bronstein relation from the principles of invariance*

For conservative isotropic scattering, equation (17) takes the form

$$I(0,\mu) = \tfrac{3}{4}F\left[\mu + \frac{1}{2\mu} \int\limits_0^1 S(\mu, \mu')\mu'\, d\mu'\right].$$ (48)

Integrating this equation over the range of μ we obtain the relation

$$J(0) = \tfrac{3}{16}F\left[1 + \int\limits_0^1\int\limits_0^1 S(\mu, \mu')\frac{\mu'}{\mu}\, d\mu'd\mu\right].$$ (49)

Similarly, multiplying equation (43) by 2μ and integrating over the range of μ we obtain

$$F = J(0)\left[1 + \int\limits_0^1\int\limits_0^1 S(\mu, \mu')\frac{\mu}{\mu'}\, d\mu'd\mu\right].$$ (50)

The identity of the quantities in brackets on the right-hand sides of equations (49) and (50) follows from the symmetry of $S(\mu, \mu')$ in μ and μ'; hence

$$\frac{J(0)}{F} = \frac{3}{16}\frac{F}{J(0)}, \quad \text{or} \quad J(0) = \frac{\sqrt{3}}{4}F,$$ (51)

which is the Hopf–Bronstein relation. Using this relation, we can bring equation (47) into the standard form

$$I(0,\mu) = \frac{\sqrt{3}}{4} F H(\mu). \tag{52}$$

We thus see how, from the principles of invariance *alone*, we can derive the *complete* solutions for the emergent radiations for the two basic problems and express them in terms of a single H-function which satisfies a non-linear integral equation. We shall see in later chapters how similar reductions can be effected for all problems in plane-parallel atmospheres formulated in Chapter I.

34. The reduction of the integral equation for S for the case $p(\cos\Theta) = \varpi_0(1+x\cos\Theta)$

For a law of scattering according to the phase function $\varpi_0(1+x\cos\Theta)$

$$p(\mu,\varphi;\mu',\varphi') = \varpi_0[1+x\mu\mu'+x(1-\mu^2)^{\frac{1}{2}}(1-\mu'^2)^{\frac{1}{2}}\cos(\varphi'-\varphi)]. \tag{53}$$

This form for the phase function and the manner in which it enters the integral equation for S suggests that we express the scattering function in the form

$$S(\mu,\varphi;\mu_0,\varphi_0)$$
$$= \varpi_0[S^{(0)}(\mu,\mu_0)+x(1-\mu^2)^{\frac{1}{2}}(1-\mu_0^2)^{\frac{1}{2}}S^{(1)}(\mu,\mu_0)\cos(\varphi_0-\varphi)], \tag{54}$$

where, as the notation indicates, $S^{(0)}$ and $S^{(1)}$ are functions of μ and μ_0 only.

Substituting for p and S according to equations (53) and (54) in the integral equation for S and making use of the relations

$$\frac{1}{2\pi} \int_0^{2\pi} \cos m(\varphi'-\varphi_0)\cos n(\varphi-\varphi')\, d\varphi' = 0 \qquad\qquad \text{if } m \neq n,$$
$$= \tfrac{1}{2}\cos m(\varphi-\varphi_0) \quad \text{if } m = n \neq 0,$$
$$= 1 \qquad\qquad \text{if } m = n = 0, \tag{55}$$

we find that the equations for $S^{(0)}$ and $S^{(1)}$ separate and that

$$\left(\frac{1}{\mu}+\frac{1}{\mu_0}\right)S^{(0)}(\mu,\mu_0) = 1-x\mu\mu_0+\tfrac{1}{2}\varpi_0 \int_0^1 (1+x\mu\mu'')S^{(0)}(\mu'',\mu_0)\frac{d\mu''}{\mu''}+$$
$$+\tfrac{1}{2}\varpi_0 \int_0^1 S^{(0)}(\mu,\mu')(1+x\mu'\mu_0)\frac{d\mu'}{\mu'}+$$
$$+\tfrac{1}{4}\varpi_0^2 \int_0^1\int_0^1 S^{(0)}(\mu,\mu')(1-x\mu'\mu'')S^{(0)}(\mu'',\mu_0)\frac{d\mu'}{\mu'}\frac{d\mu''}{\mu''}, \tag{56}$$

and

$$\left(\frac{1}{\mu}+\frac{1}{\mu_0}\right)S^{(1)}(\mu,\mu_0) = \left\{1+\tfrac{1}{4}x\varpi_0\int_0^1\frac{d\mu''}{\mu''}(1-\mu''^2)S^{(1)}(\mu'',\mu_0)\right\}\times$$

$$\times\left\{1+\tfrac{1}{4}x\varpi_0\int_0^1\frac{d\mu'}{\mu'}(1-\mu'^2)S^{(1)}(\mu,\mu')\right\}. \quad (57)$$

34.1. *The reduction of the equation for $S^{(0)}$*

On examination, it is seen that equation (56) can be written in the form

$$\left(\frac{1}{\mu}+\frac{1}{\mu_0}\right)S^{(0)}(\mu,\mu_0)$$

$$= \left\{1+\tfrac{1}{2}\varpi_0\int_0^1 S^{(0)}(\mu'',\mu_0)\frac{d\mu''}{\mu''}\right\}\left\{1+\tfrac{1}{2}\varpi_0\int_0^1 S^{(0)}(\mu,\mu')\frac{d\mu'}{\mu'}\right\} -$$

$$-x\left\{\mu_0-\tfrac{1}{2}\varpi_0\int_0^1 S^{(0)}(\mu'',\mu_0)\,d\mu''\right\}\left\{\mu-\tfrac{1}{2}\varpi_0\int_0^1 S^{(0)}(\mu,\mu')\,d\mu'\right\}. \quad (58)$$

From this equation it follows that $S^{(0)}(\mu,\mu_0)$ can be expressed in the form

$$\left(\frac{1}{\mu}+\frac{1}{\mu_0}\right)S^{(0)}(\mu,\mu_0) = \psi(\mu)\psi(\mu_0)-x\,\phi(\mu)\phi(\mu_0), \quad (59)$$

where

$$\psi(\mu) = 1+\tfrac{1}{2}\varpi_0\int_0^1 S^{(0)}(\mu,\mu')\frac{d\mu'}{\mu'}$$

and

$$\phi(\mu) = \mu-\tfrac{1}{2}\varpi_0\int_0^1 S^{(0)}(\mu,\mu')\,d\mu'. \quad (60)$$

Now, substituting for $S^{(0)}$ from equation (59) back into equations (60), we obtain the pair of equations

$$\psi(\mu) = 1+\tfrac{1}{2}\varpi_0\mu\psi(\mu)\int_0^1\frac{\psi(\mu')}{\mu+\mu'}d\mu'-\tfrac{1}{2}x\varpi_0\mu\phi(\mu)\int_0^1\frac{\phi(\mu')}{\mu+\mu'}d\mu' \quad (61)$$

and

$$\phi(\mu) = \mu-\tfrac{1}{2}\varpi_0\mu\psi(\mu)\int_0^1\frac{\psi(\mu')}{\mu+\mu'}\mu'\,d\mu'+\tfrac{1}{2}x\varpi_0\mu\phi(\mu)\int_0^1\frac{\phi(\mu')}{\mu+\mu'}\mu'\,d\mu'. \quad (62)$$

The solutions of these equations will be given in Chapter VI, § 46.

34.2. *The expression of $S^{(1)}$ in terms of an H-function*

Returning to equation (57), we observe that $S^{(1)}(\mu, \mu_0)$ is expressible in the form

$$\left(\frac{1}{\mu}+\frac{1}{\mu_0}\right)S^{(1)}(\mu, \mu_0) = H^{(1)}(\mu)H^{(1)}(\mu_0), \tag{63}$$

where

$$H^{(1)}(\mu) = 1+\tfrac{1}{4}x\varpi_0 \int_0^1 \frac{d\mu'}{\mu'}(1-\mu'^2)S^{(1)}(\mu, \mu'). \tag{64}$$

According to equations (63) and (64), $H(\mu)$ satisfies the non-linear integral equation

$$H^{(1)}(\mu) = 1+\tfrac{1}{4}x\varpi_0 \mu H^{(1)}(\mu) \int_0^1 \frac{(1-\mu'^2)}{\mu+\mu'} H^{(1)}(\mu')\, d\mu'. \tag{65}$$

35. The reduction of the integral equation for S for the case $p(\cos\Theta) = \tfrac{3}{4}(1+\cos^2\Theta)$

For scattering according to Rayleigh's phase function (Chap. I, eqs. [99] and [100])

$$p(\mu, \varphi; \mu', \varphi') = \tfrac{3}{8}[3-\mu^2-\mu'^2+3\mu^2\mu'^2+$$
$$+4\mu\mu'(1-\mu^2)^{\frac{1}{2}}(1-\mu'^2)^{\frac{1}{2}}\cos(\varphi'-\varphi)+(1-\mu^2)(1-\mu'^2)\cos 2(\varphi'-\varphi)]. \tag{66}$$

We accordingly express the scattering function in the form

$$S(\mu, \varphi; \mu_0, \varphi_0) = \tfrac{3}{8}[S^{(0)}(\mu, \mu_0)-4\mu\mu_0(1-\mu^2)^{\frac{1}{2}}(1-\mu_0^2)^{\frac{1}{2}}S^{(1)}(\mu, \mu_0)\cos(\varphi_0-\varphi)+$$
$$+(1-\mu^2)(1-\mu_0^2)S^{(2)}(\mu, \mu_0)\cos 2(\varphi_0-\varphi)], \tag{67}$$

and find that $S^{(0)}$, $S^{(1)}$, and $S^{(2)}$ satisfy the independent integral equations

$$\left(\frac{1}{\mu}+\frac{1}{\mu_0}\right)S^{(0)}(\mu, \mu_0) = p^{(0)}(\mu, \mu_0)+\frac{3}{16} \int_0^1 p^{(0)}(\mu, \mu'')S^{(0)}(\mu'', \mu_0)\frac{d\mu''}{\mu''}+$$

$$+\frac{3}{16} \int_0^1 S^{(0)}(\mu, \mu')p^{(0)}(\mu', \mu_0)\frac{d\mu'}{\mu'}+$$

$$+\frac{9}{256} \int_0^1\int_0^1 S^{(0)}(\mu, \mu')p^{(0)}(\mu', \mu'')S^{(0)}(\mu'', \mu_0)\frac{d\mu'}{\mu'}\frac{d\mu''}{\mu''}, \tag{68}$$

$$\left(\frac{1}{\mu}+\frac{1}{\mu_0}\right)S^{(1)}(\mu, \mu_0) = \left\{1+\frac{3}{8} \int_0^1 \mu''^2(1-\mu''^2)S^{(1)}(\mu'', \mu_0)\frac{d\mu''}{\mu''}\right\} \times$$

$$\times \left\{1+\frac{3}{8} \int_0^1 \mu'^2(1-\mu'^2)S^{(1)}(\mu, \mu')\frac{d\mu'}{\mu'}\right\} \tag{69}$$

and

$$\left(\frac{1}{\mu}+\frac{1}{\mu_0}\right)S^{(2)}(\mu,\mu_0) = \left\{1+\frac{3}{32}\int_0^1 (1-\mu''^2)^2 S^{(2)}(\mu'',\mu_0)\frac{d\mu''}{\mu''}\right\} \times$$

$$\times\left\{1+\frac{3}{32}\int_0^1 (1-\mu'^2)^2 S^{(2)}(\mu,\mu')\frac{d\mu'}{\mu'}\right\}, \quad (70)$$

where in equation (68) we have written

$$p^{(0)}(\mu,\mu') = 3-\mu^2-\mu'^2+3\mu^2\mu'^2 = \tfrac{1}{3}(3-\mu^2)(3-\mu'^2)+\tfrac{8}{3}\mu^2\mu'^2. \quad (71)$$

Of the three integral equations, (68)–(70), only the first need be considered in any detail, since it is at once apparent that the solutions for the other two must be of the forms

$$\left(\frac{1}{\mu}+\frac{1}{\mu_0}\right)S^{(1)}(\mu,\mu_0) = H^{(1)}(\mu)H^{(1)}(\mu_0) \quad (72)$$

and

$$\left(\frac{1}{\mu}+\frac{1}{\mu_0}\right)S^{(2)}(\mu,\mu_0) = H^{(2)}(\mu)H^{(2)}(\mu_0), \quad (73)$$

where $H^{(1)}(\mu)$ and $H^{(2)}(\mu)$ satisfy the integral equations

$$H^{(1)}(\mu) = 1+\tfrac{3}{8}\mu H^{(1)}(\mu)\int_0^1 \frac{\mu'^2(1-\mu'^2)}{\mu+\mu'}H^{(1)}(\mu')\,d\mu' \quad (74)$$

and

$$H^{(2)}(\mu) = 1+\tfrac{3}{32}\mu H^{(2)}(\mu)\int_0^1 \frac{(1-\mu'^2)^2}{\mu+\mu'}H^{(2)}(\mu')\,d\mu'. \quad (75)$$

Returning to equation (68) we find, on examination, that this equation can be written in the form

$$\left(\frac{1}{\mu}+\frac{1}{\mu_0}\right)S^{(0)}(\mu,\mu_0) = \tfrac{1}{3}\left\{3-\mu^2+\frac{3}{16}\int_0^1 (3-\mu'^2)S^{(0)}(\mu,\mu')\frac{d\mu'}{\mu'}\right\} \times$$

$$\times\left\{3-\mu_0^2+\frac{3}{16}\int_0^1 (3-\mu''^2)S^{(0)}(\mu'',\mu_0)\frac{d\mu''}{\mu''}\right\} +$$

$$+\tfrac{8}{3}\left\{\mu^2+\frac{3}{16}\int_0^1 \mu'^2 S^{(0)}(\mu,\mu')\frac{d\mu'}{\mu'}\right\}\left\{\mu_0^2+\frac{3}{16}\int_0^1 \mu''^2 S^{(0)}(\mu'',\mu_0)\frac{d\mu''}{\mu''}\right\}. \quad (76)$$

From equation (76) it follows that $S^{(0)}(\mu,\mu_0)$ can be expressed in the form

$$\left(\frac{1}{\mu}+\frac{1}{\mu_0}\right)S^{(0)}(\mu,\mu_0) = \tfrac{1}{3}\psi(\mu)\psi(\mu_0)+\tfrac{8}{3}\phi(\mu)\phi(\mu_0), \quad (77)$$

where

$$\psi(\mu) = 3-\mu^2+\frac{3}{16}\int_0^1 (3-\mu'^2)S^{(0)}(\mu,\mu')\frac{d\mu'}{\mu'}$$

and
$$\phi(\mu) = \mu^2 + \frac{3}{16} \int_0^1 \mu'^2 S^{(0)}(\mu,\mu') \frac{d\mu'}{\mu'}. \tag{78}$$

Now, substituting for $S^{(0)}$ according to equation (77) in equations (78), we obtain the integral equations

$$\psi(\mu) = 3 - \mu^2 + \tfrac{1}{16}\mu\psi(\mu) \int_0^1 \frac{\psi(\mu')}{\mu+\mu'} (3-\mu'^2) \, d\mu' +$$

$$+ \tfrac{1}{2}\mu\phi(\mu) \int_0^1 \frac{\phi(\mu')}{\mu+\mu'} (3-\mu'^2) \, d\mu' \tag{79}$$

and

$$\phi(\mu) = \mu^2 + \tfrac{1}{16}\mu\psi(\mu) \int_0^1 \frac{\psi(\mu')}{\mu+\mu'} \mu'^2 \, d\mu' + \tfrac{1}{2}\mu\phi(\mu) \int_0^1 \frac{\phi(\mu')}{\mu+\mu'} \mu'^2 \, d\mu'. \tag{80}$$

The solutions of these equations will be given in Chapter VI, § 44.

The reduction of the integral equation for S into independent systems of equations for functions of one variable only, which we have effected for two special cases (other than the case of isotropic scattering), is possible quite generally by expanding the phase functions, $p(\mu,\varphi;\mu',\varphi')$, in spherical harmonics. But we shall not carry out this reduction for the general case here, as an analogous reduction for the more difficult problem of diffuse reflection and transmission by atmospheres of finite optical thicknesses will be considered in Chapter VII (§ 53).

36. The principles of invariance when the polarization of the radiation field is taken into account

We have already seen in Chapter I (§ 17, see particularly eqs. [212] and [231]) that when the polarization of the radiation field is taken into account, the equation of transfer becomes a vector equation for I (whose components are the Stokes parameters) in which a phase-matrix, $P(\mu,\varphi;\mu',\varphi')$ plays, *identically*, the same role as the phase-function, $p(\mu,\varphi;\mu',\varphi')$, in the more conventional problems. Consequently, all the equations of §§ 29, 30, and 32 will continue to be valid also when polarization is taken into account, if I is regarded as a vector and p and S as matrices. Thus, the integral equation for the scattering matrix, $S(\mu,\varphi;\mu_0,\varphi_0)$ (cf. Chap. I, eq. [230]) will be identical in form with equation (28): only S and p must be replaced by matrices S and P. Similarly, the integral equation relating the law of darkening in the problem with a constant net flux and the scattering matrix governing the law of diffuse reflection can be written down in exact analogy with equation (36). In the case of Rayleigh scattering, the equation in question is, for example (cf. Chap. I, eqs. [221] and [227]),

$$I(0,\mu) = \tfrac{1}{2} \int_0^1 P^{(0)}(\mu,\mu'')I(0,\mu'')\,d\mu'' +$$

$$+ \frac{1}{4} \int_0^1 \int_0^1 S^{(0)}(\mu,\mu')P^{(0)}(\mu',\mu'')I(0,\mu'')\frac{d\mu'}{\mu'}\,d\mu'', \quad (81)$$

where
$$I = (I_l, I_r), \quad (82)$$

$$P^{(0)}(\mu,\mu') = \frac{3}{4}\binom{2(1-\mu^2)(1-\mu'^2)+\mu^2\mu'^2 \quad \mu^2}{\mu'^2 \qquad\qquad\qquad 1}, \quad (83)$$

and $S^{(0)}(\mu,\mu')$ is a matrix of two rows and columns which is the azimuth independent part of the scattering matrix.

BIBLIOGRAPHICAL NOTES

We owe to Ambarzumian the first introduction of a principle of invariance in the treatment of transfer problems—

1. V. A. AMBARZUMIAN, C.R. (Doklady), Sci. U.R.S.S. **38**, 257 (1943).
2. V. A. AMBARZUMIAN, Russian Astronomical Journal, **19**, 1 (1942).
3. V. A. AMBARZUMIAN, Journal of Physics of the Academy of Sciences of U.S.S.R. **8**, 65 (1944).

The integral equation for the scattering function (§ 30) is derived by Ambarzumian in refs. 2 and 3 (though by a method somewhat different from the one we have adopted in the text).

The formulation of a complete set of principles of invariance and their systematic integration into a general theory are subsequent developments. They are contained in—

4. S. CHANDRASEKHAR, Astrophys. J. **105**, 164, 441 (1947).

The present chapter is largely based on ref. 4.

The principle of invariance arising from the asymptotic solution at infinity, in conservative cases (§§ 29.3 and 33.3), is considered here for the first time: it arose in a discussion between the writer and Dr. Louis G. Henyey.

In view of the very important role which principles of invariance play in the modern theory of radiative transfer, it is of interest to recall that the basic ideas underlying the formulation of the various integral equations in this chapter (and in Chap. VII) resemble those which were introduced by Sir George Stokes and Lord Rayleigh in their treatment of the reflection and transmission of light by piles of plates.

5. SIR GEORGE STOKES, Mathematical and Physical Papers of Sir George Stokes, iv (Cambridge, 1904), p. 145.
6. LORD RAYLEIGH, Scientific Papers of Lord Rayleigh, vi (Cambridge, 1920), p. 492.

V

THE *H*-FUNCTIONS

37. Introduction

THE discussion of the principles of invariance in Chapter IV has disclosed the importance for the theory of radiative transfer, of non-linear integral equations of the form (Chap. IV, eqs. [42], [65], [74], and [75])

$$H(\mu) = 1 + \mu H(\mu) \int_0^1 \frac{\Psi(\mu')}{\mu + \mu'} H(\mu')\, d\mu', \tag{1}$$

where the *characteristic function*, $\Psi(\mu)$, is an even polynomial in μ satisfying the condition

$$\int_0^1 \Psi(\mu)\, d\mu \leqslant \tfrac{1}{2}. \tag{2}$$

For the standard problems in isotropic scattering the angular distributions of the emergent radiations are expressed directly in terms of H-functions (Chap. IV, eqs. [45] and [52]). But for more general laws of scattering, the principles of invariance lead to simultaneous non-linear integral equations (cf. Chap. IV, eqs. [61], [62] and [79], [80]) which, in appearance, are far more complex than equation (1); nevertheless we shall show that the solutions of these equations can also be expressed in terms of H-functions. Thus the H-functions play an important part in the theory of radiative transfer in semi-infinite atmospheres. In view of this, we shall devote this chapter to the study of these functions.

The plan of this chapter is as follows:

In § 38 we establish certain elementary integral properties of the H-functions. In § 39 we consider the relation of the function

$$H(\mu) = \frac{1}{\mu_1 \cdots \mu_n} \frac{\prod_{i=1}^{n} (\mu + \mu_i)}{\prod_{\alpha}(1 + k_\alpha \mu)}, \tag{3}$$

defined in terms of the zeros, μ_i, of $P_{2n}(\mu)$ and the positive, non-vanishing roots, k_α, of the associated *characteristic equation*

$$1 = 2 \sum_{j=1}^{n} \frac{a_j \Psi(\mu_j)}{1 - k^2 \mu_j^2}, \tag{4}$$

to the solution of equation (1). In § 40 we obtain, explicitly, the solution of equation (1), and in § 41 we show how we can solve, numerically, for the *H*-functions, in practice. And, finally, in § 42, we tabulate the solutions for the case

$$\Psi(\mu) = \text{constant} = \tfrac{1}{2}\varpi_0 \tag{5}$$

for various values of the albedo ϖ_0.

38. Integral properties of the *H*-functions

In this section we shall suppose that a solution of equation (1) exists which is bounded in the interval $0 \leqslant \mu \leqslant 1$. The various moments

$$\alpha_n = \int_0^1 H(\mu)\mu^n \, d\mu \quad (n \geqslant 0) \tag{6}$$

then exist and are finite.

THEOREM 1. $\displaystyle\int_0^1 H(\mu)\Psi(\mu) \, d\mu = 1 - \left[1 - 2\int_0^1 \Psi(\mu) \, d\mu\right]^{\frac{1}{2}}. \tag{7}$

PROOF. Multiplying the equation satisfied by $H(\mu)$ by $\Psi(\mu)$ and integrating over the range of μ, we have

$$\int_0^1 H(\mu)\Psi(\mu) \, d\mu$$

$$= \int_0^1 \Psi(\mu) \, d\mu + \int_0^1 \int_0^1 \frac{\mu}{\mu + \mu'} H(\mu)\Psi(\mu)H(\mu')\Psi(\mu') \, d\mu \, d\mu'. \tag{8}$$

Interchanging μ and μ' in the double integral in equation (8) and taking the average of the two equations, we obtain

$$\int_0^1 H(\mu)\Psi(\mu) \, d\mu$$

$$= \int_0^1 \Psi(\mu) \, d\mu + \tfrac{1}{2}\int_0^1 \int_0^1 H(\mu)\Psi(\mu)H(\mu')\Psi(\mu') \, d\mu \, d\mu', \tag{9}$$

or, alternatively,

$$\frac{1}{2}\left[\int_0^1 H(\mu)\Psi(\mu) \, d\mu\right]^2 - \int_0^1 H(\mu)\Psi(\mu) \, d\mu + \int_0^1 \Psi(\mu) \, d\mu = 0. \tag{10}$$

Solving this equation for the integral in question, we have

$$\int_0^1 H(\mu)\Psi(\mu) \, d\mu = 1 \pm \left[1 - 2\int_0^1 \Psi(\mu) \, d\mu\right]^{\frac{1}{2}}. \tag{11}$$

The ambiguity in the sign in equation (11) can be removed by the consideration that the integral on the left-hand side must uniformly converge to zero when $\Psi(\mu)$ tends to zero uniformly in the interval $(0, 1)$.

This requires us to choose the negative sign in equation (11) and the result stated follows.

COROLLARY 1. A necessary condition that $H(\mu)$ be real is

$$\int_0^1 \Psi(\mu)\, d\mu \leqslant \tfrac{1}{2}. \tag{12}$$

This is, of course, an immediate consequence of the theorem.

The physical meaning of the limitation (12) is interesting: it is equivalent to the condition that, on each scattering, more radiation be not emitted than was incident. Further, it may be noted that we have the equality sign in (12) only in the *conservative case*, i.e. only when there is no true absorption and the efficiency of scattering is unity.

COROLLARY 2. An alternative form of the integral equation satisfied by $H(\mu)$ is

$$\frac{1}{H(\mu)} = \left[1 - 2 \int_0^1 \Psi(\mu)\, d\mu \right]^{\frac{1}{2}} + \int_0^1 \frac{\mu'\,\Psi(\mu')}{\mu+\mu'} H(\mu')\, d\mu'. \tag{13}$$

PROOF. This can be proved as follows: Using the result of Theorem 1 and recalling the integral equation satisfied by $H(\mu)$, we have

$$H(\mu) \int_0^1 \frac{\mu'\,\Psi(\mu')}{\mu+\mu'} H(\mu')\, d\mu' = H(\mu) \int_0^1 H(\mu')\,\Psi(\mu') \left[1 - \frac{\mu}{\mu+\mu'} \right] d\mu'$$

$$= H(\mu) \left\{ 1 - \left[1 - 2 \int_0^1 \Psi(\mu)\, d\mu \right]^{\frac{1}{2}} \right\} - H(\mu) + 1$$

$$= 1 - H(\mu) \left[1 - 2 \int_0^1 \Psi(\mu)\, d\mu \right]^{\frac{1}{2}}, \tag{14}$$

which is equivalent to equation (13).

COROLLARY 3. In the conservative case, when

$$\int_0^1 \Psi(\mu)\, d\mu = \tfrac{1}{2}, \tag{15}$$

the results of Theorem 1 and Corollary 2 take the simple forms

$$\int_0^1 H(\mu)\Psi(\mu)\, d\mu = 1, \tag{16}$$

and

$$\frac{1}{H(\mu)} = \int_0^1 \frac{\mu'\,\Psi(\mu')}{\mu+\mu'} H(\mu')\, d\mu'. \tag{17}$$

THEOREM 2.

$$\left[1-2\int_0^1 \Psi(\mu)\,d\mu\right]^{\frac{1}{2}}\int_0^1 H(\mu)\Psi(\mu)\mu^2\,d\mu+$$

$$+\frac{1}{2}\left[\int_0^1 H(\mu)\Psi(\mu)\mu\,d\mu\right]^2=\int_0^1 \Psi(\mu)\mu^2\,d\mu.\quad(18)$$

PROOF. To prove equation (18) we multiply the equation defining $H(\mu)$ by $\Psi(\mu)\mu^2$ and integrate over the range of μ. We obtain

$$\int_0^1 H(\mu)\Psi(\mu)\mu^2\,d\mu$$

$$=\int_0^1 \Psi(\mu)\mu^2\,d\mu+\int_0^1\int_0^1 \frac{\mu^3}{\mu+\mu'}H(\mu)\Psi(\mu)H(\mu')\Psi(\mu')\,d\mu d\mu'$$

$$=\int_0^1 \Psi(\mu)\mu^2\,d\mu+\frac{1}{2}\int_0^1\int_0^1 \frac{\mu^3+\mu'^3}{\mu+\mu'}H(\mu)\Psi(\mu)H(\mu')\Psi(\mu')\,d\mu d\mu'$$

$$=\int_0^1 \Psi(\mu)\mu^2\,d\mu+\frac{1}{2}\int_0^1\int_0^1 (\mu^2-\mu\mu'+\mu'^2)H(\mu)\Psi(\mu)H(\mu')\Psi(\mu')\,d\mu d\mu'$$

$$=\int_0^1 \Psi(\mu)\mu^2\,d\mu+\left[\int_0^1 H(\mu)\Psi(\mu)\mu^2\,d\mu\right]\left[\int_0^1 H(\mu)\Psi(\mu)\,d\mu\right]-$$

$$-\frac{1}{2}\left[\int_0^1 H(\mu)\Psi(\mu)\mu\,d\mu\right]^2.\quad(19)$$

Using Theorem 1, and after some rearranging of the terms, we obtain

$$\left[1-2\int_0^1 \Psi(\mu)\,d\mu\right]^{\frac{1}{2}}\int_0^1 H(\mu)\Psi(\mu)\mu^2\,d\mu+\frac{1}{2}\left[\int_0^1 H(\mu)\Psi(\mu)\mu\,d\mu\right]^2$$

$$=\int_0^1 \Psi(\mu)\mu^2\,d\mu.\quad(20)$$

COROLLARY 1. For the conservative case, (15), we have the further integral

$$\int_0^1 H(\mu)\Psi(\mu)\mu\,d\mu=\left[2\int_0^1 \Psi(\mu)\mu^2\,d\mu\right]^{\frac{1}{2}}.\quad(21)$$

COROLLARY 2. For the conservative case, still another form of the integral equation for $H(\mu)$ is

$$\frac{\mu}{H(\mu)} = \left[2\int_0^1 \Psi(\mu)\mu^2\,d\mu\right]^{\frac{1}{2}} - \int_0^1 \frac{\mu'^2\Psi(\mu')}{\mu+\mu'}H(\mu')\,d\mu'. \qquad (22)$$

This follows from equations (17) and (21):

$$\frac{\mu}{H(\mu)} = \int_0^1 \frac{\mu}{\mu+\mu'}\mu'\,\Psi(\mu')H(\mu')\,d\mu'$$

$$= \int_0^1 \left(1 - \frac{\mu'}{\mu+\mu'}\right)\mu'\,\Psi(\mu')H(\mu')\,d\mu'$$

$$= \left[2\int_0^1 \Psi(\mu)\mu^2\,d\mu\right]^{\frac{1}{2}} - \int_0^1 \frac{\mu'^2\Psi(\mu')}{\mu+\mu'}H(\mu')\,d\mu'. \qquad (23)$$

It is now seen that equations (16) and (21) represent a generalization of the Hopf–Bronstein relation for the conservative isotropic case; for, in this latter case,

$$\Psi(\mu) = \text{constant} = \tfrac{1}{2}, \qquad (24)$$

and the integrals (16) and (21) become

$$\int_0^1 H(\mu)\,d\mu = 2 \qquad (25)$$

and

$$\int_0^1 H(\mu)\mu\,d\mu = \frac{2}{\sqrt{3}}. \qquad (26)$$

Equation (25) is clearly necessary for Chapter IV, equation (47), to be consistent with itself; and the ratio of equations (25) and (26) expresses the Hopf–Bronstein relation.

THEOREM 3. When the characteristic function $\Psi(\mu)$ has the form

$$\Psi(\mu) = a + b\mu^2, \qquad (27)$$

where a and b are two constants,† we have the relations

$$\alpha_0 = 1 + \tfrac{1}{2}(a\alpha_0^2 + b\alpha_1^2) \qquad (28)$$

and

$$(a+b\mu^2)\int_0^1 \frac{H(\mu')}{\mu+\mu'}\,d\mu' = \frac{H(\mu)-1}{\mu H(\mu)} - b(\alpha_1 - \mu\alpha_0), \qquad (29)$$

where α_0 and α_1 are the moments of order zero and one of $H(\mu)$.

---

† The condition $\int_0^1 \Psi(\mu)\,d\mu \leqslant \tfrac{1}{2}$ requires that $a + \tfrac{1}{3}b \leqslant \tfrac{1}{2}$.

PROOF. To prove the relation (28) we simply integrate the equation satisfied by $H(\mu)$. We find

$$\alpha_0 = 1 + \int\limits_0^1 \int\limits_0^1 \frac{a+b\mu'^2}{\mu+\mu'} \mu H(\mu)H(\mu')\,d\mu d\mu'$$

$$= 1 + \frac{1}{2} \int\limits_0^1 \int\limits_0^1 (a+b\mu\mu')H(\mu)H(\mu')\,d\mu d\mu'$$

$$= 1 + \tfrac{1}{2}(a\alpha_0^2 + b\alpha_1^2). \tag{30}$$

The relation (29) can be established in the following manner:

$$a \int\limits_0^1 \frac{H(\mu')}{\mu+\mu'}\,d\mu' = \int\limits_0^1 \frac{a+b\mu'^2}{\mu+\mu'} H(\mu')\,d\mu' - b \int\limits_0^1 \frac{\mu'^2}{\mu+\mu'} H(\mu')\,d\mu'$$

$$= \frac{H(\mu)-1}{\mu H(\mu)} - b \int\limits_0^1 \left(\mu'-\mu+\frac{\mu^2}{\mu+\mu'}\right)H(\mu')\,d\mu'$$

$$= \frac{H(\mu)-1}{\mu H(\mu)} - b(\alpha_1 - \mu\alpha_0) - b\mu^2 \int\limits_0^1 \frac{H(\mu')}{\mu+\mu'}\,d\mu', \tag{31}$$

from which the relation stated follows.

39. The relation of the *H*-function defined in terms of the Gaussian division and characteristic roots to the solution of the integral equation (1)

The principal result to be established in this section is that the *H*-function defined as in equation (3) is the unique solution of the equation

$$H(\mu) = 1 + \mu H(\mu) \sum_{j=1}^n \frac{a_j \Psi(\mu_j)}{\mu+\mu_j} H(\mu_j), \tag{32}$$

which is regular and non-zero for $\mu \geqslant 0$.

Rewriting equation (32) in the form

$$\frac{1}{H(\mu)} = 1 - \mu \sum_{j=1}^n \frac{q_j H(\mu_j)}{\mu+\mu_j}, \tag{33}$$

where

$$q_j = a_j \Psi(\mu_j) \quad (j = 1,...,n), \tag{34}$$

we conclude that $H(\mu)$ is a rational function whose zeros are at $\mu = -\mu_j$ $(j = 1,...,n)$. We may therefore write

$$H(\mu) = \frac{(-1)^n}{\mu_1 \cdots \mu_n} \frac{P(-\mu)}{R(-\mu)}, \tag{35}$$

where

$$P(\mu) = \prod_{i=1}^{n} (\mu - \mu_i), \tag{36}$$

and $R(\mu)$ is a polynomial of degree at most n. Actually, $R(\mu)$ will be of degree n unless

$$\frac{1}{H(\mu)} \to 0 \quad \text{as} \quad \mu \to \infty, \quad \text{i.e.} \quad \sum_{j=1}^{n} q_j H(\mu_j) = 1. \tag{37}$$

And this last condition can be satisfied only in conservative cases. For, multiplying equation (32) by $a_j \Psi(\mu_j)$ and summing over j, we find, as in the proof of Theorem 1 in § 38, that (cf. eq. [10])

$$\tfrac{1}{2}\Big[\sum_{j=1}^{n} q_j H(\mu_j) \Big]^2 - \Big[\sum_{j=1}^{n} q_j H(\mu_j) \Big] + \sum_{j=1}^{n} q_j = 0, \tag{38}$$

or

$$\sum_{j=1}^{n} q_j H(\mu_j) = 1 - \Big[1 - 2\sum_{j=1}^{n} a_j \Psi(\mu_j) \Big]^{\tfrac{1}{2}}; \tag{39}$$

so that $\sum q_j H(\mu_j) = 1$, only when $\sum q_j \Psi(\mu_j) = \tfrac{1}{2}$.

Since $R(\mu)$ is a polynomial of degree at most n, we can write, using Lagrange's interpolation formula:

$$R(\mu) = \frac{(-1)^n}{\mu_1 \cdots \mu_n} P(\mu) + \mu P(\mu) \sum_{j=1}^{n} \frac{R(\mu_j)}{\mu_j(\mu - \mu_j)P'(\mu_j)}, \tag{40}$$

where, it will be noticed, that we have arranged that $R(0) = 1$ as required by the condition $H(0) = 1$ (cf. eqs. [32] and [35]).

From equations (35) and (40) we have

$$\frac{1}{H(\mu)} = 1 + (-1)^n \mu \mu_1 \cdots \mu_n \sum_{j=1}^{n} \frac{R(\mu_j)}{\mu_j(\mu + \mu_j)P'(\mu_j)}. \tag{41}$$

On the other hand, from equations (33) and (35), we have

$$\frac{1}{H(\mu)} = 1 - \frac{(-1)^n}{\mu_1 \cdots \mu_n} \mu \sum_{j=1}^{n} \frac{q_j P(-\mu_j)}{(\mu + \mu_j)R(-\mu_j)}. \tag{42}$$

Now, comparing equations (41) and (42), we conclude that it is necessary and sufficient that

$$R(\mu_j)R(-\mu_j) = -\frac{q_j \mu_j}{\mu_1^2 \cdots \mu_n^2} P(-\mu_j)P'(\mu_j) \quad (j = 1,...,n). \tag{43}$$

Returning to equation (33), we have, by repeated applications of this equation:

$$\left\{1-\frac{1}{H(\mu)}\right\}\left\{1-\frac{1}{H(-\mu)}\right\}$$

$$= \mu^2 \sum_{j=1}^{n} \sum_{i=1}^{n} \frac{q_j\,H(\mu_j)}{\mu+\mu_j}\,\frac{q_i\,H(\mu_i)}{\mu-\mu_i}$$

$$= \mu \sum_{j=1}^{n} \sum_{i=1}^{n} \frac{q_j q_i\,H(\mu_j)H(\mu_i)}{\mu_i+\mu_j}\left(\frac{\mu_j}{\mu+\mu_j}+\frac{\mu_i}{\mu-\mu_i}\right)$$

$$= \mu \sum_{j=1}^{n} \frac{q_j\,H(\mu_j)}{\mu+\mu_j}\left\{\mu_j \sum_{i=1}^{n}\frac{q_i\,H(\mu_i)}{\mu_i+\mu_j}\right\}+\mu \sum_{i=1}^{n}\frac{q_i\,H(\mu_i)}{\mu-\mu_i}\left\{\mu_i\sum_{j=1}^{n}\frac{q_j\,H(\mu_j)}{\mu_i+\mu_j}\right\}$$

$$= \mu \sum_{j=1}^{n} \frac{q_j\,H(\mu_j)}{\mu+\mu_j}\left\{1-\frac{1}{H(\mu_j)}\right\}+\mu\sum_{i=1}^{n}\frac{q_i\,H(\mu_i)}{\mu-\mu_i}\left\{1-\frac{1}{H(\mu_i)}\right\}$$

$$= 1-\frac{1}{H(\mu)}+1-\frac{1}{H(-\mu)}-\mu\sum_{j=1}^{n}\frac{q_j}{\mu+\mu_j}-\mu\sum_{i=1}^{n}\frac{q_i}{\mu-\mu_i}. \tag{44}$$

Hence
$$\frac{1}{H(\mu)H(-\mu)}=1-2\mu^2\sum_{j=1}^{n}\frac{a_j\,\Psi(\mu_j)}{\mu^2-\mu_j^2}. \tag{45}$$

By arguments similar to those used in establishing the identity in Chapter III, § 26.4, we can express

$$T(\mu) = 1-2\mu^2\sum_{j=1}^{n}\frac{a_j\,\Psi(\mu_j)}{\mu^2-\mu_j^2}, \tag{46}$$

in terms of the non-vanishing roots, $\pm k_\alpha$, of the characteristic equation

$$1 = 2\sum_{j=1}^{n}\frac{a_j\,\Psi(\mu_j)}{1-k^2\mu_j^2}. \tag{47}$$

Equation (47) admits $2n$ distinct roots $\pm k_\alpha$ $(\alpha = 1,...,n)$ except in conservative cases when $k^2 = 0$ is a root. Restricting ourselves, in the first instance, to non-conservative cases, we observe that

$$\prod_{j=1}^{n}(\mu^2-\mu_j^2)T(\mu) \tag{48}$$

is a polynomial of degree $2n$ in μ which vanishes for

$$\pm\frac{1}{k_\alpha}\quad(\alpha = 1,...,n). \tag{49}$$

Consequently, $T(\mu)$ cannot differ from

$$\frac{\prod\limits_{\alpha=1}^{n}(1-k_\alpha^2\mu^2)}{\prod\limits_{j=1}^{n}(\mu^2-\mu_j^2)} \tag{50}$$

by more than a constant factor. The constant of proportionality can be found from a comparison of (46) and (50) at $\mu = 0$. In this manner we find:

$$T(\mu) = (-1)^n\mu_1^2...\mu_n^2\frac{\prod\limits_{\alpha}(1-k_\alpha^2\mu^2)}{\prod\limits_{j=1}^{n}(\mu^2-\mu_j^2)}. \tag{51}$$

It can be readily verified that in the form (51) the equation is also valid in conservative cases when one of the roots $k_\alpha^2 = 0$.

Returning to equation (45) we can now write

$$H(\mu)H(-\mu) = \frac{1}{\mu_1^2...\mu_n^2}\frac{P(\mu)P(-\mu)}{\prod\limits_{\alpha}(1-k_\alpha^2\mu^2)}. \tag{52}$$

On the other hand, according to equation (35),

$$H(\mu)H(-\mu) = \frac{1}{\mu_1^2...\mu_n^2}\frac{P(\mu)P(-\mu)}{R(\mu)R(-\mu)}. \tag{53}$$

Hence
$$R(\mu) = \prod\limits_{\alpha}(1\mp k_\alpha\mu), \tag{54}$$

where $(1-k_\alpha\mu)$ or $(1+k_\alpha\mu)$, but not both, is a factor of $R(\mu)$.

With $R(\mu)$ given by equation (54), (35) in fact provides a solution of the original equation; for it gives (cf. eq. [45])

$$R(\mu)R(-\mu) = \frac{P(\mu)P(-\mu)}{\mu_1^2...\mu_n^2}\left[1-2\mu^2\sum\limits_{j=1}^{n}\frac{q_j}{\mu^2-\mu_j^2}\right], \tag{55}$$

from which the validity of equation (43) follows.

The solution of equation (32) must therefore be of the form

$$\frac{1}{\mu_1...\mu_n}\frac{\prod\limits_{i=1}^{n}(\mu+\mu_i)}{\prod\limits_{\alpha}(1\pm k_\alpha\mu)}. \tag{56}$$

If we now require that the solution be regular and non-zero for all $\mu \geqslant 0$, it is evident that we must always choose the factor $(1+k_\alpha\mu)$ in (56). We have thus proved:

THEOREM 4. Consider the equation

$$H(\mu) = 1+\mu H(\mu)\sum\limits_{j=1}^{n}\frac{a_j\,\Psi(\mu_j)}{\mu+\mu_j}H(\mu_j), \tag{57}$$

where $a_{\pm j}$ $(j = 1,...,n,\ a_j = a_{-j})$ and $\mu_{\pm j}$ $(j = 1,...,n,\ \mu_{-j} = -\mu_j)$ are the weights and divisions appropriate to a quadrature formula† in the interval $(-1, +1)$; the unique solution of this equation which is regular and non-zero for $\mu \geqslant 0$, is

$$H(\mu) = \frac{1}{\mu_1 \cdots \mu_n} \frac{\prod\limits_{i=1}^{n} (\mu + \mu_i)}{\prod\limits_{\alpha=1}^{n} (1 + k_\alpha \mu)}, \tag{58}$$

where the k_α's $(\alpha = 1,...,n)$ are the non-negative roots of the associated characteristic equation

$$1 = 2 \sum_{j=1}^{n} \frac{a_j \Psi'(\mu_j)}{1 - k^2 \mu_j^2}. \tag{59}$$

From Theorem 4 it would appear that the exact solution of the integral equation (1) can be obtained as a limit by making the division (μ_j) finer and finer. In any event, the theorem suggests a simple practical method for solving for the *H*-functions numerically. For, starting with an approximate solution for $H(\mu)$ (in the third approximation, for example), we can determine the exact *H*-functions by a process of iteration applied to the integral equation which it must satisfy. This is the method which will be described in greater detail in § 41.

39.1. *The representation of the solution of equation* (32) *as a complex integral*

Starting with the identity

$$\frac{1}{H(z)H(-z)} = T(z) = 1 - 2z^2 \sum_{j=1}^{n} \frac{a_j \Psi'(\mu_j)}{z^2 - \mu_j^2} = \frac{\prod\limits_{\alpha=1}^{n} (1 - k_\alpha^2 z^2)}{\prod\limits_{j=1}^{n} (1 - z^2/\mu_j^2)}, \tag{60}$$

we can obtain a representation of $H(z)$ as a complex integral. For this purpose consider the integral

$$K(z) = \frac{1}{2\pi i} \int\limits_{-i\infty}^{i\infty} \log T(w) \frac{z\, dw}{w^2 - z^2}, \tag{61}$$

taken along the entire imaginary axis. Defined in this manner, it is evident that $K(z)$ is regular for $R(z) > 0$.

---

† It is clearly not necessary to restrict ourselves here to Gauss's quadrature formula as we had no occasion to use the particular properties of the Gaussian division in our demonstration.

By evaluating the residue at the pole on the right

$$\frac{1}{2\pi i}\int_{-i\infty}^{i\infty} \log\left(1+\frac{w}{a}\right)\frac{z\,dw}{w^2-z^2} = -\tfrac{1}{2}\log\left(1+\frac{z}{a}\right), \tag{62}$$

if $R(z) > 0$ and $R(a) > 0$. Similarly, by evaluating the residue at the pole on the left, we have

$$\frac{1}{2\pi i}\int_{-i\infty}^{i\infty} \log\left(1-\frac{w}{a}\right)\frac{z\,dw}{w^2-z^2} = -\tfrac{1}{2}\log\left(1+\frac{z}{a}\right), \tag{63}$$

again, provided $R(z) > 0$ and $R(a) > 0$. Hence

$$\frac{1}{2\pi i}\int_{-i\infty}^{i\infty} \log\left(1-\frac{w^2}{a^2}\right)\frac{z\,dw}{w^2-z^2} = -\log\left(1+\frac{z}{a}\right). \tag{64}$$

Using this result, we have

$$\begin{aligned}
K(z) &= \frac{1}{2\pi i}\int_{-i\infty}^{i\infty}\left\{\sum_\alpha \log(1-k_\alpha^2 w^2) - \sum_j \log\left(1-\frac{w^2}{\mu_j^2}\right)\right\}\frac{z\,dw}{w^2-z^2} \\
&= -\sum_\alpha \log(1+k_\alpha z) + \sum_j \log\left(1+\frac{z}{\mu_j}\right) \\
&= \log H(z). \tag{65}
\end{aligned}$$

Hence
$$\log H(z) = \frac{1}{2\pi i}\int_{-i\infty}^{i\infty} \log T(w)\frac{z\,dw}{w^2-z^2}, \tag{66}$$

which is the required representation.

In § 40 we shall see that the solution of the integral equation (1), which is bounded in the interval $0 \leqslant \mu \leqslant 1$, is again given by an integral of the form (66) in which $T(z)$, now given by equation (60), is replaced by

$$T(z) = 1 - 2z^2\int_0^1 \frac{\Psi(\mu)\,d\mu}{z^2-\mu^2}. \tag{67}$$

40. The explicit solution of the integral equation satisfied by $H(\mu)$

The following general discussion of the integral equation satisfied by $H(\mu)$ is due to M. M. Crum.

Considering the integral equation (1) in the complex z-plane, we shall write it in the form

$$\frac{1}{H(z)} = 1 - \int_0^1 \frac{z}{z+\mu}H(\mu)\Psi(\mu)\,d\mu. \tag{68}$$

The characteristic functions which occur in the astrophysical contexts are even, real, non-negative, and bounded in the interval $0 \leqslant \mu \leqslant 1$, and satisfy the conditions

$$\int_0^1 |\Psi(\mu)| \, d\mu \leqslant \tfrac{1}{2} \quad \text{and} \quad \int_0^\delta |\Psi(\mu)| \, d\mu \to 0 \quad \text{as} \quad \delta \to 0. \tag{69}$$

Also, we shall be interested only in solutions which are bounded in the interval $0 \leqslant \mu \leqslant 1$:

$$|H(\mu)| < M \quad (0 \leqslant \mu \leqslant 1). \tag{70}$$

Let $H(z)$ be a solution of equation (68) under the conditions (69) and (70). Then $1/H(z)$ is analytic in the entire z-plane slit between -1 to 0. Except for $-1 < z < 1$, we therefore have, by repeated application of equation (68) (cf. § 39, eq. [44])

$$\left\{1 - \frac{1}{H(z)}\right\}\left\{1 - \frac{1}{H(-z)}\right\}$$

$$= \int_0^1\int_0^1 \frac{z}{z+\mu} \frac{z}{z-\mu'} H(\mu)\Psi(\mu)H(\mu')\Psi(\mu') \, d\mu \, d\mu'$$

$$= \int_0^1\int_0^1 H(\mu)\Psi(\mu)H(\mu')\Psi(\mu') \frac{z}{\mu+\mu'}\left(\frac{\mu}{z+\mu}+\frac{\mu'}{z-\mu'}\right) d\mu \, d\mu'$$

$$= z\int_0^1 \frac{d\mu}{z+\mu} H(\mu)\Psi(\mu)\mu \int_0^1 \frac{d\mu'}{\mu+\mu'} H(\mu')\Psi(\mu') +$$

$$\qquad\qquad + z\int_0^1 \frac{d\mu'}{z-\mu'} H(\mu')\Psi(\mu')\mu' \int_0^1 \frac{d\mu}{\mu+\mu'} H(\mu)\Psi(\mu)$$

$$= z\int_0^1 \frac{\Psi(\mu)}{z+\mu} H(\mu)\left\{1 - \frac{1}{H(\mu)}\right\} d\mu + z\int_0^1 \frac{\Psi(\mu')}{z-\mu'} H(\mu')\left\{1 - \frac{1}{H(\mu')}\right\} d\mu'$$

$$= 1 - \frac{1}{H(z)} + 1 - \frac{1}{H(-z)} - z\int_0^1 \frac{\Psi(\mu)}{z+\mu} \, d\mu - z\int_0^1 \frac{\Psi(\mu')}{z-\mu'} \, d\mu'. \tag{71}$$

Hence $\quad \dfrac{1}{H(z)H(-z)} = 1 - 2z^2 \displaystyle\int_0^1 \frac{\Psi(\mu)}{z^2-\mu^2} \, d\mu = T(z) \quad \text{(say)}. \tag{72}$

Still excepting the case $-1 < z < 1$, substituting for $1/H(z)$ according to equation (72), we can rewrite equation (68) in the form

$$H(-z)\left[1 - \int_0^1 \frac{z\Psi(\mu)}{z+\mu}\,d\mu - \int_0^1 \frac{z\Psi(\mu)}{z-\mu}\,d\mu\right] = 1 - \int_0^1 \frac{z\Psi(\mu)}{z+\mu}\,H(\mu)\,d\mu, \quad (73)$$

or, after some minor rearranging of the terms, we have

$$H(-z)\left[1 - \int_0^1 \frac{z\Psi(\mu)}{z-\mu}\,d\mu\right] = 1 + \int_0^1 \frac{z\Psi(\mu)}{z+\mu}[H(-z) - H(\mu)]\,d\mu. \quad (74)$$

Writing z for $-z$ in this equation, we obtain

$$H(z)\left[1 - \int_0^1 \frac{z\Psi(\mu)}{z+\mu}\,d\mu\right] = 1 + \int_0^1 \frac{z\Psi(\mu)}{z-\mu}[H(z) - H(\mu)]\,d\mu. \quad (75)$$

Since both sides of this equation are analytic for $R(z) > 0$ the equation is valid also for $0 < z < 1$.

Conversely, equations (72) and (75) together imply that $H(\mu)$ is a solution of equation (68).

It will now be shown that under the conditions (69) and (70), the unique solution which is analytic in the half-plane $R(z) > 0$ is given by (cf. § 39, eq. [66])

$$\log H(z) = \frac{1}{2\pi i} \int_{-i\infty}^{+i\infty} \log T(w) \frac{z}{w^2 - z^2}\,dw. \quad (76)$$

In conservative cases, this is the *only* solution which is bounded in the interval $0 \leqslant z \leqslant 1$. But in non-conservative cases it is possible that there is another solution which is also bounded in $(0 \leqslant z \leqslant 1)$, but which has a pole on the real axis for a value of $z > 1$. We shall describe later the relation of this other solution to (76).

The function $T(z)$ is even and regular in the plane slit from -1 to $+1$. More particularly, writing $z^2 = u + iv$, we have

$$T(z) = 1 - 2\int_0^1 \Psi(\mu)\,d\mu - 2\int_0^1 \frac{\Psi(\mu)\mu^2}{u+iv-\mu^2}\,d\mu$$

$$= 1 - 2\int_0^1 \Psi(\mu)\,d\mu - 2\int_0^1 \frac{(u-\mu^2)-iv}{(u-\mu^2)^2+v^2}\mu^2\Psi(\mu)\,d\mu. \quad (77)$$

From this equation it follows that the imaginary part of $T(z)$ can vanish only when z^2 is real. The zeros of $T(z)$, if any, must therefore be on the real or the imaginary axis or at infinity. When $v = 0$, equation (77) becomes

$$T(z) = 1 - 2 \int_0^1 \Psi(\mu) \, d\mu - 2 \int_0^1 \frac{\mu^2 \Psi(\mu)}{u - \mu^2} \, d\mu \quad (z^2 = u = \text{real}). \quad (78)$$

The expression on the right-hand side is monotonic, increasing for $u < 0$ and $u > 1$. Therefore, in conservative cases, the only zero of $T(z)$ is at infinity; and in non-conservative cases $T(\infty) > 0$ and there *may* be a pair of real zeros at $\pm 1/k$ (say) where $1/k = \sqrt{u_0} > 1$. Thus, summarizing, we have

$$T(z) \sim -\frac{\text{positive constant}}{z^2} \quad \left(z \to \infty \quad \text{and} \quad \int_0^1 \Psi(\mu) \, d\mu = \tfrac{1}{2} \right) \quad (79)$$

and

in non-conservative $\left(\int_0^1 \Psi(\mu) \, d\mu < \tfrac{1}{2} \right)$ cases $T(z)$ is positive

and bounded on the entire imaginary axis and may have   (80)
a pair of zeros $\pm 1/k$ $(k > 1)$ on the real axis.

Otherwise $T(z)$ is non-zero in the plane slit from -1 to $+1$. Also, (§ 38, eq. [7])

$$T(\infty) = 1 - 2 \int_0^1 \Psi(\mu) \, d\mu = \left[1 - \int_0^1 H(\mu) \Psi(\mu) \, d\mu \right]^2 = \left[\frac{1}{H(\infty)} \right]^2. \quad (81)$$

Let

$$B(z) = T(z)(1 - z^2)^n. \quad \begin{pmatrix} n = 0 \text{ in non-conservative cases,} \\ n = 1 \text{ in conservative cases.} \end{pmatrix} \quad (82)$$

Defined in this manner $B(z)$ is even, analytic, and non-zero on the imaginary axis (and at infinity) except perhaps at $z = 0$; since $B(iv) \to 1$ as $v \to 0$, $\log B(iv)$ is bounded.

For $R(z) > 0$, we define

$$\log K(z) = \frac{1}{2\pi i} \int_{-i\infty}^{i\infty} \log B(w) \frac{z}{w^2 - z^2} \, dw, \quad (83)$$

where $\log B(iv) \to 0$ as $v \to 0$. The integral defines a function which is analytic for $R(z) > 0$. And we can obtain an analytic continuation of $K(z)$ across the imaginary axis by deforming the contour. Thus, when z is in the second quadrant we integrate along the contour indicated in Fig. 9 (a). Then

$$\log K(z) = \frac{1}{2\pi i} \int_{-i\infty}^{i\infty} \log B(w) \frac{z}{w^2 - z^2} \, dw - \tfrac{1}{2} \log B(-z) - \tfrac{1}{2} \log B(z)$$

$$= -\log K(-z) - \log B(z) - 2v\pi i, \quad (84)$$

since $R(-z) > 0$. If z passes from the fourth to the third quadrant we deform the contour as in Fig. 9 (b) and we get the same result (84). Thus $K(z)$ is single valued for all z and

$$\frac{1}{K(z)K(-z)} = B(z). \tag{85}$$

Now let
$$H(z) = K(z)(1+z)^n. \tag{86}$$

Then
$$\frac{1}{H(z)H(-z)} = T(z); \tag{87}$$

i.e. $H(z)$ satisfies equation (72).

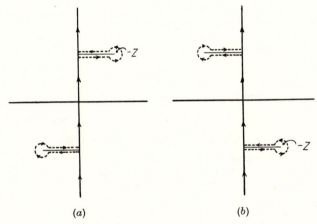

FIG. 9 (a) and (b). The deformed contours for the analytic continuation of $K(z)$ (eq. [83]) across the imaginary axis.

Next, using the result expressed by equation (64) (cf. eqs. [82] and [86]),

$$\log H(z) = \frac{1}{2\pi i} \int_{-i\infty}^{i\infty} \{\log T(w) + n\log(1-w^2)\} \frac{z}{w^2-z^2} dw + n\log(1+z)$$

$$= \frac{1}{2\pi i} \int_{-i\infty}^{i\infty} \log T(w) \frac{z}{w^2-z^2} dw. \tag{88}$$

It remains to prove that $H(z)$ defined in this manner satisfies equation (75): we should then have proved that $H(z)$ is a solution of equation (68).

Now if $z = x+iy = re^{i\theta}$, then from the definition of $K(z)$, for $x > 0$,

$$\log K(z) = O\left\{\int_0^\infty \frac{r\,dv}{|r^2e^{2i\theta}+v^2|}\right\}$$

$$= O\left\{\int_0^\infty \frac{dv_1}{|e^{2i\theta}+v_1^2|}\right\} \quad (v = rv_1)$$

$$= O(1)+O\left\{\int_{-2}^2 \frac{dv_1}{|v_1+ie^{i\theta}|}\right\}$$

$$= O(1)+O\left\{\int_{-2}^2 \frac{dv_1}{|v_1-\sin\theta|+\cos\theta}\right\}$$

$$= O(1)+O\left\{\int_0^2 \frac{dv_2}{v_2+\cos\theta}\right\} \quad (v_2 = v_1-\sin\theta)$$

$$= O(1)+O(\log\sec\theta). \tag{89}$$

In other words, there exist constants κ and λ such that

$$|\log|K(z)|\,| \leqslant |\log K(z)| \leqslant \kappa+\lambda\log\sec\theta,$$

or
$$K(z) = O(\sec\theta)^\lambda \quad \text{and} \quad 1/K(z) = O(\sec\theta)^\lambda. \tag{90}$$

Hence, for large enough ν (cf. eq. [86])

$$H(z) = O\{(1+r^n)\sec^\nu\theta\}$$

and
$$\frac{1}{H(z)} = O(\sec^\nu\theta). \tag{91}$$

Also, when $r \to 0$ with $|\theta| \leqslant \frac{1}{4}\pi$, $K(z) \to 1$. For,

$$\log K(z) = O\left\{\int_0^\infty |\log B(iv)|\frac{r\,dv}{r^2+v^2}\right\}$$

$$= O\left\{\int_0^\delta |\log B(iv)|\frac{r\,dv}{r^2+v^2}\right\}+O\left\{\int_\delta^\infty \frac{r\,dv}{r^2+v^2}\right\}$$

$$= o\left\{\int_0^\delta \frac{r\,dv}{r^2+v^2}\right\}+O\left(r\int_\delta^\infty \frac{dv}{v^2}\right)$$

$$= o(1)+O\left(\frac{r}{\delta}\right) = o(1), \tag{92}$$

by choosing first δ and then r small enough. Hence

$$H(z) \to 1 \quad \text{as} \quad r \to 0 \quad (|\theta| \leqslant \tfrac{1}{4}\pi). \tag{93}$$

Now let (cf. eq. [75])

$$\phi(z) = H(z)\left\{1 - \int_0^1 \frac{z\Psi(\mu)}{z+\mu}\,d\mu\right\} - 1 - \int_0^1 \frac{z\Psi(\mu)}{z-\mu}[H(z)-H(\mu)]\,d\mu. \quad (94)$$

From this equation we obtain, by using the relation (87):

$$\phi(-z) = H(-z)\left\{T(z) + \int_0^1 \frac{z}{z+\mu}\Psi(\mu)\,d\mu\right\} -$$

$$- 1 - \int_0^1 \frac{z}{z+\mu}[H(-z)-H(\mu)]\Psi(\mu)\,d\mu$$

$$= \frac{1}{H(z)} - 1 + \int_0^1 \frac{z}{z+\mu}H(\mu)\Psi(\mu)\,d\mu. \quad (95)$$

From equation (95) it is apparent that $\phi(z)$ is analytic for $x \leqslant 0$ except perhaps at $z = 0$; and for $x < 0$ (cf. eq. [91])

$$\phi(z) = O(|\sec\theta|^\nu) + O(1) = O(|\sec\theta|^\nu). \quad (96)$$

On the other hand, from equation (94) we conclude that $\phi(z)$ is analytic for $x > 0$, and, for $y \neq 0$, we have, by (91)

$$\phi(z) = O\left\{(1+r^n)\sec^\nu\theta\left(1 + \int_0^1 \left|\frac{r\,\Psi(\mu)}{y}\right|\,d\mu\right)\right\}$$

$$= O\{(1+r^n)\sec^\nu\theta\,\mathrm{cosec}\,\theta\} = O\{(1+r^n)\mathrm{cosec}^\nu 2\theta\} \quad (97)$$

if we choose $\nu \geqslant 1$.

According to equation (97) for $r = \frac{1}{2}R$ and for $r = 2R$

$$\left|\left(1 - 2^4\frac{z^4}{R^4}\right)^\nu\left(1 - \frac{z^4}{2^4 R^4}\right)^\nu\phi(z)\right| < C(1+R^n), \quad (98)$$

where C is a constant independent of R.† By the maximum-modulus

† For $r = \frac{1}{2}R$, for example, the quantity on the left-hand side of (98) is

$$|(1-e^{4i\theta})^\nu|\left|\left(1-\frac{e^{4i\theta}}{2^8}\right)^\nu\right||\phi(z)| = O(1)|(1-e^{4i\theta})^\nu||\phi(z)|.$$

But
$$|(1-e^{4i\theta})| = |\{1-(\cos 2\theta + i\sin 2\theta)^2\}|$$
$$= |\{1-\cos^2 2\theta + \sin^2 2\theta - 2i\cos 2\theta\sin 2\theta\}|$$
$$= |2\sin 2\theta(\sin 2\theta - i\cos 2\theta)|$$
$$= |2\sin 2\theta|.$$

A similar reduction clearly holds for $r = 2R$.

theorem† the inequality (98) holds also for $r = R$, so that

$$\phi(z) = O(1+r^n). \qquad (99)$$

Hence, $\phi(z)$ is analytic at $z = 0$ and is a polynomial. But from (93), (95), and (96) $\phi(-z)$ is bounded for $\theta = \tfrac{1}{4}\pi$ and tends to zero as $r \to 0$. Hence $\phi(z) \equiv 0$ and equations (94) and (95) state that $H(z)$ is a solution of equation (68).

It remains to find the general solution, $H_1(z)$, of equation (68) which is bounded for $0 \leqslant z \leqslant 1$. By (68), $1/H_1(z)$ is analytic except for $-1 \leqslant z < 0$ and bounded for $x \geqslant 0$. Also equation (72) holds, so that if

$$\psi(z) = \frac{H_1(z)}{H(z)}, \qquad (100)$$

then by equations (72) and (87)

$$\psi(z)\psi(-z) = 1. \qquad (101)$$

Since $\psi(z)$ is meromorphic‡ for $x > 0$, except perhaps at $z = 0$, $\phi(z)$ is meromorphic everywhere except perhaps at $z = 0$; but $\psi(z) \to 1$ as $z \to 0$ with $x \geqslant 0$ (since by [68] both H_1 and $H \to 1$ as $z \to 0$), hence also as $z \to 0$ with $x < 0$; hence $\psi(z)$ is analytic at $z = 0$. But it is analytic at infinity because by (72) and (87), $H(z)$ and $H_1(z)$ have poles of the same order. Hence $\psi(z)$ is rational and

$$\psi(z) = \frac{\prod\limits_{\alpha'} (1-z/z_{\alpha'})}{\prod\limits_{\alpha} (1-z/z_{\alpha})}. \qquad (102)$$

By (101)

$$\prod_{\alpha}\left(1-\frac{z^2}{z_{\alpha'}^2}\right) = \prod_{\alpha}\left(1-\frac{z^2}{z_{\alpha}^2}\right). \qquad (103)$$

We can therefore suppose that $z_{\alpha} = -z_{\alpha}'$. In this manner we conclude that

$$H_1(z) = H(z) \prod_{\alpha}\left(\frac{1+z/z_{\alpha}}{1-z/z_{\alpha}}\right). \qquad (104)$$

Since both $H(z)$ and $H_1(z)$ are bounded and non-zero for $0 \leqslant z \leqslant 1$, none of the z_{α}'s can be on the line $-1 \leqslant z \leqslant 1$; also the z_{α}'s are the poles of $H_1(z)$ and $H(-z)$ and, according to equation (72), must be the zeros of $T(z)$. But we have already shown that in conservative cases there are no zeros of $T(z)$ in the finite part of the plane slit between -1 to 1, while in non-conservative cases we can have, at most, a pair of zeros $\pm 1/k$

† The maximum-modulus theorem states that if $|f(z)| \leqslant M$ on a simple closed contour C, then $|f| < M$ at all interior points of D unless $|f(z)|$ is a constant, when $|f(z)| = M$ throughout the domain D (cf. E. C. Titchmarsh, *The Theory of Functions*, chap. v, Oxford, 1932).

‡ A function is said to be meromorphic in a region if it is analytic in the region except for a finite number of poles.

$(k < 1)$. Hence, in conservative cases, equation (76) represents the unique solution of the integral equation (1) which is bounded in the interval $0 \leqslant z \leqslant 1$. In non-conservative cases, if the equation $T(z) = 0$ admits the roots $\pm 1/k$ $(0 < k < 1)$, then there is a further solution given by

$$H_1(z) = H(z)\frac{1+kz}{1-kz}. \tag{105}$$

Conversely, if $-1/k$ is a pole of $H(z)$, then $H_1(z)$ defined by equation (105) represents a solution of (68). For

$$\int_0^1 \frac{z}{z+\mu} H_1(\mu)\Psi(\mu)\, d\mu$$

$$= \int_0^1 \frac{z(1+k\mu)}{(z+\mu)(1-k\mu)} H(\mu)\Psi(\mu)\, d\mu$$

$$= \frac{1-kz}{1+kz}\int_0^1 \frac{z}{z+\mu} H(\mu)\Psi(\mu)\, d\mu + \frac{2kz}{1+kz}\int_0^1 \frac{H(\mu)}{1-k\mu} \Psi(\mu)\, d\mu$$

$$= \frac{1-kz}{1+kz}\left\{1 - \frac{1}{H(z)}\right\} + \frac{2kz}{1+kz}\left\{1 - \frac{1}{H(-1/k)}\right\}$$

$$= 1 - \frac{1-kz}{1+kz}\frac{1}{H(z)} = 1 - \frac{1}{H_1(z)}, \tag{106}$$

which completes the verification.

41. A practical method for evaluating the *H*-functions

While the representation of $H(\mu)$ as a definite integral given in § 40 can be used for the evaluation of the *H*-functions, it is found that, in practice, it is more convenient to solve for the *H*-functions, directly, by a process of iteration applied to the integral equation. We shall briefly outline this iteration method:

First, we may observe that the integral equation in the form (eq. [13])

$$\frac{1}{H(\mu)} = \left[1 - 2\int_0^1 \Psi(\mu)\, d\mu\right]^{\frac{1}{2}} + \int_0^1 \frac{\mu'\Psi(\mu')}{\mu+\mu'}H(\mu')\, d\mu', \tag{107}$$

is more suitable for purposes of iteration than the equation in the original form. The solution can, in general, be started with the third approximation for $H(\mu)$ in terms of the Gaussian division and characteristic roots, though in conservative cases it is preferable to start with the fourth approximation.

With some experience the successive iterations can be performed quite rapidly. Thus the iterations need not be carried out explicitly for every tabulated value: it is sufficient to evaluate the iterates at some 'strategic' points and then predict the intermediate values by interpolating among the differences between the successive iterates. Also, when the *H*-functions are evaluated for various values of a parameter (e.g. the albedo ϖ_0 in isotropic scattering) we can often 'correct', in advance, the third approximations by interpolating among the differences between the approximate solutions and the exact solutions for neighbouring values of the parameter.

A satisfactory check on the accuracy reached at any stage of the iteration is provided by evaluating the integral

$$\int_0^1 H(\mu)\Psi(\mu)\, d\mu$$

numerically and comparing it with its exact value (eq. [7])

$$1 - \left[1 - 2\int_0^1 \Psi(\mu)\, d\mu \right]^{\frac{1}{2}}.$$

Using the method described, Mrs. Frances H. Breen and the writer have evaluated over forty *H*-functions in the context of various problems (cf. § 42 below, Chap. VI, §§ 44 and 46 and Chap. X, §§ 68 and 70).

42. The *H*-functions for problems in isotropic scattering

In Chapters III and IV we have shown that the angular distributions of the emergent radiations for the standard problems of radiative transfer under conditions of isotropic scattering can be expressed in terms of *H*-functions with

$$\Psi(\mu) = \text{constant} = \tfrac{1}{2}\varpi_0. \tag{108}$$

Thus, the law of diffuse reflection is given by

$$I(0,\mu;\mu_0) = \tfrac{1}{4}\varpi_0\, F\, \frac{\mu_0}{\mu+\mu_0}\, H(\mu)H(\mu_0), \tag{109}$$

and in the conservative case ($\varpi_0 = 1$) the law of darkening in the problem with a constant net flux is

$$I(0,\mu) = \frac{\sqrt{3}}{4}\, FH(\mu). \tag{110}$$

The *H*-functions (computed by Mrs. Frances H. Breen and the writer) for various values of ϖ_0 are given in Table XI.

Table XI

The H-Functions obtained as Solutions of the Exact Integral Equations which they satisfy

μ	$\varpi_0 = 0 \cdot 1$	$\varpi_0 = 0 \cdot 2$	$\varpi_0 = 0 \cdot 3$	$\varpi_0 = 0 \cdot 4$	$\varpi_0 = 0 \cdot 5$	$\varpi_0 = 0 \cdot 6$	$\varpi_0 = 0 \cdot 7$
0 . .	1·00000	1·00000	1·00000	1·00000	1·00000	1·00000	1·00000
0·05 . .	1·00783	1·01608	1·02484	1·03422	1·04439	1·05544	1·06780
0·10 . .	1·01238	1·02562	1·03989	1·05535	1·07241	1·09137	1·11306
0·15 . .	1·01584	1·03295	1·05155	1·07196	1·09474	1·12045	1·15036
0·20 . .	1·01864	1·03892	1·06115	1·08577	1·11349	1·14517	1·18253
0·25 . .	1·02099	1·04396	1·06930	1·09758	1·12968	1·16674	1·21095
0·30 . .	1·02300	1·04829	1·07637	1·10789	1·14391	1·18587	1·23643
0·35 . .	1·02475	1·05209	1·08259	1·11700	1·15659	1·20304	1·25951
0·40 . .	1·02630	1·05546	1·08811	1·12516	1·16800	1·21861	1·28063
0·45 . .	1·02768	1·05847	1·09308	1·13251	1·17833	1·23280	1·30003
0·50 . .	1·02892	1·06117	1·09756	1·13918	1·18776	1·24581	1·31796
0·55 . .	1·03004	1·06363	1·10164	1·14528	1·19640	1·25781	1·33459
0·60 . .	1·03106	1·06587	1·10538	1·15087	1·20436	1·26893	1·35009
0·65 . .	1·03199	1·06793	1·10881	1·15602	1·21173	1·27925	1·36457
0·70 . .	1·03284	1·06982	1·11198	1·16080	1·21858	1·28888	1·37815
0·75 . .	1·03363	1·07157	1·11491	1·16523	1·22495	1·29788	1·39090
0·80 . .	1·03436	1·07319	1·11763	1·16935	1·23091	1·30631	1·40291
0·85 . .	1·03504	1·07469	1·12017	1·17320	1·23648	1·31424	1·41425
0·90 . .	1·03567	1·07610	1·12254	1·17681	1·24171	1·32171	1·42497
0·95 . .	1·03626	1·07741	1·12476	1·18019	1·24664	1·32875	1·43512
1·00 . .	1·03682	1·07864	1·12685	1·18337	1·25128	1·33541	1·44476

Table XI (continued)

μ	$\varpi_0 = 0 \cdot 8$	$\varpi_0 = 0 \cdot 85$	$\varpi_0 = 0 \cdot 9$	$\varpi_0 = 0 \cdot 925$	$\varpi_0 = 0 \cdot 95$	$\varpi_0 = 0 \cdot 975$	$\varpi_0 = 1 \cdot 0$
0 .	1·0000	1·0000	1·0000	1·0000	1·0000	1·0000	1·0000
0·05 .	1·0820	1·0903	1·0999	1·1053	1·1117	1·1196	1·1368
0·10 .	1·1388	1·1541	1·1722	1·1828	1·1952	1·2111	1·2474
0·15 .	1·1866	1·2086	1·2349	1·2506	1·2693	1·2936	1·3508
0·20 .	1·2286	1·2570	1·2914	1·3123	1·3373	1·3703	1·4503
0·25 .	1·2663	1·3009	1·3433	1·3692	1·4008	1·4427	1·5473
0·30 .	1·3006	1·3411	1·3914	1·4224	1·4604	1·5117	1·6425
0·35 .	1·3320	1·3783	1·4363	1·4724	1·5170	1·5778	1·7364
0·40 .	1·3611	1·4129	1·4785	1·5197	1·5709	1·6414	1·8293
0·45 .	1·3881	1·4453	1·5183	1·5646	1·6224	1·7027	1·9213
0·50 .	1·4132	1·4758	1·5560	1·6073	1·6718	1·7621	2·0128
0·55 .	1·4368	1·5044	1·5918	1·6480	1·7191	1·8195	2·1037
0·60 .	1·4590	1·5315	1·6259	1·6869	1·7647	1·8753	2·1941
0·65 .	1·4798	1·5571	1·6583	1·7242	1·8086	1·9295	2·2842
0·70 .	1·4995	1·5814	1·6893	1·7600	1·8509	1·9822	2·3740
0·75 .	1·5182	1·6045	1·7190	1·7943′	1·8918	2·0334	2·4635
0·80 .	1·5358	1·6265	1·7474	1·8274	1·9313	2·0833	2·5527
0·85 .	1·5526	1·6475	1·7746	1·8592	1·9695	2·1320	2·6417
0·90 .	1·5685	1·6675	1·8008	1·8898	2·0065	2·1795	2·7306
0·95 .	1·5837	1·6867	1·8259	1·9194	2·0423	2·2258	2·8193
1·00 .	1·5982	1·7050	1·8501	1·9479	2·0771	2·2710	2·9078

An idea of the accuracy reached in the calculations can be obtained from Table XII, where a comparison is made between the values of the integral

$$\int_0^1 H(\mu)\,d\mu, \tag{111}$$

evaluated numerically with the aid of the tabulated functions and the exact values given by the formula

$$\frac{2}{\varpi_0}[1-(1-\varpi_0)^{\frac{1}{2}}]. \tag{112}$$

TABLE XII

Comparison of the Integrals $\int_0^1 H(\mu)\,d\mu$ evaluated with the Aid of the Tabulated Functions with their Exact Values

$$2[1-(1-\varpi_0)^{\frac{1}{2}}]/\varpi_0$$

ϖ_0	Iterated	Exact	ϖ_0	Iterated	Exact
0·1	1·02632	1·0263340	0·8	1·3819	1·381966
0·2	1·05572	1·0557281	0·85	1·4416	1·441651
0·3	1·08892	1·0889331	0·9	1·5194	1·519494
0·4	1·12698	1·1270167	0·925	1·5699	1·570030
0·5	1·17157	1·1715729	0·950	1·6344	1·634512
0·6	1·22512	1·2251482	0·975	1·7269	1·726946
0·7	1·29219	1·2922213	1·000	1·9999	2·000000

BIBLIOGRAPHICAL NOTES

§§ 38 and 39. The analyses in these sections are largely derived from—

1. S. CHANDRASEKHAR, *Astrophys. J.* **105**, 164 (1946) (see particularly §§ 7, 11, and 12 in this paper).

2. S. CHANDRASEKHAR, *Bulletin American Math. Soc.* **53**, 641, 1947 (see particularly §§ 24, 25, 35, and 35*a* in this paper).

§ 40. The discussion in this section follows—

3. M. M. CRUM, *The Quarterly Journal of Mathematics* (Oxford Series), **18**, 244 (1947).

§ 42. The *H*-functions tabulated in this section are taken from—

4. S. CHANDRASEKHAR and FRANCES H. BREEN, *Astrophys. J.* **106**, 143 (1947).

PROBLEMS WITH GENERAL LAWS
OF SCATTERING

43. Introduction

In this chapter we shall consider transfer problems in semi-infinite plane-parallel atmospheres on laws of scattering more general than isotropic. In these cases the principles of invariance lead to systems of integral equations which are non-linear, non-homogeneous, and of high degree. Thus, already, for the phase functions $\frac{3}{4}(1+\cos^2\Theta)$ and $\varpi_0(1+x\cos\Theta)$ we have to deal with simultaneous integral equations of order 2 (cf. Chap. IV, § 34, eqs. [61] and [62]; § 35, eqs. [79] and [80]). For more general laws of scattering we must expect to deal with systems of a higher order of complexity: for example, the case of Rayleigh scattering leads to a system of order four (cf. Chap. X, § 70, eqs. [135] and [136]). The solution or reduction of these systems of equations might well have been considered impossible had it not been for the guidance provided by the *form* of the solutions obtained in the direct solution of the equations of transfer, in the method of approximation described in Chapter III, in the context of isotropic scattering. The origin and nature of this guidance is the following.

Quite generally it is found that the system (or systems) of equations to which the equation of transfer is equivalent in the nth approximation can be treated in a manner analogous to the equations of Chapter III (cf. § 48 later in this chapter). While the details of the solution vary in individual cases (and can indeed be quite complicated) the analysis, nevertheless, shows certain broad similarities with the simple case of isotropic scattering. Thus the angular distribution of the emergent (equivalently, reflected) radiation is always described by a function for values of the argument in the interval $(0, 1)$, while the boundary conditions specify the zeros of the same function in the complementary interval $(-1, 0)$. This reciprocity, which exists in all problems, enables the elimination of the constants and the expression of the solutions in closed forms. Moreover, these solutions, apart from certain constants, involve only H-functions of the form (Chap. V, eq. [58])

$$H(\mu) = \frac{1}{\mu_1 \cdots \mu_n} \frac{\prod\limits_{i=1}^{n} (\mu+\mu_i)}{\prod\limits_{\alpha=1}^{n} (1+k_\alpha\mu)}, \tag{1}$$

where the μ_i's are the zeros of $P_{2n}(\mu)$ and the k_α's are the non-negative roots of a characteristic equation of the form

$$1 = 2 \sum_{j=1}^{n} \frac{a_j \Psi(\mu_j)}{1-k^2\mu_j^2},\tag{2}$$

where $\Psi(\mu)$ is an even polynomial in μ satisfying the condition

$$\int_0^1 \Psi(\mu)\,d\mu \leqslant \tfrac{1}{2}.\tag{3}$$

From the theorems on the H-functions proved in Chapter V, §§ 39 and 40, we can expect that in the limit of infinite approximation the H-function (1) will become the solution of the associated integral equation

$$H(\mu) = 1 + \mu H(\mu) \int_0^1 \frac{\Psi(\mu')}{\mu+\mu'} H(\mu')\,d\mu',\tag{4}$$

which is bounded in the entire half-plane $R(z) > 0$. In view of this correspondence between the exact H-functions and their rational representations in finite approximations, we can often write down the *form* of the solutions.

In this chapter we shall obtain, in the manner indicated, the solutions in terms of H-functions, of the integral equations derived from the principles of invariance, for the phase functions $\tfrac{3}{4}(1+\cos^2\Theta)$ and $\varpi_0(1+x\cos\Theta)$. We shall also tabulate the various H-functions and constants which occur in these solutions.

The exact laws of diffuse reflection derived for the cases (i) isotropic scattering, (ii) Rayleigh's phase function, and (iii) the phase function $\varpi_0(1+x\cos\Theta)$ are illustrated and contrasted in § 47.

The last sections of this chapter are devoted to a generalization of the theory to include the case of scattering according to a phase function which can be expanded in spherical harmonics.

44. The law of diffuse reflection for scattering in accordance with Rayleigh's phase function

As we have already shown in Chapter IV, § 35, the law of diffuse reflection for the case of scattering according to Rayleigh's phase function can be expressed in the form (Chap. IV, eqs. [5] and [67])

$$I(0,\mu,\varphi) = \frac{3}{32\mu} F[S^{(0)}(\mu,\mu_0) - 4\mu\mu_0(1-\mu^2)^{\tfrac{1}{2}}(1-\mu_0^2)^{\tfrac{1}{2}} S^{(1)}(\mu,\mu_0)\cos(\varphi_0-\varphi)+$$

$$+(1-\mu^2)(1-\mu_0^2)S^{(2)}(\mu,\mu_0)\cos 2(\varphi_0-\varphi)],\tag{5}$$

where (Chap. IV, eqs. [72], [73], [77], and [78])

$$\left(\frac{1}{\mu}+\frac{1}{\mu_0}\right)S^{(0)}(\mu,\mu_0) = \tfrac{1}{3}\psi(\mu)\psi(\mu_0)+\tfrac{8}{3}\phi(\mu)\phi(\mu_0), \tag{6}$$

$$\left(\frac{1}{\mu}+\frac{1}{\mu_0}\right)S^{(i)}(\mu,\mu_0) = H^{(i)}(\mu)H^{(i)}(\mu_0) \quad (i = 1, 2), \tag{7}$$

$$\psi(\mu) = 3-\mu^2+\frac{3}{16}\int_0^1 (3-\mu'^2)S^{(0)}(\mu,\mu')\frac{d\mu'}{\mu'}, \tag{8}$$

$$\phi(\mu) = \mu^2+\frac{3}{16}\int_0^1 \mu'^2 S^{(0)}(\mu,\mu')\frac{d\mu'}{\mu'}, \tag{9}$$

and $H^{(1)}(\mu)$ and $H^{(2)}(\mu)$ are H-functions defined in terms of the characteristic functions

$$\tfrac{3}{8}\mu^2(1-\mu^2) \quad \text{and} \quad \tfrac{3}{32}(1-\mu^2)^2, \tag{10}$$

respectively.

44.1. *The form of the solution for $S^{(0)}(\mu,\mu_0)$*

By substituting for $S^{(0)}$ from equation (6) in equations (8) and (9), we shall obtain the integral equations (Chap. IV, eqs. [79] and [80]) for ψ and ϕ in their normal forms. In solving systems of equations of this type we shall be guided, as we have already stated, by the form of the solutions obtained in the direct solution of the equations of transfer in the general nth approximation and the correspondence between the H-functions occurring in these solutions and the exact functions defined in terms of integral equations.

For the case on hand it would appear that the solution for $S^{(0)}(\mu,\mu_0)$ must be of the form†

$$\left(\frac{1}{\mu}+\frac{1}{\mu_0}\right)S^{(0)}(\mu,\mu_0) = H(\mu)H(\mu_0)[3-c(\mu+\mu_0)+\mu\mu_0], \tag{11}$$

where c is a constant (for the present, unspecified) and $H(\mu)$ is the unique‡ solution of the equation

$$H(\mu) = 1+\tfrac{3}{16}\mu H(\mu)\int_0^1 \frac{3-\mu'^2}{\mu+\mu'}H(\mu')\,d\mu', \tag{12}$$

which is bounded in the interval $(0 \leqslant \mu \leqslant 1)$.

---

† Cf. S. Chandrasekhar, *Astrophys. J.* **103**, 165, 1946 (eq. [189]) and ibid. **105**, 164, 1947 (eq. [246]).

‡ It is unique because we are dealing here with a conservative case:

$$\tfrac{3}{16}\int_0^1 (3-\mu^2)\,d\mu = \tfrac{1}{2}.$$

44.2. *The verification of the solution and the expression of the constant c in terms of the moments of* $H(\mu)$

The verification that the solution for $S^{(0)}(\mu, \mu_0)$ has the form (11) will consist in first evaluating $\psi(\mu)$ and $\phi(\mu)$ according to equations (8) and (9), and then requiring that when the resulting expressions for ψ and ϕ are substituted back into equation (6) we shall recover the form of the solution assumed. In general, such a procedure will lead to certain conditions which the constants introduced in the solution (such as c in the present instance) must satisfy; these conditions will serve to specify the constants and make the solution determinate.

Our first step, then, is to evaluate ψ and ϕ according to equations (8) and (9) for $S^{(0)}$ given by (11). For this purpose we need the various integral properties of the H-function satisfying equation (12). Since, in the present case

$$\Psi(\mu) = \tfrac{3}{16}(3-\mu^2), \tag{13}$$

the properties expressed by Theorems 1, 2, and 3 of Chapter V become

$$\tfrac{3}{16}\int_0^1 (3-\mu^2)H(\mu)\,d\mu = \tfrac{3}{16}(3\alpha_0-\alpha_2) = 1, \tag{14}$$

$$\alpha_0 = 1+\tfrac{3}{32}(3\alpha_0^2-\alpha_1^2), \tag{15}$$

and $$\tfrac{3}{16}(3-\mu^2)\int_0^1 \frac{H(\mu')}{\mu+\mu'}\,d\mu' = \frac{H(\mu)-1}{\mu H(\mu)}+\tfrac{3}{16}(\alpha_1-\mu\alpha_0), \tag{16}$$

where α_n denotes the moment of order n of $H(\mu)$.

A further relation among the moments, which we shall find useful, can be derived from equations (14) and (15) in the following manner:

$$32+9\alpha_2^2 = 32+(9\alpha_0-16)^2$$

$$= 288(1-\alpha_0)+81\alpha_0^2 = 27\alpha_1^2,$$

or $$\alpha_1^2 = \tfrac{1}{3}\alpha_2^2+\tfrac{32}{27}. \tag{17}$$

Turning now to the evaluation of ψ and ϕ according to equations (8), (9), and (11), and considering first ψ, we have

$$\psi(\mu) = 3-\mu^2+\tfrac{3}{16}\mu H(\mu)\int_0^1 \frac{3-\mu'^2}{\mu+\mu'}H(\mu')[3-c(\mu+\mu')+\mu\mu']\,d\mu'$$

$$= 3-\mu^2+\tfrac{3}{16}\mu H(\mu)\int_0^1 (3-\mu'^2)H(\mu')\left[\frac{3-\mu^2}{\mu+\mu'}+(\mu-c)\right]d\mu'. \tag{18}$$

With the help of equations (12) and (14) we can reduce this equation to the form

$$\psi(\mu) = 3-\mu^2+(3-\mu^2)[H(\mu)-1]+(\mu-c)\mu H(\mu),$$

or
$$\psi(\mu) = H(\mu)(3-c\mu). \tag{19}$$

Next, considering $\psi+\phi$, we have

$$\psi(\mu)+\phi(\mu) = 3+\frac{9}{16}\int_0^1 S^{(0)}(\mu,\mu')\frac{d\mu'}{\mu'}$$

$$= 3+\tfrac{9}{16}\mu H(\mu)\int_0^1 \frac{H(\mu')}{\mu+\mu'}[3-c(\mu+\mu')+\mu\mu']\,d\mu'$$

$$= 3+\tfrac{9}{16}\mu H(\mu)\int_0^1 H(\mu')\Big[\frac{3-\mu^2}{\mu+\mu'}+(\mu-c)\Big]\,d\mu'$$

$$= 3+\tfrac{9}{16}\mu H(\mu)(\mu-c)\alpha_0+3[H(\mu)-1+\tfrac{3}{16}\mu H(\mu)(\alpha_1-\mu\alpha_0)]$$

$$= 3H(\mu)+\tfrac{9}{16}\mu H(\mu)(\alpha_1-c\alpha_0), \tag{20}$$

where we have made use of equation (16) in the reductions.

From equations (19) and (20) we obtain

$$\phi(\mu) = q\mu H(\mu) \tag{21}$$

where (cf. eq. [14])

$$q = \tfrac{1}{16}[9\alpha_1-c(9\alpha_0-16)] = \tfrac{3}{16}(3\alpha_1-c\alpha_2). \tag{22}$$

Now substituting ψ and ϕ according to equations (19) and (21) in equation (16) we have

$$\Big(\frac{1}{\mu}+\frac{1}{\mu_0}\Big)S^{(0)}(\mu,\mu_0) = H(\mu)H(\mu_0)[\tfrac{1}{3}(3-c\mu)(3-c\mu_0)+\tfrac{8}{3}q^2\mu\mu_0]$$

$$= H(\mu)H(\mu_0)[3-c(\mu+\mu_0)+\tfrac{1}{3}(8q^2+c^2)\mu\mu_0]. \tag{23}$$

Comparing equations (11) and (23) we observe that we must require

$$8q^2+c^2 = 3. \tag{24}$$

This equation determines c; for, substituting for q according to equation (22), we have
$$\tfrac{1}{3}(3\alpha_1-c\alpha_2)^2+\tfrac{32}{27}c^2 = \tfrac{32}{9},$$

or
$$(\tfrac{1}{3}\alpha_2^2+\tfrac{32}{27})c^2-2\alpha_1\alpha_2 c+3(\alpha_1^2-\tfrac{32}{27}) = 0. \tag{25}$$

Using relation (17), equation (25) reduces to

$$\alpha_1^2 c^2-2\alpha_1\alpha_2 c+\alpha_2^2 = 0. \tag{26}$$

Hence
$$c = \frac{\alpha_2}{\alpha_1}. \tag{27}$$

The constant q now becomes (cf. eqs. [17] and [22])

$$q = \frac{3}{16\alpha_1}(3\alpha_1^2 - \alpha_2^2) = \frac{2}{3\alpha_1}. \tag{28}$$

Thus we have shown that, with q and c given by equations (27) and (28), equations (19) and (21) represent the solutions of the integral equations, Chapter IV, equations (79) and (80).

Finally, combining equations (5), (7), and (11), we can express the law of diffuse reflection in the form

$$I(0, \mu, \varphi) = \tfrac{3}{32}F[H(\mu)H(\mu_0)\{3 - c(\mu + \mu_0) + \mu\mu_0\} -$$

$$- 4\mu\mu_0(1 - \mu^2)^{\frac{1}{2}}(1 - \mu_0^2)^{\frac{1}{2}}H^{(1)}(\mu)H^{(1)}(\mu_0)\cos(\varphi_0 - \varphi) +$$

$$+ (1 - \mu^2)(1 - \mu_0^2)H^{(2)}(\mu)H^{(2)}(\mu_0)\cos 2(\varphi_0 - \varphi)]\frac{\mu_0}{\mu + \mu_0}, \tag{29}$$

where $H(\mu)$, $H^{(1)}(\mu)$, and $H^{(2)}(\mu)$ are defined in terms of the characteristic functions

$$\tfrac{3}{16}(3 - \mu^2), \qquad \tfrac{3}{8}\mu^2(1 - \mu^2) \quad \text{and} \quad \tfrac{3}{32}(1 - \mu^2)^2, \tag{30}$$

respectively, and c is a constant which is related to the moments of H by

$$c = \frac{\alpha_2}{\alpha_1}. \tag{31}$$

The H-functions occurring in the solution (29) have been evaluated numerically by Mrs. Frances H. Breen and the writer by the method described in Chapter V, § 41. The solutions are given in Table XIII. The various related constants are listed in Table XIV.

TABLE XIII

The Functions $H(\mu)$, $H^{(1)}(\mu)$, and $H^{(2)}(\mu)$, obtained as Solutions of the Exact Integral Equations

μ	$H(\mu)$	$H^{(1)}(\mu)$	$H^{(2)}(\mu)$	μ	$H(\mu)$	$H^{(1)}(\mu)$	$H^{(2)}(\mu)$
0 .	1·00000	1·00000	1·00000	0·55.	2·17098	1·02539	1·03586
0·05.	1·14691	1·00430	1·01145	0·60.	2·26650	1·02652	1·03679
0·10.	1·26470	1·00786	1·01724	0·65.	2·36162	1·02757	1·03761
0·15.	1·37457	1·01089	1·02134	0·70.	2·45639	1·02854	1·03836
0·20.	1·48009	1·01352	1·02448	0·75.	2·55085	1·02944	1·03904
0·25.	1·58281	1·01582	1·02700	0·80.	2·64503	1·03028	1·03966
0·30.	1·68355	1·01785	1·02909	0·85.	2·73899	1·03106	1·04024
0·35.	1·78287	1·01968	1·03085	0·90.	2·83274	1·03179	1·04076
0·40.	1·88105	1·02132	1·03236	0·95.	2·92631	1·03247	1·04125
0·45.	1·97836	1·02280	1·03368	1·00.	3·01973	1·03312	1·04170
0·50.	2·07496	1·02415	1·03483				

TABLE XIV

The Constants derived from the Exact Function, $H(\mu)$

$$\alpha_0 = 2\cdot06088 \qquad\qquad q = 0\cdot55835$$
$$\alpha_1 = 1\cdot19400 \qquad\qquad c = 0\cdot71139$$
$$\alpha_2 = 0\cdot84940$$

The discussion and the illustration of the solution (29) is postponed to § 47.

45. The law of darkening for the problem with a constant net flux and for Rayleigh's phase function

Since scattering according to Rayleigh's phase function is conservative, the problem with a constant net flux has a meaning (cf. Chap. I, § 11). The appropriate equation of transfer has, in fact, been formulated in Chapter I, § 11.2. However, to obtain the angular distribution of the emergent radiation we need not go back to the equation of transfer: we can obtain the required distribution more directly from the law of diffuse reflection derived in the preceding section and the principles of invariance. Thus, from the equation (Chap. IV, eqs. [12] and [17])

$$I(0,\mu) = \tfrac{3}{4}F\left[\mu + \frac{1}{4\pi\mu}\int_0^1\int_0^{2\pi} S(\mu,\varphi;\mu',\varphi')\mu'\,d\mu'd\varphi'\right] \tag{32}$$

and (eqs. [5] and [11])

$$\frac{1}{2\pi}\int_0^{2\pi} S(\mu,\varphi;\mu',\varphi')\,d\varphi' = \frac{3}{8}\frac{\mu\mu'}{\mu+\mu'}H(\mu)H(\mu')[3-c(\mu+\mu')+\mu\mu'], \tag{33}$$

we now obtain

$$I(0,\mu) = \tfrac{3}{4}F\left\{\mu + \tfrac{3}{16}H(\mu)\int_0^1 \frac{\mu'^2H(\mu')}{\mu+\mu'}[3-c(\mu+\mu')+\mu\mu']\,d\mu'\right\}$$

$$= \tfrac{3}{4}F\left\{\mu + \tfrac{3}{16}H(\mu)\int_0^1 \mu'^2H(\mu')\left[\frac{3-\mu'^2}{\mu+\mu'}+\mu'-c\right]d\mu'\right\}. \tag{34}$$

On the other hand, according to Chapter V, equations (21) and (22)

$$\tfrac{3}{16}\int_0^1 H(\mu)(3-\mu^2)\mu\,d\mu = \tfrac{3}{16}(3\alpha_1-\alpha_3)$$

$$= \left[\tfrac{3}{8}\int_0^1 (3-\mu^2)\mu^2\,d\mu\right]^{\frac{1}{2}} = \sqrt{0\cdot3} \tag{35}$$

and

$$\mu + \tfrac{3}{16}H(\mu)\int_0^1 \frac{\mu'^2(3-\mu'^2)}{\mu+\mu'}H(\mu')\,d\mu' = H(\mu)\sqrt{0\cdot3}. \tag{36}$$

Using this last relation in equation (34), we have

$$I(0,\mu) = \tfrac{3}{4}F[H(\mu)\sqrt{0{\cdot}3}+\tfrac{3}{16}H(\mu)(\alpha_3-c\alpha_2)], \tag{37}$$

and substituting for α_3 from equation (35), we obtain

$$I(0,\mu) = \tfrac{9}{64}F(3\alpha_1-c\alpha_2)H(\mu). \tag{38}$$

Since $c = \alpha_2/\alpha_1$, equation (38) becomes

$$I(0,\mu) = \frac{9}{64\alpha_1}F(3\alpha_1^2-\alpha_2^2)H(\mu), \tag{39}$$

or (cf. eq. [17]) $\qquad I(0,\mu) = \dfrac{1}{2\alpha_1}FH(\mu), \tag{40}$

which expresses the required law of darkening.

It should be noticed that equation (40) is consistent with itself; for

$$F = 2\int_0^1 I(0,\mu)\mu\, d\mu = \frac{1}{\alpha_1}F\int_0^1 H(\mu)\mu\, d\mu = F. \tag{41}$$

Again, since $I(0,\mu)$ is proportional to $H(\mu)$, we can combine the integrals (14) and (35) to give

$$\frac{\displaystyle\int_0^1 I(0,\mu)(3-\mu^2)\, d\mu}{\displaystyle\int_0^1 I(0,\mu)(3-\mu^2)\mu\, d\mu} = \frac{1}{\sqrt{0{\cdot}3}}. \tag{42}$$

This is the analogue of the Hopf–Bronstein relation for the problem on hand.

Using the H-functions tabulated in § 44 and Chapter V, § 42 (Table XI, case $\varpi_0 = 1$), we can now compare the exact law of darkening as expressed by equation (40) and the corresponding law (Chap. IV, eq. [52] and Chap. V, eq. [110]) for the case of isotropic scattering. This comparison is made in Table XV. On examining this table, we observe that the solution for Rayleigh's phase function shows a greater darkening towards the limb ($\mu = 0$) than the solution for isotropic scattering; also, on Rayleigh's phase function, the intensity at the centre ($\mu = 1$) is greater and the intensity at the limb ($\mu = 0$) is less than in the case of isotropic scattering for the same net flux. These differences in the two cases are clearly due to the forward-throwing character of Rayleigh's phase function.

TABLE XV

Comparison of the Laws of Darkening for an Atmosphere with a Constant Net Flux for the Cases, Isotropic Scattering and Rayleigh's Phase Function

	$I(0,\mu)/F$		$I(0,\mu)/I(0,1)$	
μ	Isotropic scattering	Rayleigh's phase function	Isotropic scattering	Rayleigh's phase function
0 . .	0·43301	0·41876	0·34390	0·33116
0·05 . .	0·49223	0·48028	0·39094	0·37981
0·10 . .	0·54013	0·52961	0·42897	0·41881
0·15 . .	0·58493	0·57562	0·46455	0·45520
0·20 . .	0·62802	0·61981	0·49878	0·49014
0·25 . .	0·67001	0·66282	0·53213	0·52416
0·30 . .	0·71123	0·70501	0·56486	0·55752
0·35 . .	0·75188	0·74660	0·59715	0·59041
0·40 . .	0·79210	0·78771	0·62909	0·62292
0·45 . .	0·83197	0·82846	0·66075	0·65514
0·50 . .	0·87156	0·86891	0·69220	0·68791
0·55 . .	0·91092	0·90912	0·72346	0·71893
0·60 . .	0·95009	0·94912	0·75456	0·75056
0·65 . .	0·98909	0·98896	0·78554	0·78206
0·70 . .	1·02796	1·02864	0·81641	0·81345
0·75 . .	1·06671	1·06820	0·84719	0·84473
0·80 . .	1·10535	1·10764	0·87788	0·87592
0·85 . .	1·14390	1·14698	0·90850	0·90703
0·90 . .	1·18238	1·18624	0·93905	0·93808
0·95 . .	1·22078	1·22543	0·96955	0·96906
1·00 . .	1·25912	1·26455	1·00000	1·00000

46. The law of diffuse reflection for scattering in accordance with the phase function $\varpi_0(1+x\cos\Theta)$

As we have already shown in Chapter IV, § 34, the law of diffuse reflection for the case of scattering according to the phase function $\varpi_0(1+x\cos\Theta)$ can be expressed in the form

$$I(0,\mu,\varphi) = \frac{\varpi_0}{4\mu} F[S^{(0)}(\mu,\mu_0)+x(1-\mu^2)^{\frac{1}{2}}(1-\mu_0^2)^{\frac{1}{2}}S^{(1)}(\mu,\mu_0)\cos(\varphi_0-\varphi)], \quad (43)$$

where

$$\left(\frac{1}{\mu}+\frac{1}{\mu_0}\right)S^{(0)}(\mu,\mu_0) = \psi(\mu)\psi(\mu_0)-x\,\phi(\mu)\phi(\mu_0), \quad (44)$$

$$\left(\frac{1}{\mu}+\frac{1}{\mu_0}\right)S^{(1)}(\mu,\mu_0) = H^{(1)}(\mu)H^{(1)}(\mu_0), \quad (45)$$

$$\psi(\mu) = 1+\tfrac{1}{2}\varpi_0\int_0^1 S^{(0)}(\mu,\mu')\frac{d\mu'}{\mu'}, \quad (46)$$

$$\phi(\mu) = \mu-\tfrac{1}{2}\varpi_0\int_0^1 S^{(0)}(\mu,\mu')\,d\mu', \quad (47)$$

and $H^{(1)}(\mu)$ is an H-function defined in terms of the characteristic function

$$\Psi'^{(1)}(\mu) = \tfrac{1}{4}x\varpi_0(1-\mu^2). \tag{48}$$

46.1. *The form of the solution for $S^{(0)}(\mu, \mu_0)$*

The elimination of $S^{(0)}$ from equations (46) and (47) with the aid of equation (44) leads to a pair of integral equations for ψ and ϕ which are equations (61) and (62) of Chapter IV. As in § 44, in solving this system of equations we shall be guided by the form of the solution obtained in the direct solution of the equation of transfer in the method of approximation of Chapter III. In the present instance the form for $S^{(0)}(\mu, \mu_0)$ suggested is†

$$\left(\frac{1}{\mu}+\frac{1}{\mu_0}\right)S^{(0)}(\mu,\mu_0) = H(\mu)H(\mu_0)[1-c(\mu+\mu_0)-x(1-\varpi_0)\mu\mu_0], \tag{49}$$

where c is a constant and $H(\mu)$ is the solution (bounded in the half-plane $R(z) > 0$) of the equation

$$H(\mu) = 1+\tfrac{1}{2}\varpi_0\mu H(\mu)\int_0^1 \frac{1+x(1-\varpi_0)\mu'^2}{\mu+\mu'}H(\mu')\,d\mu'. \tag{50}$$

46.2. *The verification of the solution and the expression of the constant c in terms of the moments of $H(\mu)$*

As in § 44.2, the verification that $S^{(0)}$ has the form (49) will consist in first evaluating ψ and ϕ according to equations (46) and (47); then requiring that when the resulting expressions for ψ and ϕ are substituted back into equation (44), we shall recover the form of the solution assumed; and finally showing that the various requirements can be met.

The evaluation of ψ and ϕ according to equations (46), (47), and (49) is straightforward if proper use is made of the integral properties of the H-functions. Since $H(\mu)$ is now defined in terms of the characteristic function

$$\Psi'(\mu) = \tfrac{1}{2}\varpi_0[1+x(1-\varpi_0)\mu^2], \tag{51}$$

we have (Chap. V, Th. 3, eqs. [28] and [29])

$$\alpha_0 = 1+\tfrac{1}{4}\varpi_0[\alpha_0^2+x(1-\varpi_0)\alpha_1^2], \tag{52}$$

and

$$[1+x(1-\varpi_0)\mu^2]\int_0^1 \frac{H(\mu')}{\mu+\mu'}\,d\mu' = \frac{H(\mu)-1}{\tfrac{1}{2}\varpi_0\mu H(\mu)} - x(1-\varpi_0)(\alpha_1-\mu\alpha_0). \tag{53}$$

† Cf. S. Chandrasekhar, *Astrophys. J.* **103**, 165, 1946 (eq. [108]).

Considering first ψ, we have

$$\psi(\mu) = 1 + \tfrac{1}{2}\varpi_0\mu H(\mu)\int_0^1 \frac{H(\mu')}{\mu+\mu'}[1-c(\mu+\mu')-x(1-\varpi_0)\mu\mu']\,d\mu'$$

$$= 1 + \tfrac{1}{2}\varpi_0\mu H(\mu)\int_0^1 H(\mu')\left[\frac{1+x(1-\varpi_0)\mu^2}{\mu+\mu'}-c-x(1-\varpi_0)\mu\right]d\mu'$$

$$= 1 + [H(\mu)-1-\tfrac{1}{2}x\varpi_0(1-\varpi_0)(\alpha_1-\mu\alpha_0)\mu H(\mu)] - $$
$$- \tfrac{1}{2}\varpi_0\mu H(\mu)[c+x(1-\varpi_0)\mu]\alpha_0$$

$$= H(\mu) - \tfrac{1}{2}\varpi_0\mu H(\mu)[c\alpha_0+x(1-\varpi_0)\alpha_1], \tag{54}$$

where we have made use of equation (53) in the reductions. Thus, $\psi(\mu)$ has the form

$$\psi(\mu) = H(\mu)(1-p\mu), \tag{55}$$

where

$$p = \tfrac{1}{2}\varpi_0[c\alpha_0+x(1-\varpi_0)\alpha_1]. \tag{56}$$

Turning next to $\phi(\mu)$, we have

$$\phi(\mu) = \mu - \tfrac{1}{2}\varpi_0\mu H(\mu)\int_0^1 \frac{\mu'H(\mu')}{\mu+\mu'}[1-c(\mu+\mu')-x(1-\varpi_0)\mu\mu']\,d\mu'$$

$$= \mu - \tfrac{1}{2}\varpi_0\mu H(\mu)\int_0^1 H(\mu')\left\{[1+x(1-\varpi_0)\mu^2]\left(1-\frac{\mu}{\mu+\mu'}\right)-\right.$$
$$\left. - \mu'[c+x(1-\varpi_0)\mu]\right\}d\mu'$$

$$= \mu + \tfrac{1}{2}\varpi_0\mu H(\mu)\{[c+x(1-\varpi_0)\mu]\alpha_1-[1+x(1-\varpi_0)\mu^2]\alpha_0\}+$$
$$+ \mu\{H(\mu)-1-\tfrac{1}{2}x\varpi_0(1-\varpi_0)(\alpha_1-\mu\alpha_0)\mu H(\mu)\}$$

$$= \mu H(\mu)(\tfrac{1}{2}\varpi_0 c\alpha_1-\tfrac{1}{2}\varpi_0\alpha_0+1). \tag{57}$$

Thus $\phi(\mu)$ is of the form

$$\phi(\mu) = q\mu H(\mu), \tag{58}$$

where

$$q = \tfrac{1}{2}[\varpi_0 c\alpha_1+(2-\varpi_0\alpha_0)]. \tag{59}$$

Now substituting for ψ and ϕ according to equations (55) and (58) in equation (44), we obtain

$$\left(\frac{1}{\mu}+\frac{1}{\mu_0}\right)S^{(0)}(\mu,\mu_0) = H(\mu)H(\mu_0)[1-p(\mu+\mu_0)-(xq^2-p^2)\mu\mu_0]. \tag{60}$$

Comparing equations (49) and (60), we observe that we must have

$$p = c = \tfrac{1}{2}\varpi_0[c\alpha_0+x(1-\varpi_0)\alpha_1] \tag{61}$$

and

$$xq^2-p^2 = xq^2-c^2 = x(1-\varpi_0). \tag{62}$$

From equation (61) we find that

$$c = x\varpi_0(1-\varpi_0)\frac{\alpha_1}{2-\varpi_0\alpha_0}. \tag{63}$$

With this value of c the constant q, given by equation (59), becomes

$$q = \frac{1}{2(2-\varpi_0\alpha_0)}[x\varpi_0^2(1-\varpi_0)\alpha_1^2+(2-\varpi_0\alpha_0)^2]$$

$$= \frac{1}{2(2-\varpi_0\alpha_0)}[4-4\varpi_0\alpha_0+\varpi_0^2\{\alpha_0^2+x(1-\varpi_0)\alpha_1^2\}]. \tag{64}$$

This last equation can be simplified by using equation (52). We find

$$q = \frac{1}{2(2-\varpi_0\alpha_0)}[4-4\varpi_0\alpha_0+4\varpi_0(\alpha_0-1)],$$

or

$$q = \frac{2(1-\varpi_0)}{2-\varpi_0\alpha_0}. \tag{65}$$

It remains to verify that q and c, as given by equations (63) and (65), satisfy the relation (62). To show that this is the case, we shall evaluate xq^2-c^2 according to equations (63) and (65). We have

$$xq^2-c^2 = \frac{x}{(2-\varpi_0\alpha_0)^2}[4(1-\varpi_0)^2-x\varpi_0^2(1-\varpi_0)^2\alpha_1^2]$$

$$= \frac{x(1-\varpi_0)}{(2-\varpi_0\alpha_0)^2}[4(1-\varpi_0)-x\varpi_0^2(1-\varpi_0)\alpha_1^2]. \tag{66}$$

But, according to equation (52),

$$-x\varpi_0^2(1-\varpi_0)\alpha_1^2 = 4\varpi_0(1-\alpha_0+\tfrac{1}{4}\varpi_0\alpha_0^2). \tag{67}$$

Using this relation in equation (66), we find

$$xq^2-c^2 = \frac{x(1-\varpi_0)}{(2-\varpi_0\alpha_0)^2}(4-4\varpi_0\alpha_0+\varpi_0^2\alpha_0^2) = x(1-\varpi_0), \tag{68}$$

which shows that q and c are related as required. Hence, with q and c given by equations (63) and (65),

$$\psi(\mu) = H(\mu)(1-c\mu) \quad \text{and} \quad \phi(\mu) = q\mu H(\mu), \tag{69}$$

represent the solution of the integral equations, Chapter IV, equations (61) and (62).

Finally, combining equations (43), (45), and (49), we can express the law of diffuse reflection in the form

$$I(0,\mu,\varphi) = \tfrac{1}{4}\varpi_0 F\Big\{H(\mu)H(\mu_0)[1-c(\mu+\mu_0)-x(1-\varpi_0)\mu\mu_0]+$$

$$+x(1-\mu^2)^{\frac{1}{2}}(1-\mu_0^2)^{\frac{1}{2}}H^{(1)}(\mu)H^{(1)}(\mu_0)\cos(\varphi_0-\varphi)\Big\}\frac{\mu_0}{\mu+\mu_0}, \tag{70}$$

TABLE XVI

The Functions $H(\mu)$ obtained as Solutions of the Exact Integral Equations they satisfy (the Case $x = 1$)

μ	$\varpi_0 = 0\cdot1$	$\varpi_0 = 0\cdot2$	$\varpi_0 = 0\cdot3$	$\varpi_0 = 0\cdot4$	$\varpi_0 = 0\cdot5$	$\varpi_0 = 0\cdot6$
0 . .	1·0000	1·0000	1·0000	1·0000	1·0000	1·0000
0·05 . .	1·0089	1·0183	1·0280	1·0383	1·0492	1·0608
0·10 . .	1·0145	1·0297	1·0459	1·0632	1·0817	1·1020
0·15 . .	1·0188	1·0388	1·0602	1·0832	1·1084	1·1361
0·20 . .	1·0224	1·0463	1·0722	1·1003	1·1311	1·1656
0·25 . .	1·0254	1·0528	1·0825	1·1151	1·1511	1·1918
0·30 . .	1·0280	1·0584	1·0916	1·1281	1·1689	1·2153
0·35 . .	1·0303	1·0634	1·0996	1·1398	1·1850	1·2366
0·40 . .	1·0324	1·0678	1·1069	1·1504	1·1996	1·2562
0·45 . .	1·0343	1·0719	1·1135	1·1600	1·2129	1·2742
0·50 . .	1·0359	1·0755	1·1194	1·1688	1·2252	1·2908
0·55 . .	1·0375	1·0788	1·1249	1·1769	1·2365	1·3063
0·60 . .	1·0389	1·0819	1·1300	1·1844	1·2470	1·3207
0·65 . .	1·0401	1·0847	1·1346	1·1913	1·2568	1·3342
0·70 . .	1·0413	1·0873	1·1389	1·1978	1·2659	1·3468
0·75 . .	1·0424	1·0897	1·1429	1·2038	1·2745	1·3587
0·80 . .	1·0434	1·0919	1·1467	1·2094	1·2825	1·3699
0·85 . .	1·0444	1·0940	1·1502	1·2147	1·2900	1·3805
0·90 . .	1·0453	1·0960	1·1535	1·2196	1·2972	1·3905
0·95 . .	1·0461	1·0978	1·1566	1·2243	1·3039	1·4000
1·00 . .	1·0469	1·0995	1·1595	1·2287	1·3103	1·4090

TABLE XVI (continued)

μ	$\varpi_0 = 0\cdot7$	$\varpi_0 = 0\cdot8$	$\varpi_0 = 0\cdot9$	$\varpi_0 = 0\cdot925$	$\varpi_0 = 0\cdot95$	$\varpi_0 = 0\cdot975$
0 . .	1·0000	1·0000	1·0000	1·0000	1·0000	1·0000
0·05 . . .	1·0735	1·0876	1·1045	1·1096	1·1153	1·1223
0·10 . .	1·1244	1·1501	1·1819	1·1917	1·2029	1·2169
0·15 . . .	1·1673	1·2038	1·2500	1·2645	1·2814	1·3027
0·20 . . .	1·2049	1·2516	1·3120	1·3313	1·3539	1·3830
0·25 . . .	1·2387	1·2951	1·3695	1·3936	1·4222	1·4593
0·30 . . .	1·2693	1·3351	1·4233	1·4523	1·4869	1·5323
0·35 . . .	1·2973	1·3722	1·4740	1·5079	1·5487	1·6026
0·40 . . .	1·3233	1·4068	1·5220	1·5608	1·6078	1·6706
0·45 . . .	1·3473	1·4392	1·5677	1·6114	1·6647	1·7365
0·50 . . .	1·3697	1·4697	1·6112	1·6598	1·7195	1·8005
0·55 . . .	1·3907	1·4985	1·6528	1·7063	1·7724	1·8627
0·60 . . .	1·4103	1·5257	1·6926	1·7510	1·8235	1·9234
0·65 . . .	1·4288	1·5515	1·7308	1·7941	1·8730	1·9826
0·70 . . .	1·4462	1·5760	1·7674	1·8356	1·9210	2·0403
0·75 . . .	1·4627	1·5993	1·8027	1·8757	1·9675	2·0967
0·80 . . .	1·4783	1·6216	1·8367	1·9144	2·0127	2·1519
0·85 . . .	1·4931	1·6428	1·8694	1·9519	2·0567	2·2058
0·90 . . .	1·5072	1·6631	1·9010	1·9882	2·0994	2·2586
0·95 . . .	1·5206	1·6825	1·9315	2·0233	2·1409	2·3103
1·00 . . .	1·5334	1·7011	1·9610	2·0574	2·1814	2·3609

where $H(\mu)$ and $H^{(1)}(\mu)$ are defined in terms of the characteristic functions

$$\tfrac{1}{2}\varpi_0[1+x(1-\varpi_0)\mu^2] \quad \text{and} \quad \tfrac{1}{4}x\varpi_0(1-\mu^2) \tag{71}$$

and

$$c = x(1-\varpi_0)\varpi_0 \frac{\alpha_1}{2-\varpi_0\alpha_0}, \tag{72}$$

α_0 and α_1 being the moments of order zero and one of $H(\mu)$.

We notice that when

$$\varpi_0 = 1, \quad c = 0, \quad \Psi(\mu) = \tfrac{1}{2} \quad \text{for } all \ x, \tag{73}$$

and the azimuth independent term becomes identical with the law of diffuse reflection of a perfectly, isotropically, scattering atmosphere.

For the case $x = 1$, the H-functions which occur in the solution (70) have been computed for various values of ϖ_0 by Mrs. Frances H. Breen and the writer; the functions $H^{(1)}$ have also been computed for various values of $x\varpi_0$. The results of these calculations are summarized in Tables XVI, XVII, and XVIII. The cases $(1+x\cos\Theta)$ and $\varpi_0(1+\cos\Theta)$ are covered by these tabulations.

47. Illustration and comparison of the laws of diffuse reflection for the cases (i) isotropic scattering, (ii) Rayleigh's phase function, and (iii) the phase function $\varpi_0(1+x\cos\Theta)$

The laws of diffuse reflection for the three cases for which we have derived exact expressions are:

I. *Isotropic scattering*:

$$I(0,\mu,\varphi;\mu_0,\varphi_0) \equiv I(0,\mu,\mu_0) = \frac{1}{4}\frac{\varpi_0}{\mu+\mu_0} H(\mu)H(\mu_0)\mu_0 F,$$

where $H(\mu)$ is defined in terms of the characteristic function

$$\Psi(\mu) = \text{constant} = \tfrac{1}{2}\varpi_0.$$

The H-functions for this problem have been tabulated in Chapter V, Table XI.

II. *Scattering in accordance with the phase function* $\varpi_0(1+x\cos\Theta)$:

$$I(0,\mu,\varphi;\mu_0,\varphi_0) = \frac{1}{4}\frac{\varpi_0}{\mu+\mu_0} \{\psi(\mu)\psi(\mu_0)-x\,\phi(\mu)\phi(\mu_0)+$$

$$+x[(1-\mu^2)^{\tfrac{1}{2}}H^{(1)}(\mu)][(1-\mu_0^2)^{\tfrac{1}{2}}H^{(1)}(\mu_0)]\cos(\varphi_0-\varphi)\}\mu_0 F,$$

where $\quad \psi(\mu) = H(\mu)(1-c\mu) \quad \text{and} \quad \phi(\mu) = q\mu H(\mu)$

TABLE XVII

The Moments α_0 and α_1 and the Constants q and c

ϖ_0	α_0	α_1	q	c
0·1 . .	1·032729	0·519588	0·949003	0·024654
0·2 . .	1·068832	0·541348	0·895740	0·048491
0·3 . .	1·109034	0·565767	0·839686	0·071260
0·4 . .	1·154378	0·593541	0·780108	0·092605
0·5 . .	1·206366	0·625686	0·715914	0·111984
0·6 . .	1·267352	0·663798	0·645375	0·128520
0·7 . .	1·341368	0·710639	0·565482	0·140649
0·8 . .	1·436535	0·771792	0·470161	0·145147
0·9 . .	1·574492	0·862276	0·343079	0·133123
0·925 .	1·623024	0·894620	0·300780	0·124451
0·950 .	1·683484	0·935277	0·249569	0·110873
0·975 .	1·767379	0·992380	0·180632	0·087387

TABLE XVIII

The Functions $H^{(1)}(\mu)$ obtained as Solutions of the Exact Integral Equations they satisfy

μ	$x\varpi_0 = 1\cdot0$	$x\varpi_0 = 0\cdot8$	$x\varpi_0 = 0\cdot6$	$x\varpi_0 = 0\cdot4$	$x\varpi_0 = 0\cdot2$	$x\varpi_0 = -0\cdot2$	$x\varpi_0 = -0\cdot4$	$x\varpi_0 = -0\cdot6$	$x\varpi_0 = -0\cdot8$	$x\varpi_0 = -1\cdot0$
0 .	1·0000	1·0000	1·0000	1·0000	1·0000	1·0000	1·0000	1·0000	1·0000	1·0000
0·05	1·0359	1·0281	1·0206	1·0135	1·0066	0·9936	0·9875	0·9815	0·9758	0·9702
0·10	1·0561	1·0436	1·0319	1·0207	1·0101	0·9903	0·9811	0·9722	0·9637	0·9555
0·15	1·0711	1·0550	1·0400	1·0259	1·0126	0·9880	0·9766	0·9657	0·9553	0·9453
0·20	1·0832	1·0642	1·0466	1·0301	1·0146	0·9862	0·9731	0·9607	0·9488	0·9375
0·25	1·0933	1·0718	1·0520	1·0335	1·0162	0·9847	0·9703	0·9566	0·9436	0·9313
0·30	1·1019	1·0783	1·0565	1·0364	1·0176	0·9835	0·9679	0·9533	0·9393	0·9261
0·35	1·1093	1·0838	1·0604	1·0388	1·0187	0·9825	0·9660	0·9504	0·9358	0·9219
0·40	1·1157	1·0886	1·0638	1·0409	1·0197	0·9816	0·9643	0·9480	0·9327	0·9182
0·45	1·1214	1·0928	1·0667	1·0428	1·0206	0·9808	0·9628	0·9459	0·9300	0·9150
0·50	1·1265	1·0966	1·0694	1·0444	1·0214	0·9801	0·9615	0·9441	0·9277	0·9122
0·55	1·1311	1·1000	1·0718	1·0459	1·0221	0·9795	0·9604	0·9424	0·9256	0·9097
0·60	1·1353	1·1031	1·0739	1·0472	1·0227	0·9789	0·9593	0·9409	0·9237	0·9075
0·65	1·1391	1·1059	1·0759	1·0484	1·0232	0·9784	0·9584	0·9396	0·9220	0·9055
0·70	1·1426	1·1085	1·0776	1·0495	1·0238	0·9780	0·9575	0·9384	0·9205	0·9037
0·75	1·1458	1·1108	1·0792	1·0505	1·0242	0·9776	0·9568	0·9373	0·9191	0·9020
0·80	1·1487	1·1130	1·0807	1·0514	1·0246	0·9772	0·9560	0·9363	0·9179	0·9005
0·85	1·1514	1·1150	1·0821	1·0523	1·0250	0·9769	0·9554	0·9354	0·9167	0·8992
0·90	1·1540	1·1168	1·0834	1·0531	1·0254	0·9765	0·9548	0·9345	0·9156	0·8979
0·95	1·1563	1·1185	1·0846	1·0538	1·0258	0·9762	0·9542	0·9338	0·9146	0·8967
1·00	1·1585	1·1201	1·0857	1·0545	1·0261	0·9760	0·9537	0·9330	0·9137	0·8956

and $H(\mu)$ and $H^{(1)}(\mu)$ are defined in terms of the characteristic functions

$$\tfrac{1}{2}\varpi_0[1+x(1-\varpi_0)\mu^2] \quad \text{and} \quad \tfrac{1}{4}x\varpi_0(1-\mu^2),$$

respectively. The constants q and c are related to the moments of $H(\mu)$ by

$$q = \frac{2(1-\varpi_0)}{2-\varpi_0\alpha_0} \quad \text{and} \quad c = x\varpi_0(1-\varpi_0)\frac{\alpha_1}{2-\varpi_0\alpha_0}.$$

For the case $x = 1$, the functions $\psi(\mu)$ and $\phi(\mu)$ are given for various values of ϖ_0 in Table XIX. The function $(1-\mu^2)^{\frac{1}{2}} H^{(1)}(\mu)$ for certain values of $x\varpi_0$ is tabulated in Table XX.

TABLE XIX

The Functions $\psi(\mu)$ and $\phi(\mu)$ for the Case $x = 1$ and Various Values of ϖ_0

μ	$\varpi_0 = 0.1$		$\varpi_0 = 0.2$		$\varpi_0 = 0.3$		$\varpi_0 = 0.4$	
	ψ	ϕ	ψ	ϕ	ψ	ϕ	ψ	ϕ
0	1·0000	0	1·0000	0	1·0000	0	1·0000	0
0·05	1·0077	0·0479	1·0158	0·0456	1·0244	0·0432	1·0335	0·0405
0·10	1·0120	0·0963	1·0247	0·0922	1·0385	0·0878	1·0533	0·0829
0·15	1·0150	0·1450	1·0312	0·1396	1·0488	0·1335	1·0682	0·1268
0·20	1·0173	0·1940	1·0362	0·1874	1·0569	0·1801	1·0799	0·1717
0·25	1·0191	0·2433	1·0400	0·2358	1·0632	0·2272	1·0892	0·2175
0·30	1·0204	0·2927	1·0430	0·2844	1·0682	0·2750	1·0968	0·2640
0·35	1·0214	0·3422	1·0453	0·3334	1·0722	0·3232	1·1029	0·3112
0·40	1·0222	0·3919	1·0471	0·3826	1·0753	0·3718	1·1078	0·3590
0·45	1·0228	0·4417	1·0485	0·4320	1·0778	0·4207	1·1117	0·4072
0·50	1·0232	0·4916	1·0494	0·4817	1·0796	0·4700	1·1147	0·4559
0·55	1·0234	0·5415	1·0501	0·5315	1·0808	0·5195	1·1170	0·5050
0·60	1·0235	0·5915	1·0504	0·5814	1·0816	0·5693	1·1186	0·5544
0·65	1·0235	0·6416	1·0505	0·6315	1·0821	0·6193	1·1196	0·6041
0·70	1·0233	0·6917	1·0504	0·6817	1·0821	0·6694	1·1201	0·6541
0·75	1·0231	0·7419	1·0501	0·7321	1·0819	0·7198	1·1202	0·7043
0·80	1·0228	0·7922	1·0496	0·7825	1·0813	0·7703	1·1198	0·7548
0·85	1·0225	0·8424	1·0489	0·8330	1·0805	0·8209	1·1191	0·8054
0·90	1·0221	0·8928	1·0481	0·8835	1·0795	0·8717	1·1180	0·8563
0·95	1·0216	0·9431	1·0472	0·9342	1·0783	0·9226	1·1166	0·9073
1·00	1·0210	0·9935	1·0462	0·9849	1·0769	0·9736	1·1149	0·9585

TABLE XIX (continued)

μ	$\varpi_0 = 0.5$		$\varpi_0 = 0.6$		$\varpi_0 = 0.7$		$\varpi_0 = 0.8$	
	ψ	ϕ	ψ	ϕ	ψ	ϕ	ψ	ϕ
0	1·0000	0	1·0000	0	1·0000	0	1·0000	0
0·05	1·0433	0·0376	1·0540	0·0342	1·0659	0·0304	1·0797	0·0256
0·10	1·0696	0·0774	1·0878	0·0711	1·1086	0·0636	1·1334	0·0541
0·15	1·0897	0·1190	1·1142	0·1100	1·1427	0·0990	1·1776	0·0849
0·20	1·1058	0·1620	1·1357	0·1505	1·1710	0·1363	1·2153	0·1177
0·25	1·1189	0·2060	1·1535	0·1923	1·1951	0·1751	1·2481	0·1522
0·30	1·1297	0·2511	1·1684	0·2353	1·2157	0·2153	1·2770	0·1883
0·35	1·1385	0·2969	1·1810	0·2793	1·2335	0·2568	1·3025	0·2258
0·40	1·1458	0·3435	1·1916	0·3243	1·2488	0·2993	1·3251	0·2646
0·45	1·1518	0·3907	1·2005	0·3700	1·2620	0·3428	1·3452	0·3045
0·50	1·1566	0·4386	1·2078	0·4165	1·2734	0·3873	1·3631	0·3455
0·55	1·1603	0·4869	1·2139	0·4637	1·2831	0·4325	1·3789	0·3875
0·60	1·1632	0·5356	1·2188	0·5114	1·2913	0·4785	1·3928	0·4304
0·65	1·1653	0·5848	1·2227	0·5597	1·2982	0·5252	1·4051	0·4741
0·70	1·1667	0·6344	1·2256	0·6084	1·3039	0·5725	1·4159	0·5187
0·75	1·1674	0·6843	1·2277	0·6576	1·3084	0·6204	1·4252	0·5640
0·80	1·1676	0·7345	1·2290	0·7073	1·3120	0·6688	1·4333	0·6099
0·85	1·1673	0·7850	1·2297	0·7573	1·3146	0·7177	1·4401	0·6565
0·90	1·1664	0·8358	1·2296	0·8076	1·3164	0·7671	1·4458	0·7037
0·95	1·1652	0·8868	1·2290	0·8583	1·3174	0·8169	1·4505	0·7515
1·00	1·1635	0·9380	1·2279	0·9093	1·3177	0·8671	1·4542	0·7998

TABLE XIX (*continued*)

μ	$\varpi_0 = 0.9$		$\varpi_0 = 0.925$		$\varpi_0 = 0.95$		$\varpi_0 = 0.975$	
	ψ	ϕ	ψ	ϕ	ψ	ϕ	ψ	ϕ
0	1·0000	0	1·0000	0	1·0000	0	1·0000	0
0·05	1·0972	0·0189	1·1027	0·0167	1·1091	0·0139	1·1174	0·0101
0·10	1·1662	0·0405	1·1768	0·0358	1·1896	0·0300	1·2063	0·0220
0·15	1·2250	0·0643	1·2409	0·0570	1·2601	0·0480	1·2857	0·0353
0·20	1·2771	0·0900	1·2981	0·0801	1·3239	0·0676	1·3588	0·0500
0·25	1·3239	0·1175	1·3502	0·1048	1·3828	0·0887	1·4274	0·0659
0·30	1·3664	0·1465	1·3980	0·1310	1·4375	0·1113	1·4921	0·0830
0·35	1·4053	0·1770	1·4422	0·1587	1·4886	0·1353	1·5536	0·1013
0·40	1·4410	0·2089	1·4831	0·1878	1·5365	0·1605	1·6122	0·1207
0·45	1·4737	0·2420	1·5211	0·2181	1·5816	0·1870	1·6682	0·1411
0·50	1·5039	0·2764	1·5565	0·2496	1·6242	0·2146	1·7218	0·1626
0·55	1·5318	0·3119	1·5895	0·2823	1·6643	0·2433	1·7732	0·1851
0·60	1·5574	0·3484	1·6203	0·3160	1·7022	0·2731	1·8226	0·2085
0·65	1·5810	0·3860	1·6490	0·3508	1·7380	0·3038	1·8699	0·2328
0·70	1·6027	0·4245	1·6757	0·3865	1·7719	0·3356	1·9155	0·2580
0·75	1·6227	0·4639	1·7006	0·4231	1·8039	0·3683	1·9593	0·2841
0·80	1·6411	0·5041	1·7238	0·4607	1·8342	0·4019	2·0014	0·3110
0·85	1·6579	0·5452	1·7454	0·4990	1·8628	0·4363	2·0420	0·3387
0·90	1·6732	0·5870	1·7655	0·5382	1·8899	0·4715	2·0809	0·3672
0·95	1·6872	0·6295	1·7841	0·5781	1·9154	0·5076	2·1185	0·3964
1·00	1·6999	0·6728	1·8014	0·6188	1·9395	0·5444	2·1546	0·4265

III. *Scattering in accordance with Rayleigh's phase function*:

$$I(0,\mu,\varphi;\mu_0,\varphi_0) = \frac{3}{32(\mu+\mu_0)}\{\tfrac{1}{3}\psi(\mu)\psi(\mu_0)+\tfrac{8}{3}\phi(\mu)\phi(\mu_0)-$$

$$-[2\mu(1-\mu^2)^{\frac{1}{2}}H^{(1)}(\mu)][2\mu_0(1-\mu_0^2)^{\frac{1}{2}}H^{(1)}(\mu_0)]\cos(\varphi_0-\varphi)+$$

$$+[(1-\mu^2)H^{(2)}(\mu)][(1-\mu_0^2)H^{(2)}(\mu_0)]\cos 2(\varphi_0-\varphi)\}\mu_0\, F,$$

where $\psi(\mu) = H(\mu)(3-c\mu)$ and $\phi(\mu) = q\mu H(\mu)$

and $H(\mu)$, $H^{(1)}(\mu)$, and $H^{(2)}(\mu)$ are defined in terms of the characteristic functions

$$\tfrac{3}{16}(3-\mu^2), \quad \tfrac{3}{8}\mu^2(1-\mu^2), \quad \text{and} \quad \tfrac{3}{32}(1-\mu^2)^2,$$

respectively. The constants c and q are related to the moments of $H(\mu)$ by

$$c = \frac{\alpha_2}{\alpha_1} \quad \text{and} \quad q = \frac{2}{3\alpha_1}.$$

The functions $\psi(\mu)$, $\phi(\mu)$, $2\mu(1-\mu^2)^{\frac{1}{2}}H^{(1)}(\mu)$, and $(1-\mu^2)H^{(2)}(\mu)$ are given in Table XXI.

In Fig. 10 we have illustrated the laws of diffuse reflection for isotropic scattering and Rayleigh's phase function, for $\mu_0 = 0.8$ and for $\varphi_0-\varphi = 0°$ and $90°$. In Fig. 11 we have illustrated the cases, isotropic scattering and $(1\pm\cos\Theta)$ for the same angle of incidence ($\mu_0 = 0.8$) and for $\varphi_0-\varphi = 0$ and $90°$.

In Fig. 12 we have similarly compared the laws of diffuse reflection for the cases (i) $\varpi_0 = 0.8$ and $x = 0$, and (ii) $\varpi_0 = 0.8$ and $x = 1$ for $\mu_0 = 0.6$ and $\varphi_0-\varphi = 0°$ and $90°$.

TABLE XX

The Function $(1-\mu^2)^{\frac{1}{2}}H^{(1)}(\mu)$ *for Various Values of* $x\varpi_0$

μ	$x\varpi_0 = 1\cdot0$	$x\varpi_0 = 0\cdot8$	$x\varpi_0 = 0\cdot6$	$x\varpi_0 = 0\cdot4$	$x\varpi_0 = 0\cdot2$
0 . .	1·0000	1·0000	1·0000	1·0000	1·0000
0·05 . .	1·0346	1·0268	1·0194	1·0122	1·0053
0·10 . .	1·0508	1·0384	1·0267	1·0156	1·0050
0·15 . .	1·0590	1·0431	1·0283	1·0143	1·0012
0·20 . .	1·0613	1·0427	1·0254	1·0093	0·9941
0·25 . .	1·0586	1·0378	1·0186	1·0007	0·9840
0·30 . .	1·0511	1·0286	1·0079	0·9886	0·9707
0·35 . .	1·0391	1·0152	0·9933	0·9731	0·9543
0·40 . .	1·0225	0·9977	0·9750	0·9540	0·9346
0·45 . .	1·0015	0·9759	0·9526	0·9312	0·9114
0·50 . .	0·9756	0·9497	0·9261	0·9045	0·8845
0·55 . .	0·9447	0·9187	0·8951	0·8735	0·8536
0·60 . .	0·9082	0·8825	0·8591	0·8378	0·8181
0·65 . .	0·8656	0·8404	0·8176	0·7967	0·7776
0·70 . .	0·8160	0·7916	0·7696	0·7495	0·7311
0·75 . .	0·7578	0·7347	0·7139	0·6949	0·6775
0·80 . .	0·6892	0·6678	0·6484	0·6309	0·6148
0·85 . .	0·6065	0·5873	0·5700	0·5543	0·5400
0·90 . .	0·5030	0·4868	0·4722	0·4590	0·4470
0·95 . .	0·3611	0·3493	0·3387	0·3291	0·3203
1·00 . .	0	0	0	0	0

TABLE XX (*continued*)

μ	$x\varpi_0 = -0\cdot2$	$x\varpi_0 = -0\cdot4$	$x\varpi_0 = -0\cdot6$	$x\varpi_0 = -0\cdot8$	$x\varpi_0 = -1\cdot0$
0 . .	1·0000	1·0000	1·0000	1·0000	1·0000
0·05 . .	0·9924	0·9862	0·9803	0·9745	0·9689
0·10 . .	0·9854	0·9762	0·9674	0·9589	0·9507
0·15 . .	0·9768	0·9656	0·9548	0·9445	0·9346
0·20 . .	0·9663	0·9534	0·9413	0·9296	0·9186
0·25 . .	0·9534	0·9395	0·9262	0·9137	0·9017
0·30 . .	0·9382	0·9234	0·9093	0·8961	0·8835
0·35 . .	0·9203	0·9049	0·8903	0·8766	0·8635
0·40 . .	0·8996	0·8838	0·8689	0·8548	0·8415
0·45 . .	0·8759	0·8598	0·8447	0·8305	0·8171
0·50 . .	0·8488	0·8327	0·8176	0·8034	0·7900
0·55 . .	0·8180	0·8021	0·7871	0·7730	0·7597
0·60 . .	0·7832	0·7675	0·7528	0·7390	0·7260
0·65 . .	0·7436	0·7283	0·7140	0·7007	0·6881
0·70 . .	0·6984	0·6838	0·6702	0·6574	0·6454
0·75 . .	0·6466	0·6328	0·6200	0·6079	0·5966
0·80 . .	0·5863	0·5736	0·5618	0·5507	0·5403
0·85 . .	0·5146	0·5033	0·4927	0·4829	0·4737
0·90 . .	0·4257	0·4162	0·4074	0·3991	0·3914
0·95 . .	0·3048	0·2980	0·2916	0·2856	0·2800
1·00 . .	0	0	0	0	0

TABLE XXI

The Functions $\psi(\mu)$, $\phi(\mu)$, $2\mu(1-\mu^2)^{\frac{1}{2}}H^{(1)}(\mu)$, and $(1-\mu^2)H^{(2)}(\mu)$

μ				$\psi(\mu)$	$\phi(\mu)$	$2\mu(1-\mu^2)^{\frac{1}{2}}H^{(1)}(\mu)$	$(1-\mu^2)H^{(2)}(\mu)$
0	.	.	.	3·00000	0	0	1·00000
0·05	.	.	.	3·39993	0·03202	0·10030	1·00892
0·10	.	.	.	3·70413	0·07061	0·20056	1·00707
0·15	.	.	.	3·97703	0·11512	0·29984	0·99836
0·20	.	.	.	4·22969	0·16528	0·39722	0·98350
0·25	.	.	.	4·46693	0·22094	0·49178	0·96281
0·30	.	.	.	4·69135	0·28200	0·58258	0·93647
0·35	.	.	.	4·90470	0·34841	0·66863	0·90457
0·40	.	.	.	5·10789	0·42011	0·74884	0·86718
0·45	.	.	.	5·30176	0·49708	0·82205	0·82436
0·50	.	.	.	5·48683	0·57928	0·88694	0·77613
0·55	.	.	.	5·66351	0·66669	0·94201	0·72252
0·60	.	.	.	5·83208	0·75930	0·98546	0·66354
0·65	.	.	.	5·99284	0·85710	1·01515	0·59922
0·70	.	.	.	6·14595	0·96007	1·02833	0·52956
0·75	.	.	.	6·29156	1·06820	1·02136	0·45458
0·80	.	.	.	6·42977	1·18148	0·98906	0·37428
0·85	.	.	.	6·56075	1·29992	0·92334	0·28866
0·90	.	.	.	6·68456	1·42349	0·80954	0·19774
0·95	.	.	.	6·80127	1·55221	0·61254	0·10152
1·00	.	.	.	6·91099	1·68606	0	0

47.1. *The intensity of the light which has been scattered once*

In the earlier treatments of the problem of diffuse reflection it was customary to consider the emergent radiation as consisting of light which has been scattered once, twice, three times, etc., in the atmosphere. Because of the excessive complexity of the resulting expressions, such calculations have never been rigorously extended beyond the stage of secondary scattering (cf. Chap. IX, § 63). It is, however, a simple matter to obtain an expression for the light which has suffered a single scattering process in the atmosphere. Thus, considering the layer of the atmosphere between τ and $\tau + d\tau$, we observe that at this depth the fraction $e^{-\tau/\mu_0}$ of the incident flux appears without having suffered any scattering process in the intervening atmosphere. The contribution to the diffuse intensity in the direction (μ, φ) at the depth τ by this reduced incident flux is

$$\tfrac{1}{4}Fe^{-\tau/\mu_0}p(\mu,\varphi;-\mu_0,\varphi_0)\frac{d\tau}{\mu}. \tag{74}$$

A fraction $e^{-\tau/\mu}$ of this amount emerges in the direction (μ, φ) without suffering any further scattering. Consequently, the contribution to the emergent intensity by light which has suffered a single scattering in the layer between τ and $\tau + d\tau$ is

$$\tfrac{1}{4} F e^{-\tau/\mu} e^{-\tau/\mu_0} p(\mu, \varphi; -\mu_0, \varphi_0) \frac{d\tau}{\mu}. \tag{75}$$

Integrating this expression over τ, we obtain the expression

$$I^{(1)}(0, \mu, \varphi; \mu_0, \varphi_0) = \frac{1}{4} \frac{1}{\mu + \mu_0} p(\mu, \varphi; -\mu_0, \varphi_0) \mu_0 F, \tag{76}$$

for the light which has been scattered only once in the atmosphere.

If in analogy with equation (76) we write the law of diffuse reflection in the form

$$I(0, \mu, \varphi; \mu_0, \varphi_0) = \frac{1}{4} \frac{1}{\mu + \mu_0} R(\mu, \varphi; \mu_0, \varphi_0) \mu_0 F, \tag{77}$$

then
$$\frac{I(0, \mu, \varphi; \mu_0, \varphi_0) - I^{(1)}(0, \mu, \varphi; \mu_0, \varphi_0)}{I(0, \mu, \varphi; \mu_0, \varphi_0)} = 1 - \frac{p(\mu, \varphi; -\mu_0, \varphi_0)}{R(\mu, \varphi; \mu_0, \varphi_0)}, \tag{78}$$

represents the fraction of the light, emergent in the direction (μ, φ), which has been scattered *more* than once in the atmosphere. The departure of $R(\mu, \varphi; \mu_0, \varphi_0)$ from $p(\mu, \varphi; -\mu_0, \varphi_0)$ is therefore a measure of the importance, for diffuse reflection, of orders of scattering higher than the first.

The relation of the function $R(\mu, \varphi; \mu_0, \varphi_0)$ to the scattering function $S(\mu, \varphi; \mu_0, \varphi_0)$ is

$$\left(\frac{1}{\mu} + \frac{1}{\mu_0} \right) S(\mu, \varphi; \mu_0, \varphi_0) = R(\mu, \varphi; \mu_0, \varphi_0). \tag{79}$$

The function $R(\mu, \varphi; \mu_0, \varphi_0)$ has therefore a somewhat simpler structure than $S(\mu, \varphi; \mu_0, \varphi_0)$, though the principles of invariance are more conveniently formulated in terms of the latter.

In Figs. 10, 11, and 12 we have indicated (according to eq. [76]) the light which has been scattered once in the atmosphere. The differences $I(0, \mu, \varphi; \mu_0, \varphi_0) - I^{(1)}(0, \mu, \varphi; \mu_0, \varphi_0)$ for the various cases are also shown. It is of interest to notice how very nearly these differences agree for the conservative cases illustrated in Fig. 10, and (to a lesser extent) for the two cases with the same albedo ($\varpi_0 = 0.8$) illustrated in Fig. 12. These agreements clearly imply that, for a given albedo, the fraction of the emergent light which has arisen from orders of scattering higher than the first is, in a first approximation, independent of the phase function: a result which is physically understandable.

Another comparison which is of interest in this connexion is to the so-called *Lambert's law* of diffuse reflection. *On this law the diffusely reflected light is isotropic in the outward hemisphere and is natural, independently of the state of polarization and the angle of incidence of the incident light.* If the surface reflects all the light which is incident on it,

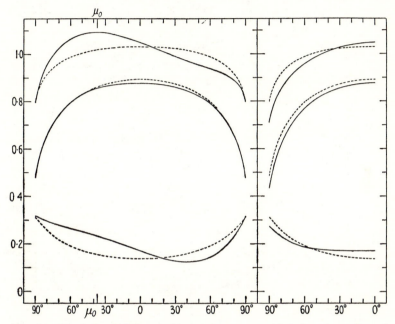

Fig. 10. The law of diffuse reflection from a semi-infinite atmosphere for conservative isotropic scattering (dashed curves) and scattering according to Rayleigh's phase function (full line curves).

The ordinates represent the intensities in the units $\mu_0 F$ and the abscissae the angles in degrees.

An angle of incidence corresponding to $\mu_0 = 0.8$ is considered and the variation of the reflected intensity in the planes $\varphi_0 - \varphi = 0°$ (the curves on the left side of the diagram) and $\varphi_0 - \varphi = 90°$ (the curves on the right side of the diagram) are illustrated.

The top set of curves represent the diffusely reflected light as given by the exact solutions of the problem; the bottom set of curves represent the light which has suffered only a single scattering process in the atmosphere; the middle set of curves (obtained by subtracting the bottom curves from the corresponding top ones) represent the result of the higher orders of scattering.

then Lambert's law of diffuse reflection can be expressed in the form

$$I(0, \mu, \varphi; \mu_0, \varphi_0) \equiv \mu_0 F. \tag{80}$$

More generally, a surface is said to reflect according to Lambert's law with an 'albedo' λ_0 if

$$I(0, \mu, \varphi; \mu_0, \varphi_0) \equiv \lambda_0 \mu_0 F. \tag{81}$$

When we are dealing with diffuse reflection from an atmosphere, it is convenient to identify λ_0 and ϖ_0 and write, for comparison, Lambert's law in the form

$$I(0, \mu, \varphi; \mu_0, \varphi_0) = \varpi_0 \mu_0 F. \tag{82}$$

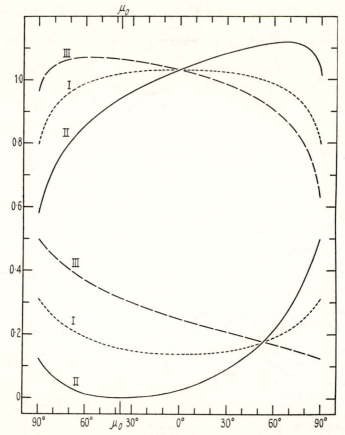

FIG. 11. The law of diffuse reflection from a semi-infinite atmosphere for conservative isotropic scattering (curves marked I) and for scattering according to the phase functions $1 + \cos \Theta$ (curves marked II) and $1 - \cos \Theta$ (curves marked III).

The ordinates and the abscissae have the same meanings as in Fig. 10.

An angle of incidence corresponding to $\mu_0 = 0.8$ is considered and the variation of the reflected intensity in the plane $\varphi_0 - \varphi = 0°$ is illustrated.

The top set of curves represent the diffusely reflected light as given by the exact solution of the problem; the bottom set of curves represent the light which has suffered only a single scattering process in the atmosphere.

It may be noted that in the exact solution of the problem, the variation of the reflected intensity in the plane $\varphi_0 - \varphi = 90°$ is the same on all three laws of scattering and agree with that given by conservative isotropic scattering for any of the planes $\varphi = \text{constant}$.

The corresponding expression for R is

$$R \text{ (Lambert)} = 4\varpi_0(\mu + \mu_0). \tag{83}$$

From Figs. 10 and 11 it is seen that, for conservative cases, Lambert's law provides a crude 'first approximation' to the exact laws of diffuse reflection.

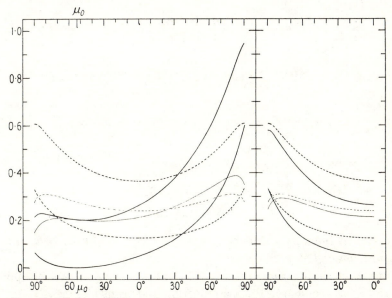

FIG. 12. The law of diffuse reflection by a semi-infinite atmosphere for isotropic scattering with an albedo 0·8 (the dashed curves) and for scattering according to the phase function $0·8(1+\cos\Theta)$.

The ordinates and the abscissae have the same meanings as in Figs. 10 and 11.

An angle of incidence corresponding to $\mu_0 = 0·6$ is considered and the variation of the reflected intensity in the planes $\varphi_0-\varphi = 0°$ (the curves on the left side of the diagram) and $\varphi_0-\varphi = 90°$ (the curves on the right side of the diagram) are illustrated.

The various sets of curves have the same meanings as explained in the legend for Fig. 10.

48. The equation of transfer for a general phase function and its solution in the nth approximation

In the first instance, it is convenient to suppose that the expansion of the phase function in Legendre polynomials consists of only a finite number of terms, say N:

$$p(\cos\Theta) = \sum_{l=0}^{N} \varpi_l P_l(\cos\Theta), \tag{84}$$

where ϖ_l $(l = 0, 1, ..., N)$ are a set of $N+1$ constants. For a phase function of this form

$$p(\mu, \varphi; \mu', \varphi') = \sum_{l=0}^{N} \varpi_l P_l[\mu\mu' + (1-\mu^2)^{\frac{1}{2}}(1-\mu'^2)^{\frac{1}{2}}\cos(\varphi'-\varphi)]. \tag{85}$$

Expanding the Legendre polynomials for the argument

$$\mu\mu' + (1-\mu^2)^{\frac{1}{2}}(1-\mu'^2)^{\frac{1}{2}}\cos(\varphi'-\varphi)$$

by the addition theorem of spherical harmonics, we have

$$p(\mu, \varphi; \mu', \varphi')$$
$$= \sum_{l=0}^{N} \varpi_l \left\{ P_l(\mu) P_l(\mu') + 2 \sum_{m=1}^{l} \frac{(l-m)!}{(l+m)!} P_l^m(\mu) P_l^m(\mu') \cos m(\varphi'-\varphi) \right\}.$$
(86)

Inverting the order of the summation on the right-hand side of this equation, we can write

$$p(\mu, \varphi; \mu', \varphi') = \sum_{m=0}^{N} (2-\delta_{0,m}) \left\{ \sum_{l=m}^{N} \varpi_l^m P_l^m(\mu) P_l^m(\mu') \right\} \cos m(\varphi'-\varphi), \quad (87)$$

where
$$\varpi_l^m = \varpi_l \frac{(l-m)!}{(l+m)!} \quad (l = m,..., N, \ 0 \leqslant m \leqslant N) \quad (88)$$

and
$$\delta_{0,m} = 1 \quad \text{if } m = 0,$$
$$= 0 \quad \text{otherwise.} \quad (89)$$

48.1. *The equation of transfer for the problem of diffuse reflection and transmission and its reduction*

For the phase function given by (87), the equation of transfer appropriate to the problem of diffuse reflection and transmission is

$$\mu \frac{dI(\tau, \mu, \varphi)}{d\tau} = I(\tau, \mu, \varphi) - \frac{1}{4\pi} \int_{-1}^{+1} \int_{0}^{2\pi} \left[\sum_{m=0}^{N} (2-\delta_{0,m}) \times \right.$$

$$\times \left\{ \sum_{l=m}^{N} \varpi_l^m P_l^m(\mu) P_l^m(\mu') \right\} \cos m(\varphi'-\varphi) \right] I(\tau, \mu', \varphi') \, d\mu' d\varphi' -$$

$$- \tfrac{1}{4} F \left[\sum_{m=0}^{N} (2-\delta_{0,m}) \left\{ \sum_{l=m}^{N} \varpi_l^m (-1)^{m+l} P_l^m(\mu) P_l^m(\mu_0) \right\} \cos m(\varphi_0 - \varphi) \right] e^{-\tau/\mu_0}.$$
(90)

This equation suggests that we expand $I(\tau, \mu, \varphi)$ in the form

$$I(\tau, \mu, \varphi) = \sum_{m=0}^{N} I^{(m)}(\tau, \mu) \cos m(\varphi_0 - \varphi). \quad (91)$$

With this substitution equation (90) splits up into the $(N+1)$ independent equations:

$$\mu \frac{dI^{(m)}(\tau, \mu)}{d\tau} = I^{(m)}(\tau, \mu) - \tfrac{1}{2} \sum_{l=m}^{N} \varpi_l^m P_l^m(\mu) \int_{-1}^{+1} P_l^m(\mu') I^{(m)}(\tau, \mu') \, d\mu' -$$

$$- \left\{ \tfrac{1}{4} (2-\delta_{0,m}) \sum_{l=m}^{N} \varpi_l^m (-1)^{m+l} P_l^m(\mu) P_l^m(\mu_0) \right\} F e^{-\tau/\mu_0}$$

$$(m = 0, 1,..., N). \quad (92)$$

48.2. *The equivalent system of linear equations in the nth approximation*

On the method of approximation described in Chapters II and III, the integrals which occur in the equation of transfer are to be replaced by sums according to Gauss's quadrature formula. In the nth approximation the equivalent system of linear equations is of order $2n$, corresponding to the division of the interval $(-1, +1)$ according to the zeros of $P_{2n}(\mu)$. Consequently, in the present context in which Legendre polynomials of maximum order N occur under the integral sign, we must seek solutions in approximations n such that

$$4n-1 > 2N. \tag{93}$$

Conversely, if we limit ourselves to solutions of the equation of transfer in the nth approximation, we are not entitled to include in an expansion of the phase function in Legendre polynomials spherical harmonics of order higher than $2n-1$.

Assuming that the condition (93) is met, we have the following system of $2n$ linear equations which is equivalent to equation (92) in the nth approximation:

$$\mu_i \frac{dI_i^{(m)}}{d\tau} = I_i^{(m)} - \tfrac{1}{2} \sum_{l=m}^{N} \varpi_l^m P_l^m(\mu_i) \sum_j a_j I_j^{(m)} P_l^m(\mu_j) -$$

$$- \left\{ \tfrac{1}{4}(2-\delta_{0,m}) \sum_{l=m}^{N} (-1)^{l+m} \varpi_l^m P_l^m(\mu_i) P_l^m(\mu_0) \right\} F e^{-\tau/\mu_0}$$

$$(i = \pm 1, ..., \pm n; \quad 0 \leqslant m \leqslant N). \tag{94}$$

48.3. *The solution of the associated homogeneous system*

Considering a particular $0 \leqslant m \leqslant N$, we shall first seek the general solution of the associated homogeneous system

$$\mu \frac{dI_i^{(m)}}{d\tau} = I_i^{(m)} - \tfrac{1}{2} \sum_{l=m}^{N} \varpi_l^m P_l^m(\mu_i) \sum_j a_j I_j^{(m)} P_l^m(\mu_j) \quad (i = \pm 1, ..., \pm n), \tag{95}$$

and add a particular integral of the non-homogeneous system (94).

To obtain the different linearly independent solutions of the system (95), we let

$$I_i^{(m)} = g_i^{(m)} e^{-k\tau} \quad (i = \pm 1, ..., \pm n), \tag{96}$$

where the $g_i^{(m)}$'s and k are constants. From equation (95) we find that $g_i^{(m)}$ must be of the form

$$g_i^{(m)} = \frac{1}{1+\mu_i k} \sum_{l=m}^{N} \varpi_l^m \xi_l^m P_l^m(\mu_i) \quad (i = \pm 1, ..., \pm n), \tag{97}$$

where the ξ_l^m's $(l = m,..., N)$ are certain constants which have to be determined in conformity with the equation

$$\sum_{l=m}^{N} \varpi_l^m \xi_l^m P_l^m(\mu_i) = \frac{1}{2} \sum_{l=m}^{N} \varpi_l^m P_l^m(\mu_i) \sum_j \frac{a_j P_l^m(\mu_j)}{1+\mu_j k} \sum_{\lambda=m}^{N} \varpi_\lambda^m \xi_\lambda^m P_\lambda^m(\mu_j). \quad (98)$$

Defining
$$D_{l,\lambda}^m(x) = \frac{1}{2} \sum_j \frac{a_j P_l^m(\mu_j) P_\lambda^m(\mu_j)}{1+\mu_j x}, \quad (99)$$

we can rewrite equation (98) in the form

$$\sum_{l=m}^{N} \varpi_l^m \xi_l^m P_l^m(\mu_i) = \sum_{l=m}^{N} \varpi_l^m P_l^m(\mu_i) \sum_{\lambda=m}^{N} \varpi_\lambda^m \xi_\lambda^m D_{l,\lambda}^m(k). \quad (100)$$

Since this equation must be valid for all i, we must require that

$$\xi_l^m = \sum_{\lambda=m}^{N} \varpi_\lambda^m \xi_\lambda^m D_{l,\lambda}^m(k) \quad (l = m,..., N). \quad (101)$$

This represents a homogeneous system of linear equations for $(N-m+1)$ unknowns. The determinant of the system must therefore vanish; this condition will lead to the characteristic equation for k; and for each root of the characteristic equation we shall have a set of ξ_l^m's which will be determined apart from a constant of proportionality. However, it is possible to obtain the characteristic equation and the ξ_l^m's without going through the procedure indicated by the following artifice:

First we observe that $D_{l,\lambda}^m(x)$ satisfies a simple recursion formula; for, writing $D_{l,\lambda}^m$ in the form

$$D_{l,\lambda}^m = \frac{1}{2} \sum_j a_j P_l^m(\mu_j) P_\lambda^m(\mu_j) \left(1 - \frac{x\mu_j}{1+x\mu_j}\right), \quad (102)$$

and remembering that, since

$$l+\lambda \leqslant 2N < 4n-1, \quad (103)$$

$$\frac{1}{2} \sum_j a_j P_l^m(\mu_j) P_\lambda^m(\mu_j) = \frac{1}{2} \int_{-1}^{+1} P_l^m(\mu) P_\lambda^m(\mu) \, d\mu$$

$$= \frac{\delta_{l,\lambda}}{2l+1} \frac{(l+m)!}{(l-m)!}, \quad (104)$$

we have

$$D_{l,\lambda}^m = \frac{\delta_{l,\lambda}}{2l+1} \frac{(l+m)!}{(l-m)!} - \frac{1}{2}x \sum_j \frac{a_j \mu_j P_l^m(\mu_j) P_\lambda^m(\mu_j)}{1+x\mu_j}$$

$$= \frac{\delta_{l,\lambda}}{2l+1} \frac{(l+m)!}{(l-m)!} - \frac{x}{2(2l+1)} \sum_j \frac{a_j P_\lambda^m(\mu_j)}{1+x\mu_j} \times$$

$$\times [(l-m+1)P_{l+1}^m(\mu_j) + (l+m)P_{l-1}^m(\mu_j)]$$

$$= \frac{\delta_{l,\lambda}}{2l+1} \frac{(l+m)!}{(l-m)!} - \frac{x}{2l+1}[(l-m+1)D_{l+1,\lambda}^m + (l+m)D_{l-1,\lambda}^m]. \quad (105)$$

Hence

$$(2l+1)D_{l,\lambda}^m(x) = \frac{(l+m)!}{(l-m)!}\delta_{l,\lambda} - x[(l-m+1)D_{l+1,\lambda}^m(x) + (l+m)D_{l-1,\lambda}^m(x)].$$

$$(106)$$

Using the recursion formula (106) in equation (101), we have

$$(2l+1)\xi_l^m$$

$$= \sum_{\lambda=m}^N \varpi_\lambda^m \xi_\lambda^m \left\{ \frac{(l+m)!}{(l-m)!}\delta_{l,\lambda} - k[(l-m+1)D_{l+1,\lambda}^m(k) + (l+m)D_{l-1,\lambda}^m(k)] \right\},$$

$$(107)$$

or, again, using equation (101), we have (cf. eq. [88])

$$(2l+1)\xi_l^m = \varpi_l \xi_l^m - k[(l-m+1)\xi_{l+1}^m + (l+m)\xi_{l-1}^m].$$

$$(108)$$

Alternatively, we can write

$$\xi_{l+1}^m = -\frac{2l+1-\varpi_l}{k(l-m+1)}\xi_l^m - \frac{l+m}{l-m+1}\xi_{l-1}^m \quad (l = m,...,N-1). \quad (109)\dagger$$

Equation (109) determines the ξ's apart from a constant of proportionality which we shall make determinate by the choice

$$\xi_m^m = 1.$$

$$(110)$$

With this choice of ξ_m^m, the remaining ξ_l^m's are uniquely determined. Thus

$$\xi_{m+1}^m = -\frac{2m+1-\varpi_m}{k},$$

$$\xi_{m+2}^m = \frac{(2m+1-\varpi_m)(2m+3-\varpi_{m+1})}{2k^2} - \frac{2m+1}{2}, \quad \text{etc.} \quad (111)$$

It is to be particularly noted that the ξ_l^m's determined in this fashion are, in general, functions of k. To emphasize this dependence we shall sometimes write $\xi_l^m(k)$.

The characteristic equation for k now follows from equation (101) by letting $l = m$. Thus

$$1 = \sum_{\lambda=m}^N D_{m,\lambda}^m(k)\xi_\lambda^m(k)\varpi_\lambda^m,$$

$$(112)$$

or, more explicitly,

$$1 = \frac{1}{2}\sum_j \frac{a_j}{1+\mu_j k}\left\{ \sum_{\lambda=m}^N \varpi_\lambda^m \xi_\lambda^m(k) P_\lambda^m(\mu_j) P_m^m(\mu_j) \right\}.$$

$$(113)$$

By virtue of equation (109), equation (113) is of order n in k^2 and admits, in general, $2n$ distinct non-vanishing roots which must occur in pairs as

$$k_{\pm\alpha}^m \quad (\alpha = 1,...,n \text{ and } k_\alpha^m = -k_{-\alpha}^m).$$

$$(114)$$

† It should be noted that, in applying this equation for the case $l = m$, we must put $\xi_{m-1}^m = 0$ since the ξ's are not defined for $l < m$.

The $2n$ linearly independent solutions of the homogeneous system associated with equation (94) are therefore

$$I_i^{(m)} = \text{constant} \sum_{l=m}^{N} \xi_l^m(k_\alpha^m)\varpi_l^m \frac{P_l^m(\mu_i)}{1+\mu_i k_\alpha^m} e^{-k_\alpha^m \tau}$$

$$(\alpha = \pm 1,...,\pm n;\ i = \pm 1,...,\pm n;\ 0 \leqslant m \leqslant N). \quad (115)$$

In conservative cases, when $\varpi_0 = 1$, $k^2 = 0$ will be a root of the characteristic equation for $m = 0$; for

$$\tfrac{1}{2} \sum_j \sum_{\lambda=m}^{N} a_j\, \varpi_\lambda \xi_\lambda P_\lambda(\mu_j) = \tfrac{1}{2} \sum a_j\, \varpi_0 = 1. \quad (116)$$

We shall then have only $(2n-2)$ non-vanishing roots, and (115) provides only $(2n-2)$ independent integrals. On the other hand, in conservative cases the equation of transfer admits the further integral (cf. Chap. I, § 10, eq. [81])

$$I_i^{(0)} = \text{constant}[(1-\tfrac{1}{3}\varpi_1)\tau+\mu_i+Q]. \quad (117)$$

To avoid repetition we shall suppose that $\varpi_0 \neq 1$, unless it is explicitly stated to the contrary.

48.4. *A particular integral of the non-homogeneous system* (94)

To complete the solution of the system of equations (94) we need a particular integral. This can be obtained in the following manner:

Setting

$$I_i^{(m)} = \tfrac{1}{4}(2-\delta_{0,m})h_i^{(m)} F e^{-\tau/\mu_0} \quad (i = \pm 1,...,\pm n) \quad (118)$$

in equation (94), we verify that the constants $h_i^{(m)}$ must be expressible in the form

$$h_i^{(m)} = \sum_{l=m}^{N} \varpi_l^m \gamma_l^m \frac{P_l^m(\mu_i)}{1+\mu_i/\mu_0}, \quad (119)$$

where the γ_l^m's are certain constants to be determined in accordance with the relation (cf. eq. [101])

$$\gamma_l^m = (-1)^{l+m}P_l^m(\mu_0)+\sum_{\lambda=m}^{N} \varpi_\lambda^m \gamma_\lambda^m D_{l,\lambda}^m\!\left(\frac{1}{\mu_0}\right) \quad (l = m,...,N). \quad (120)$$

Using the recursion formula (106) satisfied by the $D_{l,\lambda}^m$'s, we derive:

$$(2l+1)\gamma_l^m = (-1)^{l+m}(2l+1)P_l^m(\mu_0)+\sum_{\lambda=m}^{N} \varpi_\lambda^m \gamma_\lambda^m \times$$

$$\times \left\{\frac{(l+m)!}{(l-m)!}\delta_{l,\lambda}-\frac{1}{\mu_0}\left[(l-m+1)D_{l+1,\lambda}^m\!\left(\frac{1}{\mu_0}\right)+(l+m)D_{l-1,\lambda}^m\!\left(\frac{1}{\mu_0}\right)\right]\right\} \quad (121)$$

or, again, using equation (120), we have

$$(2l+1)\gamma_l^m = (-1)^{l+m}(2l+1)P_l^m(\mu_0)+\varpi_l\gamma_l^m-$$

$$-\frac{1}{\mu_0}\{(l-m+1)[\gamma_{l+1}^m-(-1)^{l+m+1}P_{l+1}^m(\mu_0)]+$$

$$+(l+m)[\gamma_{l-1}^m-(-1)^{l+m-1}P_{l-1}^m(\mu_0)]\}. \quad (122)$$

The terms in the spherical harmonics cancel and we are left with

$$(2l+1)\gamma_l^m = \varpi_l\gamma_l^m-\frac{1}{\mu_0}[(l-m+1)\gamma_{l+1}^m+(l+m)\gamma_{l-1}^m]. \quad (123)$$

Comparing this relation among the γ_l^m's with equation (108), we observe that the γ_l^m's satisfy a recursion formula of the same form as the ξ_l^m's. Hence we should have

$$\gamma_l^m = \gamma_m^m(\mu_0)\xi_l^m\left(\frac{1}{\mu_0}\right) \quad (l = m,...,N), \quad (124)$$

where $\gamma_m^m(\mu_0)$ is a constant of proportionality which can be determined from equation (120) by letting $l = m$ and remembering that $\xi_m^m = 1$. We have

$$\gamma_m^m(\mu_0) = P_m^m(\mu_0)+\gamma_m^m \sum_{\lambda=m}^{N} \varpi_\lambda^m\xi_\lambda^m\left(\frac{1}{\mu_0}\right)D_{m,\lambda}^m\left(\frac{1}{\mu_0}\right), \quad (125)$$

or

$$\gamma_m^m(\mu_0) = \frac{P_m^m(\mu_0)}{1-\sum_{\lambda=m}^{N} \varpi_\lambda^m\xi_\lambda^m(1/\mu_0)D_{m,\lambda}^m(1/\mu_0)}. \quad (126)$$

From the relation of the denominator in this expression for $\gamma_m^m(\mu_0)$ to the characteristic equation (112), we conclude by arguments similar to those leading to the identity established in Chapter V, equations (46) and (51), that

$$\gamma_m^m(\mu_0) = P_m^m(\mu_0)H^{(m)}(\mu_0)H^{(m)}(-\mu_0), \quad (127)$$

where $H^{(m)}(\mu)$ is defined as usual in terms of the positive roots of the characteristic equation.

The required particular integral is therefore

$$I_i^{(m)} = \tfrac{1}{4}(2-\delta_{0,m})P_m^m(\mu_0)H^{(m)}(\mu_0)H^{(m)}(-\mu_0)Fe^{-\tau/\mu_0}\times$$

$$\times\sum_{l=m}^{N} \varpi_l^m\xi_l^m\left(\frac{1}{\mu_0}\right)\frac{P_l^m(\mu_i)}{1+\mu_i/\mu_0} \quad (i = \pm 1,...,\pm n; \ 0 \leqslant m \leqslant N). \quad (128)$$

48.5. *The general solution of the system of equations* (94)

Combining equations (115) and (128) we can write the general solution of the system (94) in the form

$$I_i^{(m)} = \tfrac{1}{4}(2-\delta_{0,m})FP_m^m(\mu_0)\Bigg\{ \sum_{\alpha=-n}^{+n} \frac{L_\alpha^m e^{-k_\alpha^m \tau}}{1+\mu_i k_\alpha^m}\Bigg[\sum_{l=m}^{N} \varpi_l^m \xi_l^m(k_\alpha^m)P_l^m(\mu_i)\Bigg] +$$

$$+ \frac{H^{(m)}(\mu_0)H^{(m)}(-\mu_0)}{1+\mu_i/\mu_0} e^{-\tau/\mu_0}\Bigg[\sum_{l=m}^{N} \varpi_l^m \xi_l^m\Big(\frac{1}{\mu_0}\Big)P_l^m(\mu_i)\Bigg] \Bigg\} \quad (i = \pm 1,...,\pm n),$$

$$(129)\dagger$$

where the L_α^m's are $2n(N+1)$ constants of integration.

The constants of integration in the solution (129) must be found from the boundary conditions. Thus, if we are considering the problem of diffuse reflection by a semi-infinite atmosphere, we must suppress all the terms which are unbounded as $\tau \to \infty$. The remaining constants will then be determined by the conditions

$$I_{-i}^{(m)} = 0 \quad \text{at} \quad \tau = 0 \quad \text{for} \quad i = 1,...,n \quad \text{and} \quad m = 0,...,N, \quad (130)$$

at the surface. On the other hand, if we are dealing with the standard problem in the theory of diffuse reflection and transmission, then the $2n(N+1)$ constants must be determined from the conditions (cf. Chap. I, eqs. [127] and [128])

$$I_{-i}^{(m)} = 0 \quad \text{at} \quad \tau = 0 \quad \text{for} \quad i = 1,...,n \quad \text{and} \quad m = 0,...,N,$$

and $\quad I_{+i}^{(m)} = 0 \quad \text{at} \quad \tau = \tau_1 \quad \text{for} \quad i = 1,...,n \quad \text{and} \quad m = 0,...,N. \quad (131)$

48.6. *The problem with a constant net flux in conservative cases*

In this case the radiation field has no azimuth dependent terms and the solution appropriate to the problem is

$$I_i = \tfrac{3}{4}F\Bigg\{ \sum_{\alpha=1}^{n-1} \frac{L_\alpha e^{-k_\alpha \tau}}{1+\mu_i k_\alpha}\Bigg[\sum_{l=0}^{N} \varpi_l \xi_l(k_\alpha)P_l(\mu_i)\Bigg] + (1-\tfrac{1}{3}\varpi_1)\tau + \mu_i + L_n\Bigg\}, \quad (132)$$

where the n constants L_α are to be determined from the boundary conditions

$$I_{-i} = 0 \quad \text{for} \quad i = 1,...,n \quad \text{at} \quad \tau = 0. \quad (133)$$

Equation (109) which determines the ξ's reduces in this case, $m = 0$, to

$$\xi_{l+1} = -\frac{2l+1-\varpi_l}{k(l+1)}\xi_l - \frac{l}{l+1}\xi_{l-1} \quad (l = 0,...,N-1). \quad (134)$$

---

† It should be noted that in this equation, in the summation over α, there is no term with $\alpha = 0$ and that $k_{-\alpha}^m = -k_\alpha^m$. Also, in conservative cases, for $m = 0$, α runs only from $-(n-1)$ to $(n-1)$; instead, we shall have the term

$$L_0^0[(1-\tfrac{1}{3}\varpi_1)\tau+\mu_i]+L_n^0$$

in the solution.

In particular (since $\varpi_0 = 1$),

$$\xi_0 = 1, \qquad \xi_1 = 0, \qquad \xi_2 = -\tfrac{1}{2}, \qquad \xi_3 = \frac{5-\varpi_2}{6k}, \quad \text{etc.} \quad (135)$$

The characteristic equation which determines the k's is

$$1 = \frac{1}{2} \sum_j \frac{a_j}{1+\mu_j k} \left\{ \sum_{l=0}^{N} \varpi_l \xi_l(k) P_l(\mu_j) \right\}. \qquad (136)$$

48.7. *The solution for the phase function* $1+\varpi_1 P_1(\cos\Theta)+\varpi_2 P_2(\cos\Theta)$

As an illustration of the general theory, we shall consider the conservative case

$$p(\cos\Theta) = 1+\varpi_1 P_1(\cos\Theta)+\varpi_2 P_2(\cos\Theta). \qquad (137)$$

For this phase function the solution has the form

$$I(\tau,\mu,\varphi) = I^{(0)}(\tau,\mu)+I^{(1)}(\tau,\mu)\cos(\varphi_0-\varphi)+I^{(2)}(\tau,\mu)\cos 2(\varphi_0-\varphi). \qquad (138)$$

Considering first $I^{(0)}(\tau,\mu)$, we have the characteristic equation

$$1 = \frac{1}{2} \sum_j \frac{a_j[1-\tfrac{1}{2}\varpi_2 P_2(\mu_j)]}{1+\mu_j k}, \qquad (139)$$

or, somewhat differently,

$$1 = \tfrac{3}{4}\varpi_2 \sum_{j=1}^{n} \left(\frac{4+\varpi_2}{3\varpi_2}-\mu_j^2 \right) \frac{a_j}{1-\mu_j^2 k^2}. \qquad (140)$$

Equation (140) admits only $(n-1)$ distinct non-vanishing roots for k^2. We must accordingly make use of the integral (117) in writing the general solution for $I_i^{(0)}$. Thus we have

$$I_i^{(0)} = \frac{3\varpi_2}{16} F\left[\left(\frac{4+\varpi_2}{3\varpi_2}-\mu_i^2 \right) \sum_{\alpha=-n+1}^{n-1} \frac{L_\alpha^0 e^{-k_\alpha^0 \tau}}{1+\mu_i k_\alpha^0} + L_0^0(\tau+\mu_i)+L_n^0+ \right.$$
$$\left. + \left(\frac{4+\varpi_2}{3\varpi_2}-\mu_i^2 \right) \frac{H^{(0)}(\mu_0)H^{(0)}(-\mu_0)}{1+\mu_i/\mu_0} e^{-\tau/\mu_0} \right] \quad (i = \pm 1,...,\pm n). \qquad (141)$$

Considering next the solution for $I^{(1)}(\tau,\mu)$, we have

$$\xi_1^1 = 1 \quad \text{and} \quad \xi_2^1 = -\frac{3-\varpi_1}{k}. \qquad (142)$$

The characteristic equation (113) becomes

$$1 = \tfrac{1}{2}\varpi_1 \sum_{j=1}^{n} \frac{a_j(1-\mu_j^2)}{1-\mu_j^2 k^2} \left[1+\frac{\varpi_2(3-\varpi_1)}{\varpi_1}\mu_j^2 \right]. \qquad (143)$$

The corresponding solution for $I_i^{(1)}$ is

$$I_i^{(1)} = \tfrac{1}{4}\varpi_1 F(1-\mu_i^2)^{\frac{1}{2}}(1-\mu_0^2)^{\frac{1}{2}}\left\{ \sum_{\alpha=-n}^{+n} \frac{L_\alpha^1 e^{-k_\alpha^1 \tau}}{1+\mu_i k_\alpha^1}\left[1 - \frac{\varpi_2(3-\varpi_1)}{\varpi_1}\frac{\mu_i}{k_\alpha^1}\right] + \right.$$

$$\left. + \frac{H^{(1)}(\mu_0)H^{(1)}(-\mu_0)}{1+\mu_i/\mu_0}e^{-\tau/\mu_0}\left[1 - \frac{\varpi_2(3-\varpi_1)}{\varpi_1}\mu_i\mu_0\right]\right\}$$

$$(i = \pm 1,...,\pm n). \quad (144)$$

And, finally, the solution for $I_i^{(2)}$ is

$$I_i^{(2)} = \frac{3\varpi_2}{16} F(1-\mu_i^2)(1-\mu_0^2)\left\{ \sum_{\alpha=-n}^{+n} \frac{L_\alpha^2 e^{-k_\alpha^2 \tau}}{1+\mu k_\alpha^2} + \frac{H^{(2)}(\mu_0)H^{(2)}(-\mu_0)}{1+\mu_i/\mu_0}e^{-\tau/\mu_0}\right\}$$

$$(i = \pm 1,...,\pm n), \quad (145)$$

the characteristic equation being

$$1 = \tfrac{3}{8}\varpi_2 \sum_{j=1}^{n} \frac{a_j(1-\mu_j^2)^2}{1-\mu_j^2 k^2}. \quad (146)$$

For the standard problems in semi-infinite plane-parallel atmospheres, the constants in the solutions (141), (144), and (145) can be eliminated and the angular distributions of the emergent radiations expressed in closed forms. Indeed, for the phase function (137), the integral equations derived from the principles of invariance can be solved and expressed in terms of H-functions. The solutions are given below:

48.8. *The exact solutions for the standard problems in semi-infinite atmospheres for the case* $P(\cos\Theta) = 1+\varpi_1 P_1(\cos\Theta)+\varpi_2 P_2(\cos\Theta)$

By solving the integral equations derived from the principles of invariance, we find that the law of diffuse reflection is expressible in the form

$$I(0,\mu,\varphi;\mu_0,\varphi_0)$$

$$= \frac{\mu_0 F}{4(\mu_0+\mu)}\left\{\frac{9\varpi_2}{4+\varpi_2}\left[\tfrac{1}{4}\varpi_2\,\psi(\mu)\psi(\mu_0)+\phi(\mu)\phi(\mu_0)\right]+\right.$$

$$+\left[\varpi_1\,\theta(\mu)\theta(\mu_0)-3\varpi_2\,\sigma(\mu)\sigma(\mu_0)\right](1-\mu^2)^{\frac{1}{2}}(1-\mu_0^2)^{\frac{1}{2}}\cos(\varphi_0-\varphi)+$$

$$\left.+\tfrac{3}{4}\varpi_2(1-\mu^2)(1-\mu_0^2)H^{(2)}(\mu)H^{(2)}(\mu_0)\cos 2(\varphi_0-\varphi)\right\}, \quad (147)$$

where

$$\psi(\mu) = H^{(0)}(\mu)\left(\frac{4+\varpi_2}{3\varpi_2}-c\mu\right), \qquad \phi(\mu) = q\mu H^{(0)}(\mu),$$

$$\theta(\mu) = H^{(1)}(\mu)(1-\gamma\mu), \qquad \sigma(\mu) = \kappa\mu H^{(1)}(\mu), \quad (148)$$

and $H^{(0)}(\mu)$, $H^{(1)}(\mu)$, and $H^{(2)}(\mu)$ are defined in terms of the characteristic functions

$$\frac{3\varpi_2}{8}\left(\frac{4+\varpi_2}{3\varpi_2}-\mu^2\right), \qquad \tfrac{1}{4}(1-\mu^2)[\varpi_1+\varpi_2(3-\varpi_1)\mu^2],$$

and
$$\tfrac{3}{16}\varpi_2(1-\mu^2)^2, \tag{149}$$

respectively. Further, the constants, c, q, γ, and κ are given by

$$c=\frac{\alpha_2^{(0)}}{\alpha_1^{(0)}}; \qquad \gamma=\frac{\varpi_2(3-\varpi_1)(\alpha_1^{(1)}-\alpha_3^{(1)})}{4-\varpi_1(\alpha_0^{(1)}-\alpha_2^{(1)})},$$

$$q=\frac{2}{3\alpha_1^{(0)}}; \qquad \kappa=\frac{4}{3}\frac{(3-\varpi_1)}{4-\varpi_1(\alpha_0^{(1)}-\alpha_2^{(1)})}, \tag{150}$$

where $\alpha_n^{(0)}$ and $\alpha_n^{(1)}$ denote the moments of order n of $H^{(0)}(\mu)$ and $H^{(1)}(\mu)$ respectively.

Similarly, the law of darkening for the problem with a constant net flux is given by

$$I(0,\mu)=\frac{F}{2\alpha_1^{(0)}}H^{(0)}(\mu). \tag{151}$$

It will be noticed that in the foregoing solutions the azimuth independent terms do not involve ϖ_1 in any way; that for the case $\varpi_1=3$ the term in $\cos(\varphi_0-\varphi)$ in the law of diffuse reflection also becomes independent of ϖ_2; and, finally, that for $\varpi_1=0$ and $\varpi_2=\tfrac{1}{2}$ the solutions reduce to those given in §§ 44 and 45 for the case of Rayleigh's phase function.

BIBLIOGRAPHICAL NOTES

§§ 44 and 45. Problems of radiative transfer in semi-infinite plane-parallel atmospheres and allowing for an anisotropy of the scattered radiation according to Rayleigh's phase function are considered in—

1. S. Chandrasekhar, *Astrophys. J.* **100**, 117 (1944).
2. —— ibid. **103**, 165, 1946 (Section II of this paper).
3. —— ibid. **105**, 164, 1947 (Section V of this paper).
4. —— and Frances H. Breen, ibid. **105**, 435, 1947 (Section I of this paper).

In references 1 and 2 the standard problems of constant net flux and diffuse reflection are considered in the method of approximation of Chapter III. In reference 2 the elimination of the constants is performed and the angular distributions of the emergent radiations are expressed in closed forms. The exact solution of these problems will be found in reference **3**. The H-functions tabulated in § 45 are reproduced from reference **4**.

§ 46. The law of diffuse reflection in accordance with the phase function $\varpi_0(1 + x \cos \Theta)$ is considered in—

5. V. A. AMBARZUMIAN, *Journal of Physics of the Academy of Sciences of U.S.S.R.* **8**, 65 (1944).

6. S. CHANDRASEKHAR, *Astrophys. J.* **103**, 165, 1946 (Section I of this paper).

7. —— ibid. **105**, 164, 1946 (Section IV of this paper).

8. —— and FRANCES H. BREEN, ibid. **107**, 216 (1948).

In reference 6 the problem is considered in the general nth approximation; the elimination of the constants is also effected here. In reference 7 the solution of the integral equations derived from the principles of invariance are reduced to H-functions. And finally, in reference 8, the H-functions which occur in the solution of this problem are tabulated. Reference 5 is of historical interest. We have already commented on it in the Bibliographical Notes for the preceding chapter.

§ 47. The writer is indebted to Mrs. Frances H. Breen for the computations included in this section.

A discussion along similar lines will be found in—

9. H. C. VAN DE HULST, *The Atmospheres of the Earth and Planets* (edited by G. P. Kuiper, University of Chicago Press, 1949, Chicago), Chapter IV (see particularly pp. 92–9).

§ 48. The analysis in this section is in the main derived from—

10. S. CHANDRASEKHAR, *Astrophys. J.* **104**, 191 (1946).

PRINCIPLES OF INVARIANCE (*continued*)

49. Introduction

IN Chapters III–VI we have considered various transfer problems in semi-infinite plane-parallel atmospheres and shown how, in all cases, it is possible to solve the equation of transfer by a method of approximation which can be made to give solutions of any desired accuracy. However, in some ways, the most striking aspect of the theory which has been presented is the role which the principles of invariance of Chapter IV have played in the developments: they have enabled the exact solution of the problem of the angular distribution of the emergent radiation and, in some instances, have even circumvented the explicit use of the equation of transfer (cf. Chap. VI, § 45). Now this method of using integral equations derived from principles of invariance is not limited to problems in semi-infinite atmospheres. Principles of invariance can be formulated also for transfer problems in atmospheres of finite optical thicknesses. Indeed, in some respects, the method shows itself to its best advantage in these latter contexts, as only by its means has it been possible to solve exactly, for the first time, a large class of problems, long considered impossible. In this chapter we shall formulate these general principles and derive their most immediate consequences. The chapter also includes a discussion of certain related matters, such as the principle of reciprocity.

50. The principles of invariance

Let a parallel beam of radiation of net flux πF_0 per unit area normal to itself be incident on a plane-parallel atmosphere of optical thickness τ_1 in the direction $(-\mu_0, \varphi_0)$. Let the intensity diffusely reflected in the direction (μ, φ) be denoted by $I(0, \mu, \varphi)$ $(0 \leqslant \mu \leqslant 1)$; similarly, let the intensity diffusely transmitted in the direction $(-\mu, \varphi)$, below the surface $\tau = \tau_1$, be denoted by $I(\tau_1, -\mu, \varphi)$. We shall express these reflected and transmitted intensities in terms of a *scattering function*

$$S(\tau_1; \mu, \varphi; \mu_0, \varphi_0)$$

and a *transmission function* $T(\tau_1; \mu, \varphi; \mu_0, \varphi_0)$ in the forms (cf. Chap. I, eq. [120])

$$I(0, \mu, \varphi) = \frac{F}{4\mu} S(\tau_1; \mu, \varphi; \mu_0, \varphi_0)$$

and
$$I(\tau_1, -\mu, \varphi) = \frac{F}{4\mu} T(\tau_1; \mu, \varphi; \mu_0, \varphi_0). \tag{1}$$

It will be noticed that we have included τ_1 explicitly as a parameter in the definitions of S and T to emphasize their dependence on the optical thickness of the atmosphere.

Now, considering the radiation field in such an atmosphere, we distinguish between the reduced incident flux of amount

$$\pi F e^{-\tau/\mu_0} \tag{2}$$

in the direction $(-\mu_0, \varphi_0)$ and the diffuse radiation field characterized by the intensity $I(\tau, \mu, \varphi)$. To distinguish further between the outward $(0 \leqslant \mu \leqslant 1)$ and the inward $(-1 \leqslant \mu < 0)$ directed radiations, we shall write

$$I(\tau, +\mu, \varphi) \quad (0 \leqslant \mu \leqslant 1) \tag{3}$$

and

$$I(\tau, -\mu, \varphi) \quad (0 < \mu \leqslant 1). \tag{4}$$

With these definitions we can formulate the following four principles:

I. *The intensity, $I(\tau, +\mu, \varphi)$, in the outward direction at any level τ results from the reflection of the reduced incident flux $\pi F e^{-\tau/\mu_0}$ and the diffuse radiation $I(\tau, -\mu', \varphi')$ $(0 < \mu' \leqslant 1)$ incident on the surface τ, by the atmosphere of optical thickness $(\tau_1 - \tau)$, below τ.*

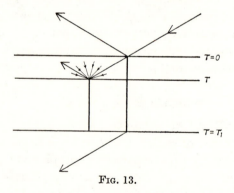

Fɪɢ. 13.

The mathematical expression of this principle is clearly (see Fig. 13)

$$I(\tau, +\mu, \varphi) = \frac{F}{4\mu} e^{-\tau/\mu_0} S(\tau_1 - \tau; \mu, \varphi; \mu_0, \varphi_0) +$$

$$+ \frac{1}{4\pi\mu} \int_0^1 \int_0^{2\pi} S(\tau_1 - \tau; \mu, \varphi; \mu', \varphi') I(\tau, -\mu', \varphi') \, d\mu' d\varphi'. \tag{5}$$

II. *The intensity, $I(\tau, -\mu, \varphi)$, in the inward direction at any level τ results from the transmission of the incident flux by the atmosphere of optical thickness τ, above the surface τ, and the reflection by this same* *surface of the diffuse radiation* $I(\tau, +\mu', \varphi')$ $(0 \leqslant \mu' \leqslant 1)$ *incident on it from below.*

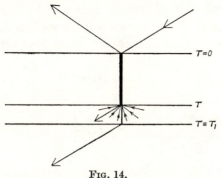

FIG. 14.

The mathematical expression of this principle is (see Fig. 14)

$$I(\tau, -\mu, \varphi) = \frac{F}{4\mu} T(\tau; \mu, \varphi; \mu_0, \varphi_0) +$$

$$+ \frac{1}{4\pi\mu} \int_0^1 \int_0^{2\pi} S(\tau; \mu, \varphi; \mu', \varphi') I(\tau, +\mu', \varphi') \, d\mu' \, d\varphi'. \quad (6)$$

III. *The diffuse reflection of the incident light by the entire atmosphere is equivalent to the reflection by the part of the atmosphere of optical thickness* τ, *above the level* τ, *and the transmission by this same atmosphere of the diffuse radiation* $I(\tau, +\mu', \varphi')$ $(0 \leqslant \mu' \leqslant 1)$ *incident on the surface* τ *from below.*

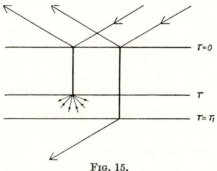

FIG. 15.

The mathematical expression of this principle is (see Fig. 15)

$$\frac{F}{4\mu} S(\tau_1; \mu, \varphi; \mu_0, \varphi_0) = \frac{F}{4\mu} S(\tau; \mu, \varphi; \mu_0, \varphi_0) + e^{-\tau/\mu} I(\tau, +\mu, \varphi) +$$

$$+ \frac{1}{4\pi\mu} \int_0^1 \int_0^{2\pi} T(\tau; \mu, \varphi; \mu', \varphi') I(\tau, +\mu', \varphi') \, d\mu' d\varphi'. \quad (7)$$

The three terms on the right-hand side represent, respectively, the contribution to the reflected intensity by the reflection of the incident flux by the part of the atmosphere above τ, the *direct* transmission of the diffuse intensity $I(\tau, +\mu, \varphi)$ already in the direction (μ, φ), and the *diffuse* transmission of the radiation field (3) incident on the surface τ from below.

IV. *The diffuse transmission of the incident light by the entire atmosphere is equivalent to the transmission of the reduced incident flux* $\pi F e^{-\tau/\mu_0}$ *and the diffuse radiation* $I(\tau, -\mu', \varphi')$ $(0 < \mu' \leqslant 1)$ *incident on the surface* τ *by the atmosphere of optical thickness* $(\tau_1 - \tau)$ *below* τ.

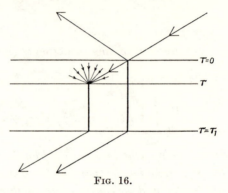

FIG. 16.

The mathematical expression of this principle is (see Fig. 16)

$$\frac{F}{4\mu} T(\tau_1; \mu, \varphi; \mu_0, \varphi_0)$$

$$= \frac{F}{4\mu} e^{-\tau/\mu_0} T(\tau_1 - \tau; \mu, \varphi; \mu_0, \varphi_0) + e^{-(\tau_1 - \tau)/\mu} I(\tau, -\mu, \varphi) +$$

$$+ \frac{1}{4\pi\mu} \int_0^1 \int_0^{2\pi} T(\tau_1 - \tau; \mu, \varphi; \mu', \varphi') I(\tau, -\mu', \varphi') \, d\mu' d\varphi'. \quad (8)$$

The three terms on the right-hand side represent, respectively, the contribution to the transmitted intensity by the reduced incident flux (2) by the transmission of the atmosphere below τ, the *direct* transmission of the diffuse intensity $I(\tau, -\mu, \varphi)$ already in the direction $(-\mu, \varphi)$, and the diffuse transmission of the radiation field (4) by the atmosphere of optical thickness $(\tau_1 - \tau)$.

It is clear that equations (5) and (6) or (7) and (8) will suffice to determine the radiation field in the atmosphere, uniquely, in terms of the scattering and the transmission functions for plane-parallel atmospheres of finite optical thicknesses.

Integral equations analogous to equations (5)–(8) for the axially symmetric radiation field in semi-infinite atmospheres with a constant net flux are (see Figs. 17, 18, and 19)

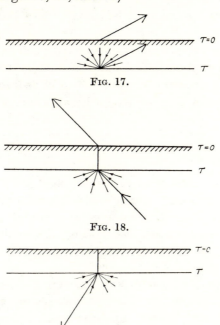

FIG. 17.

FIG. 18.

FIG. 19.

$$I(\tau, +\mu) = I(0, +\mu) + \frac{1}{2\mu} \int_0^1 S^{(0)}(\infty; \mu, \mu') I(\tau, -\mu') \, d\mu', \qquad (9)$$

$$I(0, +\mu) = e^{-\tau/\mu} I(\tau, +\mu) + \frac{1}{2\mu} \int_0^1 T^{(0)}(\tau; \mu, \mu') I(\tau, +\mu') \, d\mu', \quad (10)$$

and
$$I(\tau, -\mu) = \frac{1}{2\mu} \int_0^1 S^{(0)}(\tau; \mu, \mu') I(\tau, +\mu') \, d\mu', \qquad (11)$$

where $S^{(0)}$ and $T^{(0)}$ are the azimuth independent terms in the scattering and the transmission functions as defined in (1). Equations (9)–(11) are the mathematical expressions of the following principle:

V. *In a semi-infinite plane-parallel atmosphere with a constant net flux, the emergent radiation is invariant to the addition (or subtraction) of layers of arbitrary optical thickness to (or from) the atmosphere; also, the emergent radiation can be regarded as the result of transmission of the radiation incident on the surface τ. from below; and finally, the inward directed radiation at any level, τ, can be regarded as the reflection of the outward directed radiation by the atmosphere of optical thickness τ overlying τ.*

The first part of this principle has already been analysed in Chapter IV, §§ 29.2 and 32.

The various principles we have enunciated clarify and emphasize the remark made in Chapter I, § 13 (p. 20), that the solution of all transfer problems in plane-parallel atmospheres can be reduced to the problem of diffuse reflection and transmission by atmospheres of finite optical thicknesses.

51. Integral equations for the scattering and the transmission functions

The importance of the principles enunciated in § 50 arises from the fact that they can be used to derive a basic set of four integral equations for the scattering and the transmission functions. These integral equations are non-linear and non-homogeneous; nevertheless they provide, as we shall see, the most practical means for solving many of the fundamental problems of the subject.

The basic integral equations which govern the scattering and the transmission functions can be obtained from the principles I–IV of § 50 by differentiating with respect to τ the equations representing these principles and passing either to the limit $\tau = 0$ (Principles I and IV) or to the limit $\tau = \tau_1$ (Principles II and III). Thus, differentiating equations (5)–(8) of § 50 and making use of the boundary conditions

$$I(0, -\mu, \varphi) \equiv 0 \quad \text{and} \quad I(\tau_1, +\mu, \varphi) \equiv 0 \quad (0 < \mu \leqslant 1), \qquad (12)$$

we obtain

$$\left[\frac{dI(\tau, +\mu, \varphi)}{d\tau}\right]_{\tau=0} = \frac{F}{4\mu}\left[-\frac{1}{\mu_0} S(\tau_1; \mu, \varphi; \mu_0, \varphi_0) - \frac{\partial S(\tau_1; \mu, \varphi; \mu_0, \varphi_0)}{\partial \tau_1}\right] +$$

$$+ \frac{1}{4\pi\mu} \int_0^1 \int_0^{2\pi} S(\tau_1; \mu, \varphi; \mu', \varphi') \left[\frac{dI(\tau, -\mu', \varphi')}{d\tau}\right]_{\tau=0} d\mu' d\varphi',$$

$$\tag{13}$$

$$\left[\frac{dI(\tau, -\mu, \varphi)}{d\tau}\right]_{\tau=\tau_1} = \frac{F}{4\mu} \frac{\partial T(\tau_1; \mu, \varphi; \mu_0, \varphi_0)}{\partial \tau_1} +$$

$$+ \frac{1}{4\pi\mu} \int_0^1 \int_0^{2\pi} S(\tau_1; \mu, \varphi; \mu', \varphi') \left[\frac{dI(\tau, +\mu', \varphi')}{d\tau}\right]_{\tau=\tau_1} d\mu' d\varphi', \tag{14}$$

$$0 = \frac{F}{4\mu} \frac{\partial S(\tau_1; \mu, \varphi; \mu', \varphi')}{\partial \tau_1} + e^{-\tau_1/\mu} \left[\frac{dI(\tau, +\mu, \varphi)}{d\tau} \right]_{\tau=\tau_1} +$$

$$+ \frac{1}{4\pi\mu} \int_0^1 \int_0^{2\pi} T(\tau_1; \mu, \varphi; \mu', \varphi') \left[\frac{dI(\tau, +\mu', \varphi')}{d\tau} \right]_{\tau=\tau_1} d\mu' d\varphi', \quad (15)$$

and

$$0 = \frac{F}{4\mu} \left[-\frac{1}{\mu_0} T(\tau_1; \mu, \varphi; \mu_0, \varphi_0) - \frac{\partial T(\tau_1; \mu, \varphi; \mu_0, \varphi_0)}{\partial \tau_1} \right] +$$

$$+ e^{-\tau_1/\mu} \left[\frac{dI(\tau, -\mu, \varphi)}{d\tau} \right]_{\tau=0} + \frac{1}{4\pi\mu} \int_0^1 \int_0^{2\pi} T(\tau_1; \mu, \varphi; \mu', \varphi') \times$$

$$\times \left[\frac{dI(\tau, -\mu', \varphi')}{d\tau} \right]_{\tau=0} d\mu' d\varphi'. \quad (16)$$

The derivatives which occur in equations (13)–(16) can be found from the equation of transfer (cf. Chap. IV, eqs. [19]–[21])

$$\mu \frac{dI(\tau, \mu, \varphi)}{d\tau} = I(\tau, \mu, \varphi) - \mathfrak{J}(\tau, \mu, \varphi), \quad (17)$$

where

$$\mathfrak{J}(\tau, \mu, \varphi) = \tfrac{1}{4} F p(\mu, \varphi; -\mu_0, \varphi_0) e^{-\tau/\mu_0} +$$

$$+ \frac{1}{4\pi} \int_{-1}^1 \int_0^{2\pi} p(\mu, \varphi; \mu'', \varphi'') I(\tau, \mu'', \varphi'') d\mu'' d\varphi''. \quad (18)$$

We have (cf. eqs. [1] and [12])

$$\left[\frac{dI(\tau, +\mu, \varphi)}{d\tau} \right]_{\tau=0} = +\frac{1}{\mu} \left[\frac{F}{4\mu} S(\tau_1; \mu, \varphi; \mu_0, \varphi_0) - \mathfrak{J}(0, +\mu, \varphi) \right], \quad (19)$$

$$\left[\frac{dI(\tau, -\mu, \varphi)}{d\tau} \right]_{\tau=0} = +\frac{1}{\mu} \mathfrak{J}(0, -\mu, \varphi), \quad (20)$$

$$\left[\frac{dI(\tau, +\mu, \varphi)}{d\tau} \right]_{\tau=\tau_1} = -\frac{1}{\mu} \mathfrak{J}(\tau_1, +\mu, \varphi), \quad (21)$$

$$\left[\frac{dI(\tau, -\mu, \varphi)}{d\tau} \right]_{\tau=\tau_1} = -\frac{1}{\mu} \left[\frac{F}{4\mu} T(\tau_1; \mu, \varphi; \mu_0, \varphi_0) - \mathfrak{J}(\tau_1, -\mu, \varphi) \right]. \quad (22)$$

Using these expressions for the derivatives in equations (13)–(16), we find, after some minor rearranging of the terms, that

$$\tfrac{1}{4}F\left\{\left(\frac{1}{\mu}+\frac{1}{\mu_0}\right)S(\tau_1;\mu,\varphi;\mu_0,\varphi_0)+\frac{\partial S(\tau_1;\mu,\varphi;\mu_0,\varphi_0)}{\partial\tau_1}\right\}$$

$$=\Im(0,+\mu,\varphi)+\frac{1}{4\pi}\int_0^1\int_0^{2\pi}S(\tau_1;\mu,\varphi;\mu',\varphi')\Im(0,-\mu',\varphi')\frac{d\mu'}{\mu'}d\varphi', \quad (23)$$

$$\tfrac{1}{4}F\left\{\frac{1}{\mu}T(\tau_1;\mu,\varphi;\mu_0,\varphi_0)+\frac{\partial T(\tau_1;\mu,\varphi;\mu_0,\varphi_0)}{\partial\tau_1}\right\}$$

$$=\Im(\tau_1,-\mu,\varphi)+\frac{1}{4\pi}\int_0^1\int_0^{2\pi}S(\tau_1;\mu,\varphi;\mu',\varphi')\Im(\tau_1,+\mu',\varphi')\frac{d\mu'}{\mu'}d\varphi', \quad (24)$$

$$\tfrac{1}{4}F\frac{\partial S(\tau_1;\mu,\varphi;\mu_0,\varphi_0)}{\partial\tau_1}=e^{-\tau_1/\mu}\Im(\tau_1,+\mu,\varphi)+$$

$$+\frac{1}{4\pi}\int_0^1\int_0^{2\pi}T(\tau_1;\mu,\varphi;\mu',\varphi')\Im(\tau_1,+\mu',\varphi')\frac{d\mu'}{\mu'}d\varphi', \quad (25)$$

and

$$\tfrac{1}{4}F\left\{\frac{1}{\mu_0}T(\tau_1;\mu,\varphi;\mu_0,\varphi_0)+\frac{\partial T(\tau_1;\mu,\varphi;\mu_0,\varphi_0)}{\partial\tau_1}\right\}$$

$$=e^{-\tau_1/\mu}\Im(0,-\mu,\varphi)+\frac{1}{4\pi}\int_0^1\int_0^{2\pi}T(\tau_1;\mu,\varphi;\mu',\varphi')\Im(0,-\mu',\varphi')\frac{d\mu'}{\mu'}d\varphi'. \tag{26}$$

On the other hand, according to equations (1), (12), and (18)

$$\Im(0,\mu,\varphi)$$

$$=\tfrac{1}{4}F\left\{p(\mu,\varphi;-\mu_0,\varphi_0)+\frac{1}{4\pi}\int_0^1\int_0^{2\pi}p(\mu,\varphi;\mu'',\varphi'')S(\tau_1;\mu'',\varphi'';\mu_0,\varphi_0)\frac{d\mu''}{\mu''}d\varphi''\right\} \tag{27}$$

and

$$\Im(\tau_1,\mu,\varphi)=\tfrac{1}{4}F\left\{e^{-\tau_1/\mu_0}p(\mu,\varphi;-\mu_0,\varphi_0)+\right.$$

$$\left.+\frac{1}{4\pi}\int_0^1\int_0^{2\pi}p(\mu,\varphi;-\mu'',\varphi'')T(\tau_1;\mu'',\varphi'';\mu_0,\varphi_0)\frac{d\mu''}{\mu''}d\varphi''\right\}. \tag{28}$$

Substituting equations (27) and (28) in equations (23)–(26), we obtain

$$\left(\frac{1}{\mu}+\frac{1}{\mu_0}\right)S(\tau_1;\mu,\varphi;\mu_0,\varphi_0)+\frac{\partial S(\tau_1;\mu,\varphi;\mu_0,\varphi_0)}{\partial \tau_1}=p(\mu,\varphi;-\mu_0,\varphi_0)+$$

$$+\frac{1}{4\pi}\int_0^1\int_0^{2\pi}p(\mu,\varphi;\mu'',\varphi'')S(\tau_1;\mu'',\varphi'';\mu_0,\varphi_0)\frac{d\mu''}{\mu''}\,d\varphi''+$$

$$+\frac{1}{4\pi}\int_0^1\int_0^{2\pi}S(\tau_1;\mu,\varphi;\mu',\varphi')p(-\mu',\varphi';-\mu_0,\varphi_0)\frac{d\mu'}{\mu'}\,d\varphi'+$$

$$+\frac{1}{16\pi^2}\int_0^1\int_0^{2\pi}\int_0^1\int_0^{2\pi}S(\tau_1;\mu,\varphi;\mu',\varphi')p(-\mu',\varphi';\mu'',\varphi'')\times$$

$$\times S(\tau_1;\mu'',\varphi'';\mu_0,\varphi_0)\frac{d\mu'}{\mu'}\,d\varphi'\frac{d\mu''}{\mu''}\,d\varphi'',\quad(29)$$

$$\frac{\partial S(\tau_1;\mu,\varphi;\mu_0,\varphi_0)}{\partial \tau_1}=p(\mu,\varphi;-\mu_0,\varphi_0)\exp\left\{-\tau_1\left(\frac{1}{\mu}+\frac{1}{\mu_0}\right)\right\}+$$

$$+\frac{1}{4\pi}e^{-\tau_1/\mu}\int_0^1\int_0^{2\pi}p(\mu,\varphi;-\mu'',\varphi'')T(\tau_1;\mu'',\varphi'';\mu_0,\varphi_0)\frac{d\mu''}{\mu''}\,d\varphi''+$$

$$+\frac{1}{4\pi}e^{-\tau_1/\mu_0}\int_0^1\int_0^{2\pi}T(\tau_1;\mu,\varphi;\mu',\varphi')p(\mu',\varphi';-\mu_0,\varphi_0)\frac{d\mu'}{\mu'}\,d\varphi'+$$

$$+\frac{1}{16\pi^2}\int_0^1\int_0^{2\pi}\int_0^1\int_0^{2\pi}T(\tau_1;\mu,\varphi;\mu',\varphi')p(\mu',\varphi';-\mu'',\varphi'')\times$$

$$\times T(\tau_1;\mu'',\varphi'';\mu_0,\varphi_0)\frac{d\mu'}{\mu'}\,d\varphi'\frac{d\mu''}{\mu''}\,d\varphi'',\quad(30)$$

$$\frac{1}{\mu}T(\tau_1;\mu,\varphi;\mu_0,\varphi_0)+\frac{\partial T(\tau_1;\mu,\varphi;\mu_0,\varphi_0)}{\partial \tau_1}=e^{-\tau_1/\mu_0}p(-\mu,\varphi;-\mu_0,\varphi_0)+$$

$$+\frac{1}{4\pi}\int_0^1\int_0^{2\pi}p(-\mu,\varphi;-\mu'',\varphi'')T(\tau_1;\mu'',\varphi'';\mu_0,\varphi_0)\frac{d\mu''}{\mu''}\,d\varphi''+$$

$$+\frac{e^{-\tau_1/\mu_0}}{4\pi}\int_0^1\int_0^{2\pi}S(\tau_1;\mu,\varphi;\mu',\varphi')p(\mu',\varphi';-\mu_0,\varphi_0)\frac{d\mu'}{\mu'}\,d\varphi'+$$

$$+\frac{1}{16\pi^2}\int_0^1\int_0^{2\pi}\int_0^1\int_0^{2\pi}S(\tau_1;\mu,\varphi;\mu',\varphi')p(\mu',\varphi';-\mu'',\varphi'')\times$$

$$\times T(\tau_1;\mu'',\varphi'';\mu_0,\varphi_0)\frac{d\mu'}{\mu'}\,d\varphi'\frac{d\mu''}{\mu''}\,d\varphi'',\quad(31)$$

and

$$\frac{1}{\mu_0} T(\tau_1; \mu, \varphi; \mu_0, \varphi_0) + \frac{\partial T(\tau_1; \mu, \varphi; \mu_0, \varphi_0)}{\partial \tau_1} = e^{-\tau_1/\mu} p(-\mu, \varphi; -\mu_0, \varphi_0) +$$

$$+ \frac{e^{-\tau_1/\mu}}{4\pi} \int_0^1 \int_0^{2\pi} p(-\mu, \varphi; \mu'', \varphi'') S(\tau_1; \mu'', \varphi''; \mu_0, \varphi_0) \frac{d\mu''}{\mu''} d\varphi'' +$$

$$+ \frac{1}{4\pi} \int_0^1 \int_0^{2\pi} T(\tau_1; \mu, \varphi; \mu', \varphi') p(-\mu', \varphi'; -\mu_0, \varphi_0) \frac{d\mu'}{\mu'} d\varphi' +$$

$$+ \frac{1}{16\pi^2} \int_0^1 \int_0^{2\pi} \int_0^1 \int_0^{2\pi} T(\tau_1; \mu, \varphi; \mu', \varphi') p(-\mu', \varphi'; \mu'', \varphi'') \times$$

$$\times S(\tau_1; \mu'', \varphi''; \mu_0, \varphi_0) \frac{d\mu'}{\mu'} d\varphi' \frac{d\mu''}{\mu''} d\varphi''. \quad (32)$$

Equations (29)–(32) represent four integral equations which govern the problem of diffuse reflection and transmission by plane-parallel atmospheres of finite optical thicknesses.

Now, by simple subtraction, we can eliminate $\partial S/\partial \tau_1$ from equations (29) and (30); similarly, we can eliminate $\partial T/\partial \tau_1$ from equations (31) and (32). The resulting pair of integral equations, between S and T only, may be regarded as the expression of *the invariance of the laws of diffuse reflection and transmission to the addition (or, removal) of layers of arbitrary optical thickness to (or, from) the atmosphere at the top and the simultaneous removal (or, addition) of layers of equal optical thickness from (or, to) the atmosphere at the bottom.*

To generalize the foregoing considerations to allow for the polarization of the radiation field, we must, of course, first express the laws of diffuse reflection and transmission in terms of a scattering matrix $S(\tau_1; \mu, \varphi; \mu_0, \varphi_0)$ and a transmission matrix $T(\tau_1; \mu, \varphi; \mu_0, \varphi_0)$ (cf. Chap. I, eq. [230]). However, in view of the complete similarity of the equation of transfer for I (whose components are the Stokes parameters I_l, I_r, U, and V) and the ordinary equation of transfer (when polarization is ignored) and the entirely parallel roles which a phase-matrix $P(\mu, \varphi; \mu', \varphi')$ and a phase function $p(\mu, \varphi; \mu', \varphi')$ play in the respective equations, it is apparent that S and T will satisfy equations of *exactly* the same forms as equations (29)–(32): only, the functions S, T, and p which occur in these equations must be replaced by the matrices S, T, and P.

52. The principle of reciprocity

In § 51 we have indicated how, by subtracting equations (30) and (32) from equations (29) and (31), respectively, we can eliminate the derivatives $\partial S/\partial \tau_1$ and $\partial T/\partial \tau_1$ and obtain a pair of integral equations for S and T only. From the structure of these equations and the properties of the phase function for transposition (cf. Chap. IV, § 31, eqs. [29] and [31]) it follows that *if $S(\tau_1; \mu, \varphi; \mu_0, \varphi_0)$ and $T(\tau_1; \mu, \varphi; \mu_0, \varphi_0)$ are solutions, then so are*

$$\tilde{S}(\tau_1; \mu, \varphi; \mu_0, \varphi_0) = S(\tau_1; \mu_0, \varphi_0; \mu, \varphi),$$

and
$$\tilde{T}(\tau_1; \mu, \varphi; \mu_0, \varphi_0) = T(\tau_1; \mu_0, \varphi_0; \mu, \varphi). \tag{33}$$

For example, by transposing the variables (μ, φ) and (μ_0, φ_0), we find that the right-hand side of equation (31) becomes:

$$e^{-\tau_1/\mu} p(-\mu, \varphi; -\mu_0, \varphi_0) +$$

$$+ \frac{1}{4\pi} \int_0^1 \int_0^{2\pi} \tilde{T}(\tau_1; \mu, \varphi; \mu'', \varphi'') p(-\mu'', \varphi''; -\mu_0, \varphi_0) \frac{d\mu''}{\mu''} d\varphi'' +$$

$$+ \frac{e^{-\tau_1/\mu}}{4\pi} \int_0^1 \int_0^{2\pi} p(-\mu, \varphi; \mu', \varphi') \tilde{S}(\tau_1; \mu', \varphi'; \mu_0, \varphi_0) \frac{d\mu'}{\mu'} d\varphi' +$$

$$+ \frac{1}{16\pi^2} \int_0^1 \int_0^{2\pi} \int_0^1 \int_0^{2\pi} \tilde{T}(\tau_1; \mu, \varphi; \mu'', \varphi'') \times$$

$$\times p(-\mu'', \varphi''; \mu', \varphi') \tilde{S}(\tau_1; \mu', \varphi'; \mu_0, \varphi_0) \frac{d\mu'}{\mu'} d\varphi' \frac{d\mu''}{\mu''} d\varphi'',$$

and this is the right-hand side of equation (32) for $\tilde{S}$ and $\tilde{T}$. (The other equations transform similarly for the interchange of the variables.)

On the other hand, from equation (30) and the corresponding equation for $\partial T/\partial \tau_1$ obtained by eliminating T between equations (31) and (32), it follows that if $S = \tilde{S}$ and $T = \tilde{T}$ for a particular value of τ_1, then the derivatives, $\partial S/\partial \tau_1$ and $\partial T/\partial \tau_1$, are also symmetrical for the same value of τ_1: for example, from equation (30)

$$\frac{\partial S(\tau_1; \mu_0, \varphi_0; \mu, \varphi)}{\partial \tau_1} = p(\mu, \varphi; -\mu_0, \varphi_0) \exp\left\{-\tau_1\left(\frac{1}{\mu} + \frac{1}{\mu_0}\right)\right\} +$$

$$+ \frac{1}{4\pi} e^{-\tau_1/\mu_0} \int_0^1 \int_0^{2\pi} \tilde{T}(\tau_1; \mu, \varphi; \mu'', \varphi'') p(\mu'', \varphi''; -\mu_0, \varphi_0) \frac{d\mu''}{\mu''} d\varphi'' +$$

$$+ \frac{1}{4\pi} e^{-\tau_1/\mu} \int_0^1 \int_0^{2\pi} p(\mu, \varphi; -\mu', \varphi') \tilde{T}(\tau_1; \mu', \varphi'; \mu_0, \varphi_0) \frac{d\mu'}{\mu'} d\varphi' +$$

$$+\frac{1}{16\pi^2}\int_0^1\int_0^{2\pi}\int_0^1\int_0^{2\pi}\tilde{T}(\tau_1;\mu,\varphi;\mu'',\varphi'')p(\mu'',\varphi'';-\mu',\varphi')\times$$

$$\times\tilde{T}(\tau_1;\mu',\varphi';\mu_0,\varphi_0)\frac{d\mu'}{\mu'}\,d\varphi'\,\frac{d\mu''}{\mu''}\,d\varphi''$$

$$=\frac{\partial S(\tau_1;\mu,\varphi;\mu_0,\varphi_0)}{\partial\tau_1}\quad\text{if }\ T=\tilde{T}\ \ \text{for }\ \tau=\tau_1;\quad(34)$$

S and T will therefore continue to be symmetrical for $\tau_1+d\tau_1$. Hence, S and T will be symmetrical for all values of τ_1 *if* they are symmetrical for any particular value of τ_1.

Now from equations (29)–(32) it is apparent that, as $\tau_1\to 0$,

$$\left(\frac{1}{\mu}+\frac{1}{\mu_0}\right)S(\tau_1;\mu,\varphi;\mu_0,\varphi_0)$$

$$=p(\mu,\varphi;-\mu_0,\varphi_0)\left[1-\exp\left\{-\tau_1\left(\frac{1}{\mu}+\frac{1}{\mu_0}\right)\right\}\right]+O(\tau_1^2)$$

and

$$\left(\frac{1}{\mu}-\frac{1}{\mu_0}\right)T(\tau_1;\mu,\varphi;\mu_0,\varphi_0)=p(\mu,\varphi;\mu_0,\varphi_0)[e^{-\tau_1/\mu_0}-e^{-\tau_1/\mu}]+O(\tau_1^2).\quad(35)\dagger$$

Evidently S and T are symmetrical in (μ,φ) and (μ_0,φ_0) for $\tau_1\to 0$. According to our earlier remarks, we conclude that S and T will be symmetrical for all values of τ_1. We have thus shown that

$$\tilde{S}(\tau_1;\mu,\varphi;\mu_0,\varphi_0)\equiv S(\tau_1;\mu,\varphi;\mu_0,\varphi_0)$$

and

$$\tilde{T}(\tau_1;\mu,\varphi;\mu_0,\varphi_0)\equiv T(\tau_1;\mu,\varphi;\mu_0,\varphi_0).\quad(36)$$

In other words, *the scattering and the transmission functions are unaltered when the directions of incidence and emergence are interchanged.* This is the statement of the principle of reciprocity as it applies to the problem of diffuse reflection and transmission. It is of particular interest to notice that in the present context the principle emerges directly as a consequence of its validity for single scattering, a circumstance which clarifies the origin of the principle.

When the polarization of the radiation field is taken into account, the corresponding statement of the principle of reciprocity derives

---

† These expressions for S and T can also be derived, directly, from the consideration that as $\tau_1\to 0$, single scattering is all that need be taken into account (cf. Chap. VI, § 47.1 and Chap. IX, § 63).

similarly from the symmetry of the phase-matrix for transposition. Thus, letting

$$\boldsymbol{P}(\mu,\varphi;\mu',\varphi') = \boldsymbol{Q}\boldsymbol{\mathscr{P}}(\mu,\varphi;\mu',\varphi'), \tag{37}\dagger$$

where

$$\boldsymbol{Q} = \begin{bmatrix} 1 & 0 & 0 & 0 \\ 0 & 1 & 0 & 0 \\ 0 & 0 & 2 & 0 \\ 0 & 0 & 0 & 2 \end{bmatrix}, \tag{38}$$

we have (cf. Chap. I, eqs. [220]–[225])

$$\widetilde{\boldsymbol{\mathscr{P}}}(\mu,\varphi;\mu',\varphi') = \boldsymbol{\mathscr{P}}(\mu,\varphi;\mu',\varphi'), \tag{39}$$

where it may be recalled that $\widetilde{\boldsymbol{\mathscr{P}}}$ stands for the matrix obtained from $\boldsymbol{\mathscr{P}}$ by interchanging its rows and columns and the arguments (μ,φ) and (μ',φ'). In terms of its elements, equation (39) is equivalent to

$$\widetilde{\mathscr{P}}_{ik}(\mu,\varphi;\mu',\varphi') = \mathscr{P}_{ki}(\mu,\varphi;\mu',\varphi') \quad (i,k = l,r,U,\text{and } V). \tag{40}$$

In addition, we also have the relations

$$\widetilde{\mathscr{P}}_{ik}(\mu,\varphi;-\mu',\varphi') = \mathscr{P}_{ki}(\mu,\varphi;-\mu',\varphi') \quad (i = k \text{ and } i \neq k \neq U)$$

and $\quad \widetilde{\mathscr{P}}_{iU}(\mu,\varphi;-\mu',\varphi') = -\mathscr{P}_{Ui}(\mu,\varphi;-\mu',\varphi') \quad (i = l,r,V). \tag{41}$

Now writing

$$\boldsymbol{S}(\tau_1;\mu,\varphi;\mu_0,\varphi_0) = \boldsymbol{Q}\boldsymbol{\mathscr{S}}(\tau_1;\mu,\varphi;\mu_0,\varphi_0)$$

and $\quad \boldsymbol{T}(\tau_1;\mu,\varphi;\mu_0,\varphi_0) = \boldsymbol{Q}\boldsymbol{\mathscr{T}}(\tau_1;\mu,\varphi;\mu_0,\varphi_0), \tag{42}$

we infer, from equations for $\boldsymbol{\mathscr{S}}$ and $\boldsymbol{\mathscr{T}}$ analogous to equations (35), that for $\tau_1 \to 0$ these matrices have the same symmetries as $\boldsymbol{\mathscr{P}}(\mu,\varphi;-\mu_0,\varphi_0)$ and $\boldsymbol{\mathscr{P}}(-\mu,\varphi;-\mu_0,\varphi_0)$, respectively, for transposition. By continuation we can prove (though the details are somewhat elaborate) that these symmetry properties are preserved identically for all values of τ_1. Thus

$$\widetilde{\mathscr{S}}_{ik}(\tau_1;\mu,\varphi;\mu_0,\varphi_0) = \mathscr{S}_{ki}(\tau;\mu,\varphi;\mu_0,\varphi_0) \quad (i = k \text{ and } i \neq k \neq U),$$

$$\widetilde{\mathscr{S}}_{iU}(\tau_1;\mu,\varphi;\mu_0,\varphi_0) = -\mathscr{S}_{Ui}(\tau_1;\mu,\varphi;\mu_0,\varphi_0) \quad (i = l,r,V) \tag{43}$$

and $\quad \widetilde{\mathscr{T}}_{ik}(\tau_1;\mu,\varphi;\mu_0,\varphi_0) = \mathscr{T}_{ki}(\tau_1;\mu,\varphi;\mu_0,\varphi_0) \quad (i,k = l,r,U,V). \tag{44}$

The relationships expressed by equations (43) and (44) for the reversal of the light path by interchanging the directions of incidence and

---

† The origin of the factor $\boldsymbol{Q}$ goes back to the particular choice of I_l, I_r, U, and V as the parameters for representing polarized light. That there is a slight 'asymmetry' between I_l and I_r on the one hand and U and V on the other is apparent from the form of the linear transformation $\boldsymbol{L}(\phi)$ (Chap. I, eq. [190]) which these parameters undergo for a rotation of the axes (see also Chap. I, eq. [161]).

emergence represent the complete statement of the principle of reciprocity as it applies to the problem of diffuse reflection and transmission.

To elucidate the meaning of the relations (43) and (44), we consider the case of incidence of a parallel beam of light, plane polarized in a direction χ_0 to the meridian plane containing the direction, $(-\mu_0, \varphi_0)$, of incidence. The components of the incident flux, then, are

$$F_l = F\cos^2\chi_0, \quad F_r = F\sin^2\chi_0, \quad F_U = F\sin 2\chi_0 \quad \text{and} \quad F_V = 0. \tag{45}$$

The intensities in the three components l, r, and U, reflected in the direction (μ, φ), are therefore

$$I_l(0, \mu, \varphi) = \frac{F}{4\mu}(\mathscr{S}_{ll}\cos^2\chi_0 + \mathscr{S}_{lr}\sin^2\chi_0 + \mathscr{S}_{lU}\sin 2\chi_0),$$

$$I_r(0, \mu, \varphi) = \frac{F}{4\mu}(\mathscr{S}_{rl}\cos^2\chi_0 + \mathscr{S}_{rr}\sin^2\chi_0 + \mathscr{S}_{rU}\sin 2\chi_0),$$

$$U(0, \mu, \varphi) = \frac{F}{2\mu}(\mathscr{S}_{Ul}\cos^2\chi_0 + \mathscr{S}_{Ur}\sin^2\chi_0 + \mathscr{S}_{UU}\sin 2\chi_0). \tag{46}$$

The intensity of this reflected light in a direction making an angle χ with the meridian plane through (μ, φ) can be obtained by combining the foregoing equations according to the formula (Chap. I, eq. [161])

$$I(\chi) = I_l\cos^2\chi + I_r\sin^2\chi + \tfrac{1}{2}U\sin 2\chi. \tag{47}$$

We thus have

$$I(0; \mu, \varphi, \chi; \mu_0, \varphi_0, \chi_0)$$

$$= \frac{F}{4\mu}[\mathscr{S}_{ll}\cos^2\chi_0\cos^2\chi + \mathscr{S}_{rr}\sin^2\chi_0\sin^2\chi + \mathscr{S}_{UU}\sin 2\chi_0\sin 2\chi +$$

$$+ (\mathscr{S}_{lr}\sin^2\chi_0\cos^2\chi + \mathscr{S}_{rl}\cos^2\chi_0\sin^2\chi) +$$

$$+ (\mathscr{S}_{lU}\sin 2\chi_0\cos^2\chi + \mathscr{S}_{Ul}\cos^2\chi_0\sin 2\chi) +$$

$$+ (\mathscr{S}_{rU}\sin 2\chi_0\sin^2\chi + \mathscr{S}_{Ur}\sin^2\chi_0\sin 2\chi)]. \tag{48}$$

Interchanging the variables (μ, φ) and (μ_0, φ_0) and using the relations (43), we observe that

$$\mu I(0; \mu, \varphi, \chi; \mu_0, \varphi_0, \chi_0) = \mu_0\, I(0; \mu_0, \varphi_0, -\chi_0; \mu, \varphi, -\chi). \tag{49}$$

The reversal of the signs of χ and χ_0 when the directions of incidence and reflection are interchanged is required by the fact that the interchange reverses the sense of the measurement of the azimuthal difference $(\varphi_0 - \varphi)$ and consequently, also, the sense of χ.

For transmitted light, the equation analogous to (49) is

$$\mu I(\tau_1; -\mu, \varphi, \chi; \mu_0, \varphi_0, \chi_0) = \mu_0\, I(\tau_1; -\mu_0, \varphi_0, \chi_0; \mu, \varphi, \chi). \tag{50}$$

More generally than for incident plane-polarized light, equations (43) and (44) imply the following principle of reciprocity:

If a parallel beam of light, elliptically polarized in a direction χ_0 and characterized by an ellipticity parameter β_0, be incident in a direction $(-\mu_0, \varphi_0)$, and if the light reflected in the direction (μ, φ) be analysed

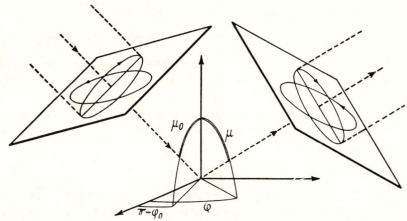

FIG. 20. Illustrating the principle of reciprocity for incident elliptically polarized light: Let light elliptically polarized and characterized by the parameters χ_0, β_0 be incident in the direction $(-\mu_0, \varphi_0)$ and let the light diffusely reflected in the direction (μ, φ) be analysed in the states of opposite polarization (χ, β) and $(\chi + \frac{1}{2}\pi, -\beta)$. In the reverse experiment, light in the state of polarization (χ, β) is incident in the direction $(-\mu, \varphi)$ and the light diffusely reflected in the direction (μ_0, φ_0) is analysed in the states of opposite polarization (χ_0, β_0) and $(\chi_0 + \frac{1}{2}\pi, -\beta_0)$ and the intensity in the former state is compared with what was reflected in the state (χ, β) in the first experiment.

in the two states of opposite polarization (χ, β) and $(\chi + \frac{1}{2}\pi, -\beta)$ then the intensity $I(0; \mu, \varphi, \chi, \beta; \mu_0, \varphi_0, \chi_0, \beta_0)$ in the component (χ, β) has the symmetry property

$$\mu I(0; \mu, \varphi, \chi, \beta; \mu_0, \varphi_0, \chi_0, \beta_0) = \mu_0 I(0; \mu_0, \varphi_0, -\chi_0, \beta_0; \mu, \varphi, -\chi, \beta). \quad (51)$$

Similarly, for the intensity transmitted in the direction $(-\mu, \varphi)$, we have

$$\mu I(\tau_1; -\mu, \varphi, \chi, \beta; \mu_0, \varphi_0, \chi_0, \beta_0) = \mu_0 I(\tau_1; -\mu_0, \varphi_0, \chi_0, \beta_0; \mu, \varphi, \chi, \beta). \quad (52)$$

According to equations (51) and (52), when the light path is reversed by interchanging the directions of incidence and emergence, light in the state of polarization in which the reflected (respectively, transmitted) light was originally resolved in the direction (μ, φ) is incident in that direction, while the light reflected (respectively, transmitted) in the direction (μ_0, φ_0) is resolved into a component having the ellipticity and the plane of polarization which the light originally incident in this direction had (see Fig. 20). This principle of reciprocity for incident elliptically polarized light follows from the symmetry properties of $\mathscr{S}$ and $\mathscr{T}$ and the formulae

$$I(\chi,\beta) = I_l(\cos^2\chi\,\cos^2\beta + \sin^2\chi\,\sin^2\beta) + I_r(\sin^2\chi\,\cos^2\beta + \cos^2\chi\,\sin^2\beta) +$$
$$+ \tfrac{1}{2}U\sin 2\chi\,\cos 2\beta + \tfrac{1}{2}V\sin 2\beta,$$

and

$$I(\chi + \tfrac{1}{2}\pi, -\beta) = I_l(\sin^2\chi\,\cos^2\beta + \cos^2\chi\,\sin^2\beta) +$$
$$+ I_r(\cos^2\chi\,\cos^2\beta + \sin^2\chi\,\sin^2\beta) - \tfrac{1}{2}U\sin 2\chi\,\cos 2\beta - \tfrac{1}{2}V\sin 2\beta, \quad (53)$$

which give the intensities of a beam characterized by the Stokes parameters I_l, I_r, U, and V, when resolved in the two states of opposite polarization (χ,β) and $(\chi + \tfrac{1}{2}\pi, -\beta)$.

Finally, we may notice that a principle of reciprocity can be formulated for the *total* intensity when a beam of *natural* light is reflected or transmitted; for then

$$I_l = I_r = \tfrac{1}{2}F \quad \text{and} \quad F_U = F_V = 0, \quad (54)$$

and the total intensities of the reflected and the transmitted light can be expressed in terms of a scattering and a transmission function which are related to the elements of $\mathscr{S}$ and $\mathscr{T}$ according to

$$S_{\text{natural}}(\tau_1; \mu, \varphi; \mu_0, \varphi_0) = \tfrac{1}{2}[\mathscr{S}_{ll} + \mathscr{S}_{lr} + \mathscr{S}_{rl} + \mathscr{S}_{rr}]_{(\tau_1;\,\mu,\varphi;\,\mu_0,\varphi_0)}$$

and

$$T_{\text{natural}}(\tau_1; \mu, \varphi; \mu_0, \varphi_0) = \tfrac{1}{2}[\mathscr{T}_{ll} + \mathscr{T}_{lr} + \mathscr{T}_{rl} + \mathscr{T}_{rr}]_{(\tau_1;\,\mu,\varphi;\,\mu_0,\varphi_0)}. \quad (55)$$

From equations (43) and (44) it now follows that

$$\tilde{S}_{\text{natural}} = S_{\text{natural}}; \quad \tilde{T}_{\text{natural}} = T_{\text{natural}}, \quad (56)$$

which is clearly a statement of a principle of reciprocity. It should, however, be emphasized that for total intensity, a principle of reciprocity of this character exists only for the case of incident natural light; in no other case is the principle valid for total intensities.

The symmetries of the scattering and the transmission functions and matrices which we have designated under 'a principle of reciprocity' derive from a theorem in geometrical optics due to Helmholtz which states that, if a ray of light (i) after any number of refractions and reflections at plane or nearly plane surfaces gives rise (among others) to a ray (e) whose intensity is a certain fraction f_{ie} of the intensity of the ray (i), then, on reversing the path of light, an incident ray (e)' will give rise (among others) to a ray (i)' whose intensity is a fraction f_{ei} of the ray (e)' such that

$$f_{ei} = f_{ie}. \quad (57)$$

For the validity of Helmholtz's theorem in geometrical optics, the surfaces if not plane must be such that their radius of curvature is large compared with the lateral extent of the ray, and this lateral extent must itself be large compared with the wave-lengths in the ray. However, in the context of problems of diffuse reflection and transmission, the following observation of Rayleigh is more relevant:

'Suppose that in any direction (i) and at any distance r from a small surface (s) reflecting in *any manner* there be situated a radiant point (A) of given intensity, and consider the intensity of the reflected vibrations at any point B situated in a direction ϵ and a distance r' from s. The theorem is to the effect that the intensity is the same as it would be at A if the radiant point were transferred to B.'

Recently, Minnaert and van de Hulst have re-examined the basic ideas underlying the theorems of Rayleigh and Helmholtz with a view to formulating a principle of as wide a range of applicability as possible. It is beyond the scope of this book to go into these discussions here: it suffices for our purposes that the principle in its entire generality can be derived from the integral equations incorporating the various principles of invariance underlying the problem of diffuse reflection and transmission by plane-parallel atmospheres.

53. The reduction of the integral equations (29)–(32) for the case in which the phase function is expressible as a series in Legendre polynomials

We shall now show how the integral equations (29)–(32) can be reduced to a basic system of equations for functions in one variable only, when the phase function has the form considered in Chapter VI, § 48, namely,

$$p(\mu,\varphi;\mu',\varphi') = \sum_{m=0}^{N} (2-\delta_{0,m})\Big\{\sum_{l=m}^{N} \varpi_l^m P_l^m(\mu)P_l^m(\mu')\Big\}\cos m(\varphi'-\varphi). \quad (58)$$

From the manner in which the phase function enters the integral equations (29)–(32), it is clear that when the phase function has the form (58), the scattering and the transmission functions have the analogous expansions

$$S(\tau_1;\mu,\varphi;\mu_0,\varphi_0) = \sum_{m=0}^{N} S^{(m)}(\tau_1;\mu,\mu_0)\cos m(\varphi_0-\varphi)$$

and $$T(\tau_1;\mu,\varphi;\mu_0,\varphi_0) = \sum_{m=0}^{N} T^{(m)}(\tau_1;\mu,\mu_0)\cos m(\varphi_0-\varphi), \quad (59)$$

where, as the notation indicates, $S^{(m)}$ and $T^{(m)}$ are functions of τ_1, μ, and μ_0 only. Substituting these expansions for S and T in equations

(29)–(32), we find that the equations for the various Fourier components separate; thus equation (29) leads to the system

$$\left(\frac{1}{\mu}+\frac{1}{\mu_0}\right)S^{(m)}(\tau_1;\mu,\mu_0)+\frac{\partial S^{(m)}(\tau_1;\mu,\mu_0)}{\partial \tau_1}$$

$$= (2-\delta_{0,m})\sum_{l=m}^{N}(-1)^{m+l}\varpi_l^m P_l^m(\mu)P_l^m(\mu_0)+$$

$$+\tfrac{1}{2}\sum_{l=m}^{N}\varpi_l^m P_l^m(\mu)\int_0^1 P_l^m(\mu'')S^{(m)}(\tau_1;\mu'',\mu_0)\frac{d\mu''}{\mu''}+$$

$$+\tfrac{1}{2}\sum_{l=m}^{N}\varpi_l^m P_l^m(\mu_0)\int_0^1 S^{(m)}(\tau_1;\mu,\mu')P_l^m(\mu')\frac{d\mu'}{\mu'}+$$

$$+\frac{1}{4(2-\delta_{0,m})}\sum_{l=m}^{N}(-1)^{m+l}\varpi_l^m\int_0^1\int_0^1 S^{(m)}(\tau_1;\mu,\mu')P_l^m(\mu')P_l^m(\mu'')\times$$

$$\times S^{(m)}(\tau_1;\mu'',\mu_0)\frac{d\mu'}{\mu'}\frac{d\mu''}{\mu''}, \qquad (60)$$

for $m = 0$, 1, etc. The right-hand side of this equation can clearly be expressed as a sum of products in the form

$$\left(\frac{1}{\mu}+\frac{1}{\mu_0}\right)S^{(m)}(\tau_1;\mu,\mu_0)+\frac{\partial S^{(m)}(\tau_1;\mu,\mu_0)}{\partial \tau_1} = (2-\delta_{0,m})\sum_{l=m}^{N}(-1)^{m+l}\varpi_l^m\times$$

$$\times\left[P_l^m(\mu)+\frac{(-1)^{m+l}}{2(2-\delta_{0,m})}\int_0^1 S^{(m)}(\tau_1;\mu,\mu')P_l^m(\mu')\frac{d\mu'}{\mu'}\right]\times$$

$$\times\left[P_l^m(\mu_0)+\frac{(-1)^{m+l}}{2(2-\delta_{0,m})}\int_0^1 P_l^m(\mu'')S^{(m)}(\tau_1;\mu'',\mu_0)\frac{d\mu''}{\mu''}\right]. \quad (61)$$

Similarly, from equations (30)–(32) we find:

$$\frac{\partial S^{(m)}(\tau_1;\mu,\mu_0)}{\partial \tau_1} = (2-\delta_{0,m})\sum_{l=m}^{N}(-1)^{m+l}\varpi_l^m\times$$

$$\times\left[e^{-\tau_1/\mu}P_l^m(\mu)+\frac{1}{2(2-\delta_{0,m})}\int_0^1 T^{(m)}(\tau_1;\mu,\mu')P_l^m(\mu')\frac{d\mu'}{\mu'}\right]\times$$

$$\times\left[e^{-\tau_1/\mu_0}P_l^m(\mu_0)+\frac{1}{2(2-\delta_{0,m})}\int_0^1 P_l^m(\mu'')T^{(m)}(\tau_1;\mu'',\mu_0)\frac{d\mu''}{\mu''}\right], \quad (62)$$

$$\frac{1}{\mu} T^{(m)}(\tau_1; \mu, \mu_0) + \frac{\partial T^{(m)}(\tau_1; \mu, \mu_0)}{\partial \tau_1}$$

$$= (2 - \delta_{0,m}) \sum_{l=m}^{N} \varpi_l^m \left[P_l^m(\mu) + \frac{(-1)^{m+l}}{2(2 - \delta_{0,m})} \int_0^1 S^{(m)}(\tau_1; \mu, \mu') P_l^m(\mu') \frac{d\mu'}{\mu'} \right] \times$$

$$\times \left[e^{-\tau_1/\mu_0} P_l^m(\mu_0) + \frac{1}{2(2 - \delta_{0,m})} \int_0^1 P_l^m(\mu'') T^{(m)}(\tau_1; \mu'', \mu_0) \frac{d\mu''}{\mu''} \right], \quad (63)$$

and

$$\frac{1}{\mu_0} T^{(m)}(\tau_1; \mu, \mu_0) + \frac{\partial T^{(m)}(\tau_1; \mu, \mu_0)}{\partial \tau_1} = (2 - \delta_{0,m}) \sum_{l=m}^{N} \varpi_l^m \times$$

$$\times \left[e^{-\tau_1/\mu} P_l^m(\mu) + \frac{1}{2(2 - \delta_{0,m})} \int_0^1 T^{(m)}(\tau_1; \mu, \mu') P_l^m(\mu') \frac{d\mu'}{\mu'} \right] \times$$

$$\times \left[P_l^m(\mu_0) + \frac{(-1)^{m+l}}{2(2 - \delta_{0,m})} \int_0^1 P_l^m(\mu'') S^{(m)}(\tau_1; \mu'', \mu_0) \frac{d\mu''}{\mu''} \right]. \quad (64)$$

If we now let

$$\psi_l^m(\tau_1; \mu) = P_l^m(\mu) + \frac{(-1)^{m+l}}{2(2 - \delta_{0,m})} \int_0^1 S^{(m)}(\tau_1; \mu, \mu') P_l^m(\mu') \frac{d\mu'}{\mu'} \quad (65)$$

and

$$\phi_l^m(\tau_1; \mu) = e^{-\tau_1/\mu} P_l^m(\mu) + \frac{1}{2(2 - \delta_{0,m})} \int_0^1 T^{(m)}(\tau_1; \mu, \mu') P_l^m(\mu') \frac{d\mu'}{\mu'}, \quad (66)$$

then, in view of the principle of reciprocity (§ 52), we can rewrite equations (61)–(64) in the forms

$$\left(\frac{1}{\mu} + \frac{1}{\mu_0} \right) S^{(m)}(\tau_1; \mu, \mu_0) + \frac{\partial S^{(m)}(\tau_1; \mu, \mu_0)}{\partial \tau_1}$$

$$= (2 - \delta_{0,m}) \sum_{l=m}^{N} (-1)^{m+l} \varpi_l^m \psi_l^m(\tau_1; \mu) \psi_l^m(\tau_1; \mu_0), \quad (67)$$

$$\frac{\partial S^{(m)}(\tau_1; \mu, \mu_0)}{\partial \tau_1} = (2 - \delta_{0,m}) \sum_{l=m}^{N} (-1)^{m+l} \varpi_l^m \phi_l^m(\tau_1; \mu) \phi_l^m(\tau; \mu_0), \quad (68)$$

$$\frac{1}{\mu} T^{(m)}(\tau_1; \mu, \mu_0) + \frac{\partial T^{(m)}(\tau_1; \mu, \mu_0)}{\partial \tau_1} = (2 - \delta_{0,m}) \sum_{l=m}^{N} \varpi_l^m \psi_l^m(\tau_1; \mu) \phi_l^m(\tau_1; \mu_0), \quad (69)$$

and

$$\frac{1}{\mu_0} T^{(m)}(\tau_1; \mu, \mu_0) + \frac{\partial T^{(m)}(\tau_1; \mu, \mu_0)}{\partial \tau_1} = (2 - \delta_{0,m}) \sum_{l=m}^{N} \varpi_l^m \phi_l^m(\tau_1; \mu) \psi_l^m(\tau_1; \mu_0). \quad (70)$$

Alternative forms of the foregoing equations are

$$\left(\frac{1}{\mu_0}+\frac{1}{\mu}\right)S^{(m)}(\tau_1;\mu,\mu_0)$$

$$= (2-\delta_{0,m})\sum_{l=m}^{N}(-1)^{m+l}\varpi_l^m[\psi_l^m(\tau_1;\mu)\psi_l^m(\tau_1;\mu_0)-\phi_l^m(\tau_1;\mu)\phi_l^m(\tau_1;\mu_0)], \quad (71)$$

$$\left(\frac{1}{\mu_0}-\frac{1}{\mu}\right)T^{(m)}(\tau_1;\mu,\mu_0)$$

$$= (2-\delta_{0,m})\sum_{l=m}^{N}\varpi_l^m[\phi_l^m(\tau_1;\mu)\psi_l^m(\tau_1;\mu_0)-\psi_l^m(\tau_1;\mu)\phi_l^m(\tau_1;\mu_0)], \quad (72)$$

$$\frac{\partial S^{(m)}(\tau_1;\mu,\mu_0)}{\partial\tau_1} = (2-\delta_{0,m})\sum_{l=m}^{N}(-1)^{m+l}\varpi_l^m\phi_l^m(\tau_1;\mu)\phi_l^m(\tau_1;\mu_0) \quad (73)$$

and

$$\left(\frac{1}{\mu_0}-\frac{1}{\mu}\right)\frac{\partial T^{(m)}(\tau_1;\mu,\mu_0)}{\partial\tau_1}$$

$$= (2-\delta_{0,m})\sum_{l=m}^{N}\varpi_l^m\left[\frac{1}{\mu_0}\psi_l^m(\tau_1;\mu)\phi_l^m(\tau_1;\mu_0)-\frac{1}{\mu}\psi_l^m(\tau_1;\mu_0)\phi_l^m(\tau_1;\mu)\right].$$
$$(74)$$

Finally substituting for $S^{(m)}(\tau;\mu,\mu')$ and $T^{(m)}(\tau_1;\mu,\mu')$ according to equations (71) and (72) back into equations (65) and (66), we obtain the following basic set of equations:

$$\psi_l^m(\tau_1;\mu) = P_l^m(\mu)+$$

$$+\tfrac{1}{2}\mu\sum_{k=m}^{N}(-1)^{l+k}\varpi_k^m\int_0^1\frac{d\mu'\,P_l^m(\mu')}{\mu+\mu'}[\psi_k^m(\tau_1;\mu)\psi_k^m(\tau_1;\mu')-\phi_k^m(\tau_1;\mu)\phi_k^m(\tau_1;\mu')]$$
$$(75)$$

and

$$\phi_l^m(\tau_1;\mu) = e^{-\tau_1/\mu}P_l^m(\mu)+$$

$$+\tfrac{1}{2}\mu\sum_{k=m}^{N}\varpi_k^m\int_0^1\frac{d\mu'\,P_l^m(\mu')}{\mu-\mu'}[\phi_k^m(\tau_1;\mu)\psi_k^m(\tau_1;\mu')-\psi_k^m(\tau_1;\mu)\phi_k^m(\tau_1;\mu')]$$

$$(m = 0, 1,...). \quad (76)$$

54. The integral equations for the case of isotropic scattering

In the case of isotropic scattering

$$\varpi_0 \neq 0 \quad \text{and} \quad \varpi_l = 0 \quad (l = 1,...) \quad (77)$$

and the equations of the preceding section become

$$X(\mu) = \psi_0^0(\tau_1; \mu) = 1 + \tfrac{1}{2} \int_0^1 S(\tau_1; \mu, \mu') \frac{d\mu'}{\mu'}, \tag{78}$$

$$Y(\mu) = \phi_0^0(\tau_1; \mu) = e^{-\tau_1/\mu} + \tfrac{1}{2} \int_0^1 T(\tau_1; \mu, \mu') \frac{d\mu'}{\mu'}, \tag{79}$$

$$\left(\frac{1}{\mu_0} + \frac{1}{\mu}\right) S(\tau_1; \mu, \mu_0) = \varpi_0 [X(\mu)X(\mu_0) - Y(\mu)Y(\mu_0)], \tag{80}$$

$$\left(\frac{1}{\mu_0} - \frac{1}{\mu}\right) T(\tau_1; \mu, \mu_0) = \varpi_0 [Y(\mu)X(\mu_0) - X(\mu)Y(\mu_0)], \tag{81}$$

$$\frac{\partial S(\tau_1; \mu, \mu_0)}{\partial \tau_1} = \varpi_0 \, Y(\mu)Y(\mu_0), \tag{82}$$

$$\left(\frac{1}{\mu_0} - \frac{1}{\mu}\right)\frac{\partial T(\tau_1; \mu, \mu_0)}{\partial \tau_1} = \varpi_0 \left[\frac{1}{\mu_0} X(\mu)Y(\mu_0) - \frac{1}{\mu} Y(\mu)X(\mu_0)\right], \tag{83}$$

$$X(\mu) = 1 + \tfrac{1}{2}\varpi_0 \mu \int_0^1 \frac{d\mu'}{\mu + \mu'}[X(\mu)X(\mu') - Y(\mu)Y(\mu')], \tag{84}$$

and $\quad Y(\mu) = e^{-\tau_1/\mu} + \tfrac{1}{2}\varpi_0 \mu \int_0^1 \frac{d\mu'}{\mu - \mu'}[Y(\mu)X(\mu') - X(\mu)Y(\mu')]. \tag{85}$

From a comparison of the foregoing equations with the corresponding equations for the problem of diffuse reflection by a semi-infinite atmosphere (Chap. IV, § 33.1, eqs. [41] and [42]) it would appear that the pair of equations

$$X(\mu) = 1 + \mu \int_0^1 \frac{\Psi(\mu')}{\mu + \mu'}[X(\mu)X(\mu') - Y(\mu)Y(\mu')]\, d\mu' \tag{86}$$

and $\quad Y(\mu) = e^{-\tau_1/\mu} + \mu \int_0^1 \frac{\Psi(\mu')}{\mu - \mu'}[Y(\mu)X(\mu') - X(\mu)Y(\mu')]\, d\mu', \tag{87}$

where $\Psi(\mu)$ is a certain characteristic function, will play the same basic rule in the theory of radiative transfer in atmospheres of finite optical thicknesses as the equation

$$H(\mu) = 1 + \mu H(\mu) \int_0^1 \frac{\Psi(\mu')}{\mu + \mu'} H(\mu')\, d\mu' \tag{88}$$

played in the theory of semi-infinite atmospheres. We shall find that this is indeed the case.

BIBLIOGRAPHICAL NOTES

§ 50. The various principles of invariance formulated in this section were first enunciated in

1. S. CHANDRASEKHAR, *Astrophys. J.* **105**, 441, 1947 (Section II of this paper). In the limited context of isotropic scattering Ambarzumian had considered, earlier, the invariance of the laws of diffuse reflection and transmission to the addition of a layer of arbitrary optical thickness to the atmosphere at $\tau = 0$ and the simultaneous removal of an equal layer from the bottom at $\tau = \tau_1$. (But this procedure leads to only one-half of the required number of equations.)

2. V. A. AMBARZUMIAN, *C.R. (Doklady) Acad. d. Sc. U.R.S.S.* **38**, 229 (1943).

§ 51. The integral equations derived in this section are contained in reference 1 quoted above.

§ 52. The earliest discussions of the principle of reciprocity are due to

3. H. VON HELMHOLTZ, *Theorie der Warme*, i. 3, § 42; see also *Helmholtz's Treatise on Physiological Optics*, edited by J. P. C. Southall (Optical Society of America, 1924), vol. i, pp. 57–90 and pp. 261–300.

4. LORD RAYLEIGH, *Scientific Papers* (Cambridge, England, 1903), iv. 480; also Lord Rayleigh, *Theory of Sound* (Macmillan, London, 1896), vol. i, §§ 107–11.

Helmholtz formulated the principle as a theorem in geometrical optics. In the context of diffuse reflection it appears that Lord Rayleigh was the first to state the principle. More recent discussions of the principle are those of

5. M. MINNAERT, *Astrophys. J.* **93**, 403 (1941).

6. H. VAN DE HULST (unpublished).

The statement of the principle in terms of the scattering and the transmission matrices (when polarization is allowed for) will be found in

7. S. CHANDRASEKHAR, *Astrophys. J.* **105**, 151, 1947 (§ 4 of this paper).

See also

8. F. PERRIN, *Journal of Chemical Physics*, **10**, 415 (1942).

The writer is not aware that a rigorous proof of the principle from the basic equations of the problem, in the manner outlined in this section, has been attempted before. It would also appear that the principle of reciprocity for elliptically polarized light as formulated in this section (p. 175) is new.

For the extension of the principle to matter waves see:

9. R. H. FOWLER, *Proc. Camb. Philos. Soc.* **25**, 193 (1929).

§§ 53 and 54. The analysis in these sections are also taken from reference 1.

VIII

THE X- AND THE Y-FUNCTIONS

55. Definitions and alternative forms of the basic equations

It will appear in the following chapters that the integral equations incorporating the various invariances of the problem of diffuse reflection and transmission can be reduced to one or more pairs of integral equations of the following standard forms:

$$X(\mu) = 1 + \mu \int_0^1 \frac{\Psi(\mu')}{\mu+\mu'}[X(\mu)X(\mu')-Y(\mu)Y(\mu')]\,d\mu' \tag{1}$$

and

$$Y(\mu) = e^{-\tau_1/\mu} + \mu \int_0^1 \frac{\Psi(\mu')}{\mu-\mu'}[Y(\mu)X(\mu')-X(\mu)Y(\mu')]\,d\mu', \tag{2}$$

where the characteristic function $\Psi(\mu)$ is an even polynomial in μ, generally, satisfying the condition

$$\int_0^1 \Psi(\mu)\,d\mu \leqslant \tfrac{1}{2}, \tag{3}$$

and τ_1 is the optical thickness of the atmosphere.

The functions X and Y defined in terms of the integral equations (1) and (2) play, in the general theory, a role similar to the H-functions in the theory of transfer in semi-infinite atmospheres. Indeed, we should require that

$$X(\mu) \to H(\mu) \quad \text{and} \quad Y(\mu) \to 0 \quad \text{as} \quad \tau_1 \to \infty, \tag{4}$$

where $H(\mu)$ satisfies the integral equation

$$H(\mu) = 1 + \mu H(\mu) \int_0^1 \frac{\Psi(\mu')}{\mu+\mu'} H(\mu')\,d\mu'. \tag{5}$$

Moreover, it will also appear that

$$X(\mu) \to 1 \quad \text{and} \quad Y(\mu) \to e^{-\tau_1/\mu} \quad \text{as} \quad \tau_1 \to 0. \tag{6}$$

In dealing with the solutions of equations (1) and (2) it is convenient to introduce the following abbreviations:

$$x_n = \int_0^1 X(\mu)\Psi(\mu)\mu^n\,d\mu, \qquad y_n = \int_0^1 Y(\mu)\Psi(\mu)\mu^n\,d\mu,$$

$$\alpha_n = \int_0^1 X(\mu)\mu^n\,d\mu \quad \text{and} \quad \beta_n = \int_0^1 Y(\mu)\mu^n\,d\mu; \tag{7}$$

i.e. x_n and y_n are the moments of order n of $X(\mu)$ and $Y(\mu)$, weighted by $\Psi(\mu)$, while α_n and β_n are the simple moments themselves.

Certain alternative forms of the basic equations which we shall find useful may also be noted here. Writing

$$\frac{\mu}{\mu+\mu'} = 1 - \frac{\mu'}{\mu+\mu'}, \quad \text{and} \quad \frac{\mu}{\mu-\mu'} = 1 + \frac{\mu'}{\mu-\mu'}, \qquad (8)$$

in equations (1) and (2), we readily find that

$$\int_0^1 \frac{\mu'\Psi(\mu')}{\mu+\mu'}[X(\mu)X(\mu') - Y(\mu)Y(\mu')]\,d\mu' = 1 - [(1-x_0)X(\mu) + y_0 Y(\mu)] \qquad (9)$$

and

$$\int_0^1 \frac{\mu'\Psi(\mu')}{\mu-\mu'}[Y(\mu)X(\mu') - X(\mu)Y(\mu')]\,d\mu'$$
$$= -e^{-\tau_1/\mu} + [y_0 X(\mu) + (1-x_0)Y(\mu)]. \qquad (10)$$

We also have

$$\int_0^1 \frac{\mu'^2\Psi(\mu')}{\mu+\mu'}[X(\mu)X(\mu') - Y(\mu)Y(\mu')]\,d\mu'$$
$$= x_1 X(\mu) - y_1 Y(\mu) - \mu + \mu[(1-x_0)X(\mu) + y_0 Y(\mu)] \qquad (11)$$

and

$$\int_0^1 \frac{\mu'^2\Psi(\mu')}{\mu-\mu'}[Y(\mu)X(\mu') - X(\mu)Y(\mu')]\,d\mu'$$
$$= y_1 X(\mu) - x_1 Y(\mu) - \mu e^{-\tau_1/\mu} + \mu[y_0 X(\mu) + (1-x_0)Y(\mu)]. \qquad (12)$$

The foregoing equations can be verified by writing

$$\frac{\mu'^2}{\mu+\mu'} = \mu' - \frac{\mu\mu'}{\mu+\mu'}, \qquad \frac{\mu'^2}{\mu-\mu'} = -\mu' + \frac{\mu\mu'}{\mu-\mu'},$$

and using equations (10) and (11).

56. Integro-differential equations for $X(\mu,\tau_1)$ and $Y(\mu,\tau_1)$

In equations (1) and (2), $0 < \tau_1 < \infty$ is, of course, to be regarded as some assigned constant. Nevertheless, it is sometimes convenient to emphasize explicitly the dependence of the solutions X and Y on τ_1. We shall then write $X(\mu,\tau_1)$ and $Y(\mu,\tau_1)$ instead of simply $X(\mu)$ and $Y(\mu)$. And, considered as functions of τ_1 also, X and Y satisfy certain integro-differential equations which are of importance. We shall state this result in the form of the following theorem:

THEOREM 1. If $X(\mu,\tau_1)$ and $Y(\mu,\tau_1)$ are solutions of equations (1) and (2) for a particular value of τ_1, then solutions for other values of τ_1 can be obtained from the integro-differential equations

$$\frac{\partial X(\mu,\tau_1)}{\partial \tau_1} = Y(\mu,\tau_1) \int_0^1 \frac{d\mu'}{\mu'} \Psi(\mu') Y(\mu',\tau_1)$$

$$= y_{-1}(\tau_1) Y(\mu,\tau_1) \tag{13}$$

and $$\frac{\partial Y(\mu,\tau_1)}{\partial \tau_1} + \frac{Y(\mu,\tau_1)}{\mu} = X(\mu,\tau_1) \int_0^1 \frac{d\mu'}{\mu'} \Psi(\mu') Y(\mu',\tau_1)$$

$$= y_{-1}(\tau_1) X(\mu,\tau_1). \tag{14}$$

PROOF. According to equations (13) and (14)

$$\mu \frac{\partial}{\partial \tau_1} \int_0^1 \frac{\Psi(\mu')}{\mu+\mu'} [X(\mu)X(\mu') - Y(\mu)Y(\mu')]\, d\mu'$$

$$= \mu \int_0^1 \frac{\Psi(\mu')}{\mu+\mu'} \Big\{ y_{-1} X(\mu) Y(\mu') + y_{-1} X(\mu') Y(\mu) -$$

$$- Y(\mu)\Big[-\frac{Y(\mu')}{\mu'} + y_{-1} X(\mu')\Big] - Y(\mu')\Big[-\frac{Y(\mu)}{\mu} + y_{-1} X(\mu)\Big] \Big\}\, d\mu'. \tag{15}$$

Hence

$$\mu \frac{\partial}{\partial \tau_1} \int_0^1 \frac{\Psi(\mu')}{\mu+\mu'} [X(\mu)X(\mu') - Y(\mu)Y(\mu')]\, d\mu' = y_{-1} Y(\mu). \tag{16}$$

Similarly,

$$\mu \frac{\partial}{\partial \tau_1} \int_0^1 \frac{\Psi(\mu')}{\mu-\mu'} [Y(\mu)X(\mu') - X(\mu)Y(\mu')]\, d\mu'$$

$$= \int_0^1 \frac{\Psi(\mu')}{\mu-\mu'}\Big[-Y(\mu)X(\mu') + \frac{\mu}{\mu'} X(\mu)Y(\mu')\Big] d\mu'. \tag{17}$$

We therefore have

$$\int_0^1 \frac{\Psi(\mu')}{\mu-\mu'} [Y(\mu)X(\mu') - X(\mu)Y(\mu')]\, d\mu' +$$

$$+ \mu \frac{\partial}{\partial \tau_1} \int_0^1 \frac{\Psi(\mu')}{\mu-\mu'} [Y(\mu)X(\mu') - X(\mu)Y(\mu')]\, d\mu' = y_{-1} X(\mu). \tag{18}$$

On the other hand, if X and Y are solutions of equations (1) and (2), we must have

$$\frac{\partial X}{\partial \tau_1} = \mu \frac{\partial}{\partial \tau_1} \int_0^1 \frac{\Psi'(\mu')}{\mu+\mu'} [X(\mu)X(\mu') - Y(\mu)Y(\mu')] \, d\mu' \qquad (19)$$

and

$$\frac{\partial Y}{\partial \tau_1} + \frac{Y}{\mu} = \int_0^1 \frac{\Psi'(\mu')}{\mu-\mu'} [Y(\mu)X(\mu') - X(\mu)Y(\mu')] \, d\mu' +$$

$$+ \mu \frac{\partial}{\partial \tau_1} \int_0^1 \frac{\Psi'(\mu')}{\mu-\mu'} [Y(\mu)X(\mu') - X(\mu)Y(\mu')] \, d\mu'. \qquad (20)$$

From equations (16), (18), (19), and (20) we now conclude that if $X(\mu, \tau_1)$ and $Y(\mu, \tau_1)$ are solutions of equations (1) and (2) for a particular value of τ_1 then

$$X(\mu, \tau_1) + y_{-1} Y(\mu, \tau_1) \, d\tau_1 \qquad (21)$$

and

$$Y(\mu, \tau_1) + \left[-\frac{Y(\mu, \tau_1)}{\mu} + y_{-1} X(\mu, \tau_1) \right] d\tau_1 \qquad (22)$$

are solutions of the same equations for an infinitesimally larger value of τ_1, namely, $\tau_1 + d\tau_1$. This proves the theorem.

COROLLARY.

$$X^2(\mu, \tau_1) - Y^2(\mu, \tau_1) = H^2(\mu) - \frac{2}{\mu} \int_{\tau_1}^{\infty} Y^2(\mu, t) \, dt. \qquad (23)$$

PROOF. Eliminating y_{-1} between equations (13) and (14) we have

$$X \frac{\partial X}{\partial \tau_1} = Y \frac{\partial Y}{\partial \tau_1} + \frac{Y^2}{\mu} \qquad (24)$$

or

$$\frac{\partial}{\partial \tau_1}(X^2 - Y^2) = \frac{2}{\mu} Y^2. \qquad (25)$$

Integrating equation (25) and remembering that

$$X(\mu, \tau_1) \to H(\mu) \quad \text{and} \quad Y(\mu, \tau_1) \to 0 \quad \text{as} \quad \tau_1 \to \infty, \qquad (26)$$

we obtain the result stated.

57. Integral properties of the X- and Y-functions

As in the case of the H-functions, there are a number of integral theorems which can be proved for functions satisfying equations (1) and (2). The theorems which follow are the analogues for the X- and Y-functions, of the theorems proved for the H-functions in Chapter V, § 38.

THEOREM 2.

$$\int\limits_0^1 X(\mu)\Psi(\mu)\,d\mu = 1-\left[1-2\int\limits_0^1 \Psi(\mu)\,d\mu+\left\{\int\limits_0^1 Y(\mu)\Psi(\mu)\,d\mu\right\}^2\right]^{\frac{1}{2}}. \quad (27)$$

PROOF. Multiplying the equation satisfied by $X(\mu)$ by $\Psi(\mu)$ and integrating over the range of μ, we have

$$x_0 = \int\limits_0^1 \Psi(\mu)\,d\mu+$$

$$+ \int\limits_0^1\int\limits_0^1 \frac{\mu}{\mu+\mu'}\Psi(\mu)\Psi(\mu')[X(\mu)X(\mu')-Y(\mu)Y(\mu')]\,d\mu d\mu'. \quad (28)$$

Interchanging μ and μ' in the double integral on the right-hand side and taking the average of the two equations, we have

$$x_0 = \int\limits_0^1 \Psi(\mu)\,d\mu+\tfrac{1}{2}\int\limits_0^1\int\limits_0^1 \Psi(\mu)\Psi(\mu')[X(\mu)X(\mu')-Y(\mu)Y(\mu')]\,d\mu d\mu'$$

$$= \int\limits_0^1 \Psi(\mu)\,d\mu+\tfrac{1}{2}(x_0^2-y_0^2). \quad (29)$$

Solving this equation for x_0 we have

$$x_0 = 1\pm\left[1-2\int\limits_0^1 \Psi(\mu)\,d\mu+y_0^2\right]^{\frac{1}{2}}. \quad (30)$$

The ambiguity in the sign in equation (30) can be removed by the consideration that the quantity on the right-hand side must uniformly converge to zero when $\Psi(\mu) \to 0$ uniformly in the interval $(0,1)$. This requires us to choose the negative sign in equation (30) and the result stated follows.

COROLLARY 1. In the conservative case

$$\int\limits_0^1 \Psi(\mu)\,d\mu = \tfrac{1}{2}, \quad (31)$$

we have $\qquad\qquad \int\limits_0^1 [X(\mu)+Y(\mu)]\Psi(\mu)\,d\mu = 1. \quad (32)$

COROLLARY 2. In case

$$\int\limits_0^1 \Psi(\mu)\,d\mu > \tfrac{1}{2}, \quad (33)$$

there exists a critical value of τ_1, say τ^*, such that for $\tau_1 > \tau^*$, the solutions become complex.

Since $Y(\mu) \to e^{-\tau_1/\mu}$ for $\tau_1 \to 0$ and $Y(\mu) \to 0$ $(0 \leqslant \mu \leqslant 1)$, as $\tau_1 \to \infty$, it is clear that there must exist a value of $\tau_1 = \tau^*$ such that

$$\int_0^1 Y(\mu)\Psi(\mu)\, d\mu = \left[2 \int_0^1 \Psi(\mu)\, d\mu - 1\right]^{\frac{1}{2}}. \tag{34}$$

For $\tau > \tau^*$, the quantity on the right-hand side of equation (27) becomes complex and the corollary stated follows.†

THEOREM 3.

$$(1-x_0)x_2 + y_0 y_2 + \tfrac{1}{2}(x_1^2 - y_1^2) = \int_0^1 \Psi(\mu)\mu^2\, d\mu. \tag{35}$$

PROOF. Multiplying equation (1) by $\Psi(\mu)\mu^2$ and integrating over the range of μ we have

$$x_2 = \int_0^1 \Psi(\mu)\mu^2\, d\mu +$$

$$+ \int_0^1 \int_0^1 \frac{\mu^3}{\mu+\mu'}\Psi(\mu)\Psi(\mu')[X(\mu)X(\mu')-Y(\mu)Y(\mu')]\, d\mu d\mu'$$

$$= \int_0^1 \Psi(\mu)\mu^2\, d\mu +$$

$$+ \tfrac{1}{2}\int_0^1 \int_0^1 (\mu^2 - \mu\mu' + \mu'^2)\Psi(\mu)\Psi(\mu')[X(\mu)X(\mu')-Y(\mu)Y(\mu')]\, d\mu d\mu'$$

$$= \int_0^1 \Psi(\mu)\mu^2\, d\mu + x_2 x_0 - y_2 y_0 - \tfrac{1}{2}(x_1^2 - y_1^2), \tag{36}$$

which is equivalent to equation (35).

COROLLARY. In the conservative case

$$y_0(x_2 + y_2) + \tfrac{1}{2}(x_1^2 - y_1^2) = \int_0^1 \Psi(\mu)\mu^2\, d\mu. \tag{37}$$

This follows from equation (35) and the Corollary 1 of Theorem 2, according to which

$$x_0 + y_0 = 1. \tag{38}$$

---

† It is apparent that the existence of a critical value of τ^* which we encounter here is related to the problem of the critical size of a pile.

THEOREM 4. When the characteristic function $\Psi(\mu)$ has the form

$$\Psi(\mu) = a + b\mu^2, \tag{39}$$

where a and b are two constants, we have the relations

$$\alpha_0 = 1 + \tfrac{1}{2}[a(\alpha_0^2 - \beta_0^2) + b(\alpha_1^2 - \beta_1^2)], \tag{40}$$

$$(a + b\mu^2) \int_0^1 \frac{d\mu'}{\mu + \mu'}[X(\mu)X(\mu') - Y(\mu)Y(\mu')]$$

$$= \frac{1}{\mu}[X(\mu) - 1] - b[(\alpha_1 - \mu\alpha_0)X(\mu) - (\beta_1 - \mu\beta_0)Y(\mu)], \tag{41}$$

and

$$(a + b\mu^2) \int_0^1 \frac{d\mu'}{\mu - \mu'}[Y(\mu)X(\mu') - X(\mu)Y(\mu')]$$

$$= \frac{1}{\mu}[Y(\mu) - e^{-\tau_1/\mu}] - b[(\beta_1 + \mu\beta_0)X(\mu) - (\alpha_1 + \mu\alpha_0)Y(\mu)]. \tag{42}$$

PROOF. To prove equation (40), we simply integrate the equation satisfied by $X(\mu)$. We find

$$\alpha_0 = 1 + \int_0^1 \int_0^1 \frac{(a + b\mu'^2)\mu}{\mu + \mu'}[X(\mu)X(\mu') - Y(\mu)Y(\mu')]\, d\mu\, d\mu'$$

$$= 1 + \tfrac{1}{2} \int_0^1 \int_0^1 (a + b\mu\mu')[X(\mu)X(\mu') - Y(\mu)Y(\mu')]\, d\mu\, d\mu'$$

$$= 1 + \tfrac{1}{2}[a(\alpha_0^2 - \beta_0^2) + b(\alpha_1^2 - \beta_1^2)]. \tag{43}$$

The relation (42) can be proved in the following manner:

$$a \int_0^1 \frac{d\mu'}{\mu - \mu'}[Y(\mu)X(\mu') - X(\mu)Y(\mu')]$$

$$= \int_0^1 \frac{a + b\mu'^2}{\mu - \mu'}[Y(\mu)X(\mu') - X(\mu)Y(\mu')]\, d\mu' +$$

$$+ b \int_0^1 \left(\mu + \mu' - \frac{\mu^2}{\mu - \mu'}\right)[Y(\mu)X(\mu') - X(\mu)Y(\mu')]\, d\mu'$$

$$= \frac{1}{\mu}[Y(\mu) - e^{-\tau_1/\mu}] + b[(\alpha_1 + \mu\alpha_0)Y(\mu) - (\beta_1 + \mu\beta_0)X(\mu)] -$$

$$- b\mu^2 \int_0^1 \frac{d\mu'}{\mu - \mu'}[Y(\mu)X(\mu') - X(\mu)Y(\mu')]. \tag{44}$$

Hence the result. Equation (41) follows quite similarly.

58. The non-uniqueness of the solution in the conservative case. The standard solution

We shall now prove the following theorem:

THEOREM 5. In the conservative case

$$\int_0^1 \Psi(\mu)\, d\mu = \tfrac{1}{2}, \tag{45}$$

The solutions of equations (1) and (2) are not unique; more particularly if $X(\mu)$ and $Y(\mu)$ are solutions, then so are

$$X(\mu)+Q\mu[X(\mu)+Y(\mu)]$$

and

$$Y(\mu)-Q\mu[X(\mu)+Y(\mu)], \tag{46}$$

where Q is an arbitrary constant.

PROOF. Writing

$$F(\mu) = X(\mu)+Q\mu[X(\mu)+Y(\mu)]$$

and

$$G(\mu) = Y(\mu)-Q\mu[X(\mu)+Y(\mu)], \tag{47}$$

we verify that

$$F(\mu)F(\mu')-G(\mu)G(\mu') = X(\mu)X(\mu')-Y(\mu)Y(\mu')+$$
$$+Q(\mu+\mu')[X(\mu)+Y(\mu)][X(\mu')+Y(\mu')]$$

and

$$G(\mu)F(\mu')-F(\mu)G(\mu') = Y(\mu)X(\mu')-X(\mu)Y(\mu')-$$
$$-Q(\mu-\mu')[X(\mu)+Y(\mu)][X(\mu')+Y(\mu')]. \tag{48}$$

Hence

$$\mu \int_0^1 \frac{\Psi(\mu')}{\mu+\mu'}[F(\mu)F(\mu')-G(\mu)G(\mu')]\, d\mu'$$
$$= \mu \int_0^1 \frac{\Psi(\mu')}{\mu+\mu'}[X(\mu)X(\mu')-Y(\mu)Y(\mu')]\, d\mu'+$$
$$+Q\mu[X(\mu)+Y(\mu)] \int_0^1 [X(\mu')+Y(\mu')]\Psi(\mu')\, d\mu'. \tag{49}$$

Using equation (1) and Corollary 1 of Theorem 2 (eq. [32]), we have

$$\mu \int_0^1 \frac{\Psi(\mu')}{\mu+\mu'}[F(\mu)F(\mu')-G(\mu)G(\mu')]\, d\mu'$$
$$= X(\mu)-1+Q\mu[X(\mu)+Y(\mu)]$$
$$= F(\mu)-1. \tag{50}$$

Similarly,

$$\mu \int_0^1 \frac{\Psi(\mu')}{\mu-\mu'}[G(\mu)F(\mu')-F(\mu)G(\mu')]\,d\mu'$$

$$= Y(\mu)-e^{-\tau_1/\mu}-Q\mu[X(\mu)+Y(\mu)] = G(\mu)-e^{-\tau_1/\mu}. \quad (51)$$

Hence $F(\mu)$ and $G(\mu)$ satisfy the same equations as $X(\mu)$ and $Y(\mu)$ and the theorem follows.

COROLLARY. The solutions derivable from a given one according to equations (47) form a one-parameter family which can be generated by any of its members.

PROOF. Let $\quad F_1(\mu) = F(\mu)+Q_1\mu[F(\mu)+G(\mu)]$

and $\quad\quad\quad G_1(\mu) = G(\mu)-Q_1\mu[F(\mu)+G(\mu)], \quad\quad (52)$

where Q_1 is an arbitrary constant. According to Theorem 5, F_1 and G_1 are also solutions of equations (1) and (2). On the other hand, since (cf. eq. [47]) $\quad\quad F(\mu)+G(\mu) = X(\mu)+Y(\mu), \quad\quad (53)$

we can express F_1 and G_1 alternatively in the forms

$$F_1(\mu) = X(\mu)+(Q+Q_1)\mu[X(\mu)+Y(\mu)]$$

and $\quad\quad G_1(\mu) = Y(\mu)-(Q+Q_1)\mu[X(\mu)+Y(\mu)]. \quad\quad (54)$

In other words, $F_1(\mu)$ and $G_1(\mu)$ can also be derived directly from $X(\mu)$ and $Y(\mu)$.

It would appear that in a given conservative case all the solutions of equations (1) and (2) which are bounded in the interval $(0 \leqslant \mu \leqslant 1)$ are included in one and only one family. In non-conservative cases, on the other hand, it would seem that the solutions which are bounded in the half-plane $R(z) > 0$ are unique. But so far no rigorous proofs have been supplied for these statements.

58.1. *Standard solutions*

In view of the ambiguity of the solutions of equations (1) and (2) in conservative cases, it would be convenient to select, in each case, a particular member of the one-parameter family of solutions as a *standard solution*.

Definition. In a conservative case we shall define the solutions which have the property

$$x_0 = \int_0^1 X(\mu)\Psi(\mu)\,d\mu = 1$$

and $\quad\quad y_0 = \int_0^1 Y(\mu)\Psi(\mu)\,d\mu = 0, \quad\quad (55)$

as the standard solutions of equations (1) and (2).

Such solutions can always be found; for if a particular X and Y do not satisfy the conditions (55) we can always find a Q such that the solutions derived from X and Y in the manner of equations (47) have the required property. Standard solutions defined in this manner have several interesting properties. We shall state them in the form of the following theorems:

THEOREM 6. The standard solutions are invariant to the increments of τ_1 according to the integro-differential equations of Theorem 1.

PROOF. Multiplying equations (13) and (14) by $\Psi(\mu)$ and integrating over the range of μ, we have

$$\frac{dx_0}{d\tau_1} = y_0 y_{-1} = 0$$

and

$$\frac{dy_0}{d\tau_1} = (x_0-1)y_{-1} = 0, \tag{56}$$

from which the theorem stated follows.

THEOREM 7. Let $X(\mu,\tau_1)$ and $Y(\mu,\tau_1)$ denote the standard solutions of equations (1) and (2) in a conservative case for a particular value of τ_1. Consider the solutions

$$F(\mu,\tau_1) = X(\mu,\tau_1)+Q\mu[X(\mu)+Y(\mu)]$$

and

$$G(\mu,\tau_1) = Y(\mu,\tau_1)-Q\mu[X(\mu)+Y(\mu)] \tag{57}$$

of equations (1) and (2) derived from X and Y and continue them for other values of τ_1 according to the equations of Theorem 1. These solutions for other values of τ_1 can, in turn, be derived from the standard solutions appropriate for these values of τ_1 with other values of Q. The quantity Q considered as a function of τ_1 in this manner satisfies the differential equation

$$\frac{d}{d\tau_1}\left(\frac{1}{Q}\right)-\frac{2y_{-1}}{Q} = -1. \tag{58}$$

PROOF. According to equations (13) and (14)

$$\frac{\partial F}{\partial \tau_1} = G \int_0^1 \frac{d\mu'}{\mu'}\Psi(\mu')G(\mu'),$$

and

$$\frac{\partial G}{\partial \tau_1}+\frac{G}{\mu} = F \int_0^1 \frac{d\mu'}{\mu'}\Psi(\mu')G(\mu'). \tag{59}$$

Now (cf. eq. [57])

$$\int_0^1 \frac{d\mu'}{\mu'}\Psi(\mu')G(\mu') = \int_0^1 \frac{d\mu'}{\mu'}\Psi(\mu')Y(\mu') - Q\int_0^1 [X(\mu')+Y(\mu')]\Psi(\mu')\,d\mu'$$

$$= y_{-1} - Q. \tag{60}$$

Hence
$$\frac{\partial F}{\partial \tau_1} = (y_{-1}-Q)G, \tag{61}$$

and
$$\frac{\partial G}{\partial \tau_1} + \frac{G}{\mu} = (y_{-1}-Q)F. \tag{62}$$

On the other hand, since X and Y remain standard solutions when continued for other values of τ_1 we must have

$$\frac{\partial F}{\partial \tau_1} = \frac{\partial X}{\partial \tau_1} + Q\mu\left(\frac{\partial X}{\partial \tau_1} + \frac{\partial Y}{\partial \tau_1}\right) + \mu(X+Y)\frac{dQ}{d\tau_1}$$

$$= y_{-1}Y + Q\mu\left(y_{-1}Y - \frac{Y}{\mu} + y_{-1}X\right) + \mu(X+Y)\frac{dQ}{d\tau_1}$$

$$= (y_{-1}-Q)Y + \mu(X+Y)\left[y_{-1}Q + \frac{dQ}{d\tau_1}\right]. \tag{63}$$

We can rewrite the foregoing equation in the form

$$\frac{\partial F}{\partial \tau_1} = (y_{-1}-Q)[Y - Q\mu(X+Y)] +$$

$$+ \mu(X+Y)\left[Q(y_{-1}-Q) + y_{-1}Q + \frac{dQ}{d\tau_1}\right],$$

or
$$\frac{\partial F}{\partial \tau_1} = (y_{-1}-Q)G + \mu(X+Y)\left(2y_{-1}Q - Q^2 + \frac{dQ}{d\tau_1}\right). \tag{64}$$

Comparing equations (61) and (64), we conclude that

$$\frac{dQ}{d\tau_1} + 2y_{-1}Q - Q^2 = 0. \tag{65}$$

A similar consideration of the equation for $\partial G/\partial \tau_1$ leads to the same equation for Q.

Equation (65) can be rewritten in the form

$$\frac{1}{Q^2}\frac{dQ}{d\tau_1} + \frac{2y_{-1}}{Q} = 1, \tag{66}$$

which is equivalent to equation (58).

The various relations (eqs. [9]–[12] and [37]) derived in §§ 55 and 57 for the solutions of equation (1) and (2) in general take particularly simple forms for standard solutions in conservative cases. We shall collect their relations in the form of the following theorem:

THEOREM 8. For the standard solutions in conservative cases we have the relations

$$x_0 = 1, \qquad y_0 = 0, \tag{67}$$

$$x_1^2 - y_1^2 = 2 \int_0^1 \Psi(\mu)\mu^2 \, d\mu, \tag{68}$$

$$\int_0^1 \frac{\mu'\Psi(\mu')}{\mu+\mu'}[X(\mu)X(\mu')-Y(\mu)Y(\mu')] \, d\mu' = 1, \tag{69}$$

$$\int_0^1 \frac{\mu'\Psi(\mu')}{\mu-\mu'}[Y(\mu)X(\mu')-X(\mu)Y(\mu')] \, d\mu' = -e^{-\tau_1/\mu}, \tag{70}$$

$$\int_0^1 \frac{\mu'^2\Psi(\mu')}{\mu+\mu'}[X(\mu)X(\mu')-Y(\mu)Y(\mu')] \, d\mu' = x_1 X(\mu)-y_1 Y(\mu)-\mu, \tag{71}$$

and

$$\int_0^1 \frac{\mu'^2\Psi(\mu')}{\mu-\mu'}[Y(\mu)X(\mu')-X(\mu)Y(\mu')] \, d\mu' = y_1 X(\mu)-x_1 Y(\mu)-\mu e^{-\tau_1/\mu}. \tag{72}$$

59. Rational representations of the X- and Y-functions in finite approximations

In the theory of radiative transfer in semi-infinite atmospheres it was found that the solutions of the equations of transfer in a finite approximation led to closed expressions for the angular distributions of the emergent radiations, involving certain H-functions defined, rationally, in terms of characteristic roots and points of the Gaussian division, and which in the limit of infinite approximation became solutions of integral equations of a standard form. In analogy with this theory, we may expect that in transfer problems in atmospheres of finite optical thicknesses also, the angular distributions of the emergent radiations can be expressed in terms of certain rational functions X and Y which, in the limit of infinite approximation, will become solutions of integral equations of the form considered in the preceding sections. Moreover, as in the case of the H-functions, we can also expect that the rational functions are, in some sense, approximations to the exact functions. It is found that all these expectations are fulfilled though the *complete* formal similarity in the properties of the H-functions defined rationally and as solutions of integral equations is not maintained in the case of the X- and Y-functions. Nevertheless, the fact that the form of the exact solutions of the integral equations incorporating the invariances of the problem can be written down, in virtue of the correspondence between the X- and Y-functions occurring in the solutions in finite approximations and the exact X- and Y-functions defined as solutions of integral equations, is of very great importance: otherwise, the high order and complex nature of the systems of equations which the principles of invariance of Chapter VII lead to might have prevented even their consideration for solution. In this section we shall indicate the manner and origin of this correspondence.

As in the case of transfer problems in semi-infinite atmospheres, the characteristic features of the solution already emerge in the simplest context of isotropic scattering. Considering, then, the problem of diffuse reflection and transmission by a plane-parallel atmosphere of optical thickness τ_1, we have (cf. Chap. III, eqs. [92] and [108])

$$I_i = \tfrac{1}{4}\varpi_0 F\left[\sum_{\alpha=-n}^{+n} \frac{L_\alpha e^{-k_\alpha \tau}}{1+\mu_i k_\alpha} + \frac{H(\mu_0)H(-\mu_0)}{1+\mu_i/\mu_0} e^{-\tau/\mu_0}\right] \quad (i = \pm 1,...,\pm n),$$
$$(73)\dagger$$

for the solution of the intensities I_i. In equation (73) the L_α's

$$(\alpha = \pm 1,...,\pm n)$$

are $2n$ constants of integration and the k_α's

$$(\alpha = \pm 1,...,\pm n \quad \text{and} \quad k_\alpha = -k_{-\alpha})$$

are the $2n$ characteristic roots which are non-vanishing in the case $\varpi_0 < 1$.

The boundary conditions appropriate to our problem are

$$I_{-i} = 0 \quad \text{at} \quad \tau = 0 \quad \text{and for} \quad i = 1,...,n$$

and
$$I_{+i} = 0 \quad \text{at} \quad \tau = \tau_1 \quad \text{and for} \quad i = 1,...,n. \tag{74}$$

In terms of the functions

$$S(\mu) = \sum_{\alpha=-n}^{+n} \frac{L_\alpha}{1-k_\alpha\mu} + \frac{H(\mu_0)H(-\mu_0)}{1-\mu/\mu_0}$$

and
$$T(\mu) = \sum_{\alpha=-n}^{+n} \frac{L_\alpha e^{-k_\alpha\tau_1}}{1+k_\alpha\mu} + \frac{H(\mu_0)H(-\mu_0)}{1+\mu/\mu_0} e^{-\tau_1/\mu_0}, \tag{75}$$

the boundary conditions (74) are equivalent to

$$S(\mu_i) = T(\mu_i) = 0 \quad (i = 1,...,n). \tag{76}$$

---

† In the summation over α in this equation there is no term with $\alpha = 0$. We shall adopt this convention throughout this section. It is, therefore, to be understood that in all summations (or products) extended over α from $-n$ to $+n$ the term with $\alpha = 0$ does not occur.

In analogy with the procedure adopted in the case of semi-infinite atmospheres we must express the reflected and the transmitted intensities, $I(0, \mu)$ and $I(\tau_1, -\mu)$ $(0 \leqslant \mu \leqslant 1)$ in terms of $S(\mu)$ and $T(\mu)$. For this purpose, we must first determine the radiation field in the atmosphere according to Chapter I, equations (64) and (65) with the aid of the source function (cf. Chap. III, eq. [98])

$$\mathfrak{J}(\tau) = \tfrac{1}{4}\varpi_0 F\left\{ \sum_{\alpha=-n}^{+n} L_\alpha e^{-k_\alpha \tau} + H(\mu_0)H(-\mu_0)e^{-\tau/\mu_0}\right\}. \tag{77}$$

We find

$$I(\tau, +\mu) = \tfrac{1}{4}\varpi_0 Fe^{+\tau/\mu}\left\{ \sum_{\alpha=-n}^{+n} \frac{L_\alpha}{1+k_\alpha\mu}\left[e^{-\tau(1+k_\alpha\mu)/\mu} - e^{-\tau_1(1+k_\alpha\mu)/\mu}\right] + \right.$$
$$\left. + \frac{H(\mu_0)H(-\mu_0)}{1+\mu/\mu_0}\left[e^{-\tau(\mu_0+\mu)/\mu\mu_0} - e^{-\tau_1(\mu_0+\mu)/\mu\mu_0}\right]\right\},$$

and

$$I(\tau, -\mu) = \tfrac{1}{4}\varpi_0 Fe^{-\tau/\mu}\left\{ \sum_{\alpha=-n}^{+n} \frac{L_\alpha}{1-k_\alpha\mu}\left[e^{\tau(1-k_\alpha\mu)/\mu} - 1\right] + \right.$$
$$\left. + \frac{H(\mu_0)H(-\mu_0)}{1-\mu/\mu_0}\left[e^{\tau(\mu_0-\mu)/\mu\mu_0} - 1\right]\right\}. \tag{78}$$

From these equations we find that the reflected and transmitted intensities are given by

$$I(0, \mu) = \tfrac{1}{4}\varpi_0 F\left\{ \sum_{\alpha=-n}^{+n} \frac{L_\alpha}{1+k_\alpha\mu} + \frac{H(\mu_0)H(-\mu_0)}{1+\mu/\mu_0} - \right.$$
$$\left. - e^{-\tau_1/\mu}\left[\sum_{\alpha=-n}^{+n} \frac{L_\alpha e^{-k_\alpha\tau_1}}{1+k_\alpha\mu} + \frac{H(\mu_0)H(-\mu_0)}{1+\mu/\mu_0}e^{-\tau_1/\mu_0}\right]\right\},$$

and

$$I(\tau_1, -\mu) = \tfrac{1}{4}\varpi_0 F\left\{ \sum_{\alpha=-n}^{+n} \frac{L_\alpha e^{-k_\alpha\tau_1}}{1-k_\alpha\mu} + \frac{H(\mu_0)H(-\mu_0)}{1-\mu/\mu_0}e^{-\tau_1/\mu_0} - \right.$$
$$\left. - e^{-\tau_1/\mu}\left[\sum_{\alpha=-n}^{+n} \frac{L_\alpha}{1-k_\alpha\mu} + \frac{H(\mu_0)H(-\mu_0)}{1-\mu/\mu_0}\right]\right\}. \tag{79}$$

According to the definitions of the functions $S(\mu)$ and $T(\mu)$, we can rewrite the foregoing expressions for $I(0, +\mu)$ and $I(\tau_1, -\mu)$ in the forms

$$I(0, \mu) = \tfrac{1}{4}\varpi_0 F[S(-\mu) - e^{-\tau_1/\mu}T(\mu)]$$

and
$$I(\tau_1, -\mu) = \tfrac{1}{4}\varpi_0 F[T(-\mu) - e^{-\tau_1/\mu}S(\mu)]. \tag{80}$$

Thus, in the case of finite atmospheres, as in the case of semi-infinite atmospheres, there is a relationship of reciprocity between the equations which express the boundary conditions and the functions which describe the emergent radiations. The present relationship is naturally not as direct as the one encountered in the context of semi-infinite atmospheres. But the relationship exemplified by equations (76) and (80) is quite general and occurs in all the problems.

59.1. *The elimination of the constants and the expression of the laws of reflection and transmission in closed forms*

In addition to the functions

$$P(\mu) = \prod_{i=1}^{n} (\mu - \mu_i) \quad \text{and} \quad R(\mu) = \prod_{\alpha=1}^{n} (1 - k_\alpha \mu) \tag{81}$$

which were used in Chapter III (eqs. [54] and [55]) we shall introduce the functions

$$W(\mu) = R(\mu)R(-\mu) = \prod_{\alpha=-n}^{+n} (1 - k_\alpha \mu) = \prod_{\alpha=1}^{n} (1 - k_\alpha^2 \mu^2), \tag{82}$$

and

$$W_\alpha(\mu) = \prod_{\substack{\beta=-n \\ \beta \neq \alpha}}^{+n} (1 - k_\beta \mu) \quad (\alpha = \pm 1, ..., \pm n). \tag{83}$$

Identities which follow from these definitions and which are useful in subsequent reductions are

$$W(\mu) = W(-\mu) \quad \text{and} \quad W_\alpha(\mu) = W_{-\alpha}(-\mu) \quad (\alpha = \pm 1, ..., \pm n). \tag{84}$$

Now, from equations (75) it follows that

$$S(\mu)W(\mu)\left(1 - \frac{\mu}{\mu_0}\right) \quad \text{and} \quad T(\mu)W(\mu)\left(1 + \frac{\mu}{\mu_0}\right), \tag{85}$$

are polynomials of degree $2n$ in μ; and according to equation (76), the μ_i's $(i = 1, ..., n)$ are zeros of both these polynomials. We may therefore write

$$S(\mu) = \frac{1}{\mu_1^2 \cdots \mu_n^2} \frac{P(\mu)}{W(\mu)(1 - \mu/\mu_0)} s(\mu)$$

and

$$T(\mu) = \frac{1}{\mu_1^2 \cdots \mu_n^2} \frac{P(\mu)}{W(\mu)(1 + \mu/\mu_0)} t(\mu), \tag{86}†$$

where $s(\mu)$ and $t(\mu)$ are polynomials of degree n in μ.

Two relations which follow immediately from equations (75) and (86) are

$$H(\mu_0)H(-\mu_0) = \frac{1}{\mu_1^2 \cdots \mu_n^2} \frac{P(\mu_0)P(-\mu_0)}{W(\mu_0)}$$

$$= \lim_{\mu \to \mu_0} \left(1 - \frac{\mu}{\mu_0}\right) S(\mu) = \frac{1}{\mu_1^2 \cdots \mu_n^2} \frac{P(\mu_0)}{W(\mu_0)} s(\mu_0),$$

† The factor $1/\mu_1^2 \cdots \mu_n^2$ has been introduced in these equations for reasons of convenience (see eqs. [87] and [88]).

and

$$H(\mu_0)H(-\mu_0)e^{-\tau_1/\mu_0} = \frac{e^{-\tau_1/\mu_0}}{\mu_1^2\ldots\mu_n^2}\frac{P(\mu_0)P(-\mu_0)}{W(\mu_0)}$$

$$= \lim_{\mu\to-\mu_0}\left(1+\frac{\mu}{\mu_0}\right)T(\mu) = \frac{1}{\mu_1^2\ldots\mu_n^2}\frac{P(-\mu_0)}{W(\mu_0)}t(-\mu_0). \quad (87)$$

Hence $$s(\mu_0) = P(-\mu_0),$$

and $$t(-\mu_0) = e^{-\tau_1/\mu_0}P(\mu_0). \quad (88)$$

We next observe that, since (cf. eqs. [75])

$$L_\alpha = \lim_{\mu\to1/k_\alpha}(1-k_\alpha\mu)S(\mu)$$

and $$L_\alpha e^{-k_\alpha\tau_1} = \lim_{\mu\to-1/k_\alpha}(1+k_\alpha\mu)T(\mu) \quad (\alpha = \pm1,\ldots,\pm n), \quad (89)$$

we must have (cf. eq. [83])

$$L_\alpha = \frac{1}{\mu_1^2\ldots\mu_n^2}\frac{P(1/k_\alpha)}{W_\alpha(1/k_\alpha)(1-1/k_\alpha\mu_0)}s(1/k_\alpha)$$

and

$$L_\alpha e^{-k_\alpha\tau_1} = \frac{1}{\mu_1^2\ldots\mu_n^2}\frac{P(-1/k_\alpha)}{W_\alpha(1/k_\alpha)(1-1/k_\alpha\mu_0)}t(-1/k_\alpha) \quad (\alpha = \pm1,\ldots,\pm n).$$
$$(90)$$

Comparing these expressions for L_α and $L_\alpha e^{-k_\alpha\tau_1}$, we observe that

$$s(1/k_\alpha) = e^{k_\alpha\tau_1}\frac{P(-1/k_\alpha)}{P(+1/k_\alpha)}t(-1/k_\alpha) \quad (\alpha = \pm1,\ldots,\pm n). \quad (91)$$

The relationship between s and t is a reciprocal one since writing $-\alpha$ in place of α in equation (91), we have

$$t(1/k_\alpha) = e^{k_\alpha\tau_1}\frac{P(-1/k_\alpha)}{P(+1/k_\alpha)}s(-1/k_\alpha) \quad (\alpha = \pm1,\ldots,\pm n). \quad (92)$$

From equations (91) and (92) we obtain the relation

$$s(1/k_\alpha)s(-1/k_\alpha)-t(1/k_\alpha)t(-1/k_\alpha) = 0 \quad (\alpha = \pm1,\ldots,\pm n). \quad (93)$$

Accordingly,

$$s(\mu)s(-\mu)-t(\mu)t(-\mu) = \text{constant } W(\mu), \quad (94)$$

since the quantity on the left-hand side is a polynomial of degree $2n$ in μ and vanishes for $\mu = \pm1/k_\alpha$ $(\alpha = 1,\ldots,n)$.

Now it can be shown (for proof, see Appendix II) that equation (91) determines $s(\mu)$ and $t(\mu)$ apart from two arbitrary constants q_0 and q_1 and that they can be expressed in the forms

$$s(\mu) = q_0 C_0(\mu)+q_1 C_1(\mu)$$

and $$t(\mu) = q_1 C_0(\mu)+q_0 C_1(\mu), \quad (95)$$

where $C_0(\mu)$ and $C_1(\mu)$ are two basic polynomials (each of degree n and suitably normalized) which are related in the same manner as $s(\mu)$ and $t(\mu)$:

$$C_0(1/k_\alpha) = e^{k_\alpha \tau_1} \frac{P(-1/k_\alpha)}{P(+1/k_\alpha)} C_1(-1/k_\alpha) \quad (\alpha = \pm 1,...,\pm n). \quad (96)$$

The particular polynomials $C_0(\mu)$ and $C_1(\mu)$ which satisfy the relations (96) and which are to be used are (cf. Appendix II, eqs. [25] and [26])

$$C_0(\mu) = \sum_{\substack{l=n,n-2,... \\ 2^{n-1} \text{ terms}}} \epsilon_l^{(0)} \frac{\prod\limits_{i=1}^{l} \prod\limits_{m=1}^{n-l} (k_{r_i}+k_{s_m})}{\prod\limits_{i=1}^{l} \prod\limits_{m=1}^{n-l} (k_{r_i}-k_{s_m})} \prod_{i=1}^{l} (1+k_{r_i}\mu) \prod_{m=1}^{n-l} \frac{1}{\lambda_{s_m}} (1-k_{s_m}\mu)$$

and

$$C_1(\mu) = (-1)^{n-1} \sum_{\substack{l=n-1,n-3.... \\ 2^{n-1} \text{ terms}}} \epsilon_l^{(1)} \frac{\prod\limits_{i=1}^{l} \prod\limits_{m=1}^{n-l} (k_{r_i}+k_{s_m})}{\prod\limits_{i=1}^{l} \prod\limits_{m=1}^{n-l} (k_{r_i}-k_{s_m})} \times$$

$$\times \prod_{i=1}^{l} (1+k_{r_i}\mu) \prod_{m=1}^{n-l} \frac{1}{\lambda_{s_m}} (1-k_{s_m}\mu), \quad (97)$$

where $r_1,...,r_l$ and $s_1,...,s_{n-l}$ are selections of l and $n-l$ distinct integers from the set $(1,2,...,n)$,

$$\epsilon_l^{(0)} = +1 \quad \text{for integers of the form } n-4m$$
$$= -1 \quad \text{for integers of the form } n-4m-2$$
$$= 0 \quad \text{otherwise,}$$

$$\epsilon_l^{(1)} = +1 \quad \text{for integers of the form } n-4m-1$$
$$= -1 \quad \text{for integers of the form } n-4m-3$$
$$= 0 \quad \text{otherwise,} \quad (98)$$

and $$\lambda_\alpha = e^{k_\alpha \tau_1} \frac{P(-1/k_\alpha)}{P(+1/k_\alpha)} \quad (\alpha = \pm 1,...,\pm n). \quad (99)$$

It is apparent that in virtue of the relation (96), $C_0(\mu)$ and $C_1(\mu)$ must, like $s(\mu)$ and $t(\mu)$, satisfy an identity of the form (94). We can, therefore, write

$$C_0(\mu)C_0(-\mu)-C_1(\mu)C_1(-\mu) = [C_0^2(0)-C_1^2(0)]W(\mu), \quad (100)$$

since $W(0) = 1$.

Now returning to equations (95) we can determine the constants q_0 and q_1 by making use of equations (88). Thus,

$$s(\mu_0) = q_0 C_0(\mu_0)+q_1 C_1(\mu_0) = P(-\mu_0)$$

and $$t(-\mu_0) = q_0 C_1(-\mu_0)+q_1 C_0(-\mu_0) = e^{-\tau_1/\mu_0}P(\mu_0). \quad (101)$$

Solving these equations for q_0 and q_1 and using (100), we find that

$$q_0 = \frac{1}{[C_0^2(0)-C_1^2(0)]W(\mu_0)}[P(-\mu_0)C_0(-\mu_0)-e^{-\tau_1/\mu_0}P(\mu_0)C_1(\mu_0)]$$

and

$$q_1 = \frac{1}{[C_0^2(0)-C_1^2(0)]W(\mu_0)}[e^{-\tau_1/\mu_0}P(\mu_0)C_0(\mu_0)-P(-\mu_0)C_1(-\mu_0)].$$

$$(102)$$

With $s(\mu)$ and $t(\mu)$ given by equations (101), equations (86) for $S(\mu)$ and $T(\mu)$ take the forms

$$S(\mu) = \frac{1}{\mu_1^2\ldots\mu_n^2}\frac{P(\mu)}{W(\mu)}\frac{\mu_0}{\mu_0-\mu}[q_0 C_0(\mu)+q_1 C_1(\mu)]$$

and

$$T(\mu) = \frac{1}{\mu_1^2\ldots\mu_n^2}\frac{P(\mu)}{W(\mu)}\frac{\mu_0}{\mu_0+\mu}[q_1 C_0(\mu)+q_0 C_1(\mu)]. \qquad (103)$$

Substituting for $S(\mu)$ and $T(\mu)$ from these equations in equations (80), we obtain, after some minor rearranging of the terms, the following expressions for the reflected and the transmitted intensities:

$$I(0,\mu) = \frac{1}{4}\frac{\varpi_0 F}{\mu_1^2\ldots\mu_n^2}\frac{1}{W(\mu)}[q_0\{P(-\mu)C_0(-\mu)-e^{-\tau_1/\mu}P(\mu)C_1(\mu)\}-$$
$$-q_1\{e^{-\tau_1/\mu}P(\mu)C_0(\mu)-P(-\mu)C_1(-\mu)\}]\frac{\mu_0}{\mu_0+\mu}$$

and

$$I(\tau_1,-\mu) = \frac{1}{4}\frac{\varpi_0 F}{\mu_1^2\ldots\mu_n^2}\frac{1}{W(\mu)}[q_1\{P(-\mu)C_0(-\mu)-e^{-\tau_1/\mu}P(\mu)C_1(\mu)\}-$$
$$-q_0\{e^{-\tau_1/\mu}P(\mu)C_0(\mu)-P(-\mu)C_1(-\mu)\}]\frac{\mu_0}{\mu_0-\mu}. \qquad (104)$$

Next substituting for q_0 and q_1 according to equations (102) we have

$$I(0,\mu) = \frac{1}{4}\frac{\varpi_0 F}{\mu_1^2\ldots\mu_n^2}\frac{1}{[C_0^2(0)-C_1^2(0)]}\frac{1}{W(\mu)W(\mu_0)}\frac{\mu_0}{\mu_0+\mu}\times$$
$$\times[\{P(-\mu)C_0(-\mu)-e^{-\tau_1/\mu}P(\mu)C_1(\mu)\}\times$$
$$\times\{P(-\mu_0)C_0(-\mu_0)-e^{-\tau_1/\mu_0}P(\mu_0)C_1(\mu_0)\}-$$
$$-\{e^{-\tau_1/\mu}P(\mu)C_0(\mu)-P(-\mu)C_1(-\mu)\}\times$$
$$\times\{e^{-\tau_1/\mu_0}P(\mu_0)C_0(\mu_0)-P(-\mu_0)C_1(-\mu_0)\}]$$

and

$$I(\tau_1,-\mu) = \frac{1}{4}\frac{\varpi_0 F}{\mu_1^2\ldots\mu_n^2}\frac{1}{[C_0^2(0)-C_1^2(0)]}\frac{1}{W(\mu)W(\mu_0)}\frac{\mu_0}{\mu-\mu_0}\times$$
$$\times[\{e^{-\tau_1/\mu}P(\mu)C_0(\mu)-P(-\mu)C_1(-\mu)\}\times$$
$$\times\{P(-\mu_0)C_0(-\mu_0)-e^{-\tau_1/\mu_0}P(\mu_0)C_1(\mu_0)\}-$$
$$-\{P(-\mu)C_0(-\mu)-e^{-\tau_1/\mu}P(\mu)C_1(\mu)\}\times$$
$$\times\{e^{-\tau_1/\mu_0}P(\mu_0)C_0(\mu_0)-P(-\mu_0)C_1(-\mu_0)\}]. \qquad (105)$$

Now let

$$X(\mu) = \frac{(-1)^n}{\mu_1 \cdots \mu_n} \frac{1}{[C_0^2(0) - C_1^2(0)]^{\frac{1}{2}}} \frac{1}{W(\mu)} [P(-\mu)C_0(-\mu) - e^{-\tau_1/\mu}P(\mu)C_1(\mu)]$$
(106)

and

$$Y(\mu) = \frac{(-1)^n}{\mu_1 \cdots \mu_n} \frac{1}{[C_0^2(0) - C_1^2(0)]^{\frac{1}{2}}} \frac{1}{W(\mu)} [e^{-\tau_1/\mu}P(\mu)C_0(\mu) - P(-\mu)C_1(-\mu)].$$
(107)

In terms of these functions the expressions (105) for the reflected and the transmitted intensities become

$$I(0,\mu) = \tfrac{1}{4}\varpi_0 F \frac{\mu_0}{\mu + \mu_0} [X(\mu)X(\mu_0) - Y(\mu)Y(\mu_0)]$$
(108)

and

$$I(\tau_1, -\mu) = \tfrac{1}{4}\varpi_0 F \frac{\mu_0}{\mu - \mu_0} [Y(\mu)X(\mu_0) - X(\mu)Y(\mu_0)].$$
(109)

Comparing equations (108) and (109) with the forms of the scattering and the transmission functions given by the principles of invariance (Chap. VII, § 54, eqs. [80] and [81]), we observe that X and Y defined as in equations (106) and (107) replace the solutions of the integral equations (1) and (2) for the case $\Psi(\mu) = \tfrac{1}{2}\varpi_0$ (cf. Chap. VII, eqs. [84] and [85]). Quite generally, we may conclude that the X- and the Y-functions defined exactly as by the equations (97), (106), and (107), but in terms of the n distinct positive roots of the characteristic equation

$$1 = 2 \sum_{j=1}^{n} \frac{a_j \Psi(\mu_j)}{1 - k^2 \mu_j^2} \qquad \left(\int_0^1 \Psi(\mu)\,d\mu < \tfrac{1}{2} \right),$$
(110)

are, in the limit of infinite approximation, to be associated with the solutions of the integral equations (1) and (2). An exception to this rule arises in conservative cases, when the solutions of equations (1) and (2) become ambiguous and form a one-parameter family and when, moreover, the equations defining $X(\mu)$ and $Y(\mu)$ in a finite approximation become indeterminate on account of two of the characteristic roots becoming coincident and vanishing. However, under these circumstances, in the finite approximations, we can still define functions $X(\mu)$ and $Y(\mu)$ in terms of the reduced number of non-vanishing positive characteristic roots; and it can be shown that these functions, in the limit of infinite approximation, must be associated with the standard solutions of the corresponding integral equations.

60. Solutions for small values of τ_1

As in the case of the H-functions, we may expect to solve for the X- and Y-functions by an iteration process in which the nth iterates are obtained by evaluating the integrals on the right-hand sides of equations (1) and (2) in terms of the $(n-1)$th iterates. Thus

$$X^{(n+1)}(\mu) = 1+\mu \int_0^1 \frac{\Psi(\mu')}{\mu+\mu'}[X^{(n)}(\mu)X^{(n)}(\mu')-Y^{(n)}(\mu)Y^{(n)}(\mu')]\,d\mu' \quad (111)$$

and

$$Y^{(n+1)}(\mu) = e^{-\tau_1/\mu}+\mu \int_0^1 \frac{\Psi(\mu')}{\mu-\mu'}[Y^{(n)}(\mu)X^{(n)}(\mu')-X^{(n)}(\mu)Y^{(n)}(\mu')]\,d\mu'. \quad (112)$$

The success of such an iteration scheme will largely depend on how well the first 'trial' functions $X^{(1)}(\mu)$ and $Y^{(1)}(\mu)$ are chosen. The rational representations of the X- and Y-functions given in § 59, in the third or the fourth approximation, could, of course, be used; for purposes of numerical iteration this is indeed quite satisfactory. However, an alternative method which appears suitable, particularly for small values of τ_1, is to start the iteration with the functions

$$X^{(1)}(\mu) = 1 \quad \text{and} \quad Y^{(1)}(\mu) = e^{-\tau_1/\mu}. \quad (113)$$

The appropriateness of these functions in our present context arises from the fact that with these expressions for X and Y we essentially recover the formulae for the reflected and the transmitted light which has suffered a single scattering process in the atmosphere.

Iterating the functions (113) in accordance with equations (111) and (112), we have

$$X^{(2)}(\mu) = 1+\mu \int_0^1 \frac{\Psi(\mu')}{\mu+\mu'}\left[1-\exp\left\{-\tau_1\left(\frac{1}{\mu}+\frac{1}{\mu'}\right)\right\}\right]d\mu' \quad (114)$$

and

$$Y^{(2)}(\mu) = e^{-\tau_1/\mu}+\mu \int_0^1 \frac{\Psi(\mu')}{\mu-\mu'}[e^{-\tau_1/\mu}-e^{-\tau_1/\mu'}]\,d\mu'. \quad (115)$$

Let

$$\Psi(\mu) = \sum_j a_j\,\mu^j. \quad (116)$$

(In actual fact only even powers of μ can occur on the right-hand side, since the characteristic function is an even polynomial in μ.)

For $\Psi(\mu)$ of the form (116) we can express $X^{(2)}(\mu)$ and $Y^{(2)}(\mu)$ in the forms

$$X^{(2)}(\mu) = 1+ \sum_j a_j\,F_{j+1}(\tau_1, -\mu)$$

and $$Y^{(2)}(\mu) = e^{-\tau_1/\mu}\Big[1 + \sum_j a_j F_{j+1}(\tau_1, \mu)\Big], \tag{117}$$

where $$F_{j+1}(\tau_1, \mu) = \mu \int_0^1 \frac{\mu'^j}{\mu - \mu'} \Big[1 - \exp\Big\{-\tau_1\Big(\frac{1}{\mu'} - \frac{1}{\mu}\Big)\Big\}\Big] d\mu'. \tag{118}$$

The functions $F_{j+1}(\tau, \mu)$ $(j = 0, 1, \ldots)$ defined as in equation (118) are the same as the functions

$$F_{j+1}(\tau_1, \mu) = \int_0^{\tau_1} e^{t/\mu} E_{j+1}(t)\, dt, \tag{119}$$

which have been studied by King, Hammad and Chapman, and van de Hulst. A proof of the identity of the definitions (118) and (119) and a discussion of the properties of these functions will be found in Appendix I. Tables of these functions for the orders 1, 3, and 5, which are most useful in applications, have been published by Chandrasekhar and Frances Breen.

60.1. *The moments of $X^{(2)}(\mu)$ and $Y^{(2)}(\mu)$*

The moments α_n, β_n, x_n, and y_n (cf. eq. [7]) of the functions $X(\mu)$ and $Y(\mu)$ are needed in several connexions. In the 'second approximation' these moments can all be expressed in terms of the functions

$$G_{n,m}(\tau_1) = \int_0^1 F_n(\tau_1, -\mu)\mu^m \frac{d\mu}{\mu^2}$$

and $$G'_{n,m}(\tau_1) = \int_0^1 e^{-\tau_1/\mu} F_n(\tau_1, \mu)\mu^m \frac{d\mu}{\mu^2}, \tag{120}$$

also studied by van de Hulst. Tables of these functions (symmetrical in the indices n and m) for $m = 1, 2, 3, 4, 5$, and 6 and $m \geqslant n$ have also been published by Chandrasekhar and Breen.

We clearly have

$$\alpha_n^{(2)} = \frac{1}{n+1} + \sum_j a_j\, G_{j+1, n+2}(\tau_1),$$

$$\beta_n^{(2)} = E_{n+2}(\tau_1) + \sum a_j\, G'_{j+1, n+2}(\tau_1),$$

$$x_n^{(2)} = \sum_j \frac{a_j}{j+n+1} + \sum_j \sum_k a_j a_k\, G_{j+1, k+n+2}(\tau_1),$$

and $$y_n^{(2)} = \sum_j a_j E_{j+n+2}(\tau_1) + \sum_j \sum_k a_j a_k\, G'_{j+1, k+n+2}(\tau_1) \tag{121}$$

60.2 *The correction of the approximate solutions*

In § 57 we have shown that the X- and Y-functions satisfy a number of integral properties. Of these, the most important relation is (eq. [27])

$$x_0 = 1 - \left[1 - 2 \int_0^1 \Psi'(\mu) \, d\mu + y_0^2 \right]^{\frac{1}{2}}. \tag{122}$$

An alternative form of this relation is

$$[1 - (x_0 + y_0)][1 - (x_0 - y_0)] = 1 - 2 \int_0^1 \Psi'(\mu) \, d\mu. \tag{123}$$

The approximate solutions (113) and (117) will not, naturally, satisfy equation (123) exactly. We shall now indicate how, on this account, we can make use of equation (123) to 'correct' the approximate solutions. For this purpose we shall write

$$X(\mu) = X^{(n)}(\mu) + \Delta(\tau_1)\mu(1 - e^{-\tau_1/\mu})$$

and

$$Y(\mu) = Y^{(n)}(\mu) + \Delta(\tau_1)\mu(1 - e^{-\tau_1/\mu}), \tag{124}$$

and determine $\Delta(\tau_1)$ by requiring that x_0 and y_0 evaluated with the aid of the functions (124) satisfy the relation (123) exactly. This procedure of adding a correction term of the form $\Delta(\tau_1)\mu(1 - e^{-\tau_1/\mu})$ to both X and Y can, in a sense, be regarded as equivalent to assuming that the sources giving rise to the radiation which has been scattered more than n times in the atmosphere are uniformly distributed.†

For $X(\mu)$ and $Y(\mu)$ defined as in equations (124), we have

$$x_0 = x_0^{(n)} + \Delta(\tau_1) \int_0^1 \Psi'(\mu)\mu(1 - e^{-\tau_1/\mu}) \, d\mu$$

and

$$y_0 = y_0^{(n)} + \Delta(\tau_1) \int_0^1 \Psi'(\mu)\mu(1 - e^{-\tau_1/\mu}) \, d\mu, \tag{125}$$

where $x_0^{(n)}$ and $y_0^{(n)}$ are the respective quantities evaluated with the functions $X^{(n)}$ and $Y^{(n)}$. Inserting equations (125) in equation (123), we

---

† This remark should not be taken too literally, since, even in the simplest case of isotropic scattering, the solutions obtained by successive iterations of the equations for X and Y are not in strict 1 : 1 correspondence with the solutions obtained by including successively the higher orders of scattering (cf. Chap. IX, § 63). And apart from this, the method of correcting by the moment relation (123) is not exactly the same as distributing the sources for the higher orders of scattering uniformly through the atmosphere.

obtain after some minor rearranging of the terms:

$$\Delta(\tau_1) = \frac{1}{2\int\limits_0^1 \Psi(\mu)\mu(1-e^{-\tau_1/\mu})\,d\mu}\left\{1-[x_0^{(n)}+y_0^{(n)}]-\frac{1-2\int\limits_0^1\Psi(\mu)\,d\mu}{1-[x_0^{(n)}-y_0^{(n)}]}\right\},$$

(126)

where it may be noted that, for $\Psi(\mu)$ given by (116),

$$\int\limits_0^1 \Psi(\mu)\mu(1-e^{-\tau_1/\mu})\,d\mu = \sum_j a_j\left\{\frac{1}{j+2}-E_{j+3}(\tau_1)\right\}.$$

(127)

In conservative cases equation (126) reduces to

$$\Delta(\tau_1) = \frac{1-[x_0^{(n)}+y_0^{(n)}]}{2\int\limits_0^1 \Psi(\mu)\mu(1-e^{-\tau_1/\mu})\,d\mu}.$$

(128)

As an illustration of the foregoing procedure of correcting the solutions $X^{(n)}$ and $Y^{(n)}$ ($n = 1$ and 2) consider the isotropic case [when $\Psi(\mu) = \frac{1}{2}\varpi_0$] and the correction to the first approximation (113). In this case

$$x_0^{(1)} = \tfrac{1}{2}\varpi_0 \quad \text{and} \quad y_0^{(1)} = \tfrac{1}{2}\varpi_0 E_2(\tau_1),$$

(129)

and

$$\Delta(\tau_1) = \frac{1}{\varpi_0[\frac{1}{2}-E_3(\tau_1)]}\left\{1-\tfrac{1}{2}\varpi_0[1+E_2(\tau_1)]-\frac{1-\varpi_0}{1-\frac{1}{2}\varpi_0[1-E_2(\tau_1)]}\right\}.$$

(130)

For the case $\varpi_0 = 1$ we have, in particular,

$$\Delta(\tau_1) = \frac{1-E_2(\tau_1)}{1-2E_3(\tau_1)}.$$

(131)

Actual numerical comparisons show that the first approximation corrected in this manner already gives accuracy of less than 1 per cent. for $\tau_1 \leqslant 0\cdot2$.† It would therefore seem that the second approximation corrected in the manner indicated will give satisfactory representations for $\tau_1 \leqslant 0\cdot5$.

For $X^{(2)}(\mu)$ and $Y^{(2)}(\mu)$ given by equations (117), the moments $x_0^{(2)}$ and $y_0^{(2)}$ are (cf. eqs. [121])

$$x_0^{(2)} = \sum_j{}' \frac{a_j}{j+1} + \sum_j\sum_k a_j a_k G_{j+1,k+2}$$

and

$$y_0^{(2)} = \sum_j a_j E_{j+2} + \sum_j\sum_k a_j a_k G'_{j+1,k+2}.$$

(132)

† For example, for the case $\varpi_0 = 1$ and $\mu = 1$ the corrected first approximation gives for X the values $1\cdot16$ and $1\cdot26$ for $\tau_1 = 0\cdot1$ and $0\cdot2$, respectively, while the true values are estimated to be $1\cdot159$ and $1\cdot263$ respectively.

The resulting expression for $\Delta(\tau_1)$ is

$$\Delta(\tau_1) = \left[1 - \left\{\sum_j a_j\left(\frac{1}{j+1} + E_{j+2}\right) + \sum_j \sum_k a_j a_k(G_{j+1,k+2} + G'_{j+1,k+2})\right\} - \right.$$

$$\left. - \frac{1 - 2\int_0^1 \Psi(\mu)\,d\mu}{1 - \left\{\sum_j a_j\left(\frac{1}{j+1} - E_{j+2}\right) + \sum_j \sum_k a_j a_k(G_{j+1,k+2} - G'_{j+1,k+2})\right\}}\right] \div$$

$$\div 2\left[\sum_j a_j\left(\frac{1}{j+2} - E_{j+3}\right)\right]. \quad (133)$$

In conservative cases the foregoing equation simplifies to

$$\Delta(\tau_1) = \frac{1 - \left\{\sum_j a_j\left(\frac{1}{j+1} + E_{j+2}\right) + \sum_j \sum_k a_j a_k(G_{j+1,k+2} + G'_{j+1,k+2})\right\}}{2\sum_j a_j\left(\frac{1}{j+2} - E_{j+3}\right)}. \quad (134)$$

As an illustration of the foregoing expressions, consider again the isotropic case. In this case the only non-vanishing a is

$$a_0 = \tfrac{1}{2}\varpi_0; \quad (135)$$

and equation (133) becomes

$$\Delta(\tau_1) = \left[1 - \tfrac{1}{2}\varpi_0\{1 + E_2 + \tfrac{1}{2}\varpi_0(G_{12} + G'_{12})\} - \right.$$

$$\left. - \frac{1 - \varpi_0}{1 - \tfrac{1}{2}\varpi_0\{1 - E_2 + \tfrac{1}{2}\varpi_0(G_{12} - G'_{12})\}}\right]\frac{1}{\varpi_0(\tfrac{1}{2} - E_3)}. \quad (136)$$

And for the case $\varpi_0 = 1$, we have, in particular,

$$\Delta(\tau_1) = \frac{1}{4}\frac{1}{\tfrac{1}{2} - E_3}(2 - 2E_2 - G_{12} - G'_{12}). \quad (137)$$

60.3. The standard solutions

In § 58 we have shown that in conservative cases the solutions of equations (1) and (2) are not unique and that they form, instead, a one-parameter family. A member of this family which plays an important role in the further developments of the theory is the standard solution defined by the property (55).

Now if $X(\mu)$ and $Y(\mu)$ are particular solutions of equations (1) and (2) (in the conservative case), then the standard solutions (distinguished by a superscript s) can be derived from them by writing

$$X^{(s)}(\mu) = X(\mu) + q\mu[X(\mu) + Y(\mu)]$$

and

$$Y^{(s)}(\mu) = Y(\mu) - q\mu[X(\mu) + Y(\mu)], \tag{138}$$

and determining the constant q by the condition (55). In this manner we find that

$$q = \frac{y_0}{x_1 + y_1}. \tag{139}$$

Since the approximate solutions, corrected as in § 60.2, satisfy the moment condition $x_0 + y_0 = 1$ exactly, it is apparent that standard solutions can be constructed from them according to equation (138) with q given by (139).

BIBLIOGRAPHICAL NOTES

§§ 55–8. The analysis in these sections are taken from

1. S. CHANDRASEKHAR, *Astrophys. J.* **107**, 48 (1948) (Section I, §§ 2–5 of this paper).

§ 59. The correspondence between the rational functions X and Y which occur in the solutions of the equations of transfer in the general nth approximation and the solutions of the integral equations considered in this chapter is treated more fully in reference 1 (§ 6). For the solution of the equation of transfer, itself, in the nth approximation, see

2. S. CHANDRASEKHAR, ibid. **106**, 152 (1947) (Sections I and II of this paper).

§ 60. The method of solution for small values of τ_1 given in this section derives from

3. H. C. VAN DE HULST, ibid. **107**, 220 (1948).

In reference 3 van de Hulst has treated the problem of diffuse reflection and transmission (under conditions of isotropic scattering) by analysing the emergent radiations in terms of the light which has been scattered once, twice, etc., in the atmosphere and consistently arranging that at each stage of the approximation the solutions have the forms required by the exact theory and the principles of invariance. The treatment in the text, however, follows

4. S. CHANDRASEKHAR, ibid. **108**, 92 (1948) and **109**, 555 (1949).

The functions F_{j+1} introduced in this section have been studied by

5. L. V. KING, *Phil. Trans. R. Soc. London*, A, **212**, 375 (1913).

6. A. HAMMAD and S. CHAPMAN, *Phil. Mag.* **28**, 99 (1930).

The properties of the functions F_{j+1}, $G_{m,n}$, and $G'_{m,n}$ are considered in sufficient detail by van de Hulst (ref. 3). Tables of these functions will be found in reference 4.

DIFFUSE REFLECTION AND TRANSMISSION

61. Introduction. Questions of uniqueness

IN this chapter we return to the problem of diffuse reflection and transmission by plane-parallel atmospheres and show how exact solutions can be found under a variety of scattering conditions. The method consists in starting with the integral equations derived from the principles of invariance in Chapter VII; reducing them to pairs of integral equations of the standard form

$$X(\mu) = 1 + \mu \int_0^1 \frac{\Psi(\mu')}{\mu + \mu'} [X(\mu)X(\mu') - Y(\mu)Y(\mu')] \, d\mu' \tag{1}$$

and

$$Y(\mu) = e^{-\tau_1/\mu} + \mu \int_0^1 \frac{\Psi(\mu')}{\mu - \mu'} [Y(\mu)X(\mu') - X(\mu)Y(\mu')] \, d\mu'; \tag{2}$$

and finally, relating in a unique manner the various constants occurring in the solutions with the moments of the X and Y functions appropriate to the problem. The theory is therefore similar, in its outlines, to the theory of diffuse reflection by semi-infinite atmospheres described in Chapter VI. However, there is one important respect in which the present theory differs from the theory of transfer in semi-infinite atmospheres, namely, that in all conservative cases of perfect scattering the integral equations derived from the principles of invariance do not suffice to characterize the physical solutions uniquely; for, as we shall see, the general solutions of the relevant equations have a single arbitrary parameter in them. The origin of this non-uniqueness in the solution is not clear; but we shall see that, in all cases, the ambiguity can be removed by appealing to the K-integral (Chap. I, § 10) which conservative problems always admit.

The reference to the question of the uniqueness of the solution in the preceding paragraph draws attention to a fact which we have so far ignored, namely, that there are, indeed, no grounds for believing that the principles of invariance, by themselves, can lead to determinate solutions of the various problems. The invariances and the equations they lead to are only *necessary* conditions; it is by no means obvious that they are also *sufficient*. It is therefore, really, a matter for some surprise that the invariances should suffice to determine the solutions, without any ambiguity, in as many cases as they do. Thus, in the theory of transfer in semi-infinite atmospheres, the equations derived from the invariances have proved sufficient to characterize the physical solutions completely. Conversely, we may conclude from this uniqueness of the mathematical solution that the physical formulation has been complete in these cases. This is a somewhat novel situation in physics; for, normally, a physical problem is formulated in terms of equations and boundary conditions and it is generally assumed that a well-defined physical problem admits only of a unique solution, i.e. one often infers the uniqueness of the mathematical solution from the completeness of the physical formulation. The situation we encounter in the solution for the angular distributions of the emergent radiations from the integral equations derived from certain immediately obvious 'kinematical' conditions on the solution is the opposite of the usual one: we have no assurance, to begin with, that the equations include all the physical conditions of the problem; yet, when the equations are found to have a unique mathematical solution, as they do in all cases except those of perfect scattering in finite atmospheres, we conclude that in all these cases the physical problem has been formulated completely.

62. The laws of diffuse reflection and transmission for isotropic scattering

The basic equations of the problem of diffuse reflections and transmission by an atmosphere scattering radiation isotropically with an albedo $\varpi_0 \leqslant 1$ have already been given in Chapter VII, § 54. They are:

$$I(0,\mu) = \frac{F}{4\mu} S(\tau_1; \mu, \mu_0); \qquad I(\tau_1, -\mu) = \frac{F}{4\mu} T(\tau_1; \mu, \mu_0), \qquad (3)$$

$$\left(\frac{1}{\mu_0} + \frac{1}{\mu}\right) S(\tau_1; \mu, \mu_0) = \varpi_0[X(\mu)X(\mu_0) - Y(\mu)Y(\mu_0)], \qquad (4)$$

$$\left(\frac{1}{\mu_0} - \frac{1}{\mu}\right) T(\tau_1; \mu, \mu_0) = \varpi_0[Y(\mu)X(\mu_0) - X(\mu)Y(\mu_0)], \qquad (5)$$

$$\frac{\partial S}{\partial \tau_1} = \varpi_0 Y(\mu)Y(\mu_0) \qquad (6)$$

and
$$\left(\frac{1}{\mu_0} - \frac{1}{\mu}\right)\frac{\partial T}{\partial \tau_1} = \varpi_0\left[\frac{1}{\mu_0}X(\mu)Y(\mu_0) - \frac{1}{\mu}Y(\mu)X(\mu_0)\right]. \qquad (7)$$

Further, the definitions of $X(\mu)$ and $Y(\mu)$ in terms of S and T are

$$X(\mu) = 1 + \frac{1}{2} \int_0^1 S(\tau_1; \mu, \mu') \frac{d\mu'}{\mu'} \tag{8}$$

and

$$Y(\mu) = e^{-\tau_1/\mu} + \frac{1}{2} \int_0^1 T(\tau_1; \mu, \mu') \frac{d\mu'}{\mu'}. \tag{9}$$

In virtue of equations (4), (5), (8), and (9) we have

$$X(\mu) = 1 + \tfrac{1}{2}\varpi_0 \mu \int_0^1 \frac{d\mu'}{\mu + \mu'} [X(\mu)X(\mu') - Y(\mu)Y(\mu')] \tag{10}$$

and

$$Y(\mu) = e^{-\tau_1/\mu} + \tfrac{1}{2}\varpi_0 \mu \int_0^1 \frac{d\mu'}{\mu - \mu'} [Y(\mu)X(\mu') - X(\mu)Y(\mu')]. \tag{11}$$

Thus X and Y satisfy integral equations of the form (1) and (2) with the characteristic function

$$\Psi(\mu) = \tfrac{1}{2}\varpi_0 = \text{constant}. \tag{12}$$

In considering the foregoing equations it is of interest to establish, first, that equations (6) and (7) are really equivalent to the integro-differential equations of Theorem 1 (Chapter VIII, § 56).

Thus, differentiating equation (8) with respect to τ_1 and using equation (6), we have

$$\frac{\partial X}{\partial \tau_1} = \tfrac{1}{2}\varpi_0 Y(\mu) \int_0^1 \frac{d\mu'}{\mu'} Y(\mu'). \tag{13}$$

Similarly, differentiating equation (9) and using equation (7), we have

$$\frac{\partial Y}{\partial \tau_1} = -\frac{1}{\mu} e^{-\tau_1/\mu} + \tfrac{1}{2}\varpi_0 \int_0^1 \frac{d\mu'}{\mu - \mu'} \left[\frac{\mu}{\mu'} X(\mu)Y(\mu') - Y(\mu)X(\mu') \right], \tag{14}$$

and combining this with equation (11) we have

$$\frac{\partial Y}{\partial \tau_1} + \frac{Y}{\mu} = \tfrac{1}{2}\varpi_0 X(\mu) \int_0^1 \frac{d\mu'}{\mu'} Y(\mu'). \tag{15}$$

It is seen that equations (13) and (15) are in agreement with equations (13) and (14) of Chapter VIII, § 56.

Finally, we may note that according to equations (3)–(5) we can express the reflected and the transmitted intensities in the forms

$$I(0,\mu) = \tfrac{1}{4}\varpi_0 F \frac{\mu_0}{\mu+\mu_0}[X(\mu)X(\mu_0)-Y(\mu)Y(\mu_0)]$$

and
$$I(\tau_1, -\mu) = \tfrac{1}{4}\varpi_0 F \frac{\mu_0}{\mu-\mu_0}[Y(\mu)X(\mu_0)-X(\mu)Y(\mu_0)]. \qquad (16)$$

62.1. *A meaning for the X- and Y-functions*

Equations (8) and (9) allow the following interpretation of the X- and Y-functions:

Let radiation from all directions, with an angular distribution

$$I_{\text{inc}}(0,\mu',\varphi') = \frac{I_0}{|\mu'|} \quad (-1 \leqslant \mu' \leqslant +1), \qquad (17)$$

be incident on the atmosphere. Then the light diffusely reflected in the direction (μ,φ) is given by (cf. Chap. I, eq. [122])

$$I_{\text{ref}}(0,\mu,\varphi) = \frac{1}{4\pi\mu} \int_0^1 \int_0^{2\pi} S(\tau_1;\mu,\mu') \frac{I_0}{\mu'} d\mu' d\varphi',$$

or
$$I_{\text{ref}}(0,\mu,\varphi) = \frac{I_0}{2\mu} \int_0^1 S(\tau_1;\mu,\mu') \frac{d\mu'}{\mu'}. \qquad (18)$$

Similarly, the light diffusely transmitted in the direction $(-\mu,\varphi)$, below τ_1, is

$$I_{\text{trans}}(\tau_1, -\mu,\varphi) = \frac{I_0}{2\mu} \int_0^1 T(\tau_1;\mu,\mu') \frac{d\mu'}{\mu'}. \qquad (19)$$

On the other hand, the intensity in the incident radiation field, already in the direction (μ,φ), is

$$I_{\text{inc}}(0,\mu,\varphi) = \frac{I_0}{\mu}, \qquad (20)$$

while in the direction $(-\mu,\varphi)$ at τ_1, the intensity of the directly transmitted light is

$$I_{\text{inc}}(0, -\mu,\varphi)e^{-\tau_1/\mu} = \frac{I_0}{\mu}e^{-\tau_1/\mu}. \qquad (21)$$

Hence
$$\frac{I_{\text{ref}}(0, +\mu,\varphi)+I_{\text{inc}}(0, +\mu,\varphi)}{I_{\text{inc}}(0, +\mu,\varphi)} = 1+\tfrac{1}{2}\int_0^1 S(\tau_1;\mu,\mu')\frac{d\mu'}{\mu'}$$

$$= X(\mu), \qquad (22)$$

and

$$\frac{I_{\text{trans}}(\tau_1, -\mu, \varphi) + e^{-\tau_1/\mu} I_{\text{inc}}(0, -\mu, \varphi)}{I_{\text{inc}}(0, -\mu, \varphi)} = e^{-\tau_1/\mu} + \tfrac{1}{2} \int_0^1 T(\tau_1; \mu, \mu') \frac{d\mu'}{\mu'}$$

$$= Y(\mu). \tag{23}$$

In other words, $X(\mu)$ and $Y(\mu)$ represent the relative changes in the intensities in the direction (μ, φ) at $\tau = 0$ and in the direction $(-\mu, \varphi)$ at $\tau = \tau_1$ respectively, in a prevailing radiation field $I_{\text{inc}}(0, \mu', \varphi') = I_0 |\mu'|^{-1}$, due to the presence of the atmosphere.†

Alternative meanings to the X- and Y-functions (which are in fact implicit in the original definitions of these functions) are that they are, respectively, proportional to the source functions at $\tau = 0$ and $\tau = \tau_1$ for an incident parallel beam of radiation in the direction μ; for, according to Chapter VII, equations (27) and (28) we have, in the case of isotropic scattering

$$\mathfrak{J}(0) = \tfrac{1}{4}\varpi_0 F\left[1 + \tfrac{1}{2}\int_0^1 S(\tau_1; \mu'', \mu_0)\frac{d\mu''}{\mu''}\right]$$

$$= \tfrac{1}{4}\varpi_0 F X(\mu_0)$$

and

$$\mathfrak{J}(\tau_1) = \tfrac{1}{4}\varpi_0 F\left[e^{-\tau_1/\mu_0} + \tfrac{1}{2}\int_0^1 T(\tau_1; \mu'', \mu_0)\frac{d\mu''}{\mu''}\right]$$

$$= \tfrac{1}{4}\varpi_0 F Y(\mu_0). \tag{24}$$

62.2. *The ambiguity in the solutions of the integral equations in the case $\varpi_0 = 1$ and its resolution by an appeal to the K-integral*

When $\varpi_0 = 1$, the equations (10) and (11) belong to the conservative class discussed in Chapter VIII, § 58 and according to Theorem 5 of that section, the solutions of these equations are not unique, the general solutions being expressible in the forms

$$X(\mu) + Q\mu[X(\mu) + Y(\mu)]$$

and

$$Y(\mu) - Q\mu[X(\mu) + Y(\mu)], \tag{25}$$

where Q is an arbitrary constant and $X(\mu)$ and $Y(\mu)$ are the standard solutions, having, for the characteristic function $\tfrac{1}{2}$, the property

$$\alpha_0 = \int_0^1 X(\mu)\, d\mu = 2 \quad \text{and} \quad \beta_0 = \int_0^1 Y(\mu)\, d\mu = 0. \tag{26}$$

---

† It should be emphasized that similar interpretations do not apply to X- and Y-functions defined, generally, in terms of a characteristic function. However, it is clear that analogous interpretations can be given to the functions ψ_i^m and ϕ_i^m introduced in Chap. VII, § 53, eqs. (65) and (66).

With the solutions (25), the expressions (16) for the emergent intensities take the forms

$$I(0, \mu) = \tfrac{1}{4}\mu_0 F\left\{\frac{1}{\mu_0+\mu}[X(\mu_0)X(\mu)-Y(\mu_0)Y(\mu)]+\right.$$
$$\left.+ Q[X(\mu_0)+Y(\mu_0)][X(\mu)+Y(\mu)]\right\}$$

and

$$I(\tau_1, -\mu) = \tfrac{1}{4}\mu_0 F\left\{\frac{1}{\mu_0-\mu}[Y(\mu_0)X(\mu)-X(\mu_0)Y(\mu)]-\right.$$
$$\left.- Q[X(\mu_0)+Y(\mu_0)][X(\mu)+Y(\mu)]\right\}. \quad (27)$$

These expressions for the emergent intensities involve the arbitrary constant Q and there is nothing in the framework of equations (3)–(11), for the case $\varpi_0 = 1$, which will remove this arbitrariness. We therefore conclude that, in this case, the various invariances considered in Chapter VII are not sufficient to determine the physical solution uniquely.

We now turn to the matter of the arbitrariness left in the solution (27) and the manner in which it is to be resolved.

The equation of transfer appropriate to the problem on hand is

$$\mu\frac{dI(\tau, \mu)}{d\tau} = I(\tau, \mu)-\tfrac{1}{2}\int_{-1}^{+1} I(\tau, \mu')\,d\mu'-\tfrac{1}{4}Fe^{-\tau/\mu_0}. \quad (28)$$

The flux and the K-integrals which this equation admits are (cf. Chap. I, §§ 8 and 10)

$$F(\tau) = 2\int_{-1}^{+1} I(\tau, \mu)\mu\,d\mu = \mu_0 F(e^{-\tau/\mu_0}-\gamma_1) \quad (29)$$

and

$$K(\tau) = \tfrac{1}{2}\int_{-1}^{+1} I(\tau, \mu)\mu^2\,d\mu = \tfrac{1}{4}\mu_0 F(-\mu_0 e^{-\tau/\mu_0}-\gamma_1\tau+\gamma_2), \quad (30)$$

where γ_1 and γ_2 are two constants.

Applying the integrals (29) and (30) at $\tau = 0$ and at $\tau = \tau_1$, we have

$$F(0) = 2\int_0^1 I(0, \mu)\mu\,d\mu = \mu_0 F(1-\gamma_1), \quad (31)$$

$$F(\tau_1) = -2\int_0^1 I(\tau_1, -\mu)\mu\,d\mu = \mu_0 F(e^{-\tau_1/\mu_0}-\gamma_1), \quad (32)$$

$$K(0) = \tfrac{1}{2}\int_0^1 I(0, \mu)\mu^2\,d\mu = \tfrac{1}{4}\mu_0 F(-\mu_0+\gamma_2), \quad (33)$$

and

$$K(\tau_1) = \tfrac{1}{2}\int_0^1 I(\tau_1, -\mu)\mu^2\,d\mu = \tfrac{1}{4}\mu_0 F(-\mu_0 e^{-\tau_1/\mu_0}-\gamma_1\tau_1+\gamma_2). \quad (34)$$

On the other hand, we can also evaluate these quantities according to the solution (27) for $I(0, \mu)$ and $I(\tau_1, -\mu)$. In this manner we shall obtain four relations between the three constants γ_1, γ_2, and Q. However, it will appear that two of these four relations are equivalent and that, in fact, they just suffice to determine all the constants uniquely.

The integrals defining $F(0)$, $F(\tau_1)$, $K(0)$, and $K(\tau_1)$ in terms of $I(0, \mu)$ and $I(\tau_1, -\mu)$ given by equations (27) can all be evaluated by using the various relations valid for standard solutions and collected under Theorem 8 in Chapter VIII, § 58. We find

$$F(0) = \mu_0 F\{1 + \tfrac{1}{2}Q(\alpha_1 + \beta_1)[X(\mu_0) + Y(\mu_0)]\}, \tag{35}$$

$$F(\tau_1) = \mu_0 F\{e^{-\tau_1/\mu_0} + \tfrac{1}{2}Q(\alpha_1 + \beta_1)[X(\mu_0) + Y(\mu_0)]\}, \tag{36}$$

$$K(0) = \tfrac{1}{4}\mu_0 F\{-\mu_0 + \tfrac{1}{2}\alpha_1 X(\mu_0) - \tfrac{1}{2}\beta_1 Y(\mu_0) + \tfrac{1}{2}Q(\alpha_2 + \beta_2)[X(\mu_0) + Y(\mu_0)]\}, \tag{37}$$

and

$$K(\tau_1) = \tfrac{1}{4}\mu_0 F\{-\mu_0 e^{-\tau_1/\mu_0} + \tfrac{1}{2}\beta_1 X(\mu_0) - \tfrac{1}{2}\alpha_1 Y(\mu_0) -$$
$$- \tfrac{1}{2}Q(\alpha_2 + \beta_2)[X(\mu_0) + Y(\mu_0)]\}, \tag{38}$$

where α_n and β_n are the moments of order n of $X(\mu)$ and $Y(\mu)$, respectively.

It is now clear that equations (31) and (35) and (32) and (36), in agreement with each other, determine

$$\gamma_1 = -\tfrac{1}{2}Q(\alpha_1 + \beta_1)[X(\mu_0) + Y(\mu_0)]. \tag{39}$$

From equations (33) and (37) we next find that

$$\gamma_2 = \tfrac{1}{2}\alpha_1 X(\mu_0) - \tfrac{1}{2}\beta_1 Y(\mu_0) + \tfrac{1}{2}Q(\alpha_2 + \beta_2)[X(\mu_0) + Y(\mu_0)]. \tag{40}$$

Finally, from equations (34) and (38) we obtain

$$-\gamma_1 \tau_1 + \gamma_2 = \tfrac{1}{2}\beta_1 X(\mu_0) - \tfrac{1}{2}\alpha_1 Y(\mu_0) - \tfrac{1}{2}Q(\alpha_2 + \beta_2)[X(\mu_0) + Y(\mu_0)]. \tag{41}$$

Now substituting for γ_1 and γ_2 according to equations (39) and (40) in equation (41), we find

$$\tfrac{1}{2}Q(\alpha_1 + \beta_1)\tau_1 = -\tfrac{1}{2}(\alpha_1 - \beta_1) - Q(\alpha_2 + \beta_2). \tag{42}$$

Hence
$$Q = - \frac{\alpha_1 - \beta_1}{(\alpha_1 + \beta_1)\tau_1 + 2(\alpha_2 + \beta_2)}. \tag{43}$$

With this determination of Q in terms of the optical thickness, τ_1, of the atmosphere and the moments of the standard solutions, we have removed the arbitrariness left by the integral equations in the solutions for the emergent intensities. It is in some ways remarkable that an explicit appeal to the K-integral is what is necessary to resolve the arbitrariness left by the integral equations. It is found that similar appeals to the K-integrals are necessary in all other cases of perfect scattering.

62.3. *The verification that Q satisfies the differential equation of Theorem* 7, § 58

It is apparent that the quantity Q as introduced in § 62.2 must satisfy the differential equation of Theorem 7, § 58. In our present context we can write this equation (Chap. VIII, § 58, eq. [58]) in the form

$$\frac{d}{d\tau_1}\left(\frac{1}{Q}\right) - \frac{\beta_{-1}}{Q} = -1, \tag{44}$$

since

$$y_{-1} = \tfrac{1}{2} \int_0^1 \frac{d\mu'}{\mu'} Y(\mu') = \tfrac{1}{2}\beta_{-1}. \tag{45}$$

We shall now show that Q as defined by equation (43) does in fact satisfy this equation.

Making use of the relation (cf. Th. 8, Chap. VIII, § 58, eq. [68])

$$\alpha_1^2 - \beta_1^2 = 4 \int_0^1 \mu^2 \, d\mu = \tfrac{4}{3}, \tag{46}$$

we first rewrite equation (43) in the form

$$\frac{1}{Q} = -\tfrac{3}{4}[(\alpha_1+\beta_1)^2\tau_1 + 2(\alpha_2+\beta_2)(\alpha_1+\beta_1)]. \tag{47}$$

Differentiating this equation, we obtain

$$\frac{d}{d\tau_1}\left(\frac{1}{Q}\right) = -\frac{3}{4}\Bigg[(\alpha_1+\beta_1)\Big\{(\alpha_1+\beta_1)+2\tau_1\frac{d}{d\tau_1}(\alpha_1+\beta_1)\Big\} +$$
$$+ 2(\alpha_2+\beta_2)\frac{d}{d\tau_1}(\alpha_1+\beta_1) + 2(\alpha_1+\beta_1)\frac{d}{d\tau_1}(\alpha_2+\beta_2)\Bigg]. \tag{48}$$

To simplify equation (48) we observe that, according to equations (13) and (15), we now have

$$\frac{\partial}{\partial\tau_1}(X+Y) = \tfrac{1}{2}\beta_{-1}(X+Y) - \frac{Y}{\mu}. \tag{49}$$

Multiplying this equation by μ^n and integrating over the range of μ, we obtain

$$\frac{d}{d\tau_1}(\alpha_n+\beta_n) = \tfrac{1}{2}\beta_{-1}(\alpha_n+\beta_n) - \beta_{n-1}. \tag{50}$$

In particular,

$$\frac{d}{d\tau_1}(\alpha_1+\beta_1) = \tfrac{1}{2}\beta_{-1}(\alpha_1+\beta_1) \tag{51}$$

(since, according to eq. [26], $\beta_0 = 0$), and

$$\frac{d}{d\tau_1}(\alpha_2+\beta_2) = \tfrac{1}{2}\beta_{-1}(\alpha_2+\beta_2)-\beta_1. \tag{52}$$

Using these relations in equation (48) we find, after some minor reductions, that

$$\frac{d}{d\tau_1}\left(\frac{1}{Q}\right) = -\tfrac{3}{4}[\beta_{-1}\{(\alpha_1+\beta_1)^2\tau_1+2(\alpha_2+\beta_2)(\alpha_1+\beta_1)\}+$$
$$+(\alpha_1+\beta_1)^2-2\beta_1(\alpha_1+\beta_1)]. \tag{53}$$

Hence (cf. eqs. [46] and [47])

$$\frac{d}{d\tau_1}\left(\frac{1}{Q}\right) = \frac{\beta_{-1}}{Q} - \tfrac{3}{4}(\alpha_1^2-\beta_1^2) = \frac{\beta_{-1}}{Q}-1. \tag{54}$$

This completes the verification.

63. Approximate solutions for small values of τ_1 in the case of isotropic scattering

The approximate solutions for small values of τ_1 for the X- and Y-functions given in Chapter VIII, § 60 (eqs. [113] and [117]), for the characteristic function (12), become

$$X^{(1)}(\mu) = 1, \qquad Y^{(1)}(\mu) = e^{-\tau_1/\mu} \tag{55}$$

and

$$X^{(2)}(\mu) = 1+\tfrac{1}{2}\varpi_0 F_1(\tau_1, -\mu),$$
$$Y^{(2)}(\mu) = e^{-\tau_1/\mu}[1+\tfrac{1}{2}\varpi_0 F_1(\tau_1,\mu)]. \tag{56}$$

These solutions can be further corrected in the manner described in Chapter VIII, § 60.2, by adding to both X and Y a term of the form

$$\Delta(\tau_1)\mu(1-e^{-\tau_1/\mu}) \tag{57}$$

where $\Delta(\tau_1)$ is given by Chapter VIII, equations (130) (first approximation) and (136) (second approximation).

We shall now exhibit the relationship of the solutions (55) and (56) to the approximate solutions given in earlier treatments of the subject in which the emergent intensities were analysed in terms of light which has been scattered once, twice, etc. Thus, considering the layer of the atmosphere between τ and $\tau+d\tau$, we have for the contribution to the diffuse intensity in the direction μ, by the scattering of the reduced incident flux $\pi F e^{-\tau/\mu_0}$ at this level (cf. Chap. VI, eq. [74]),

$$\tfrac{1}{4}\varpi_0 F e^{-\tau/\mu_0}\frac{d\tau}{\mu}. \tag{58}$$

Of this intensity, the fractions $e^{-\tau/\mu}$ and $e^{-(\tau_1-\tau)/\mu}$ will emerge in the directions μ and $-\mu$, from $\tau = 0$ and $\tau = \tau_1$, respectively. The contributions to the reflected and the transmitted intensities by light which has suffered a single scattering process in the layer between τ and $\tau + d\tau$, are, therefore,

$$\tfrac{1}{4}\varpi_0 \, F e^{-\tau/\mu_0} e^{-\tau/\mu} \frac{d\tau}{\mu} \quad \text{and} \quad \tfrac{1}{4}\varpi_0 \, F e^{-\tau/\mu_0} e^{-(\tau_1-\tau)/\mu} \frac{d\tau}{\mu}. \tag{59}$$

Integrating these expressions over τ from 0 to τ_1, we obtain

$$I^{(1)}(0, +\mu) = \tfrac{1}{4}\varpi_0 \, F \frac{\mu_0}{\mu+\mu_0} \left[1 - \exp\left\{ -\tau_1\left(\frac{1}{\mu} + \frac{1}{\mu_0} \right) \right\} \right]$$

and
$$I^{(1)}(\tau_1, -\mu) = \tfrac{1}{4}\varpi_0 \, F \frac{\mu_0}{\mu-\mu_0} \left[e^{-\tau_1/\mu} - e^{-\tau_1/\mu_0} \right], \tag{60}$$

for the intensities of the reflected and the transmitted light which has been scattered once in the atmosphere.

Comparing equations (16) and (60), we observe that the first approximation (55) is equivalent to a treatment in which only light which has been scattered once in the atmosphere is included. Similarly, it can be shown that the second approximation (56) is equivalent to a treatment in which light which has been scattered twice is also included.

A further comparison of interest arises when we consider the interpretation of the X- and Y-functions given in § 62.1. On this interpretation, the case when radiation from all directions with the angular distribution (17) is incident on the atmosphere is considered. In this case the contributions to diffuse reflection and transmission by light which has been scattered only once in the atmosphere are

$$I_{\text{ref}}^{(1)}(0, +\mu) = \tfrac{1}{2}\varpi_0 I_0 \int_0^1 \frac{d\mu'}{\mu'} \int_0^{\tau_1} \frac{d\tau}{\mu} e^{-\tau/\mu'} e^{-\tau/\mu}$$

and
$$I_{\text{trans}}^{(1)}(\tau_1, -\mu) = \tfrac{1}{2}\varpi_0 I_0 \int_0^1 \frac{d\mu'}{\mu'} \int_0^{\tau_1} \frac{d\tau}{\mu} e^{-\tau/\mu'} e^{-(\tau_1-\tau)/\mu}. \tag{61}$$

Performing the integrations over τ we have

$$I_{\text{ref}}^{(1)}(0, +\mu) = \tfrac{1}{2}\varpi_0 I_0 \int_0^1 \frac{d\mu'}{\mu+\mu'} \left[1 - \exp\left\{ -\tau_1\left(\frac{1}{\mu} + \frac{1}{\mu} \right) \right\} \right]$$

and
$$I_{\text{trans}}^{(1)}(\tau_1, -\mu) = \tfrac{1}{2}\varpi_0 I_0 \, e^{-\tau_1/\mu} \int_0^1 \frac{d\mu'}{\mu-\mu'} \left[1 - \exp\left\{ -\tau_1\left(\frac{1}{\mu'} - \frac{1}{\mu} \right) \right\} \right], \tag{62}$$

or recalling the definition of the function $F_{j+1}(\tau_1, \mu)$ (Chap. VIII, eq. [118]), we have

$$I_{\text{ref}}^{(1)}(0, +\mu) = \tfrac{1}{2}\varpi_0 \frac{I_0}{\mu} F_1(\tau_1, -\mu)$$

and

$$I_{\text{trans}}^{(1)}(\tau_1, -\mu) = \tfrac{1}{2}\varpi_0 \frac{I_0}{\mu} e^{-\tau_1/\mu} F_1(\tau_1, \mu). \tag{63}$$

Accordingly (cf. eqs. [22] and [23])

$$\frac{I_{\text{ref}}^{(1)}(0, +\mu) + I_{\text{inc}}(0, +\mu)}{I_{\text{inc}}(0, +\mu)} = 1 + \tfrac{1}{2}\varpi_0 F_1(\tau_1, -\mu) = X^{(2)}(\mu)$$

and

$$\frac{I_{\text{trans}}^{(1)}(\tau_1, -\mu) + e^{-\tau_1/\mu} I_{\text{inc}}(0, -\mu)}{I_{\text{inc}}(0, -\mu)} = e^{-\tau_1/\mu}[1 + \tfrac{1}{2}\varpi_0 F_1(\tau_1, \mu)] = Y^{(2)}(\mu). \tag{64}$$

Thus $X^{(2)}(\mu)$ and $Y^{(2)}(\mu)$ represent the relative changes in the intensities in the direction $+\mu$ at $\tau = 0$ and in the direction $-\mu$ at $\tau = \tau_1$, respectively, due to light which has suffered a single scattering process in the atmosphere, when radiation from all directions with an angular distribution $I^{(0)}/|\mu'|$, incident on the atmosphere, is present.

63.1. *The approximate solution for the conservative isotropic case*

In § 62.2 we have shown that in the case $\varpi_0 = 1$ the solutions for X and Y, appropriate to the problem, are

$$X^*(\mu) = X^{(s)}(\mu) + Q\mu[X^{(s)}(\mu) + Y^{(s)}(\mu)]$$

and

$$Y^*(\mu) = Y^{(s)}(\mu) - Q\mu[X^{(s)}(\mu) + Y^{(s)}(\mu)], \tag{65}$$

where $X^{(s)}(\mu)$ and $Y^{(s)}(\mu)$ are the standard solutions and

$$Q = -\frac{(\alpha_1^{(s)} - \beta_1^{(s)})}{[\alpha_1^{(s)} + \beta_1^{(s)}]\tau_1 + 2[\alpha_2^{(s)} + \beta_2^{(s)}]}. \tag{66}$$

(In the foregoing equations we have used a superscript s to distinguish the standard solutions from the others.)

When the standard solutions are themselves derived from another set $X(\mu)$ and $Y(\mu)$ (according to Chap. VIII, eqs. [138] and [139]), then the required solutions can be expressed in terms of these latter solutions in the forms (cf. Chap. VIII, eqs. [54])

$$X^*(\mu) = X(\mu) + (Q+q)\mu[X(\mu) + Y(\mu)]$$

and

$$Y^*(\mu) = Y(\mu) - (Q+q)\mu[X(\mu) + Y(\mu)], \tag{67}$$

where
$$X^{(s)}(\mu) = X(\mu) + q\mu[X(\mu) + Y(\mu)],$$

$$Y^{(s)}(\mu) = Y(\mu) - q\mu[X(\mu) + Y(\mu)] \tag{68}$$

and
$$q = \frac{y_0}{x_1 + y_1} = \frac{\beta_0}{\alpha_1 + \beta_1}, \tag{69}$$

the moments now referring to $X(\mu)$ and $Y(\mu)$.

Since (cf. eqs. [68])

$$\alpha_j^{(s)} = \alpha_j + q(\alpha_{j+1} + \beta_{j+1})$$

and
$$\beta_j^{(s)} = \beta_j - q(\alpha_{j+1} + \beta_{j+1}), \tag{70}$$

the expression (66) for Q becomes

$$Q = -\frac{\alpha_1 - \beta_1 + 2q(\alpha_2 + \beta_2)}{(\alpha_1 + \beta_1)\tau_1 + 2(\alpha_2 + \beta_2)}. \tag{71}$$

Combining this with (69), we have

$$Q + q = \frac{\beta_0\tau_1 - (\alpha_1 - \beta_1)}{(\alpha_1 + \beta_1)\tau_1 + 2(\alpha_2 + \beta_2)}. \tag{72}$$

The solutions $X^*(\mu)$ and $Y^*(\mu)$ appropriate to the problem of diffuse reflection and transmission are, therefore,

$$X^*(\mu) = X(\mu) + \frac{\beta_0\tau_1 - (\alpha_1 - \beta_1)}{(\alpha_1 + \beta_1)\tau_1 + 2(\alpha_2 + \beta_2)}\mu[X(\mu) + Y(\mu)]$$

and
$$Y^*(\mu) = Y(\mu) - \frac{\beta_0\tau_1 - (\alpha_1 - \beta_1)}{(\alpha_1 + \beta_1)\tau_1 + 2(\alpha_2 + \beta_2)}\mu[X(\mu) + Y(\mu)]. \tag{73}$$

If we now use in equations (73) the solutions in the corrected second approximation, we shall obtain solutions which will *exactly* satisfy both the integrals of the problem, namely, the flux and the K-integrals.†

64. Diffuse reflection and transmission on Rayleigh's phase function

We have already shown in Chapter VII, § 53, how the integral equations governing the angular distributions of the reflected and the

---

† In this connexion it is of interest to note that, by writing the reflected and the transmitted intensities in the forms given by equations (16) (with $\varpi_0 = 1$), we can readily show that the flux and the K-integrals are equivalent to the conditions

$$\alpha_0 + \beta_0 = 2 \quad \text{and} \quad \tau_1 = \frac{\alpha_1 - \beta_1}{\beta_0};$$

and it may be verified that, with the solutions for X and Y in the forms given by eqs. (73), the latter condition is satisfied in virtue of the former.

transmitted radiations from an atmosphere scattering according to a general phase function, expressible as a series in Legendre polynomials, can be reduced to independent systems of equations in one variable only.

In the case of scattering according to Rayleigh's phase function, we can express the scattering and the transmission functions in the forms (cf. Chap. IV, eq. [67])

$$S(\tau_1; \mu, \varphi; \mu_0, \varphi_0) = \tfrac{3}{8}[S^{(0)}(\mu, \mu_0) - 4\mu\mu_0(1-\mu^2)^{\frac{1}{2}}(1-\mu_0^2)^{\frac{1}{2}} \times$$
$$\times S^{(1)}(\mu, \mu_0)\cos(\varphi_0-\varphi) + (1-\mu^2)(1-\mu_0^2)S^{(2)}(\mu, \mu_0)\cos 2(\varphi_0-\varphi)] \quad (74)$$

and

$$T(\tau_1; \mu, \varphi; \mu_0, \varphi_0) = \tfrac{3}{8}[T^{(0)}(\mu, \mu_0) + 4\mu\mu_0(1-\mu^2)^{\frac{1}{2}}(1-\mu_0^2)^{\frac{1}{2}} \times$$
$$\times T^{(1)}(\mu, \mu_0)\cos(\varphi_0-\varphi) + (1-\mu^2)(1-\mu_0^2)T^{(2)}(\mu, \mu_0)\cos 2(\varphi_0-\varphi)], \quad (75)$$

and the functions of the different orders (distinguished by the superscripts) satisfy independent systems of equations. Of these systems, the two governing the functions of order one and two are directly reducible to X- and Y-functions. Thus (cf. Chap. IV, eqs. [72] and [73])

$$\left(\frac{1}{\mu_0}+\frac{1}{\mu}\right)S^{(i)}(\mu, \mu_0) = X^{(i)}(\mu)X^{(i)}(\mu_0) - Y^{(i)}(\mu)Y^{(i)}(\mu_0)$$

and

$$\left(\frac{1}{\mu_0}-\frac{1}{\mu}\right)T^{(i)}(\mu, \mu_0) = Y^{(i)}(\mu)X^{(i)}(\mu_0) - X^{(i)}(\mu)Y^{(i)}(\mu_0) \quad (i = 1, 2), \quad (76)$$

where $X^{(1)}$, $Y^{(1)}$, and $X^{(2)}$, $Y^{(2)}$ are defined in terms of the characteristic functions

$$\tfrac{3}{8}\mu^2(1-\mu^2) \quad \text{and} \quad \tfrac{3}{32}(1-\mu^2)^2, \quad (77)$$

respectively. These terms require, therefore, no further consideration.

Turning to the functions $S^{(0)}(\mu, \mu_0)$ and $T^{(0)}(\mu, \mu_0)$ of order zero, we find that these functions must be expressible in the forms (cf. Chap. IV, eqs. [77] and [78])

$$\left(\frac{1}{\mu_0}+\frac{1}{\mu}\right)S^{(0)}(\mu, \mu_0) = \tfrac{1}{3}[\psi(\mu)\psi(\mu_0)-\chi(\mu)\chi(\mu_0)]+$$
$$+\tfrac{8}{3}[\phi(\mu)\phi(\mu_0)-\zeta(\mu)\zeta(\mu_0)]$$

and

$$\left(\frac{1}{\mu_0}-\frac{1}{\mu}\right)T^{(0)}(\mu, \mu_0) = \tfrac{1}{3}[\chi(\mu)\psi(\mu_0)-\psi(\mu)\chi(\mu_0)]+$$
$$+\tfrac{8}{3}[\zeta(\mu)\phi(\mu_0)-\phi(\mu)\zeta(\mu_0)], \quad (78)$$

where $\qquad \psi(\mu) = 3 - \mu^2 + \dfrac{3}{16} \displaystyle\int_0^1 (3 - \mu'^2) S^{(0)}(\mu, \mu') \dfrac{d\mu'}{\mu'},$

$$\phi(\mu) = \mu^2 + \frac{3}{16} \int_0^1 \mu'^2 S^{(0)}(\mu, \mu') \frac{d\mu'}{\mu'},$$

$$\chi(\mu) = (3 - \mu^2) e^{-\tau_1/\mu} + \frac{3}{16} \int_0^1 (3 - \mu'^2) T^{(0)}(\mu, \mu') \frac{d\mu'}{\mu'},$$

and $\qquad \zeta(\mu) = \mu^2 e^{-\tau_1/\mu} + \dfrac{3}{16} \displaystyle\int_0^1 \mu'^2 T^{(0)}(\mu, \mu') \dfrac{d\mu'}{\mu'}.$ $\qquad$ (79)

Further, we must also have

$$\frac{\partial S^{(0)}}{\partial \tau_1} = \tfrac{1}{3}\chi(\mu)\chi(\mu_0) + \tfrac{8}{3}\zeta(\mu)\zeta(\mu_0)$$

and

$$\left(\frac{1}{\mu_0} - \frac{1}{\mu}\right)\frac{\partial T^{(0)}}{\partial \tau_1} = \frac{1}{\mu_0}\left[\tfrac{1}{3}\psi(\mu)\chi(\mu_0) + \tfrac{8}{3}\phi(\mu)\zeta(\mu_0)\right] -$$
$$- \frac{1}{\mu}\left[\tfrac{1}{3}\chi(\mu)\psi(\mu_0) + \tfrac{8}{3}\zeta(\mu)\phi(\mu_0)\right]. \quad (80)$$

64.1. *The form of the solutions for $S^{(0)}(\mu, \mu_0)$ and $T^{(0)}(\mu, \mu_0)$*

By substituting for $S^{(0)}$ and $T^{(0)}$ according to equations (78) in equations (79) we shall obtain a simultaneous system of integral equations, of order four, for ψ, ϕ, χ, and ζ. In solving systems of equations of this type, we shall be guided, as in the case of the solution of similar systems in the semi-infinite case, by the form of the solutions obtained in the direct solution of the equations of transfer, and the correspondence enunciated in Chapter VIII, § 59, between the X- and Y-functions occurring in these solutions and the exact functions defined in terms of integral equations.

For the case on hand, it would appear that the solutions for $S^{(0)}(\mu, \mu_0)$ and $T^{(0)}(\mu, \mu_0)$ must be of the forms

$$\left(\frac{1}{\mu_0} + \frac{1}{\mu}\right) S^{(0)}(\mu, \mu_0) = X(\mu)X(\mu_0)[3 - c_1(\mu + \mu_0) + \mu\mu_0] -$$
$$- Y(\mu)Y(\mu_0)[3 + c_1(\mu + \mu_0) + \mu\mu_0] - c_2(\mu + \mu_0)[X(\mu)Y(\mu_0) + Y(\mu)X(\mu_0)]$$

and

$$\left(\frac{1}{\mu_0} - \frac{1}{\mu}\right) T^{(0)}(\mu, \mu_0) = Y(\mu)X(\mu_0)[3 + c_1(\mu - \mu_0) - \mu\mu_0] -$$
$$- X(\mu)Y(\mu_0)[3 - c_1(\mu - \mu_0) - \mu\mu_0] + c_2(\mu - \mu_0)[X(\mu)X(\mu_0) + Y(\mu)Y(\mu_0)],$$
$$(81)$$

where c_1 and c_2 are certain constants unspecified for the present, and $X(\mu)$ and $Y(\mu)$ are the *standard solutions* of the equations

$$X(\mu) = 1 + \tfrac{3}{16}\mu \int_0^1 \frac{3-\mu'^2}{\mu+\mu'} [X(\mu)X(\mu')-Y(\mu)Y(\mu')]\,d\mu'$$

and

$$Y(\mu) = e^{-\tau_1/\mu} + \tfrac{3}{16}\mu \int_0^1 \frac{3-\mu'^2}{\mu-\mu'} [Y(\mu)X(\mu')-X(\mu)Y(\mu')]\,d\mu', \qquad (82)$$

having the property

$$\tfrac{3}{16} \int_0^1 (3-\mu^2)X(\mu)\,d\mu = \tfrac{3}{16}(3\alpha_0 - \alpha_2) = 1$$

and

$$\int_0^1 (3-\mu^2)Y(\mu)\,d\mu = (3\beta_0 - \beta_2) = 0. \qquad (83)\dagger$$

64.2. *Verification of the solution and a relation between the constants* c_1 *and* c_2

The verification that the solutions for $S^{(0)}$ and $T^{(0)}$ have the forms (81) will consist in first evaluating ψ, ϕ, χ, and ζ according to equations (79), then requiring that when the resulting expressions for ψ, ϕ, etc., are substituted back into equations (78) we shall recover the form of the solutions assumed. In general, such a procedure will lead to certain conditions which the constants introduced in the solution (such as c_1 and c_2 in the present instance) must satisfy. We shall see that, in the particular case under discussion, the conditions derived in the manner indicated do not suffice to determine c_1 and c_2 without an ambiguity and an arbitrariness. This is a further example of the non-uniqueness of the solution, in conservative cases, of the integral equations incorporating the invariances of the problem. But, again, as in the case of conservative isotropic scattering, an appeal to the integrals of the problem resolves the ambiguity and the arbitrariness.

Our first step, then, is to evaluate ψ, ϕ, χ, and ζ according to equations (79) for $S^{(0)}$ and $T^{(0)}$ given by equations (81). The evaluation of the integrals defining ψ, ϕ, etc., is fairly straightforward if appropriate use is made of the various integral properties of equations (82). It may be noted that, in addition to equations (83), use must also be made of the relations (cf. Th. 4, Chap. VIII, eqs. [40]–[42])

$$\alpha_0 = 1 + \tfrac{3}{32}[3(\alpha_0^2 - \beta_0^2) - (\alpha_1^2 - \beta_1^2)] \qquad (84)$$

† Cf. S. Chandrasekhar, *Astrophys. J.* **106**, 152, 1947 (eqs. [221] and [222]).

and

$$(3-\mu^2) \int_0^1 \frac{d\mu'}{\mu+\mu'} [X(\mu)X(\mu')-Y(\mu)Y(\mu')]$$

$$= \frac{X(\mu)-1}{\frac{3}{16}\mu} + (\alpha_1-\mu\alpha_0)X(\mu)-(\beta_1-\mu\beta_0)Y(\mu)$$

and

$$(3-\mu^2) \int_0^1 \frac{d\mu'}{\mu-\mu'} [Y(\mu)X(\mu')-X(\mu)Y(\mu')]$$

$$= \frac{Y(\mu)-e^{-\tau_1/\mu}}{\frac{3}{16}\mu} + (\beta_1+\mu\beta_0)X(\mu)-(\alpha_1+\mu\alpha_0)Y(\mu). \quad (85)$$

Evaluating ψ, ϕ, etc., in the manner indicated, we find that

$$\psi(\mu) = (3-c_1\mu)X(\mu)-c_2\mu Y(\mu),$$

$$\chi(\mu) = (3+c_1\mu)Y(\mu)+c_2\mu X(\mu),$$

$$\phi(\mu) = \mu[q_1 X(\mu)-q_2 Y(\mu)] \text{ and } \zeta(\mu) = \mu[q_2 X(\mu)-q_1 Y(\mu)], \quad (86)$$

where

$$q_1 = \tfrac{3}{16}(3\alpha_1-c_1\alpha_2-c_2\beta_2) \quad \text{and} \quad q_2 = \tfrac{3}{16}(3\beta_1+c_1\beta_2+c_2\alpha_2). \quad (87)$$

Using the expressions (86) for ψ, χ, etc., we next evaluate $S^{(0)}$ and $T^{(0)}$ according to equations (78). We find

$$\left(\frac{1}{\mu_0}+\frac{1}{\mu}\right)S^{(0)}(\mu,\mu_0)$$

$$= X(\mu)X(\mu_0)[3-c_1(\mu+\mu_0)+\tfrac{1}{3}\{c_1^2-c_2^2+8(q_1^2-q_2^2)\}\mu\mu_0]-$$

$$-Y(\mu)Y(\mu_0)[3+c_1(\mu+\mu_0)+\tfrac{1}{3}\{c_1^2-c_2^2+8(q_1^2-q_2^2)\}\mu\mu_0]-$$

$$-c_2(\mu+\mu_0)[X(\mu)Y(\mu_0)+Y(\mu)X(\mu_0)]$$

and

$$\left(\frac{1}{\mu_0}-\frac{1}{\mu}\right)T^{(0)}(\mu,\mu_0)$$

$$= Y(\mu)X(\mu_0)[3+c_1(\mu-\mu_0)-\tfrac{1}{3}\{c_1^2-c_2^2+8(q_1^2-q_2^2)\}\mu\mu_0]-$$

$$-X(\mu)Y(\mu_0)[3-c_1(\mu-\mu_0)-\tfrac{1}{3}\{c_1^2-c_2^2+8(q_1^2-q_2^2)\}\mu\mu_0]+$$

$$+c_2(\mu-\mu_0)[X(\mu)X(\mu_0)+Y(\mu)Y(\mu_0)]. \quad (88)$$

A comparison of equations (81) and (88) now shows that among the constants c_1, c_2, q_1, and q_2 we must require that there exist the relation (cf. Chap. VI, eq. [24])

$$c_1^2-c_2^2+8(q_1^2-q_2^2) = 3. \quad (89)$$

Substituting for q_1 and q_2 according to equations (87) in (89) we obtain

$$32(c_1^2-c_2^2)+9[(c_1+c_2)(\alpha_2+\beta_2)-3(\alpha_1-\beta_1)]\times$$

$$\times[(c_1-c_2)(\alpha_2-\beta_2)-3(\alpha_1+\beta_1)]-96 = 0. \quad (90)$$

After some minor rearranging of the terms, the foregoing equation can be reduced to the form

$$[32+9(\alpha_2^2-\beta_2^2)](c_1^2-c_2^2)-27(\alpha_1+\beta_1)(\alpha_2+\beta_2)(c_1+c_2)-$$
$$-27(\alpha_1-\beta_1)(\alpha_2-\beta_2)(c_1-c_2)+81(\alpha_1^2-\beta_1^2)-96 = 0. \quad (91)$$

On the other hand, according to equations (83) and (84),

$$32+9(\alpha_2^2-\beta_2^2) = 32+(9\alpha_0-16)^2-81\beta_0^2$$
$$= 288(1-\alpha_0)+81(\alpha_0^2-\beta_0^2) = 27(\alpha_1^2-\beta_1^2). \quad (92)$$

Equation (91) therefore becomes

$$(\alpha_1^2-\beta_1^2)(c_1^2-c_2^2)-(\alpha_1+\beta_1)(\alpha_2+\beta_2)(c_1+c_2)-$$
$$-(\alpha_1-\beta_1)(\alpha_2-\beta_2)(c_1-c_2)+(\alpha_2^2-\beta_2^2) = 0. \quad (93)$$

Hence

$$[(\alpha_1+\beta_1)(c_1+c_2)-(\alpha_2-\beta_2)][(\alpha_1-\beta_1)(c_1-c_2)-(\alpha_2+\beta_2)] = 0. \quad (94)$$

It is apparent that one of the two factors in equation (94) must vanish. But within the framework of the equations satisfied by ψ, ϕ, χ, and ζ it is impossible to decide which of the two it must be; and in either case we shall have only one relation between the two constants c_1 and c_2. The problem is therefore characterized by an ambiguity and an arbitrariness. We shall now show how this can be resolved.

64.3. *The resolution of the ambiguity and the arbitrariness in the solution*

Since scattering according to Rayleigh's phase function is conservative, the problem admits both the flux and the K-integrals. The emergent values of F and K must therefore be given by equations of the form (cf. eqs. [31]–[34])

$$F(0) = \mu_0 F(1-\gamma_1); \qquad F(\tau_1) = \mu_0 F(e^{-\tau_1/\mu_0}-\gamma_1), \quad (95)$$
$$K(0) = \tfrac{1}{4}\mu_0 F(-\mu_0+\gamma_2),$$

and

$$K(\tau_1) = \tfrac{1}{4}\mu_0 F(-\mu_0 e^{-\tau_1/\mu_0}-\gamma_1\tau_1+\gamma_2), \quad (96)$$

where γ_1 and γ_2 are constants.

It is evident that only the azimuth independent terms in the intensity will contribute to F and K. We have, accordingly, to evaluate $F(0)$, etc., for the emergent intensities (cf. eqs. [74], [75], and [81])

$$I^{(0)}(0,\mu) = \tfrac{1}{2}\mu_0 F\Big\{\frac{3}{16}\frac{3-\mu^2}{\mu_0+\mu}[X(\mu_0)X(\mu)-Y(\mu_0)Y(\mu)]-$$

$$-\tfrac{3}{16}X(\mu_0)[(c_1-\mu)X(\mu)+c_2 Y(\mu)]-\tfrac{3}{16}Y(\mu_0)[c_2 X(\mu)+(c_1+\mu)Y(\mu)]\Big\}$$

and

$$I^{(0)}(\tau_1, -\mu) = \tfrac{1}{2}\mu_0 F\left\{\frac{3}{16}\frac{3-\mu^2}{\mu_0-\mu}[Y(\mu_0)X(\mu)-X(\mu_0)Y(\mu)]+\right.$$
$$\left.+\tfrac{3}{16}X(\mu_0)[c_2 X(\mu)+(c_1+\mu)Y(\mu)]+\tfrac{3}{16}Y(\mu_0)[(c_1-\mu)X(\mu)+c_2 Y(\mu)]\right\}. \tag{97}$$

With $I^{(0)}(0, \mu)$ and $I^{(0)}(\tau_1, -\mu)$ given by equations (97) the integrals defining $F(0)$, etc., can all be evaluated quite simply by using the various relations given in Chapter VIII, Theorem 8, and remembering that in the present case

$$x_1 = \tfrac{3}{16}(3\alpha_1-\alpha_3) \quad \text{and} \quad y_1 = \tfrac{3}{16}(3\beta_1-\beta_3). \tag{98}$$

We thus find

$$F(0) = \mu_0 F\{1-\tfrac{3}{16}X(\mu_0)(c_1\alpha_1+c_2\beta_1-\alpha_2)-\tfrac{3}{16}Y(\mu_0)(c_1\beta_1+c_2\alpha_1+\beta_2)\},$$
$$F(\tau_1) = \mu_0 F\{e^{-\tau_1/\mu_0}-\tfrac{3}{16}X(\mu_0)(c_1\beta_1+c_2\alpha_1+\beta_2)-$$
$$-\tfrac{3}{16}Y(\mu_0)(c_1\alpha_1+c_2\beta_1-\alpha_2)\}, \quad (99)$$
$$K(0) = \tfrac{1}{4}\mu_0 F\{-\mu_0-\tfrac{3}{16}X(\mu_0)(c_1\alpha_2+c_2\beta_2-3\alpha_1)-$$
$$-\tfrac{3}{16}Y(\mu_0)(c_1\beta_2+c_2\alpha_2+3\beta_1)\},$$

and

$$K(\tau_1) = \tfrac{1}{4}\mu_0 F\{-\mu_0 e^{-\tau_1/\mu_0}+\tfrac{3}{16}X(\mu_0)(c_1\beta_2+c_2\alpha_2+3\beta_1)+$$
$$+\tfrac{3}{16}Y(\mu_0)(c_1\alpha_2+c_2\beta_2-3\alpha_1)\}. \tag{100}$$

Comparing the reflected and the transmitted fluxes given by equations (99) with those given by the flux integral (cf. eqs. [95]), we find that

$$\gamma_1 = \tfrac{3}{16}X(\mu_0)(c_1\alpha_1+c_2\beta_1-\alpha_2)+\tfrac{3}{16}Y(\mu_0)(c_1\beta_1+c_2\alpha_1+\beta_2), \tag{101}$$

and also that

$$\gamma_1 = \tfrac{3}{16}X(\mu_0)(c_1\beta_1+c_2\alpha_1+\beta_2)+\tfrac{3}{16}Y(\mu_0)(c_1\alpha_1+c_2\beta_1-\alpha_2). \tag{102}$$

We must therefore require that

$$c_1\alpha_1+c_2\beta_1-\alpha_2 = c_1\beta_1+c_2\alpha_1+\beta_2,$$

or
$$(\alpha_1-\beta_1)(c_1-c_2)-(\alpha_2+\beta_2) = 0; \tag{103}$$

but this is one of the factors in equation (94). The appeal to the flux integral has therefore removed the ambiguity and decided which of the two factors in equation (94) must be set equal to zero.

In view of equation (103), we can combine equations (101) and (102) to give
$$\gamma_1 = \tfrac{3}{32}[(c_1+c_2)(\alpha_1+\beta_1)-(\alpha_2-\beta_2)][X(\mu_0)+Y(\mu_0)]. \tag{104}$$

Next from equations (96) and (100) we find that

$$\gamma_2 = -\tfrac{3}{16}X(\mu_0)(c_1\alpha_2+c_2\beta_2-3\alpha_1)-\tfrac{3}{16}Y(\mu_0)(c_1\beta_2+c_2\alpha_2+3\beta_1), \tag{105}$$

and

$$-\gamma_1\tau_1+\gamma_2 = +\tfrac{3}{16}X(\mu_0)(c_1\beta_2+c_2\alpha_2+3\beta_1)+\tfrac{3}{16}Y(\mu_0)(c_1\alpha_2+c_2\beta_2-3\alpha_1).$$

(106)

Now, substituting for γ_1 and γ_2 according to equations (104) and (105) in equation (106), we obtain

$$[(\alpha_1+\beta_1)(c_1+c_2)-(\alpha_2-\beta_2)]\tau_1 = -2[(\alpha_2+\beta_2)(c_1+c_2)-3(\alpha_1-\beta_1)],$$

or, solving for (c_1+c_2), we have

$$c_1+c_2 = \frac{(\alpha_2-\beta_2)\tau_1+6(\alpha_1-\beta_1)}{(\alpha_1+\beta_1)\tau_1+2(\alpha_2+\beta_2)}.$$

(107)

Since we have already shown that (cf. eq. [103])

$$c_1-c_2 = \frac{\alpha_2+\beta_2}{\alpha_1-\beta_1},$$

(108)

the solution to the problem is completed.

64.4. *The law of diffuse reflection and transmission*

Combining the results of the preceding paragraphs, we have the following expressions for the reflected and the transmitted intensities:

$$I(0,\mu,\varphi;\mu_0,\varphi_0)$$

$$= \frac{3}{32(\mu+\mu_0)}\{\tfrac{1}{3}[\psi(\mu)\psi(\mu_0)-\chi(\mu)\chi(\mu_0)]+\tfrac{8}{3}[\phi(\mu)\phi(\mu_0)-\zeta(\mu)\zeta(\mu_0)]-$$

$$-4\mu\mu_0(1-\mu^2)^{\frac{1}{2}}(1-\mu_0^2)^{\frac{1}{2}}[X^{(1)}(\mu)X^{(1)}(\mu_0)-Y^{(1)}(\mu)Y^{(1)}(\mu_0)]\cos(\varphi_0-\varphi)+$$

$$+(1-\mu^2)(1-\mu_0^2)[X^{(2)}(\mu)X^{(2)}(\mu_0)-Y^{(2)}(\mu)Y^{(2)}(\mu_0)]\cos 2(\varphi_0-\varphi)\}\mu_0\,F$$

and

$$I(\tau_1,-\mu,\varphi;\mu_0,\varphi_0)$$

$$= \frac{3}{32(\mu-\mu_0)}\{\tfrac{1}{3}[\chi(\mu)\psi(\mu_0)-\psi(\mu)\chi(\mu_0)]+\tfrac{8}{3}[\zeta(\mu)\phi(\mu_0)-\phi(\mu)\zeta(\mu_0)]+$$

$$+4\mu\mu_0(1-\mu^2)^{\frac{1}{2}}(1-\mu_0^2)^{\frac{1}{2}}[Y^{(1)}(\mu)X^{(1)}(\mu_0)-X^{(1)}(\mu)Y^{(1)}(\mu_0)]\cos(\varphi_0-\varphi)+$$

$$+(1-\mu^2)(1-\mu_0^2)[Y^{(2)}(\mu)X^{(2)}(\mu_0)-X^{(2)}(\mu)Y^{(2)}(\mu_0)]\cos 2(\varphi_0-\varphi)\}\mu_0\,F,$$

where

$$\psi(\mu) = (3-c_1\mu)X(\mu)-c_2\mu Y(\mu),$$

$$\chi(\mu) = (3+c_1\mu)Y(\mu)+c_2\mu X(\mu),$$

$$\phi(\mu) = \mu[q_1 X(\mu)-q_2 Y(\mu)], \qquad \zeta(\mu) = \mu[q_2 X(\mu)-q_1 Y(\mu)],$$

$X^{(1)}$, $Y^{(1)}$ and $X^{(2)}$, $Y^{(2)}$ are defined in terms of the characteristic functions

$$\tfrac{3}{8}\mu^2(1-\mu^2) \quad \text{and} \quad \tfrac{3}{32}(1-\mu^2)^2,$$

respectively, and X and Y are the standard solutions for the characteristic function

$$\tfrac{3}{16}(3-\mu^2).$$

The constants c_1, c_2, q_1, and q_2 are related to the moments of X and Y by

$$c_1+c_2 = \frac{(\alpha_2-\beta_2)\tau_1+6(\alpha_1-\beta_1)}{(\alpha_1+\beta_1)\tau_1+2(\alpha_2+\beta_2)}; \qquad c_1-c_2 = \frac{\alpha_2+\beta_2}{\alpha_1-\beta_1},$$

$$q_1 = \tfrac{3}{16}(3\alpha_1-c_1\alpha_2-c_2\beta_2) \quad \text{and} \quad q_2 = \tfrac{3}{16}(3\beta_1+c_1\beta_2+c_2\alpha_2).$$

65. Diffuse reflection and transmission for scattering in accordance with the phase function $\varpi_0(1+x\cos\Theta)$

For the case of scattering in accordance with the phase function $\varpi_0(1+x\cos\Theta)$, we can express the scattering and the transmission functions in the forms (cf. Chap. IV, eqs. [54]–[60] and Chap. VI, eqs. [43]–[48])

$$S(\tau_1;\mu,\varphi;\mu_0,\varphi_0) = \varpi_0[S^{(0)}(\mu,\mu_0)+x(1-\mu^2)^{\frac{1}{2}}(1-\mu_0^2)^{\frac{1}{2}}S^{(1)}(\mu,\mu_0)\cos(\varphi_0-\varphi)]$$

and

$$T(\tau_1;\mu,\varphi;\mu_0,\varphi_0)$$
$$= \varpi_0[T^{(0)}(\mu,\mu_0)+x(1-\mu^2)^{\frac{1}{2}}(1-\mu_0^2)^{\frac{1}{2}}T^{(1)}(\mu,\mu_0)\cos(\varphi_0-\varphi)], \quad (109)$$

where

$$\left.\begin{aligned}
\left(\frac{1}{\mu_0}+\frac{1}{\mu}\right)S^{(0)}(\mu,\mu_0) &= \psi(\mu)\psi(\mu_0)-\chi(\mu)\chi(\mu_0)- \\
&\qquad -x[\phi(\mu)\phi(\mu_0)-\zeta(\mu)\zeta(\mu_0)], \\
\left(\frac{1}{\mu_0}-\frac{1}{\mu}\right)T^{(0)}(\mu,\mu_0) &= \chi(\mu)\psi(\mu_0)-\psi(\mu)\chi(\mu_0)+ \\
&\qquad +x[\zeta(\mu)\phi(\mu_0)-\phi(\mu)\zeta(\mu_0)],
\end{aligned}\right\} \quad (110)$$

$$\left.\begin{aligned}
\left(\frac{1}{\mu_0}+\frac{1}{\mu}\right)S^{(1)}(\mu,\mu_0) &= X^{(1)}(\mu)X^{(1)}(\mu_0)-Y^{(1)}(\mu)Y^{(1)}(\mu_0), \\
\left(\frac{1}{\mu_0}-\frac{1}{\mu}\right)T^{(1)}(\mu,\mu_0) &= Y^{(1)}(\mu)X^{(1)}(\mu_0)-X^{(1)}(\mu)Y^{(1)}(\mu_0),
\end{aligned}\right\} \quad (111)$$

$$\left.\begin{aligned}
\psi(\mu) &= 1+\tfrac{1}{2}\varpi_0\int_0^1 S^{(0)}(\mu,\mu')\frac{d\mu'}{\mu'}, \\
\phi(\mu) &= \mu-\tfrac{1}{2}\varpi_0\int_0^1 S^{(0)}(\mu,\mu')\,d\mu', \\
\chi(\mu) &= e^{-\tau_1/\mu}+\tfrac{1}{2}\varpi_0\int_0^1 T^{(0)}(\mu,\mu')\frac{d\mu'}{\mu'}, \\
\zeta(\mu) &= \mu e^{-\tau_1/\mu}+\tfrac{1}{2}\varpi_0\int_0^1 T^{(0)}(\mu,\mu')\,d\mu',
\end{aligned}\right\} \quad (112)$$

and $X^{(1)}(\mu)$ and $Y^{(1)}(\mu)$ are defined in terms of the characteristic function $\frac{1}{4}x\varpi_0(1-\mu^2)$.

The functions of order one require no further consideration as they are directly reducible to X- and Y-functions. On the other hand, according to equations (110) and (112) the functions ψ, ϕ, χ, and ζ, of order zero, satisfy a simultaneous system of integral equations of order four. We shall now briefly indicate how these equations can be solved.

65.1. *The form of the solutions for $S^{(0)}(\mu, \mu_0)$ and $T^{(0)}(\mu, \mu_0)$*

From the form of the solutions obtained in the direct solution of the equation of transfer in the nth approximation, it appears that the scattering and the transmission functions must have the forms†

$$\left(\frac{1}{\mu_0}+\frac{1}{\mu}\right)S^{(0)}(\mu,\mu_0) = X(\mu)X(\mu_0)[1-c_1(\mu+\mu_0)-x(1-\varpi_0)\mu\mu_0]-$$
$$-Y(\mu)Y(\mu_0)[1+c_1(\mu+\mu_0)-x(1-\varpi_0)\mu\mu_0]-$$
$$-c_2(\mu+\mu_0)[X(\mu)Y(\mu_0)+Y(\mu)X(\mu_0)]$$

and

$$\left(\frac{1}{\mu_0}-\frac{1}{\mu}\right)T^{(0)}(\mu,\mu_0) = Y(\mu)X(\mu_0)[1+c_1(\mu-\mu_0)+x(1-\varpi_0)\mu\mu_0]-$$
$$-X(\mu)Y(\mu_0)[1-c_1(\mu-\mu_0)+x(1-\varpi_0)\mu\mu_0]+$$
$$+c_2(\mu-\mu_0)[X(\mu)X(\mu_0)+Y(\mu)Y(\mu_0)], \quad (113)$$

where $X(\mu)$ and $Y(\mu)$ are defined in terms of the characteristic function

$$\Psi(\mu) = \tfrac{1}{2}\varpi_0[1+x(1-\varpi_0)\mu^2], \tag{114}$$

and c_1 and c_2 are two constants.

65.2. *Verification of the solution and the evaluation of the constants c_1 and c_2 in terms of the moments of $X(\mu)$ and $Y(\mu)$*

The verification that the solutions for $S^{(0)}$ and $T^{(0)}$ have the forms (113) proceeds as in § 64:

We first evaluate ψ, ϕ, etc., according to equations (112) for $S^{(0)}$ and $T^{(0)}$ given by equations (113), making proper use of the integral properties of the X- and Y-functions for the characteristic function (114). We find

$$\psi(\mu) = (1-q_0\mu)X(\mu)-p_0\mu Y(\mu),$$

$$\chi(\mu) = (1+q_0\mu)Y(\mu)+p_0\mu X(\mu),$$

$$\phi(\mu) = \mu[q_1 X(\mu)+p_1 Y(\mu)] \text{ and } \zeta(\mu) = \mu[p_1 X(\mu)+q_1 Y(\mu)], \tag{115}$$

† Cf. S. Chandrasekhar, *Astrophys. J.* **106**, 152, 1947 (eqs. [276] and [277]).

where
$$q_0 = \tfrac{1}{2}\varpi_0[c_1\alpha_0 + c_2\beta_0 + x(1-\varpi_0)\alpha_1],$$
$$p_0 = \tfrac{1}{2}\varpi_0[c_1\beta_0 + c_2\alpha_0 - x(1-\varpi_0)\beta_1],$$
$$q_1 = \tfrac{1}{2}[\varpi_0(c_1\alpha_1 + c_2\beta_1) + (2-\varpi_0\alpha_0)],$$

and
$$p_1 = \tfrac{1}{2}\varpi_0[c_1\beta_1 + c_2\alpha_1 + \beta_0]. \tag{116}$$

Using the expressions (115) for ψ, χ, etc., in equation (110) for $S^{(0)}$, we obtain

$$\left(\frac{1}{\mu_0} + \frac{1}{\mu}\right) S^{(0)}(\mu, \mu_0)$$
$$= X(\mu)X(\mu_0)[1 - q_0(\mu+\mu_0) + \{q_0^2 - p_0^2 - x(q_1^2 - p_1^2)\}\mu\mu_0] -$$
$$- Y(\mu)Y(\mu_0)[1 + q_0(\mu+\mu_0) + \{q_0^2 - p_0^2 - x(q_1^2 - p_1^2)\}\mu\mu_0] -$$
$$- p_0(\mu+\mu_0)[X(\mu)Y(\mu_0) + Y(\mu)X(\mu_0)]. \tag{117}$$

The agreement of this expression for $S^{(0)}$ with that assumed in (113) requires that
$$q_0 = c_1, \qquad p_0 = c_2 \tag{118}$$

and
$$x(q_1^2 - p_1^2) - (q_0^2 - p_0^2) = x(1-\varpi_0). \tag{119}$$

A similar consideration of $T^{(0)}$ leads to these same conditions.

According to equations (116) and (118)
$$c_1 = \tfrac{1}{2}\varpi_0[c_1\alpha_0 + c_2\beta_0 + x(1-\varpi_0)\alpha_1]$$

and
$$c_2 = \tfrac{1}{2}\varpi_0[c_1\beta_0 + c_2\alpha_0 - x(1-\varpi_0)\beta_1]. \tag{120}$$

Solving these equations for c_1 and c_2, we obtain
$$c_1 = q_0 = x\varpi_0(1-\varpi_0)\frac{(2-\varpi_0\alpha_0)\alpha_1 - \varpi_0\beta_0\beta_1}{(2-\varpi_0\alpha_0)^2 - \varpi_0^2\beta_0^2}$$

and
$$c_2 = p_0 = x\varpi_0(1-\varpi_0)\frac{-(2-\varpi_0\alpha_0)\beta_1 + \varpi_0\beta_0\alpha_1}{(2-\varpi_0\alpha_0)^2 - \varpi_0^2\beta_0^2}. \tag{121}$$

Substituting these values for c_1 and c_2 in the equations for q_1 and p_1 (eqs. [116]), and making use of the relation
$$\alpha_0 = 1 + \tfrac{1}{4}\varpi_0[\alpha_0^2 - \beta_0^2 + x(1-\varpi_0)(\alpha_1^2 - \beta_1^2)], \tag{122}$$

derived from Chapter VIII, Theorem 4, for the characteristic function (114), we find that

$$q_1 = \frac{2(1-\varpi_0)(2-\varpi_0\alpha_0)}{(2-\varpi_0\alpha_0)^2 - \varpi_0^2\beta_0^2} \quad \text{and} \quad p_1 = \frac{2(1-\varpi_0)\varpi_0\beta_0}{(2-\varpi_0\alpha_0)^2 - \varpi_0^2\beta_0^2}. \tag{123}$$

It can now be verified that with q_0, p_0, q_1, and p_1 given by equations (121) and (123), equation (119) is satisfied, identically, in virtue of the relation (122).

65.3. *The law of diffuse reflection and transmission*

Combining the results of the preceding paragraphs, we have the following expressions for the reflected and the transmitted intensities:

$$I(0, \mu, \varphi; \mu_0, \varphi_0)$$

$$= \frac{\varpi_0}{4(\mu+\mu_0)} \{\psi(\mu)\psi(\mu_0)-\chi(\mu)\chi(\mu_0)-x[\phi(\mu)\phi(\mu_0)-\zeta(\mu)\zeta(\mu_0)]+$$

$$+x(1-\mu^2)^{\frac{1}{2}}(1-\mu_0^2)^{\frac{1}{2}}[X^{(1)}(\mu)X^{(1)}(\mu_0)-Y^{(1)}(\mu)Y^{(1)}(\mu_0)]\cos(\varphi_0-\varphi)\}\mu_0\,F,$$

and

$$I(\tau_1, -\mu, \varphi; \mu_0, \varphi_0)$$

$$= \frac{\varpi_0}{4(\mu-\mu_0)} \{\chi(\mu)\psi(\mu_0)-\psi(\mu)\chi(\mu_0)+x[\zeta(\mu)\phi(\mu_0)-\phi(\mu)\zeta(\mu_0)]+$$

$$+x(1-\mu^2)^{\frac{1}{2}}(1-\mu_0^2)^{\frac{1}{2}}[Y^{(1)}(\mu)X^{(1)}(\mu_0)-X^{(1)}(\mu)Y^{(1)}(\mu_0)]\cos(\varphi_0-\varphi)\}\mu_0\,F,$$

where
$$\psi(\mu) = (1-c_1\mu)X(\mu)-c_2\mu Y(\mu),$$

$$\chi(\mu) = (1+c_1\mu)Y(\mu)+c_2\mu X(\mu),$$

$$\phi(\mu) = \mu[q_1 X(\mu)+p_1 Y(\mu)], \qquad \zeta(\mu) = \mu[p_1 X(\mu)+q_1 Y(\mu)],$$

and (X, Y) and $(X^{(1)}, Y^{(1)})$ are defined in terms of the characteristic functions

$$\tfrac{1}{2}\varpi_0[1+x(1-\varpi_0)\mu^2] \quad \text{and} \quad \tfrac{1}{4}x\varpi_0(1-\mu^2),$$

respectively. The constants c_1, c_2, q_1, and p_1 are related to the moments of X and Y by

$$c_1 = x\varpi_0(1-\varpi_0)\frac{(2-\varpi_0\alpha_0)\alpha_1-\varpi_0\beta_0\beta_1}{(2-\varpi_0\alpha_0)^2-\varpi_0^2\beta_0^2},$$

$$c_2 = x\varpi_0(1-\varpi_0)\frac{-(2-\varpi_0\alpha_0)\beta_1+\varpi_0\beta_0\alpha_1}{(2-\varpi_0\alpha_0)^2-\varpi_0^2\beta_0^2},$$

$$q_1 = \frac{2(1-\varpi_0)(2-\varpi_0\alpha_0)}{(2-\varpi_0\alpha_0)^2-\varpi_0^2\beta_0^2} \quad \text{and} \quad p_1 = \frac{2(1-\varpi_0)\varpi_0\beta_0}{(2-\varpi_0\alpha_0)^2-\varpi_0^2\beta_0^2}.$$

66. Illustrations of the laws of diffuse reflection and transmission

In Figs. 21 and 22 the laws of diffuse reflection and transmission under conditions of isotropic scattering and albedos $\varpi_0 = 1$ and 0.9 are illustrated. The solutions in the form obtained in § 62 were used in the calculations, though lacking solutions of the exact X- and Y-equations, the rational approximations of Chapter VIII, § 59 were used.

For comparison, the solutions for $\tau_1 = 0.25$ and 0.5 were also obtained from the development given in § 63.

It appears that for $\tau_1 \leqslant 0.5$ the corrected second approximations for the X- and Y-functions constructed in the manner explained in Chapter VIII, § 60.2 give an accuracy of about one part in a thousand: this

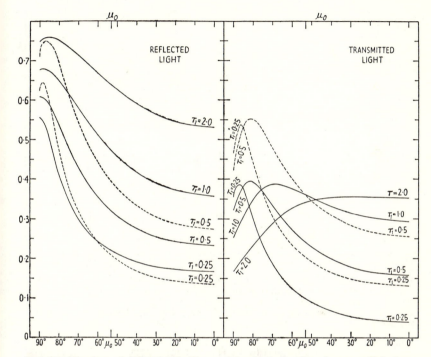

FIG. 21. The law of diffuse reflection and transmission by plane-parallel atmospheres of finite optical thickness and under conditions of conservative isotropic scattering.

The ordinates represent the intensity in the unit $\mu_0 F$ and the abscissae the angle in degrees.

An angle of incidence corresponding to $\mu_0 = 0.6$ is considered and the angular distribution of the reflected light (the curves on the left side of the diagram) and the transmitted light (the curves on the right side of the diagram) are illustrated for various values of the optical thickness τ_1.

The full-line curves have been derived from the rational approximations of the X- and Y-functions; and the dashed curves have been obtained from the solution developed for small values of τ_1.

accuracy should be sufficient for most problems. However, exact solutions of the relevant X- and Y-equations for various values of τ_1 are being obtained at the Watson Scientific Computing Laboratory (New York) by a process of numerical iteration (in the manner of the H-equations). When these tables of the X- and Y-functions become available, a variety of problems, including that of the formation of absorption lines in the process of diffuse reflection, will receive their exact solutions.

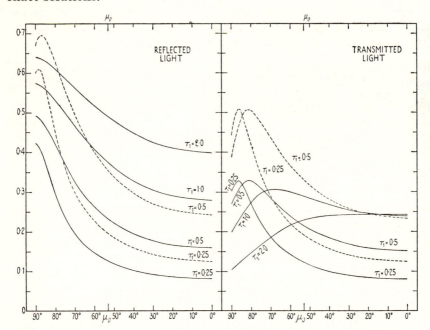

FIG. 22. The law of diffuse reflection and transmission by plane-parallel atmospheres of finite optical thicknesses and under conditions of isotropic scattering with an albedo $\varpi_0 = 0.9$.

The disposition of the diagram is the same as in Fig. 21; the angle of incidence considered is also the same ($\mu_0 = 0.6$).

BIBLIOGRAPHICAL NOTES

§ 62. The analysis in this section is taken from

1. S. CHANDRASEKHAR, *Astrophys. J.* **107**, 48 (1948) (Section II of this paper).

For the interpretation of the X- and Y-functions given in § 62.1 see

2. H. VAN DE HULST, ibid. **107**, 220 (1948).

§ 63. See van de Hulst, reference 2, and also

3. S. CHANDRASEKHAR, ibid. **108**, 92 (1948).

§ 64. See

4. S. CHANDRASEKHAR, ibid. **107**, 188 (1948) (Section III of this paper),

and

5. S. CHANDRASEKHAR, ibid. **106**, 152 (1948) (Section III of this paper).

§ 65. See, Chandrasekhar, reference 4 (Section IV) and reference 5 (Section IV).

RAYLEIGH SCATTERING AND SCATTERING BY PLANETARY ATMOSPHERES

67. Introduction

In the preceding chapters we have seen how transfer problems in plane-parallel atmospheres can be solved in a general nth approximation and how, further, exact solutions for the angular distributions of the emergent radiations can be found, most simply, by appealing to certain general principles of invariance. In the formulation of these latter principles (Chaps. IV and VII) we have indicated, in each case, how the state of polarization of the radiation field should be taken into account in a physically correct theory; but, so far, we have not given explicit solutions for any problem in which scattering has been considered as a linear transformation of the Stokes parameters characterizing the incident light. In this chapter we proceed to the consideration of these more difficult problems: more particularly we shall consider the basic problems in the theory of transfer in plane-parallel atmospheres, in the context of Rayleigh scattering.

The plan of this chapter is as follows:

In § 68 the axially symmetric problem with a constant net flux is considered, first in a general nth approximation and then in the limit of infinite approximation as far as the angular distribution and the state of polarization of the emergent radiation are concerned. In § 69, as a preliminary to the discussion of the problems of diffuse reflection and transmission, the equation of transfer is reduced to certain basic equations and explicit solutions for the azimuth dependent terms in the scattering and the transmission matrices are obtained. § 70 is devoted to the problem of diffuse reflection by a semi-infinite atmosphere and § 71 to the more general problem of diffuse reflection and transmission by atmospheres of finite optical thicknesses. In § 72 the *planetary problem*, when there is a diffusely reflecting surface at $\tau = \tau_1$, is considered and reduced to the *standard problem* when no such surface is present. In § 73 the solutions obtained in §§ 71 and 72 are illustrated; this section includes a brief discussion of the applications of the theory of diffuse reflection and transmission, developed in this chapter, to the long-standing problem of the illumination and polarization of the sunlit sky. And finally in § 74 the generalizations necessary to allow for a 'depolarization effect' (cf. Chap. I, § 18) are indicated.

68. The problem with a constant net flux. The radiative equilibrium of an electron scattering atmosphere

The equation of transfer for this problem has already been given in Chapter I, § 17.3, equation (227). Rewriting this equation, separately, for the intensities $I_l(\tau, \mu)$ and $I_r(\tau, \mu)$ in the directions parallel, respectively, perpendicular to the meridian plane containing the direction μ, we have

$$\mu \frac{dI_l}{d\tau} = I_l - \tfrac{3}{8} \left\{ 2 \int_{-1}^{+1} I_l(\tau, \mu')(1 - \mu'^2) \, d\mu' + \mu^2 \int_{-1}^{+1} I_l(\tau, \mu')(3\mu'^2 - 2) \, d\mu' + \right.$$
$$\left. + \mu^2 \int_{-1}^{+1} I_r(\tau, \mu') \, d\mu' \right\}$$

and
$$\mu \frac{dI_r}{d\tau} = I_r - \tfrac{3}{8} \left\{ \int_{-1}^{+1} I_l(\tau, \mu')\mu'^2 \, d\mu' + \int_{-1}^{+1} I_r(\tau, \mu') \, d\mu' \right\}. \tag{1}$$

We require to solve these equations with the boundary conditions, Chapter I, equations (228).

In terms of the quantities J and K, defined in the usual manner, we can rewrite the equations of transfer (1) in the following forms:

$$\mu \frac{dI_l}{d\tau} = I_l - \tfrac{3}{4}\{2(J_l - K_l) + \mu^2(3K_l - 2J_l + J_r)\}$$

and
$$\mu \frac{dI_r}{d\tau} = I_r - \tfrac{3}{4}(J_r + K_l). \tag{2}$$

These equations admit the flux and the K-integral for the total intensity $I_l + I_r$; for, multiplying equations (2) by $\tfrac{1}{2} d\mu$ and $\tfrac{1}{2}\mu \, d\mu$ and integrating over the range of μ, we obtain

$$\left. \begin{array}{l} \dfrac{1}{4} \dfrac{dF_l}{d\tau} = \tfrac{3}{4}K_l - \tfrac{1}{4}J_r, \\[2mm] \dfrac{1}{4} \dfrac{dF_r}{d\tau} = \tfrac{1}{4}J_r - \tfrac{3}{4}K_l, \end{array} \right\} \quad \text{and} \quad \begin{array}{l} \dfrac{dK_l}{d\tau} = \tfrac{1}{4}F_l, \\[2mm] \dfrac{dK_r}{d\tau} = \tfrac{1}{4}F_r. \end{array} \tag{3}$$

Accordingly,

$$\frac{d}{d\tau}(F_l + F_r) = 0 \quad \text{and} \quad \frac{d}{d\tau}(K_l + K_r) = \tfrac{1}{4}(F_l + F_r). \tag{4}$$

Hence
$$F_l + F_r = \text{constant} = F \text{ (say)},$$

and
$$K_l + K_r = \tfrac{1}{4}F(\tau + Q), \tag{5}$$

where Q is a constant.

68.1. *The general solution of the equations of transfer in the nth approximation*

Following the general method of replacing integrals by the corresponding Gauss sums, we have the following system of $4n$ linear equations in the nth approximation:

$$\mu_i \frac{dI_{l,i}}{d\tau} = I_{l,i} - \tfrac{3}{8}[2\sum a_j(1-\mu_j^2)I_{l,j} + \mu_i^2\{\sum a_j(3\mu_j^2-2)I_{l,j} + \sum a_j I_{r,j}\}]$$

$$(i = \pm 1,...,\pm n)$$

and $\quad \mu_i \dfrac{dI_{r,i}}{d\tau} = I_{r,i} - \tfrac{3}{8}(\sum a_j I_{r,j} + \sum a_j \mu_j^2 I_{l,j}) \quad (i = \pm 1,...,\pm n),$ $\qquad$ (6)

where $I_{l,i}$ and $I_{r,i}$ denote $I_l(\tau,\mu_i)$ and $I_r(\tau,\mu_i)$ and the rest of the symbols have their usual meanings.

In solving equations (6) we shall first find the different linearly independent solutions and later, by combining them, obtain the general solution.

First, we seek solutions of equations (6) of the forms

$$I_{l,i} = g_i e^{-k\tau} \quad \text{and} \quad I_{r,i} = h_i e^{-k\tau} \quad (i = \pm 1,...,\pm n), \qquad (7)$$

where the g_i's, h_i's, and k are constants. Substituting these forms for $I_{l,i}$ and $I_{r,i}$ in equations (6), we obtain

$$g_i(1+\mu_i k) = \tfrac{3}{8}[2\sum a_j(1-\mu_j^2)g_j + \mu_i^2\{\sum a_j(3\mu_j^2-2)g_j + \sum a_j h_j\}]$$

and $\qquad h_i(1+\mu_i k) = \tfrac{3}{8}[\sum a_j \mu_j^2 g_j + \sum a_j h_j]. \qquad (8)$

Equations (8) imply that g_i and h_i must be expressible in the forms

$$g_i = \frac{\alpha\mu_i^2 + \beta}{1+\mu_i k} \quad \text{and} \quad h_i = \frac{\gamma}{1+\mu_i k} \quad (i = \pm 1,...,\pm n), \qquad (9)$$

where α, β, and γ are certain constants, independent of i. Combining equations (8) and (9) we have

$$\alpha\mu_i^2 + \beta = \tfrac{3}{8}[2\{\alpha(D_2-D_4)+\beta(D_0-D_2)\}+$$

$$+\mu_i^2\{\alpha(3D_4-2D_2)+\beta(3D_2-2D_0)+\gamma D_0\}]$$

and $\qquad \gamma = \tfrac{3}{8}(\alpha D_4 + \beta D_2 + \gamma D_0), \qquad (10)$

where we have introduced the quantities D_0, D_2, and D_4 according to the formula (cf. Chap. III, eq. [18])

$$D_m = \sum \frac{a_j \mu_j^m}{1+\mu_j k}. \qquad (11)$$

Since (10) is valid for all i, we must require that

$$\tfrac{8}{3}\alpha = \alpha(3D_4-2D_2)+\beta(3D_2-2D_0)+\gamma D_0, \tag{12}$$

$$\tfrac{4}{3}\beta = \alpha(D_2-D_4)+\beta(D_0-D_2), \tag{13}$$

and $$\tfrac{8}{3}\gamma = \alpha D_4+\beta D_2+\gamma D_0. \tag{14}$$

Equations (12), (13), and (14), together, represent a system of homogeneous linear equations for α, β, and γ. The determinant of this system must, therefore, vanish. Hence

$$\begin{vmatrix} 3D_4-2D_2-\tfrac{8}{3} & 3D_2-2D_0 & D_0 \\ D_2-D_4 & D_0-D_2-\tfrac{4}{3} & 0 \\ D_4 & D_2 & D_0-\tfrac{8}{3} \end{vmatrix} = 0. \tag{15}$$

By adding or subtracting suitable multiples of the rows or columns we can transform the determinant on the left-hand side, successively, into

$$\begin{vmatrix} 3D_4-2D_2-\tfrac{8}{3} & 3D_2-2D_0 & D_0 \\ D_2 & D_0-\tfrac{4}{3} & D_0-\tfrac{8}{3} \\ D_4 & D_2 & D_0-\tfrac{8}{3} \end{vmatrix}$$

$$= \begin{vmatrix} -\tfrac{8}{3} & -\tfrac{8}{3} & \tfrac{8}{3} \\ D_2 & D_0-\tfrac{4}{3} & D_0-\tfrac{8}{3} \\ D_4 & D_2 & D_0-\tfrac{8}{3} \end{vmatrix} = \begin{vmatrix} 0 & 0 & \tfrac{8}{3} \\ D_2+D_0-\tfrac{8}{3} & 2(D_0-2) & D_0-\tfrac{8}{3} \\ D_4+D_0-\tfrac{8}{3} & D_2+D_0-\tfrac{8}{3} & D_0-\tfrac{8}{3} \end{vmatrix}. \tag{16}$$

Hence $$(D_2+D_0-\tfrac{8}{3})^2-2(D_0-2)(D_4+D_0-\tfrac{8}{3}) = 0. \tag{17}$$

Now, the D's satisfy the recursion relations (Chap. III, eqs. [21] and [22])

$$D_{2j} = \frac{1}{k^2}\left(D_{2j-2}-\frac{2}{2j-1}\right)$$

and $$D_{2j-1} = -kD_{2j}. \tag{18}$$

In particular,

$$D_2 = \frac{1}{k^2}(D_0-2) \quad \text{and} \quad D_4 = \frac{1}{k^2}(D_2-\tfrac{2}{3}) = \frac{1}{k^4}(D_0-2)-\frac{2}{3k^2}. \tag{19}$$

A relation which follows from these equations is

$$D_4(D_0-2) = D_2(D_2-\tfrac{2}{3}). \tag{20}$$

Using this last relation in equation (17), we have

$$(D_2+D_0-\tfrac{8}{3})^2-2D_2(D_2-\tfrac{2}{3})-2(D_0-2)(D_0-\tfrac{8}{3}) = 0. \tag{21}$$

On simplifying this equation further, we find

$$-D_2^2-D_0^2+2D_2 D_0-4D_2+4D_0-\tfrac{32}{9} = 0,$$

or $$(D_0-D_2)^2-4(D_0-D_2)+\tfrac{32}{9} = 0. \tag{22}$$

Hence $$(D_0-D_2-\tfrac{8}{3})(D_0-D_2-\tfrac{4}{3}) = 0. \tag{23}$$

In other words, either

$$D_0 - D_2 = \sum \frac{a_j(1-\mu_j^2)}{1+\mu_j k} = \frac{8}{3} \quad \text{(case 1)} \tag{24}$$

or,

$$D_0 - D_2 = \sum \frac{a_j(1-\mu_j^2)}{1+\mu_j k} = \frac{4}{3} \quad \text{(case 2)}. \tag{25}$$

Equations (24) and (25) can be written alternatively in the forms

$$\sum_{j=1}^{n} \frac{a_j(1-\mu_j^2)}{1-\mu_j^2 k^2} = \frac{4}{3} \quad \text{(case 1)} \tag{26}$$

and

$$\sum_{j=1}^{n} \frac{a_j(1-\mu_j^2)}{1-\mu_j^2 k^2} = \frac{2}{3} \quad \text{(case 2)}; \tag{27}$$

and k^2 must be a root of either of these two equations.

Equation (26) is of order $2n$ in k and admits $2n$ distinct non-vanishing roots which occur in pairs as, $k_{\pm\alpha}$, and

$$k_{+\alpha} = -k_{-\alpha} \quad (\alpha = 1,...,n). \tag{28}$$

On the other hand, equation (27), though of order $2n$ in k, admits only $2(n-1)$ distinct non-vanishing roots, since $k^2 = 0$ is a root.† However, the $(2n-2)$ roots for k also occur in pairs, which we shall denote by $\kappa_{\pm\beta}$, and

$$\kappa_\beta = -\kappa_{-\beta} \quad (\beta = 1,...,n-1), \tag{29}$$

to distinguish them from the roots of equation (26).

Case 1: k^2 *a root of equation* (26). In this case $D_0 - D_2 = \frac{8}{3}$; and from the relations (19) we readily find that

$$D_0 = \frac{2}{3}\frac{4k^2-3}{k^2-1}; \qquad D_2 = \frac{2}{3(k^2-1)}; \qquad D_4 = \frac{2}{3k^2}\frac{2-k^2}{k^2-1}. \tag{30}$$

With these values of D_0, D_2, and D_4, equations (13) and (14) become

$$\alpha = -k^2\beta \quad \text{and} \quad \gamma = -(k^2-1)\beta. \tag{31}$$

Accordingly, equations (6) allow $2n$ linearly independent integrals of the form

$$\left.\begin{array}{l} I_{l,i} = \text{constant}(1-k_\alpha\mu_i)e^{-k_\alpha\tau} \\[2mm] I_{r,i} = -\text{constant}\,\dfrac{k_\alpha^2-1}{1+k_\alpha\mu_i}e^{-k_\alpha\tau} \end{array}\right\} \quad \begin{array}{c} (i = \pm1,...,\pm n \\ \text{and} \\ \alpha = \pm1,...,\pm n). \end{array} \tag{32}$$

Case 2: κ^2 *a root of equation* (27). In this case $D_0 - D_2 = \frac{4}{3}$; and equations (19) now give

$$D_0 = \frac{2}{3}\frac{2\kappa^2-3}{\kappa^2-1}; \qquad D_2 = D_4 = -\frac{2}{3(\kappa^2-1)}. \tag{33}$$

---

† Note that $\sum\limits_{j=1}^{n} a_j(1-\mu^2) = \frac{2}{3}$.

With these values of D_0, D_2, and D_4 it is seen that equation (13) is satisfied identically, while the consideration of equations (12) and (14) shows that

$$\alpha = -\beta \quad \text{and} \quad \gamma = 0. \tag{34}$$

Accordingly, equations (6) admit $(2n-2)$ further linearly independent integrals of the form

$$I_{l,i} = \text{constant} \; \frac{1-\mu_i^2}{1+\mu_i \kappa_\beta} e^{-\kappa_\beta \tau} \quad (i = \pm 1,..., \pm n, \; \beta = \pm 1,..., \pm n \mp 1),$$

and

$$I_{r,i} \equiv 0 \quad (i = \pm 1,..., \pm n). \tag{35}$$

To complete the solution we verify that equations (6) also admit the solution

$$I_{l,i} = I_{r,i} = b(\tau + \mu_i + Q) \quad (i = \pm 1,..., \pm n), \tag{36}$$

where b and Q are constants.

Combining the solutions (32), (35), and (36), we can write the general solution of equations (6) in the form

$$I_{l,i} = b\Bigg\{ \tau + \mu_i + Q + (1-\mu_i^2) \sum_{\beta=-n+1}^{n-1} \frac{L_\beta e^{-\kappa_\beta \tau}}{1+\mu_i \kappa_\beta} +$$

$$+ \sum_{\alpha=-n}^{+n} M_\alpha (1-k_\alpha \mu_i) e^{-k_\alpha \tau} \Bigg\} \quad (i = \pm 1,..., \pm n)$$

and

$$I_{r,i} = b\Bigg\{ \tau + \mu_i + Q - \sum_{\alpha=-n}^{+n} M_\alpha \frac{(k_\alpha^2 - 1)}{1+\mu_i k_\alpha} e^{-k_\alpha \tau} \Bigg\} \quad (i = \pm 1,..., \pm n), \tag{37}$$

where $L_{\pm\beta}$ $(\beta = 1,..., n-1)$, $M_{\pm\alpha}$ $(\alpha = 1,..., n)$, b, and Q are $4n$ constants of integration.

68.2. *The solution satisfying the necessary boundary conditions*

For the problem on hand the boundary conditions are (Chap. I, eq. [228]) that none of the I_i's increase more rapidly than e^τ as $\tau \to \infty$ and that there is no radiation incident on $\tau = 0$. The first of these conditions requires that in the general solution (37) we omit all terms in $\exp(+k_\alpha \tau)$ and $\exp(+k_\beta \tau)$. Thus

$$I_{l,i} = b\Bigg\{ \tau + \mu_i + Q + (1-\mu_i^2) \sum_{\beta=1}^{n-1} \frac{L_\beta e^{-\kappa_\beta \tau}}{1+\mu_i \kappa_\beta} + \sum_{\alpha=1}^{n} M_\alpha (1-k_\alpha \mu_i) e^{-k_\alpha \tau} \Bigg\}$$

$$(i = \pm 1,..., \pm n)$$

and

$$I_{r,i} = b\Bigg\{ \tau + \mu_i + Q - \sum_{\alpha=1}^{n} M_\alpha \frac{k_\alpha^2 - 1}{1+\mu_i k_\alpha} e^{-k_\alpha \tau} \Bigg\} \quad (i = \pm 1,..., \pm n). \tag{38}$$

Next, the absence of any radiation incident on $\tau = 0$ requires that

$$I_{l,i} = I_{r,i} = 0 \quad \text{at} \quad \tau = 0 \quad \text{and for} \quad i = -1,...,-n, \qquad (39)$$

or, according to equations (38),

$$(1-\mu_i^2) \sum_{\beta=1}^{n-1} \frac{L_\beta}{1-\mu_i \kappa_\beta} + \sum_{\alpha=1}^{n} M_\alpha(1+k_\alpha \mu_i) - \mu_i + Q = 0 \quad (i = 1,...,n)$$

and

$$-\sum_{\alpha=1}^{n} \frac{M_\alpha(k_\alpha^2-1)}{1-\mu_i k_\alpha} - \mu_i + Q = 0 \quad (i = 1,...,n). \qquad (40)$$

These are the $2n$ equations which determine the $2n$ constants L_β ($\beta = 1,...,n-1$) and M_α ($\alpha = 1,...,n$), and Q. The constant b is left arbitrary and is related to the constant net flux in the atmosphere, as we shall presently see.

Defining the net fluxes F_l and F_r in terms of the corresponding Gauss sums we have, according to equations (38),

$$F_l = 2b\left\{\tfrac{2}{3} + \sum_{\beta=1}^{n-1} L_\beta[D_1(\kappa_\beta) - D_3(\kappa_\beta)]e^{-\kappa_\beta \tau} - \tfrac{2}{3}\sum_{\alpha=1}^{n} M_\alpha k_\alpha e^{-k_\alpha \tau}\right\}$$

and

$$F_r = 2b\left\{\tfrac{2}{3} - \sum_{\alpha=1}^{n} M_\alpha(k_\alpha^2-1)D_1(k_\alpha)e^{-k_\alpha \tau}\right\}. \qquad (41)$$

On the other hand, from equations (18), (30), and (33) we have

$$D_1(k_\alpha) = -k_\alpha D_2(k_\alpha) = -\frac{2}{3}\frac{k_\alpha}{k_\alpha^2-1},$$

and

$$D_1(\kappa_\beta) - D_3(\kappa_\beta) = -\kappa_\beta[D_2(\kappa_\beta) - D_4(\kappa_\beta)] = 0. \qquad (42)$$

Hence

$$F_l = \tfrac{4}{3}b\left(1 - \sum_{\alpha=1}^{n} M_\alpha k_\alpha e^{-k_\alpha \tau}\right)$$

and

$$F_r = \tfrac{4}{3}b\left(1 + \sum_{\alpha=1}^{n} M_\alpha k_\alpha e^{-k_\alpha \tau}\right). \qquad (43)$$

From the two preceding equations we infer the constancy of the net flux. More particularly,

$$F = F_l + F_r = \tfrac{8}{3}b = \text{constant}. \qquad (44)$$

We can, therefore, rewrite the solution (38) in the form

$$I_{l,i} = \tfrac{3}{8}F\left\{\tau + \mu_i + Q + (1-\mu_i^2)\sum_{\beta=1}^{n-1} \frac{L_\beta e^{-\kappa_\beta \tau}}{1+\mu_i \kappa_\beta} + \sum_{\alpha=1}^{n} M_\alpha(1-k_\alpha \mu_i)e^{-k_\alpha \tau}\right\},$$

$$I_{r,i} = \tfrac{3}{8}F\left\{\tau + \mu_i + Q - \sum_{\alpha=1}^{n} \frac{M_\alpha(k_\alpha^2-1)}{1+\mu_i k_\alpha}e^{-k_\alpha \tau}\right\} \quad (i = \pm 1,..., \pm n).$$

$$(45)$$

In terms of the foregoing solutions for $I_{l,i}$ and $I_{r,i}$ we can readily establish the following formulae:

$$J_l = \tfrac{1}{2}\sum a_j I_{l,j} = \tfrac{3}{8}F\left(\tau + Q + \tfrac{2}{3}\sum_{\beta=1}^{n-1} L_\beta e^{-\kappa_\beta \tau} + \sum_{\alpha=1}^{n} M_\alpha e^{-k_\alpha \tau}\right),$$

$$K_l = \tfrac{1}{2}\sum a_j \mu_j^2 I_{l,j} = \tfrac{1}{8}F\left(\tau + Q + \sum_{\alpha=1}^{n} M_\alpha e^{-k_\alpha \tau}\right),$$

$$J_r = \tfrac{3}{8}F\left(\tau + Q - \tfrac{1}{3}\sum_{\alpha=1}^{n} M_\alpha(4k_\alpha^2 - 3)e^{-k_\alpha \tau}\right),$$

and

$$K_r = \tfrac{1}{8}F\left(\tau + Q - \sum_{\alpha=1}^{n} M_\alpha e^{-k_\alpha \tau}\right). \tag{46}$$

The source functions $\mathfrak{J}_l(\tau, \mu)$ and $\mathfrak{J}_r(\tau, \mu)$ for I_l and I_r are (cf. eqs. [2])

$$\mathfrak{J}_l(\tau, \mu) = \tfrac{3}{4}[2(J_l - K_l) + \mu^2(3K_l - 2J_l + J_r)]$$

and

$$\mathfrak{J}_r(\tau, \mu) = \tfrac{3}{4}(J_r + K_l). \tag{47}$$

Now, substituting for J_l, K_l, J_r, and K_r from equations (46), we have

$$\mathfrak{J}_l(\tau, \mu) = \tfrac{3}{8}F\left\{\tau + Q + (1-\mu^2)\sum_{\beta=1}^{n-1} L_\beta e^{-\kappa_\beta \tau} + \sum_{\alpha=1}^{n} M_\alpha(1 - k_\alpha^2 \mu^2)e^{-k_\alpha \tau}\right\}$$

and

$$\mathfrak{J}_r(\tau, \mu) = \tfrac{3}{8}F\left\{\tau + Q - \sum_{\alpha=1}^{n} M_\alpha(k_\alpha^2 - 1)e^{-k_\alpha \tau}\right\}. \tag{48}$$

In terms of these source functions, the radiation field in the atmosphere can be determined in accordance with Chapter I, equations (68) and (69). In particular, for the emergent radiations, we find

$$I_l(0, \mu) = \tfrac{3}{8}F\left\{\mu + Q + (1-\mu^2)\sum_{\beta=1}^{n-1} \frac{L_\beta}{1 + \kappa_\beta \mu} + \sum_{\alpha=1}^{n} M_\alpha(1 - k_\alpha \mu)\right\}$$

and

$$I_r(0, \mu) = \tfrac{3}{8}F\left\{\mu + Q - \sum_{\alpha=1}^{n} \frac{M_\alpha(k_\alpha^2 - 1)}{1 + k_\alpha \mu}\right\}. \tag{49}$$

68.3. *The characteristic roots and the constants of integration in the third approximation*

Before we proceed further it is of interest to note the values of the characteristic roots and the constants of integration, in the third approximation. They are

$$k_1 = 3\cdot458589; \qquad k_2 = 1\cdot327570; \qquad k_3 = 1\cdot046766,$$

$$\kappa_1 = 2\cdot718381; \qquad \kappa_2 = 1\cdot118216,$$

$$L_1 = -0\cdot1402646; \qquad L_2 = -0\cdot06791696; \qquad Q = 0\cdot705927,$$

$$M_1 = +0\cdot00718392; \qquad M_2 = +0\cdot01861255; \qquad M_3 = -0\cdot0328664. \tag{50}$$

68.4. *The elimination of the constants and the expression of $I_l(0, \mu)$ and $I_r(0, \mu)$ in terms of H-functions*

According to equations (40) and (49), we can express the boundary conditions and the angular distributions of the emergent intensities in terms of the functions

$$S_l(\mu) = (1-\mu^2) \sum_{\beta=1}^{n-1} \frac{L_\beta}{1-\mu\kappa_\beta} + \sum_{\alpha=1}^{n} M_\alpha(1+k_\alpha\mu) - \mu + Q \qquad (51)$$

and

$$S_r(\mu) = - \sum_{\alpha=1}^{n} \frac{M_\alpha(k_\alpha^2-1)}{1-\mu k_\alpha} - \mu + Q. \qquad (52)$$

Thus,

$$S_l(\mu_i) = S_r(\mu_i) = 0 \quad (i = 1,...,n) \qquad (53)$$

and

$$I_l(0, \mu) = \tfrac{3}{8}FS_l(-\mu); \qquad I_r(0, \mu) = \tfrac{3}{8}FS_r(-\mu). \qquad (54)$$

Equations (53) and (54) provide further examples of the reciprocity between the equations representing the boundary conditions and the equations governing the angular distributions of the emergent radiations, which we have referred to in Chapter VI, § 43.

We shall now show how explicit expressions for $S_l(\mu)$ and $S_r(\mu)$ can be obtained without solving for the constants of integration.

First, we shall define the functions

$$R(\mu) = \prod_{\alpha=1}^{n} (1-k_\alpha\mu); \qquad R_\alpha(\mu) = \prod_{a\neq\alpha} (1-k_a\mu),$$

$$\rho(\mu) = \prod_{\beta=1}^{n-1} (1-\kappa_\beta\mu) \quad \text{and} \quad \rho_\beta(\mu) = \prod_{b\neq\beta} (1-\kappa_b\mu). \qquad (55)$$

Considering, now, the function $S_l(\mu)$, we see that $\rho(\mu)S_l(\mu)$ is a polynomial of degree n in μ which vanishes for $\mu = \mu_i$ ($i = 1,...,n$). We must accordingly have a relation of the form

$$S_l(\mu) = qk_1...k_n(-1)^n \frac{P(\mu)}{\rho(\mu)}$$

$$= qk_1...k_n \mu_1...\mu_n H_l(-\mu), \qquad (56)\dagger$$

where q is a constant, $P(\mu) = \prod (\mu-\mu_i)$, and $H_l(\mu)$ is defined in terms of the roots of the characteristic equation (27) (cf. Chap. V, eq. [3]).

Considering next $S_r(\mu)$, we observe that we must have a proportionality of the form

$$R(\mu)S_r(\mu) \propto P(\mu)(\mu-c),$$

where c is a constant, since the quantity on the left-hand side is a polynomial of degree $(n+1)$ in μ and has the zeros $\mu = \mu_i$ ($i = 1,...,n$).

† We have introduced the factor $k_1...k_n$ in equation (56) for convenience in later reductions.

The constant of proportionality can be found by comparing the coefficients of the highest power of μ on either side. Thus we find that

$$S_r(\mu) = (-1)^{n+1}k_1...k_n\frac{P(\mu)}{R(\mu)}(\mu-c)$$

$$= -k_1...k_n\,\mu_1...\mu_n\,H_r(-\mu)(\mu-c), \qquad (57)$$

where $H_r(\mu)$ is defined in terms of the roots of the characteristic equation (26).

Using the relation $k_1...k_n\,\mu_1...\mu_n = \frac{1}{\sqrt{2}},$ (58)†

between the roots of equation (26) we can rewrite equations (56) and (57) in the forms

$$S_l(\mu) = \frac{q}{\sqrt{2}}H_l(-\mu) \quad \text{and} \quad S_r(\mu) = -\frac{1}{\sqrt{2}}H_r(-\mu)(\mu-c). \qquad (59)$$

To determine the constants q and c we proceed in the following manner:

Setting $\mu = +1$, respectively, -1, in equation (51), we have (cf. eq. [59])

$$S_l(+1) = \sum_{\alpha=1}^{n}M_\alpha(1+k_\alpha)-1+Q = \frac{q}{\sqrt{2}}H_l(-1)$$

and $$S_l(-1) = \sum_{\alpha=1}^{n}M_\alpha(1-k_\alpha)+1+Q = \frac{q}{\sqrt{2}}H_l(+1). \qquad (60)$$

Next, putting $\mu = 0$ in equation (52), we have (cf. eq. [59])

$$S_r(0) = Q - \sum_{\alpha=1}^{n}M_\alpha(k_\alpha^2-1) = \frac{c}{\sqrt{2}}. \qquad (61)$$

---

† This relation follows most readily from the characteristic equation written in the form (cf. Chap. III, § 25.3)

$$\sum_{j=0}^{n}p_{2j}\,\Delta_{2j} = 0,$$

where the p_{2j}'s are the coefficients of μ^{2j} in the Legendre polynomial $P_{2n}(\mu)$ and

$$\Delta_{2j} = \sum_{i=1}^{n}\frac{a_i(1-\mu_i^2)\mu_i^{2j}}{1-\mu_i^2 k^2}.$$

The Δ's defined in this manner satisfy the recursion formula

$$\Delta_{2j} = \frac{1}{k^2}\Big(\Delta_{2j-2}-\frac{2}{4j^2-1}\Big).$$

For the characteristic equation (26), $\Delta_0 = \frac{4}{3}$ and $\Delta_2 = 2/3k^2$. From the recursion formula we therefore conclude that Δ_{2n} starts with $2/3k^{2n}$. The equation for k must accordingly have the form

$$\tfrac{4}{3}p_0 k^{2n}+...+\tfrac{2}{3}p_{2n} = 0.$$

Hence, $$k_1^2...k_n^2 = (-1)^n\frac{p_{2n}}{2p_0} = \frac{1}{2\mu_1^2...\mu_n^2}.$$

Combining equations (60) and (61) appropriately, we obtain the following relations:

$$\sum_{\alpha=1}^{n} M_\alpha = \frac{q}{\sqrt{2}} a_l - Q; \qquad \sum_{\alpha=1}^{n} M_\alpha k_\alpha = \frac{q}{\sqrt{2}} b_l + 1,$$

and

$$\sum_{\alpha=1}^{n} M_\alpha k_\alpha^2 = \frac{1}{\sqrt{2}} (q a_l - c), \qquad (62)$$

where we have used the abbreviations

$$a_l = \tfrac{1}{2}[H_l(-1) + H_l(+1)] \quad \text{and} \quad b_l = \tfrac{1}{2}[H_l(-1) - H_l(+1)]. \quad (63)$$

Now from equations (52) and (57) we obtain the following expression for M_α:

$$M_\alpha = (-1)^n k_1 ... k_n \frac{P(1/k_\alpha)}{R_\alpha(1/k_\alpha)(k_\alpha^2 - 1)} \left(\frac{1}{k_\alpha} - c \right) \quad (\alpha = 1,...,n). \quad (64)$$

We may therefore write

$$\sum_{\alpha=1}^{n} M_\alpha k_\alpha^m = \xi_{m-1} - c\xi_m, \qquad (65)$$

where

$$\xi_m = (-1)^n k_1 ... k_n \sum_{\alpha=1}^{n} \frac{P(1/k_\alpha) k_\alpha^m}{R_\alpha(1/k_\alpha)(k_\alpha^2 - 1)}. \qquad (66)$$

To carry out the summation on the right-hand side of equation (66) we introduce the function

$$f_m(x) = \sum_{\alpha=1}^{n} \frac{P(1/k_\alpha) k_\alpha^m}{R_\alpha(1/k_\alpha)(k_\alpha^2 - 1)} R_\alpha(x) \quad (m = -1, 0, 1, 2), \qquad (67)$$

and express ξ_m in terms of it. Thus (cf. eq. [58]),

$$\xi_m = (-1)^n k_1 ... k_n f_m(0) = \frac{(-1)^n}{\mu_1 ... \mu_n \sqrt{2}} f_m(0). \qquad (68)$$

Now $f_m(x)$ defined as in equation (67) is a polynomial of degree $(n-1)$ in x, which takes the values

$$P(1/k_\alpha) k_\alpha^m / (k_\alpha^2 - 1)$$

for $x = 1/k_\alpha$ ($\alpha = 1,...,n$). In other words

$$(1-x^2) f_m(x) - x^{2-m} P(x) = 0 \quad \text{for} \quad x = k_\alpha^{-1} \quad \text{and} \quad \alpha = 1,...,n. \quad (69)$$

The polynomial on the left-hand side of equation (69) must therefore divide $R(x)$. We must, accordingly, have a relation of the form

$$(1-x^2) f_m(x) - x^{2-m} P(x) = R(x) \Phi(x), \qquad (70)$$

where $\Phi(x)$ is a polynomial of degree 3, 2, 1, and 1, for $m = -1, 0, 1,$ and 2, respectively. To determine $\Phi(x)$ more explicitly, we must consider each case separately. We shall illustrate the procedure for the case $m = -1$.

For $m = -1$, equation (70) becomes

$$(1-x^2)f_{-1}(x)-x^3 P(x) = R(x)(Ax^3+Bx^2+Cx+D), \qquad (71)$$

where A, B, C, and D are certain constants to be determined. The constants A and B follow from a comparison of the coefficients of x^{n+3} and x^{n+2} on either side of equation (71). Thus

$$A = \frac{(-1)^{n+1}}{k_1...k_n} \quad \text{and} \quad B = \frac{(-1)^n}{k_1...k_n}\left(\sum_{j=1}^{n}\mu_j - \sum_{\alpha=1}^{n}\frac{1}{k_\alpha}\right). \qquad (72)$$

Next, putting $x = +1$ and -1 in equation (71), we have

$$A+B+C+D = -\frac{P(1)}{R(1)} = -(-1)^n\mu_1...\mu_n H_r(-1)$$

and

$$-A+B-C+D = +\frac{P(-1)}{R(-1)} = (-1)^n\mu_1...\mu_n H_r(+1). \qquad (73)$$

These equations determine the remaining constants C and D. In particular

$$D = f_{-1}(0) = (-1)^{n+1}\mu_1...\mu_n b_r + \frac{(-1)^{n+1}}{k_1...k_n}\left(\sum_{j=1}^{n}\mu_j - \sum_{\alpha=1}^{n}\frac{1}{k_\alpha}\right), \qquad (74)$$

where, in analogy with (63), we have introduced the abbreviations

$$a_r = \tfrac{1}{2}[H_r(-1)+H_r(+1)] \quad \text{and} \quad b_r = \tfrac{1}{2}[H_r(-1)-H_r(+1)]. \qquad (75)$$

From equations (68) and (74) we now find

$$\xi_{-1} = -\frac{b_r}{\sqrt{2}}-\left(\sum_{j=1}^{n}\mu_j - \sum_{\alpha=1}^{n}\frac{1}{k_\alpha}\right). \qquad (76)$$

The evaluations of ξ_0, ξ_1, and ξ_2 proceed along similar lines. We find

$$\xi_0 = -\frac{a_r}{\sqrt{2}}+1, \quad \xi_1 = -\frac{b_r}{\sqrt{2}} \quad \text{and} \quad \xi_2 = \frac{1}{\sqrt{2}}(1-a_r). \qquad (77)$$

With the substitutions (65), (76), and (77), equations (62) become

$$\frac{1}{\sqrt{2}}(ca_r-b_r)-c-\left(\sum_{j=1}^{n}\mu_j - \sum_{\alpha=1}^{n}\frac{1}{k_\alpha}\right) = \frac{q}{\sqrt{2}}a_l-Q, \qquad (78)$$

$$-a_r+cb_r = qb_l \quad \text{and} \quad -b_r+ca_r = qa_l. \qquad (79)$$

Solving equations (79) for q and c, we obtain

$$q = \frac{a_r^2-b_r^2}{a_l b_r-a_r b_l} \quad \text{and} \quad c = \frac{a_r a_l-b_r b_l}{a_l b_r-a_r b_l}. \qquad (80)$$

Also, from equations (78) and (79), we have

$$Q = c + \sum_{j=1}^{n} \mu_j - \sum_{\alpha=1}^{n} \frac{1}{k_\alpha}. \tag{81}$$

With the determination of q and c, we have completed the elimination of the constants in the solution for the emergent radiations. We have (cf. eqs. [54] and [59])

$$I_l(0,\mu) = \tfrac{3}{8} F \frac{q}{\sqrt{2}} H_l(\mu) \quad \text{and} \quad I_r(0,\mu) = \tfrac{3}{8} F \frac{1}{\sqrt{2}} H_r(\mu)(\mu+c). \tag{82}$$

68.5. *Relations between the constants q and c*

According to equations (63), (75), and (80), we have

$$1-c^2 = -\frac{(a_r^2-b_r^2)(a_l^2-b_l^2)}{(a_l b_r - a_r b_l)^2} = -\frac{H_l(+1)H_l(-1)H_r(+1)H_r(-1)}{(a_l b_r - a_r b_l)^2}$$

and

$$q^2 = \frac{[H_r(+1)H_r(-1)]^2}{(a_l b_r - a_r b_l)^2}. \tag{83}$$

On the other hand, from the identities (cf. Chap. V, eq. [45])

$$\frac{1}{H_l(\mu)H_l(-\mu)} = 1 - \tfrac{3}{2}\mu^2 \sum_{j=1}^{n} \frac{a_j(1-\mu_j^2)}{\mu^2-\mu_j^2}$$

and

$$\frac{1}{H_r(\mu)H_r(-\mu)} = 1 - \tfrac{3}{4}\mu^2 \sum_{j=1}^{n} \frac{a_j(1-\mu_j^2)}{\mu^2-\mu_j^2}, \tag{84}$$

we have the relations

$$H_l(+1)H_l(-1) = -2 \quad \text{and} \quad H_r(+1)H_r(-1) = 4. \tag{85}$$

Hence,

$$1-c^2 = \frac{8}{(a_l b_r - a_r b_l)^2}, \qquad q^2 = \frac{16}{(a_l b_r - a_r b_l)^2}, \tag{86}$$

and

$$q^2 = 2(1-c^2). \tag{87}$$

Since equation (87) is independent of the order of the approximation, we conclude that it represents an *exact relation* of the problem.

A further relation between the constants q and c which follows from the equation

$$1+c = \frac{(a_l-b_l)(a_r+b_r)}{a_l b_r - a_r b_l} = \frac{H_l(+1)H_r(-1)}{a_l b_r - a_r b_l}, \tag{88}$$

is

$$(1+c)H_r(+1) = \frac{H_l(+1)H_r(+1)H_r(-1)}{a_l b_r - a_r b_l} = q H_l(+1). \tag{89}$$

This last relation between q and c is a necessary one, since symmetry clearly demands that

$$I_l(0,1) = \tfrac{3}{8}F \frac{q}{\sqrt{2}} H_l(1) = I_r(0,1) = \tfrac{3}{8}F \frac{1}{\sqrt{2}} H_r(1)(1+c). \qquad (90)$$

68.6. *Passage to the limit of infinite approximation and the exact solutions for $I_l(0,\mu)$ and $I_r(0,\mu)$*

From the theory of the H-functions (Chap. V) it is known that the H-function defined rationally in terms of the positive roots of the characteristic equation

$$1 = 2\sum_{j=1}^{n} \frac{a_j \Psi(\mu_j)}{1-k^2\mu_j^2}, \qquad (91)$$

must, in the limit of infinite approximation, be associated with the solution of the integral equation

$$H(\mu) = 1 + \mu H(\mu) \int_0^1 \frac{\Psi(\mu')}{\mu+\mu'} H(\mu')\, d\mu', \qquad (92)$$

which is bounded in the entire half-plane $R(z) > 0$.

We may therefore conclude from equations (82) that the exact solutions for $I_l(0,\mu)$ and $I_r(0,\mu)$ must be expressible in the forms

$$I_l(0,\mu) = \tfrac{3}{8}F \frac{q}{\sqrt{2}} H_l(\mu) \quad \text{and} \quad I_r(0,\mu) = \tfrac{3}{8}F \frac{1}{\sqrt{2}} H_r(\mu)(\mu+c), \qquad (93)$$

where $H_l(\mu)$ and $H_r(\mu)$ are defined in terms of the characteristic functions

$$\Psi_l(\mu) = \tfrac{3}{4}(1-\mu^2) \quad \text{and} \quad \Psi_r(\mu) = \tfrac{3}{8}(1-\mu^2), \qquad (94)$$

respectively, and q and c are two constants related in the manner (cf. eq. [87])

$$q^2 = 2(1-c^2). \qquad (95)$$

A further relation between the constants q and c, which will make the solution determinate, can be obtained from the flux integral

$$F = 2\int_0^1 [I_l(0,\mu)+I_r(0,\mu)]\mu\, d\mu. \qquad (96)$$

Thus, with the solutions for $I_l(0,\mu)$ and $I_r(0,\mu)$ given by (93), we must have

$$\frac{3}{4\sqrt{2}}(q\alpha_1+A_2+cA_1) = 1, \qquad (97)$$

where α_n and A_n are the moments of order n of $H_l(\mu)$ and $H_r(\mu)$, respectively. Alternately, we could also use the relation

$$qH_l(1) = (1+c)H_r(1), \qquad (98)$$

required by symmetry.

By using the various integral properties of the functions $H_l(\mu)$ and $H_r(\mu)$, it can be shown that (cf. § 70.2 below)

$$q = 2\frac{4(A_1+2\alpha_1)-3(A_0\alpha_1+\alpha_0 A_1)}{3(A_1^2+2\alpha_1^2)}$$

and

$$c = \frac{8(A_1-\alpha_1)+3(2\alpha_1\alpha_0-A_1 A_0)}{3(A_1^2+2\alpha_1^2)}, \tag{99}$$

represent the unique solutions for q and c.

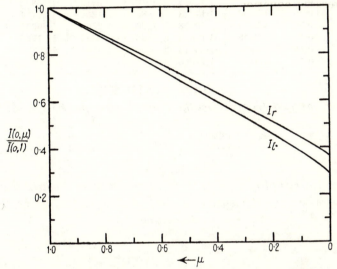

Fig. 23. The laws of darkening in the two states of polarization for an electron scattering atmosphere: l refers to the component polarized with the electric vector in the meridian plane and r refers to the component with the electric vector at right angles to the meridian plane.

68.7. *The exact laws of darkening in the two states of polarization. The degree of polarization of the emergent radiation*

The functions $H_l(\mu)$ and $H_r(\mu)$ have been evaluated numerically by Mrs. Frances H. Breen and the writer by the method described in Chapter V, § 41. The solutions are given in Table XXII. The various related constants are listed in Table XXIII. The intensities $I_l(0,\mu)$ and $I_r(0,\mu)$ and the degree of polarization,

$$\delta = \frac{I_r(0,\mu)-I_l(0,\mu)}{I_r(0,\mu)+I_l(0,\mu)}, \tag{100}$$

of the emergent radiation derived from the tabulated functions and constants are given in Table XXIV. The different laws of darkening in the two states of polarization are further illustrated in Fig. 23.

TABLE XXII
The Functions $H_l(\mu)$ and $H_r(\mu)$ obtained as solutions of the Exact Integral Equations

μ	$H_l(\mu)$	$H_r(\mu)$	μ	$H_l(\mu)$	$H_r(\mu)$
0	1·0000	1·00000	0·55	2·4319	1·22532
0·05	1·1814	1·05737	0·60	2·5486	1·23320
0·10	1·3255	1·09113	0·65	2·6649	1·24042
0·15	1·4596	1·11703	0·70	2·7807	1·24705
0·20	1·5884	1·13816	0·75	2·8962	1·25318
0·25	1·7137	1·15594	0·80	3·0113	1·25886
0·30	1·8367	1·17128	0·85	3·1262	1·26414
0·35	1·9579	1·18468	0·90	3·2408	1·26906
0·40	2·0778	1·19654	0·95	3·3552	1·27366
0·45	2·1966	1·20713	1·00	3·4695	1·27797
0·50	2·3146	1·21668			

TABLE XXIII
The Constants derived from the Exact Functions, $H_l(\mu)$ and $H_r(\mu)$

$$\alpha_0 = 2\cdot29767 \qquad A_1 = 0\cdot61733$$
$$A_0 = 1\cdot19736 \qquad q = 0\cdot68989$$
$$\alpha_1 = 1\cdot34864 \qquad c = 0\cdot87294$$

TABLE XXIV
The Exact Laws of darkening in the Two States of Polarization for an Electron-scattering Atmosphere; Degree of Polarization of the Emergent Radiation

μ	I_l/F	I_r/F	I/F	Law of darkening in state of polarization l	r	Degree of polarization
0	0·18294	0·23147	0·41441	0·28823	0·36470	0·11713
0·05	0·21613	0·25877	0·47490	0·34053	0·40771	0·08979
0·10	0·24247	0·28150	0·52397	0·38203	0·44352	0·07448
0·15	0·26702	0·30299	0·57001	0·42070	0·47739	0·06311
0·20	0·29057	0·32381	0·61439	0·45782	0·51019	0·05410
0·25	0·31350	0·34420	0·65770	0·49394	0·54231	0·04667
0·30	0·33599	0·36429	0·70029	0·52939	0·57397	0·04041
0·35	0·35817	0·38417	0·74234	0·56432	0·60529	0·03502
0·40	0·38010	0·40388	0·78398	0·59888	0·63634	0·03033
0·45	0·40184	0·42346	0·82530	0·63313	0·66719	0·02619
0·50	0·42343	0·44294	0·86637	0·66714	0·69788	0·02252
0·55	0·44489	0·46233	0·90722	0·70095	0·72844	0·01923
0·60	0·46624	0·48165	0·94789	0·73459	0·75888	0·01627
0·65	0·48750	0·50092	0·98842	0·76809	0·78924	0·01358
0·70	0·50869	0·52013	1·02882	0·80147	0·81950	0·011123
0·75	0·52981	0·53930	1·06911	0·83475	0·84971	0·008880
0·80	0·55087	0·55844	1·10931	0·86794	0·87986	0·006818
0·85	0·57189	0·57754	1·14943	0·90105	0·90996	0·004919
0·90	0·59286	0·59661	1·18947	0·93409	0·94001	0·003155
0·95	0·61379	0·61566	1·22945	0·96707	0·97002	0·001522
1·00	0·63469	0·63469	1·26938	1·00000	1·00000	0

According to the results given in Table XXIV, the emergent intensities in the two states of polarization, while equal at the centre ($\mu = 1$), differ by about 25 per cent. at the limb ($\mu = 0$). Correspondingly, the degree of polarization varies from zero at the centre to 11·7 per cent. at the limb. In stellar atmospheres of early type (spectral type B0 and earlier), in which scattering by free electrons is believed to play a predominant role in governing the transfer of radiation, we should expect to detect this polarization during the eclipse of such stars, at phases close to the primary minimum. Preliminary observations by W. A. Hiltner† would appear to indicate that the effect, of the predicted amount, is present.

69. The reduction of the equation of transfer for the problem of diffuse reflection and transmission

The equation of transfer appropriate to the problem of diffuse reflection and transmission, for Rayleigh scattering, has been given in Chapter I, § 17.4, equation (231). In this section we shall reduce this vector equation into a simpler set of equations.

First, we observe that since the phase-matrix (cf. Chap. I, eqs. [220]–[224]) is reducible with respect to the last row and column, the Stokes parameter V will be scattered independently of the others and, indeed, according to the phase function $\frac{3}{2}\cos\Theta$. Thus, the equation of transfer for V is

$$\mu\frac{dV(\tau,\mu,\varphi)}{d\tau} = V(\tau,\mu,\varphi) -$$

$$-\frac{3}{8\pi}\int_{-1}^{+1}\int_{0}^{2\pi}[\mu\mu'+(1-\mu^2)^{\frac{1}{2}}(1-\mu'^2)^{\frac{1}{2}}\cos(\varphi'-\varphi)]V(\tau,\mu',\varphi)\,d\mu'd\varphi' -$$

$$-\tfrac{3}{8}F_V[-\mu\mu_0+(1-\mu_0^2)^{\frac{1}{2}}(1-\mu^2)^{\frac{1}{2}}\cos(\varphi_0-\varphi)]e^{-\tau/\mu_0}, \quad (101)$$

or, writing V in the form

$$V(\tau,\mu,\varphi) = \tfrac{3}{8}F_V[-\mu\mu_0\,V^{(0)}(\tau,\mu)+(1-\mu^2)^{\frac{1}{2}}(1-\mu_0^2)^{\frac{1}{2}}V^{(1)}(\tau,\mu)\cos(\varphi_0-\varphi)], \quad (102)$$

we have the pair of equations

$$\mu\frac{dV^{(0)}(\tau,\mu)}{d\tau} = V^{(0)}(\tau,\mu)-\tfrac{3}{4}\int_{-1}^{+1}\mu'^2V^{(0)}(\tau,\mu')\,d\mu'-e^{-\tau/\mu_0}$$

and $\quad \mu\dfrac{dV^{(1)}(\tau,\mu)}{d\tau} = V^{(1)}(\tau,\mu)-\tfrac{3}{8}\displaystyle\int_{-1}^{+1}(1-\mu'^2)V^{(1)}(\tau,\mu')\,d\mu'-e^{-\tau/\mu_0}. \quad (103)$

† *Astrophys. J.* **106**, 231 (1947).

Next, considering I_l, I_r, and U as the components of a *three*-dimensional vector I, we have the equation of transfer

$$\mu\frac{dI(\tau,\mu,\varphi)}{d\tau} = I(\tau,\mu,\varphi) - \frac{1}{4\pi}\int_{-1}^{+1}\int_{0}^{2\pi} P(\mu,\varphi;\mu',\varphi')I(\tau,\mu',\varphi')\,d\mu'd\varphi' -$$
$$-\tfrac{1}{4}P(\mu,\varphi;-\mu_0,\varphi_0)Fe^{-\tau/\mu_0}, \quad (104)$$

where
$$F = (F_l, F_r, U), \quad (105)$$

and $P(\mu,\varphi;\mu',\varphi')$ denotes the matrix defined exactly as in Chapter I, equations (220)–(224) but with the last row and column deleted. Thus

$$P(\mu,\varphi;\mu',\varphi')$$
$$= Q[P^{(0)}(\mu,\mu')+(1-\mu^2)^{\frac{1}{2}}(1-\mu'^2)^{\frac{1}{2}}P^{(1)}(\mu,\varphi;\mu',\varphi')+P^{(2)}(\mu,\varphi;\mu',\varphi')],$$
$$(106)$$

where
$$Q = \begin{pmatrix} 1 & 0 & 0 \\ 0 & 1 & 0 \\ 0 & 0 & 2 \end{pmatrix}, \quad (107)$$

$$P^{(0)}(\mu,\mu') = \frac{3}{4}\begin{pmatrix} 2(1-\mu^2)(1-\mu'^2)+\mu^2\mu'^2 & \mu^2 & 0 \\ \mu'^2 & 1 & 0 \\ 0 & 0 & 0 \end{pmatrix},$$

$$P^{(1)}(\mu,\varphi;\mu',\varphi') = \frac{3}{4}\begin{pmatrix} 4\mu\mu'\cos(\varphi'-\varphi) & 0 & 2\mu\sin(\varphi'-\varphi) \\ 0 & 0 & 0 \\ -2\mu'\sin(\varphi'-\varphi) & 0 & \cos(\varphi'-\varphi) \end{pmatrix},$$

and

$$P^{(2)}(\mu,\varphi;\mu',\varphi')$$
$$= \frac{3}{4}\begin{pmatrix} \mu^2\mu'^2\cos 2(\varphi'-\varphi) & -\mu^2\cos 2(\varphi'-\varphi) & \mu^2\mu'\sin 2(\varphi'-\varphi) \\ -\mu'^2\cos 2(\varphi'-\varphi) & \cos 2(\varphi'-\varphi) & -\mu'\sin 2(\varphi'-\varphi) \\ -\mu\mu'^2\sin 2(\varphi'-\varphi) & \mu\sin 2(\varphi'-\varphi) & \mu\mu'\cos 2(\varphi'-\varphi) \end{pmatrix}. \quad (108)$$

Now writing

$$I(\tau,\mu,\varphi) = I^{(0)}(\tau,\mu)+(1-\mu^2)^{\frac{1}{2}}(1-\mu_0^2)^{\frac{1}{2}}I^{(1)}(\tau,\mu,\varphi)+I^{(2)}(\tau,\mu,\varphi), \quad (109)$$

where $I^{(0)}$ is azimuth independent and $I^{(1)}$ and $I^{(2)}$ contain terms in $(\varphi_0-\varphi)$ of periods 2π and π only, we have the three equations

$$\mu\frac{dI^{(0)}(\tau,\mu)}{d\tau} = I^{(0)}(\tau,\mu)-\tfrac{1}{2}\int_{-1}^{+1} P^{(0)}(\mu,\mu')I^{(0)}(\tau,\mu')\,d\mu'-\tfrac{1}{4}P^{(0)}(\mu,\mu_0)Fe^{-\tau/\mu_0},$$
$$(110)$$

$$\mu\frac{dI^{(1)}}{d\tau} = I^{(1)}-\frac{Q}{4\pi}\int_{-1}^{+1}\int_{0}^{2\pi}(1-\mu'^2)P^{(1)}(\mu,\varphi;\mu',\varphi')I^{(1)}(\tau,\mu',\varphi')\,d\mu'd\varphi'-$$
$$-\tfrac{1}{4}QP^{(1)}(\mu,\varphi;-\mu_0,\varphi_0)Fe^{-\tau/\mu_0}, \quad (111)$$

and

$$\mu\frac{d\boldsymbol{I}^{(2)}}{d\tau} = \boldsymbol{I}^{(2)} - \frac{Q}{4\pi}\int\limits_{-1}^{+1}\int\limits_{0}^{2\pi}\boldsymbol{P}^{(2)}(\mu,\varphi;\mu',\varphi')\boldsymbol{I}^{(2)}(\tau,\mu',\varphi')\,d\mu'd\varphi' -$$
$$-\tfrac{1}{4}Q\boldsymbol{P}^{(2)}(\mu,\varphi;-\mu_0,\varphi_0)Fe^{-\tau/\mu_0}. \quad (112)$$

By writing equations (111) and (112) in full, we find that, among the various coefficients in the Fourier expansions of I_l, I_r, and U, there are certain simple relations of proportionality. Thus, if $I_l^{(1)}$, $I_r^{(1)}$, $U^{(-1)}$ and $I_l^{(-1)}$, $I_r^{(-1)}$, $U^{(1)}$ are the coefficients of $\cos(\varphi_0-\varphi)$ and $\sin(\varphi_0-\varphi)$, respectively, in the Fourier expansions in $(\varphi_0-\varphi)$ of I_l, I_r, and U, then

$$I_l^{(1)}(\tau,\mu) \equiv \mu U^{(1)}(\tau,\mu); \qquad I_l^{(-1)}(\tau,\mu) \equiv -\mu U^{(-1)}(\tau,\mu);$$
$$I_r^{(1)}(\tau,\mu) \equiv I_r^{(-1)}(\tau,\mu) \equiv 0. \quad (113)$$

Similarly, among the coefficients of $\cos 2(\varphi_0-\varphi)$ and $\sin 2(\varphi_0-\varphi)$, there are the relations

$$I_l^{(2)}(\tau,\mu) \equiv -\mu^2 I_r^{(2)}(\tau,\mu) \equiv \tfrac{1}{2}\mu U^{(2)}(\tau,\mu)$$

and

$$I_l^{(-2)}(\tau,\mu) \equiv -\mu^2 I_r^{(-2)}(\tau,\mu) \equiv -\tfrac{1}{2}\mu U^{(-2)}(\tau,\mu). \quad (114)$$

The origin of these relationships can be traced to the following, directly verifiable, properties of the matrices $\boldsymbol{P}^{(1)}$ and $\boldsymbol{P}^{(2)}$:

$$\frac{1}{4\pi}\int\limits_{0}^{2\pi}Q\boldsymbol{P}^{(1)}(\mu,\varphi;\mu',\varphi')Q\boldsymbol{P}^{(1)}(\mu',\varphi';-\mu_0,\varphi_0)\,d\varphi'$$
$$= Q\boldsymbol{P}^{(1)}(\mu,\varphi;-\mu_0,\varphi_0)\tfrac{3}{8}(1+2\mu'^2)$$

and

$$\frac{1}{4\pi}\int\limits_{0}^{2\pi}Q\boldsymbol{P}^{(2)}(\mu,\varphi;\mu',\varphi')Q\boldsymbol{P}^{(2)}(\mu',\varphi';-\mu_0,\varphi_0)\,d\varphi'$$
$$= Q\boldsymbol{P}^{(2)}(\mu,\varphi;-\mu_0,\varphi_0)\tfrac{3}{16}(1+\mu'^2)^2. \quad (115)$$

Consequently, equations (111) and (112) allow solutions of the form

$$\boldsymbol{I}^{(1)}(\tau,\mu,\varphi) = \tfrac{1}{4}Q\boldsymbol{P}^{(1)}(\mu,\varphi;-\mu_0,\varphi_0)F\phi^{(1)}(\tau,\mu)$$

and

$$\boldsymbol{I}^{(2)}(\tau,\mu,\varphi) = \tfrac{1}{4}Q\boldsymbol{P}^{(2)}(\mu,\varphi;-\mu_0,\varphi_0)F\phi^{(2)}(\tau,\mu), \quad (116)$$

where $\phi^{(1)}$ and $\phi^{(2)}$ are scalars depending on τ and μ only; for, with the substitutions (116), $\tfrac{1}{4}Q\boldsymbol{P}^{(1)}(\mu,\varphi;-\mu_0,\varphi_0)F$ and $\tfrac{1}{4}Q\boldsymbol{P}^{(2)}(\mu,\varphi;-\mu_0,\varphi_0)F$ will occur as factors in every term of equations (111) and (112), respectively, after the integration over φ' has been carried out, and will leave the scalar equations

$$\mu\frac{d\phi^{(1)}}{d\tau} = \phi^{(1)} - \tfrac{3}{8}\int\limits_{-1}^{+1}(1-\mu'^2)(1+2\mu'^2)\phi^{(1)}(\tau,\mu')\,d\mu' - e^{-\tau/\mu_0}$$

and

$$\mu\frac{d\phi^{(2)}}{d\tau} = \phi^{(2)} - \tfrac{3}{16}\int\limits_{-1}^{+1}(1+\mu'^2)^2\phi^{(2)}(\tau,\mu')\,d\mu - e^{-\tau/\mu_0}, \quad (117)$$

for determining $\phi^{(1)}$ and $\phi^{(2)}$. It can now be verified that equations (116) imply the relationships (113) and (114).

Now equations (103) and (117) are of the general form

$$\mu \frac{df(\tau,\mu)}{d\tau} = f(\tau,\mu) - \int_{-1}^{+1} \Psi(\mu')f(\tau,\mu')\,d\mu' - e^{-\tau/\mu_0}, \qquad (118)$$

where $\Psi(\mu)$ is an even polynomial in μ. Equation (118) can be solved, quite readily, by the standard procedure of replacing integrals by sums. Indeed, it is evident that the analysis will parallel, exactly, the treatment of the equation

$$\mu \frac{dI(\tau,\mu)}{d\tau} = I(\tau,\mu) - \tfrac{1}{2}\varpi_0 \int_{-1}^{+1} I(\tau,\mu')\,d\mu' - \frac{\varpi_0}{4} Fe^{-\tau/\mu_0}, \qquad (119)$$

given in Chapter III, § 26, and Chapter VIII, § 59, with the only non-essential difference that the characteristic equation appropriate for the solution of equation (118) is

$$1 = 2 \sum_{j=1}^{n} \frac{a_j\,\Psi(\mu_j)}{1-k^2\mu_j^2}, \qquad (120)$$

instead of Chapter III, equation (84). The angular dependence of f at the boundaries of the atmosphere can, therefore, be reduced to H-functions (in the semi-infinite case) or X- and Y-functions (in the finite case). From the correspondences enunciated in Chapters V and VIII, between the H-functions and the X- and Y-functions, defined, rationally, in terms of characteristic roots and those defined in terms of integral equations, we conclude that the exact solutions for f at $\tau = 0$ and $\tau = \tau_1$ (in the finite case) will be given by

$$f(0,\mu) = [X(\mu)X(\mu_0) - Y(\mu)Y(\mu_0)]\frac{\mu_0}{\mu+\mu_0},$$

and

$$f(\tau_1,-\mu) = [Y(\mu)X(\mu_0) - X(\mu)Y(\mu_0)]\frac{\mu_0}{\mu-\mu_0}, \qquad (121)$$

where X and Y are defined as solutions of integral equations of the form considered in Chapter VIII.

Returning to equations (116) and (117), we now observe that this reduction of the solutions for $I^{(1)}$ and $I^{(2)}$ means that if we express the scattering and the transmission matrices in the form

$$S(\tau_1; \mu, \varphi; \mu_0, \varphi_0)$$
$$= Q[\tfrac{3}{4}S^{(0)}(\tau_1; \mu, \mu_0) + (1-\mu^2)^{\frac{1}{2}}(1-\mu_0^2)^{\frac{1}{2}}S^{(1)}(\tau_1; \mu, \varphi; \mu_0, \varphi_0) +$$
$$+ S^{(2)}(\tau_1; \mu, \varphi; \mu_0, \varphi_0)]$$

and

$$T(\tau_1; \mu, \varphi; \mu_0, \varphi_0)$$
$$= Q[\tfrac{3}{4}T^{(0)}(\tau_1; \mu, \mu_0) + (1-\mu^2)^{\frac{1}{2}}(1-\mu_0^2)^{\frac{1}{2}} T^{(1)}(\tau_1; \mu, \varphi; \mu_0, \varphi_0) +$$
$$+ T^{(2)}(\tau_1; \mu, \varphi; \mu_0, \varphi_0)], \quad (122)$$

then

$$\left(\frac{1}{\mu_0} + \frac{1}{\mu}\right) S^{(i)} = P^{(i)}(\mu, \varphi; -\mu_0, \varphi_0)[X^{(i)}(\mu)X^{(i)}(\mu_0) - Y^{(i)}(\mu)Y^{(i)}(\mu_0)]$$

and

$$\left(\frac{1}{\mu_0} - \frac{1}{\mu}\right) T^{(i)} = P^{(i)}(-\mu, \varphi; -\mu_0, \varphi_0)[Y^{(i)}(\mu)X^{(i)}(\mu_0) - X^{(i)}(\mu)Y^{(i)}(\mu_0)]$$

$$(i = 1, 2) \quad (123)$$

where $X^{(1)}$, $Y^{(1)}$ and $X^{(2)}$, $Y^{(2)}$ are defined in terms of the characteristic functions

$$\Psi^{(1)}(\mu) = \tfrac{3}{8}(1-\mu^2)(1+2\mu^2) \quad \text{and} \quad \Psi^{(2)}(\mu) = \tfrac{3}{16}(1+\mu^2)^2, \quad (124)$$

respectively. In the semi-infinite case, the corresponding expressions for $S^{(1)}$ and $S^{(2)}$ are

$$\left(\frac{1}{\mu_0} + \frac{1}{\mu}\right) S^{(i)} = P^{(i)}(\mu, \varphi; -\mu_0, \varphi_0)H^{(i)}(\mu)H^{(i)}(\mu_0) \quad (i = 1, 2), \quad (125)$$

where $H^{(1)}$ and $H^{(2)}$ are defined with respect to the same characteristic functions (124).

Similarly, according to equations (102) and (103), the law of diffuse reflection and transmission of the Stokes parameter V can also be reduced to X- and Y-functions, and expressed in terms of a scattering and a transmission function in the form

$$\left(\frac{1}{\mu_0} + \frac{1}{\mu}\right) S_V = \tfrac{3}{2}\{-\mu\mu_0[X_v(\mu)X_v(\mu_0) - Y_v(\mu)Y_v(\mu_0)] +$$
$$+ (1-\mu^2)^{\frac{1}{2}}(1-\mu_0^2)^{\frac{1}{2}}[X_r(\mu)X_r(\mu_0) - Y_r(\mu)Y_r(\mu_0)]\cos(\varphi_0-\varphi)\}$$

and

$$\left(\frac{1}{\mu_0} - \frac{1}{\mu}\right) T_V = \tfrac{3}{2}\{\mu\mu_0[Y_v(\mu)X_v(\mu_0) - X_v(\mu)Y_v(\mu_0)] +$$
$$+ (1-\mu^2)^{\frac{1}{2}}(1-\mu_0^2)^{\frac{1}{2}}[Y_r(\mu)X_r(\mu_0) - X_r(\mu)Y_r(\mu_0)]\cos(\varphi_0-\varphi)\}, \quad (126)$$

where X_v, Y_v and X_r, Y_r are defined in terms of the characteristic functions

$$\Psi_v = \tfrac{3}{4}\mu^2 \quad \text{and} \quad \Psi_r(\mu) = \tfrac{3}{8}(1-\mu^2), \quad (127)$$

respectively. In the semi-infinite case the corresponding expression for S_V is

$$\left(\frac{1}{\mu_0} + \frac{1}{\mu}\right) S_V = \tfrac{3}{2}[-\mu\mu_0 H_v(\mu)H_v(\mu_0) +$$
$$+ (1-\mu^2)^{\frac{1}{2}}(1-\mu_0^2)^{\frac{1}{2}}H_r(\mu)H_r(\mu_0)\cos(\varphi_0-\varphi)], \quad (128)$$

where H_v and H_r are defined in terms of the same characteristic functions (127).

The explicit solutions we have found for all terms in the scattering and the transmission matrices, except those independent of the azimuth in I_l and I_r, can also be derived from the integral equations governing S and T; but a derivation along these lines will require rather more elaborate considerations than the one we have given, starting, directly, from the equation of transfer.

It remains to consider the solution of equation (110); but this equation presents a problem of a different order of difficulty and will be taken up in the following sections.

70. The law of diffuse reflection by a semi-infinite atmosphere for Rayleigh scattering

Since the matrix $P^{(0)}(\mu, \mu')$ (cf. eq. [108]) is reducible with respect to U, it is clear that equation (110) represents only a pair of equations for the azimuth independent terms $I_l^{(0)}(\tau, \mu)$ and $I_r^{(0)}(\tau, \mu)$ in the Fourier expansion of $I_l(\tau, \mu, \varphi)$ and $I_r(\tau, \mu, \varphi)$ in $(\varphi_0 - \varphi)$. It is therefore convenient to regard $I_l^{(0)}$ and $I_r^{(0)}$ as the components of a two-dimensional vector, $I^{(0)}$, and rewrite equation (110) in the form

$$\mu \frac{dI^{(0)}(\tau, \mu)}{d\tau} = I^{(0)}(\tau, \mu) - \tfrac{3}{8} \int_{-1}^{+1} J(\mu, \mu') I^{(0)}(\tau, \mu') \, d\mu' - \tfrac{3}{16} J(\mu, \mu_0) F e^{-\tau/\mu_0},$$

(129)

where, now, $$F = (F_l, F_r)$$ (130)

and $$J(\mu, \mu') = \begin{pmatrix} 2(1-\mu^2)(1-\mu'^2) + \mu^2 \mu'^2 & \mu^2 \\ \mu'^2 & 1 \end{pmatrix}.$$ (131)

In § 68 we have already considered the homogeneous equation associated with equation (129). The solution can now be completed by finding a particular integral. It is then found that the constants can be eliminated from the solutions for the emergent radiations obtained in this manner. However, the reductions are quite involved and the integral equations derived from the principles of invariance provide a more suitable starting-point for the present discussion.

Expressing, then, the reflected intensities $I_l^{(0)}(0, \mu)$ and $I_r^{(0)}(0, \mu)$ in terms of a scattering matrix, $S^{(0)}(\mu, \mu_0)$ (with *two* rows and columns) in the form

$$I^{(0)}(0, \mu) = \begin{pmatrix} I_l^{(0)}(0, \mu) \\ I_r^{(0)}(0, \mu) \end{pmatrix} = \frac{3}{16\mu} S^{(0)}(\mu, \mu_0) \begin{pmatrix} F_l \\ F_r \end{pmatrix},$$ (132)

we can write the integral equation for $S^{(0)}$ in analogy with Chapter IV,

§ 30, equation (28), by replacing $p(\mu, \varphi; \mu', \varphi')$ and $S(\mu, \varphi; \mu', \varphi')$ by $\frac{3}{4}J(\mu, \mu')$ and $\frac{3}{4}S^{(0)}(\mu, \mu')$, respectively. Since $S^{(0)}$ and J are azimuth independent, the integration over φ can be effected. The resulting equation for $S^{(0)}$ can be written most compactly by using the following notation:

Let the 'product' $[A, B]_{\mu, \mu'}$ of two matrices $A(\mu, \mu')$ and $B(\mu, \mu')$ be defined by the formula

$$[A, B]_{\mu, \mu'} = \tfrac{3}{8} \int\limits_0^1 A(\mu, \mu'') B(\mu'', \mu') \frac{d\mu''}{\mu''}, \tag{133}$$

where, under the integral sign, the ordinary matrix product is intended. With this product notation, the equation satisfied by $S^{(0)}$ is

$$\left(\frac{1}{\mu_0} + \frac{1}{\mu}\right) S^{(0)} = J + [J, S^{(0)}] + [S^{(0)}, J] + [[S^{(0)}, J], S^{(0)}]. \tag{134}$$

On writing equation (134) in full for the different matrix elements, we find that $S^{(0)}$ must be expressible in the form

$$\left(\frac{1}{\mu'} + \frac{1}{\mu}\right) S^{(0)}(\mu, \mu') = \begin{pmatrix} \psi(\mu) & 2^{\frac{1}{2}}\phi(\mu) \\ \chi(\mu) & 2^{\frac{1}{2}}\zeta(\mu) \end{pmatrix} \begin{pmatrix} \psi(\mu') & \chi(\mu') \\ 2^{\frac{1}{2}}\phi(\mu') & 2^{\frac{1}{2}}\zeta(\mu') \end{pmatrix}, \tag{135}$$

where

$$\psi(\mu) = \mu^2 + \frac{3}{8} \int\limits_0^1 \frac{d\mu'}{\mu'} [\mu'^2 S_{ll}^{(0)}(\mu, \mu') + S_{lr}^{(0)}(\mu, \mu')],$$

$$\phi(\mu) = 1 - \mu^2 + \frac{3}{8} \int\limits_0^1 \frac{d\mu'}{\mu'} (1 - \mu'^2) S_{ll}^{(0)}(\mu, \mu'),$$

$$\chi(\mu) = 1 + \frac{3}{8} \int\limits_0^1 \frac{d\mu'}{\mu'} [\mu'^2 S_{rl}^{(0)}(\mu, \mu') + S_{rr}^{(0)}(\mu, \mu')],$$

and

$$\zeta(\mu) = \frac{3}{8} \int\limits_0^1 \frac{d\mu'}{\mu'} (1 - \mu'^2) S_{rl}^{(0)}(\mu, \mu'). \tag{136}$$

70.1. The form of the solution for $S^{(0)}(\mu, \mu')$

By substituting for $S_{ll}^{(0)}$, etc., according to equation (135) in equations (136) we shall obtain a simultaneous system of integral equations of order four for ψ, ϕ, χ, and ζ. In solving this system of equations we shall be guided, as in earlier instances, by the form of the solutions obtained in the direct solution of the equations of transfer in the general nth approximation. In the present case the forms for $S_{ll}^{(0)}$, etc., suggested are

$$\left(\frac{1}{\mu_0}+\frac{1}{\mu}\right)S_{ll}^{(0)}(\mu,\mu_0) = 2H_l(\mu)H_l(\mu_0)[1-c(\mu+\mu_0)+\mu\mu_0],$$

$$\left(\frac{1}{\mu_0}+\frac{1}{\mu}\right)S_{lr}^{(0)}(\mu,\mu_0) = qH_l(\mu)H_r(\mu_0)(\mu+\mu_0),$$

$$\left(\frac{1}{\mu_0}+\frac{1}{\mu}\right)S_{rl}^{(0)}(\mu,\mu_0) = qH_r(\mu)H_l(\mu_0)(\mu+\mu_0),$$

and $\quad\left(\frac{1}{\mu_0}+\frac{1}{\mu}\right)S_{rr}^{(0)}(\mu,\mu_0) = H_r(\mu)H_r(\mu_0)[1+c(\mu+\mu_0)+\mu\mu_0],$ (137)

where q and c are constants and $H_l(\mu)$ and $H_r(\mu)$ are H-functions defined in terms of the characteristic functions

$$\Psi_l^*(\mu) = \tfrac{3}{4}(1-\mu^2) \quad \text{and} \quad \Psi_r^*(\mu) = \tfrac{3}{8}(1-\mu^2), \tag{138}$$

respectively.†

70.2. *Verification of the solution and the expression of the constants q and c in terms of the moments of $H_l(\mu)$ and $H_r(\mu)$*

The verification that the solution for $S^{(0)}$ has the form given by equations (137) will consist in first evaluating ψ, ϕ, etc., according to equations (136) and (137); then, requiring that when the resulting expressions for ψ, ϕ, etc., are substituted back into equation (135), we shall recover the form of the solution assumed; and finally showing that the various requirements can be met.

The evaluation of ψ, ϕ, χ, and ζ according to equations (136) and (137) is straightforward if proper use is made of the integral properties of the H-functions. For the functions, defined in terms of the characteristic functions (138), we have (Chap. V, Th. 1, eq. [7], and Th. 3, eqs. [27]–[29])

$$\tfrac{3}{4}\int_0^1 H_l(\mu)(1-\mu^2)\,d\mu = \tfrac{3}{4}(\alpha_0-\alpha_2) = 1, \tag{139}$$

$$\tfrac{3}{8}\int_0^1 H_r(\mu)(1-\mu^2)\,d\mu = 1-\frac{1}{\sqrt{2}}, \tag{140}$$

$$\alpha_0 = 1+\tfrac{3}{8}(\alpha_0^2-\alpha_1^2); \qquad A_0 = 1+\tfrac{3}{16}(A_0^2-A_1^2), \tag{141}$$

$$(1-\mu^2)\int_0^1 \frac{H_l(\mu')}{\mu+\mu'}\,d\mu' = \frac{H_l(\mu)-1}{\tfrac{3}{4}\mu H_l(\mu)}+(\alpha_1-\mu\alpha_0), \tag{142}$$

---

† Cf. S. Chandrasekhar, *Astrophys. J.* **105**, 151, 1947 (eq. [65]).

and
$$(1-\mu^2) \int_0^1 \frac{H_r(\mu')}{\mu+\mu'} d\mu' = \frac{H_r(\mu)-1}{\frac{3}{8}\mu H_r(\mu)} + (A_1-\mu A_0), \tag{143}$$

where α_n and A_n denote the moments of order n of $H_l(\mu)$ and $H_r(\mu)$, respectively.

Considering first $\phi(\mu)$, we have

$$\phi(\mu) = 1-\mu^2+\tfrac{3}{4}\mu H_l(\mu) \int_0^1 \frac{d\mu'}{\mu+\mu'} (1-\mu'^2)H_l(\mu')[1-c(\mu+\mu')+\mu\mu']$$

$$= 1-\mu^2+\tfrac{3}{4}\mu H_l(\mu) \int_0^1 d\mu'(1-\mu'^2)H_l(\mu')\left[\mu-c+\frac{1-\mu^2}{\mu+\mu'}\right]$$

$$= 1-\mu^2+\mu(\mu-c)H_l(\mu)+(1-\mu^2)[H_l(\mu)-1], \tag{144}$$

where, in the reductions, we have made use of the integral equation satisfied by $H_l(\mu)$ and the relation (139). Hence

$$\phi(\mu) = H_l(\mu)(1-c\mu). \tag{145}$$

Considering next $\phi+\psi$, we have (cf. eqs. [136] and [137])

$$\phi(\mu)+\psi(\mu)$$

$$= 1+\frac{3}{8} \int_0^1 \frac{d\mu'}{\mu'} [S_{ll}^{(0)}(\mu,\mu')+S_{lr}^{(0)}(\mu,\mu')].$$

$$= 1+\tfrac{3}{8}\mu H_l(\mu) \int_0^1 \frac{d\mu'}{\mu+\mu'} [2H_l(\mu')\{1-c(\mu+\mu')+\mu\mu'\}+qH_r(\mu')(\mu+\mu')]$$

$$= 1+\tfrac{3}{8}q\mu H_l(\mu)A_0+\tfrac{3}{4}\mu H_l(\mu) \int_0^1 d\mu' H_l(\mu')\left[\mu-c+\frac{1-\mu^2}{\mu+\mu'}\right]$$

$$= 1+\tfrac{3}{8}q\mu H_l(\mu)A_0+\tfrac{3}{4}\mu(\mu-c)H_l(\mu)\alpha_0+\tfrac{3}{4}\mu(1-\mu^2)H_l(\mu) \int_0^1 \frac{H_l(\mu')\,d\mu'}{\mu+\mu'}. \tag{146}$$

Using equation (142) in the last step of the preceding reduction, we find

$$\phi(\mu)+\psi(\mu) = H_l(\mu)+\mu H_l(\mu)[\tfrac{3}{8}qA_0+\tfrac{3}{4}(\alpha_1-\alpha_0 c)]. \tag{147}$$

According to equations (145) and (147), we can now write

$$\psi(\mu) = q'\mu H_l(\mu), \tag{148}$$

where
$$q' = \tfrac{3}{8}qA_0+\tfrac{3}{4}(\alpha_1-\alpha_0 c)+c. \tag{149}$$

Turning next to $\zeta(\mu)$ we have

$$\zeta(\mu) = \tfrac{3}{8}q\mu H_r(\mu) \int_0^1 H_l(\mu')(1-\mu'^2)\, d\mu', \tag{150}$$

or, using equation (139), we obtain

$$\zeta(\mu) = \tfrac{1}{2}q\mu H_r(\mu). \tag{151}$$

Finally, considering $\zeta(\mu)+\chi(\mu)$, we have

$$\zeta(\mu)+\chi(\mu)$$

$$= 1+\tfrac{3}{8} \int_0^1 \frac{d\mu'}{\mu'} \left[S_{rl}^{(0)}(\mu,\mu') + S_{rr}^{(0)}(\mu,\mu') \right]$$

$$= 1+\tfrac{3}{8}\mu H_r(\mu) \int_0^1 \frac{d\mu'}{\mu+\mu'} \left[qH_l(\mu')(\mu+\mu')+H_r(\mu')\{1+c(\mu+\mu')+\mu\mu'\} \right]$$

$$= 1+\tfrac{3}{8}q\mu H_r(\mu)\alpha_0+\tfrac{3}{8}\mu(\mu+c)H_r(\mu)A_0+\tfrac{3}{8}\mu(1-\mu^2)H_r(\mu)\int_0^1 \frac{H_r(\mu')}{\mu+\mu'}\, d\mu'. \tag{152}$$

Using equations (143) and (151), we find, after some minor reductions, that

$$\chi(\mu) = H_r(\mu)(1+c'\mu), \tag{153}$$

where

$$c' = \tfrac{3}{8}q\alpha_0+\tfrac{3}{8}(A_1+A_0 c)-\tfrac{1}{2}q. \tag{154}$$

Now substituting for ϕ, ψ, ζ, and χ according to equations (145), (148), (151), and (153) in equation (135) and comparing the resulting expressions for $S_{ll}^{(0)}$, etc., with those assumed (eqs. [137]), we find that we must have

$$q = q', \quad c = c', \quad \text{and} \quad q^2 = 2(1-c^2). \tag{155}$$

According to equations (149), (154), and (155), we must have

$$(3A_0-8)q-2(3\alpha_0-4)c+6\alpha_1 = 0$$

and

$$(3\alpha_0-4)q+(3A_0-8)c+3A_1 = 0. \tag{156}$$

Solving these equations, we find

$$q = -6\frac{\alpha_1(3A_0-8)+A_1(3\alpha_0-4)}{(3A_0-8)^2+2(3\alpha_0-4)^2}$$

and

$$c = +3\frac{2\alpha_1(3\alpha_0-4)-A_1(3A_0-8)}{(3A_0-8)^2+2(3\alpha_0-4)^2}. \tag{157}$$

It remains to prove that q and c as given by equations (157) are compatible with the remaining condition

$$q^2 = 2(1-c^2). \tag{158}$$

To prove that this is the case, we evaluate $2(1-c^2)-q^2$ according to equations (157) and find

$$2(1-c^2)-q^2 = 2\frac{[(3A_0-8)^2-9A_1^2]-2[9\alpha_1^2-(3\alpha_0-4)^2]}{(3A_0-8)^2+2(3\alpha_0-4)^2}. \tag{159}$$

On the other hand, using equation (141) we find

$$(3A_0-8)^2-9A_1^2 = 9(A_0^2-A_1^2)-48A_0+64 = 16$$

and

$$9\alpha_1^2-(3\alpha_0-4)^2 = 9(\alpha_1^2-\alpha_0^2)+24\alpha_0-16 = 8. \tag{160}$$

The right-hand side of equation (159) therefore vanishes and q and c are indeed related, as required.

This completes the verification.

It may be noted that, since we have incidentally shown that

$$(3A_0-8)^2+2(3\alpha_0-4)^2 = 9(A_1^2+2\alpha_1^2), \tag{161}$$

we can rewrite the expressions for q and c in the forms

$$q = 2\frac{4(A_1+2\alpha_1)-3(A_0\alpha_1+\alpha_0 A_1)}{3(A_1^2+2\alpha_1^2)}$$

and

$$c = \frac{8(A_1-\alpha_1)+3(2\alpha_1\alpha_0-A_1 A_0)}{3(A_1^2+2\alpha_1^2)}. \tag{162}\dagger$$

70.3. *The law of diffuse reflection*

Combining the results of the preceding paragraphs with the solutions for the azimuth dependent terms given in § 69, we can write the law of diffuse reflection in the form

$$\boldsymbol{I}(0,\mu,\varphi) = \begin{pmatrix} I_l \\ I_r \\ U \end{pmatrix} = \frac{1}{4\mu}Q\boldsymbol{S}(\mu,\varphi;\mu_0,\varphi_0)\begin{pmatrix} F_l \\ F_r \\ F_U \end{pmatrix}, \tag{163}$$

---

† It will be observed that the expressions for q and c we have derived in this section agree with those quoted in § 68.6 (eqs. [99]). It can be verified now that the conditions imposed on q and c in § 68.6 (eqs. [95], [97], and [98]) are all satisfied:

We have already seen that the first of these conditions, namely that $q^2 = 2(1-c^2)$, is satisfied. Considering the condition (97), we find on evaluating $q\alpha_1+A_2+cA_1$ according to equations (162) that

$$q\alpha_1+A_2+cA_1 = \tfrac{8}{3}+(A_2-A_0),$$

which on using equation (140) becomes identical with the relation to be established. Equation (98) can be similarly established by using the relations

$$\frac{1}{H_l(1)} = 1+\tfrac{3}{4}(\alpha_1-\alpha_0) \quad \text{and} \quad \frac{1}{H_r(1)} = 1+\tfrac{3}{8}(A_1-A_0),$$

which directly follow from equations (142) and (143) by setting $\mu = 1$.

where

$$\left(\frac{1}{\mu_0}+\frac{1}{\mu}\right)S(\mu,\varphi;\mu_0,\varphi_0) = \frac{3}{4}\begin{pmatrix} \psi(\mu) & 2^{\frac{1}{2}}\phi(\mu) & 0 \\ \chi(\mu) & 2^{\frac{1}{2}}\zeta(\mu) & 0 \\ 0 & 0 & 0 \end{pmatrix}\begin{pmatrix} \psi(\mu_0) & \chi(\mu_0) & 0 \\ 2^{\frac{1}{2}}\phi(\mu_0) & 2^{\frac{1}{2}}\zeta(\mu_0) & 0 \\ 0 & 0 & 0 \end{pmatrix} +$$

$$+\frac{3}{4}\begin{pmatrix} -4\mu\mu_0\cos(\varphi_0-\varphi) & 0 & 2\mu\sin(\varphi_0-\varphi) \\ 0 & 0 & 0 \\ 2\mu_0\sin(\varphi_0-\varphi) & 0 & \cos(\varphi_0-\varphi) \end{pmatrix} \times$$

$$\times (1-\mu^2)^{\frac{1}{2}}(1-\mu_0^2)^{\frac{1}{2}}H^{(1)}(\mu)H^{(1)}(\mu_0)+$$

$$+\frac{3}{4}\begin{pmatrix} \mu^2\mu_0^2\cos 2(\varphi_0-\varphi) & -\mu^2\cos 2(\varphi_0-\varphi) & -\mu^2\mu_0\sin 2(\varphi_0-\varphi) \\ -\mu_0^2\cos 2(\varphi_0-\varphi) & \cos 2(\varphi_0-\varphi) & \mu_0\sin 2(\varphi_0-\varphi) \\ -\mu\mu_0^2\sin 2(\varphi_0-\varphi) & \mu\sin 2(\varphi_0-\varphi) & -\mu\mu_0\cos 2(\varphi_0-\varphi) \end{pmatrix} \times$$

$$\times H^{(2)}(\mu)H^{(2)}(\mu_0), \quad (164)$$

$$\psi(\mu) = q\mu H_l(\mu); \qquad \phi(\mu) = H_l(\mu)(1-c\mu),$$

$$\chi(\mu) = H_r(\mu)(1+c\mu); \qquad \zeta(\mu) = \tfrac{1}{2}q\mu H_r(\mu), \quad (165)$$

and $H_l(\mu)$, $H_r(\mu)$, $H^{(1)}(\mu)$, and $H^{(2)}(\mu)$ are defined in terms of the characteristic functions

$$\tfrac{3}{4}(1-\mu^2), \quad \tfrac{3}{8}(1-\mu^2), \quad \tfrac{3}{8}(1-\mu^2)(1+2\mu^2) \quad \text{and} \quad \tfrac{3}{16}(1+\mu^2)^2, \quad (166)$$

respectively; further, the constants q and c are related to the moments α_n and A_n of $H_l(\mu)$ and $H_r(\mu)$ by

$$q = 2\frac{4(A_1+2\alpha_1)-3(A_0\alpha_1+\alpha_0 A_1)}{3(A_1^2+2\alpha_1^2)}$$

and

$$c = \frac{8(A_1-\alpha_1)+3(2\alpha_1\alpha_0-A_1 A_0)}{3(A_1^2+2\alpha_1^2)}. \quad (167)$$

In case a partially elliptically polarized beam is incident, we must consider in addition to I_l, I_r, and U the parameter V. This is, however, scattered independently of the others and the law of its reflection is given by (cf. eq. [128])

$$V(0,\mu,\varphi) = \frac{F_V}{4\mu}S_V(\mu,\varphi;\mu_0,\varphi_0) = \frac{3}{8(\mu_0+\mu)}\{-\mu\mu_0 H_v(\mu)H_v(\mu_0)+$$

$$+(1-\mu^2)^{\frac{1}{2}}(1-\mu_0^2)^{\frac{1}{2}}H_r(\mu)H_r(\mu_0)\cos(\varphi_0-\varphi)\}\mu_0 F_V, \quad (168)$$

where $H_r(\mu)$ is the same H-function as occurs in equations (165) and $H_v(\mu)$ is defined in terms of the characteristic function $\tfrac{3}{4}\mu^2$.

The H-functions occurring in the solutions (164) and (168) have been evaluated numerically by Mrs. Frances H. Breen and the writer by the method of Chapter V, § 41. The functions $H_l(\mu)$ and $H_r(\mu)$ have already been tabulated in § 68 (Table XXII). The remaining functions, $H_v(\mu)$,

$H^{(1)}(\mu)$, and $H^{(2)}(\mu)$, are given in Table XXV. This table also includes the functions $\psi(\mu)$, $\phi(\mu)$, $\chi(\mu)$, $\zeta(\mu)$, $(1-\mu^2)^{\frac{1}{2}}H^{(1)}(\mu)$ and $H^{(2)}(\mu)$.

TABLE XXV

The Functions ψ, ϕ, χ, ζ, $(1-\mu^2)^{\frac{1}{2}}H^{(1)}$, $H^{(2)}$, H_v, and $H^{(1)}$

μ	ψ	ϕ	χ	ζ	$(1-\mu^2)^{\frac{1}{2}}H^{(1)}$	$H^{(2)}$	H_v	$H^{(1)}$
0	0	1·00000	1·00000	0	1·00000	1·00000	1·00000	1·00000
0·05	0·04075	1·12988	1·10352	0·01824	1·07167	1·04967	1·02031	1·07301
0·10	0·09144	1·20976	1·18638	0·03764	1·11602	1·08621	1·03834	1·12164
0·15	0·15105	1·26850	1·26329	0·05780	1·14837	1·11762	1·05458	1·16151
0·20	0·21916	1·31108	1·33687	0·07852	1·17155	1·14552	1·06934	1·19571
0·25	0·29557	1·33973	1·40821	0·09969	1·18685	1·17075	1·08284	1·22577
0·30	0·38014	1·35569	1·47801	0·12121	1·19487	1·19383	1·09525	1·25256
0·35	0·47276	1·35971	1·54664	0·14303	1·19599	1·21508	1·10671	1·27674
0·40	0·57338	1·35228	1·61435	0·16510	1·19030	1·23476	1·11735	1·29872
0·45	0·68195	1·33374	1·68132	0·18738	1·17774	1·25308	1·12724	1·31882
0·50	0·79843	1·30437	1·74772	0·20984	1·15816	1·27019	1·13647	1·33733
0·55	0·92277	1·26432	1·81362	0·23247	1·13118	1·28624	1·14510	1·35444
0·60	1·05497	1·21375	1·87911	0·25523	1·09624	1·30132	1·15320	1·37030
0·65	1·19501	1·15279	1·94425	0·27812	1·05256	1·31554	1·16082	1·38507
0·70	1·34286	1·08153	2·00907	0·30112	0·99899	1·32895	1·16799	1·39886
0·75	1·49852	1·00003	2·07365	0·32421	0·93381	1·34166	1·17476	1·41179
0·80	1·66198	0·90836	2·13799	0·34739	0·85435	1·35371	1·18116	1·42392
0·85	1·83321	0·80655	2·20213	0·37065	0·75611	1·36515	1·18722	1·43533
0·90	2·01223	0·69468	2·26609	0·39398	0·63033	1·37601	1·19297	1·44608
0·95	2·19902	0·57276	2·32990	0·41738	0·45471	1·38638	1·19843	1·45625
1·00	2·39357	0·44083	2·39356	0·44083	0	1·39625	1·20362	1·46586

70.4. *The law of diffuse reflection of an incident beam of natural light*

In practical applications, greatest interest is attached to the diffuse reflection of an incident beam of natural light. In this case

$$F_l = F_r = \tfrac{1}{2}F \quad \text{and} \quad F_V = F_U = 0. \tag{169}$$

From the general solution given in § 70.3 we now obtain

$$I_l(0,\mu,\varphi) = \frac{3}{32(\mu+\mu_0)}\{\psi(\mu)[\psi(\mu_0)+\chi(\mu_0)]+2\phi(\mu)[\phi(\mu_0)+\zeta(\mu_0)]-$$
$$-4\mu\mu_0(1-\mu^2)^{\frac{1}{2}}(1-\mu_0^2)^{\frac{1}{2}}H^{(1)}(\mu)H^{(1)}(\mu_0)\cos(\varphi_0-\varphi)-$$
$$-\mu^2(1-\mu_0^2)H^{(2)}(\mu)H^{(2)}(\mu_0)\cos 2(\varphi_0-\varphi)\}\mu_0\,F,$$

$$I_r(0,\mu,\varphi) = \frac{3}{32(\mu+\mu_0)}\{\chi(\mu)[\psi(\mu_0)+\chi(\mu_0)]+2\zeta(\mu)[\phi(\mu_0)+\zeta(\mu_0)]+$$
$$+(1-\mu_0^2)H^{(2)}(\mu)H^{(2)}(\mu_0)\cos 2(\varphi_0-\varphi)\}\mu_0\,F,$$

and

$$U(0,\mu,\varphi) = \frac{3}{16(\mu+\mu_0)}\{2(1-\mu^2)^{\frac{1}{2}}(1-\mu_0^2)^{\frac{1}{2}}\mu_0\,H^{(1)}(\mu)H^{(1)}(\mu_0)\sin(\varphi_0-\varphi)+$$
$$+\mu(1-\mu_0^2)H^{(2)}(\mu)H^{(2)}(\mu_0)\sin 2(\varphi_0-\varphi)\}\mu_0\,F. \tag{170}$$

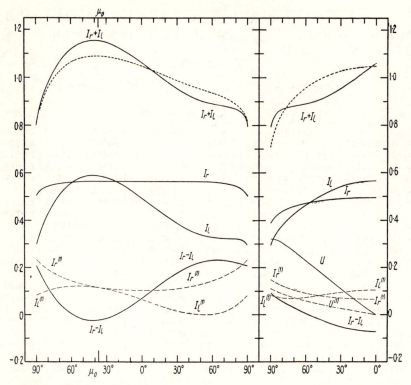

Fig. 24. The law of diffuse reflection by a semi-infinite atmosphere on Rayleigh scattering.

The ordinates represent the intensities in the unit $\mu_0 F$ and the abscissae the angle in degrees.

An angle of incidence corresponding to $\mu_0 = 0.8$ is considered and the variation of the reflected intensities in the planes $\varphi_0 - \varphi = 0°$ (the curves on the left side of the diagram) and $\varphi_0 - \varphi = 90°$ (the curves on the right side of the diagram) are illustrated. The intensities I_l and I_r (in directions parallel and perpendicular to the meridian plane containing the direction of the reflected light) the total intensity $I_l + I_r$ and the difference of intensities $I_r - I_l$ are all shown; and in the plane $\varphi_0 - \varphi = 90°$ the variation of U is also included.

The intensities $I_l^{(1)}$, $I_r^{(1)}$ and $U^{(1)}$ resulting from light which has suffered only a single scattering process in the atmosphere are also shown (the thin dashed curves).

The total intensity $I_l + I_r$ predicted by the present exact theory of Rayleigh scattering is compared with what would be expected (shown by the heavy dashed curves) on a theory which does not take into account the state of polarization of the radiation field but allows for an anisotropy of the scattered radiation according to Rayleigh's phase function.

The corresponding expressions for the intensities which represent light which has suffered a single scattering process in the atmosphere can be obtained from equations (170) by letting

$$\psi(\mu) \to \mu^2, \qquad \phi(\mu) \to 1 - \mu^2, \qquad \chi(\mu) \to 1, \qquad \zeta(\mu) \to 0,$$

$$H^{(1)}(\mu) \to 1 \quad \text{and} \quad H^{(2)}(\mu) \to 1. \tag{171}$$

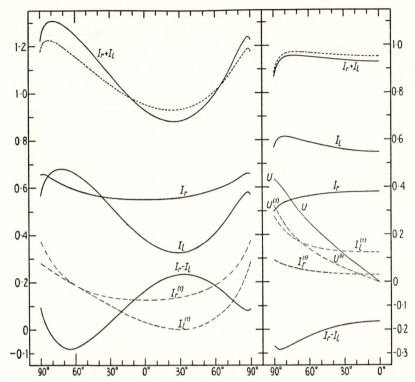

FIG. 25. Illustrates the same phenomenon as Fig. 24 but for an angle of incidence corresponding to $\mu_0 = 0.5$. (Notice that in this diagram—unlike the other diagrams—the scale of ordinates on the right side of the diagram is shifted relative to that on the left side.)

Thus,

$$I_l^{(1)}(0, \mu, \varphi) = \frac{3}{32(\mu+\mu_0)} \{\mu^2(1+\mu_0^2) + 2(1-\mu^2)(1-\mu_0^2) -$$

$$- 4\mu\mu_0(1-\mu^2)^{\frac{1}{2}}(1-\mu_0^2)^{\frac{1}{2}} \cos(\varphi_0-\varphi) - \mu^2(1-\mu_0^2)\cos 2(\varphi_0-\varphi)\}\mu_0 F,$$

$$I_r^{(1)}(0, \mu, \varphi) = \frac{3}{32(\mu+\mu_0)} \{1+\mu_0^2 + (1-\mu_0^2)\cos 2(\varphi_0-\varphi)\}\mu_0 F,$$

and

$$U^{(1)}(0, \mu, \varphi) = \frac{3}{16(\mu+\mu_0)} \{2(1-\mu^2)^{\frac{1}{2}}(1-\mu_0^2)^{\frac{1}{2}}\mu_0 \sin(\varphi_0-\varphi) +$$

$$+ \mu(1-\mu_0^2)\sin 2(\varphi_0-\varphi)\}\mu_0 F. \quad (172)$$

In Figs. 24, 25, and 26 we have illustrated the law of diffuse reflection as expressed by equations (170) for angles of incidence corresponding to $\mu_0 = 0.8$ (Fig. 24), $\mu_0 = 0.5$ (Fig. 25), and $\mu_0 = 0.2$ (Fig. 26). The variations of I_l, I_r, I_r+I_l, and I_r-I_l in the principal plane ($\varphi_0-\varphi = 0$ and π) containing the direction of incidence, and in the plane ($\varphi_0-\varphi = \pm\frac{1}{2}\pi$)

at right angles to the direction of incidence, are shown. In the principal plane $U \equiv 0$; this is clearly required by symmetry, since in the plane containing the direction of incidence the plane of polarization must be along the direction l or r. However, in the plane $\varphi_0 - \varphi = \pm \frac{1}{2}\pi$, $U \neq 0$ and the variation of U is also shown. The intensities of light which has

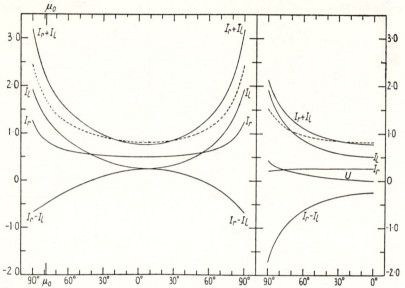

Fig. 26. Illustrates the same phenomenon as Figs. 24 and 25 but for an angle of incidence corresponding to $\mu_0 = 0.2$. (The results for single scattering are, however, not included in this diagram.)

suffered only a single scattering in the atmosphere are shown in Figs. 24 and 25.

From Figs. 24, 25, and 26 it is seen that I_r shows very much less dependence on angle than I_l. This is particularly true in the meridian plane, for moderate angles of incidence: for, while I_l shows a strong variation, I_r is nearly independent of angle. These results are physically understandable since light polarized at right angles to the plane of scattering is isotropically scattered, while light polarized parallel to the plane of scattering is scattered in accordance with the phase function $3 \cos^2 \Theta$. It will also be noticed that in the principal plane, contrary to what would be expected on single scattering, there is a reversal of the sign of polarization, the polarization vanishing at two points. This vanishing of the polarization at two points is related to the phenomenon of the 'neutral points' exhibited in the polarization of the sky (cf. § 73).

A further comparison of interest is that of the total intensities given by the present exact treatment of Rayleigh scattering and that given on the assumption of scattering according to Rayleigh's phase-function (Chap. VI, §§ 44 and 47). This comparison is made in Figs. 24, 25, and 26. It is seen that the differences between the two curves are quite large; this emphasizes the importance of treating scattering, correctly, as a linear transformation of the Stokes parameters of the incident light.

71. The law of diffuse reflection and transmission for Rayleigh scattering

We have already given in § 69 the solutions for the azimuth dependent terms in the scattering and the transmission matrices (eqs. [122] and [123]). It remains to specify the azimuth independent terms $I_l^{(0)}(0, \mu)$, $I_r^{(0)}(0, \mu)$, $I_l^{(0)}(\tau_1, -\mu)$, and $I_r^{(0)}(\tau_1, -\mu)$. Expressing these intensities in terms of two matrices $S^{(0)}$ and $T^{(0)}$ (with two rows and columns) in the forms

$$\begin{pmatrix} I_l^{(0)}(0, \mu) \\ I_r^{(0)}(0, \mu) \end{pmatrix} = \frac{3}{16\mu} S^{(0)}(\mu, \mu_0) \begin{pmatrix} F_l \\ F_r \end{pmatrix}; \qquad \begin{pmatrix} I_l^{(0)}(\tau_1, -\mu) \\ I_r^{(0)}(\tau_1, -\mu) \end{pmatrix} = \frac{3}{16\mu} T^{(0)}(\mu, \mu_0) \begin{pmatrix} F_l \\ F_r \end{pmatrix},$$
(173)

we have, in the product notation of § 70 (eq. [133]), the following analogues of Chapter VII, equations (29)–(32):

$$\left(\frac{1}{\mu_0}+\frac{1}{\mu}\right)S^{(0)} + \frac{\partial S^{(0)}}{\partial \tau_1} = J + [J, S^{(0)}] + [S^{(0)}, J] + [[S^{(0)}, J], S^{(0)}],$$

$$\frac{\partial S^{(0)}}{\partial \tau_1} = \exp\left\{-\tau_1\left(\frac{1}{\mu}+\frac{1}{\mu_0}\right)\right\}J + e^{-\tau_1/\mu}[J, T^{(0)}] + e^{-\tau_1/\mu_0}[T^{(0)}, J] +$$
$$+ [[T^{(0)}, J], T^{(0)}],$$

$$\frac{1}{\mu_0} T^{(0)} + \frac{\partial T^{(0)}}{\partial \tau_1} = e^{-\tau_1/\mu}J + e^{-\tau_1/\mu}[J, S^{(0)}] + [T^{(0)}, J] + [[T^{(0)}, J], S^{(0)}]$$

and

$$\frac{1}{\mu} T^{(0)} + \frac{\partial T^{(0)}}{\partial \tau_1} = e^{-\tau_1/\mu_0}J + [J, T^{(0)}] + e^{-\tau_1/\mu_0}[S^{(0)}, J] + [[S^{(0)}, J], T^{(0)}]. \quad (174)$$

A discussion of these equations shows that $S^{(0)}$ and $T^{(0)}$ must be expressible in the forms

$$\left(\frac{1}{\mu'}+\frac{1}{\mu}\right)S^{(0)}(\mu, \mu') = \begin{pmatrix} \psi(\mu) & 2^{\frac{1}{2}}\phi(\mu) \\ \chi(\mu) & 2^{\frac{1}{2}}\zeta(\mu) \end{pmatrix} \begin{pmatrix} \psi(\mu') & \chi(\mu') \\ 2^{\frac{1}{2}}\phi(\mu') & 2^{\frac{1}{2}}\zeta(\mu') \end{pmatrix} -$$
$$- \begin{pmatrix} \xi(\mu) & 2^{\frac{1}{2}}\eta(\mu) \\ \sigma(\mu) & 2^{\frac{1}{2}}\theta(\mu) \end{pmatrix} \begin{pmatrix} \xi(\mu') & \sigma(\mu') \\ 2^{\frac{1}{2}}\eta(\mu') & 2^{\frac{1}{2}}\theta(\mu') \end{pmatrix},$$

and

$$\left(\frac{1}{\mu'}-\frac{1}{\mu}\right)T^{(0)}(\mu,\mu')\begin{pmatrix}\xi(\mu) & 2^{\frac{1}{2}}\eta(\mu)\\ \sigma(\mu) & 2^{\frac{1}{2}}\theta(\mu)\end{pmatrix}\begin{pmatrix}\psi(\mu') & \chi(\mu')\\ 2^{\frac{1}{2}}\phi(\mu') & 2^{\frac{1}{2}}\zeta(\mu')\end{pmatrix}-$$
$$-\begin{pmatrix}\psi(\mu) & 2^{\frac{1}{2}}\phi(\mu)\\ \chi(\mu) & 2^{\frac{1}{2}}\zeta(\mu)\end{pmatrix}\begin{pmatrix}\xi(\mu') & \sigma(\mu')\\ 2^{\frac{1}{2}}\eta(\mu') & 2^{\frac{1}{2}}\theta(\mu')\end{pmatrix}, \quad (175)$$

where

$$\psi(\mu) = \mu^2+\frac{3}{8}\int_0^1\frac{d\mu'}{\mu'}[\mu'^2 S_{ll}^{(0)}(\mu,\mu')+S_{lr}^{(0)}(\mu,\mu')],$$

$$\phi(\mu) = 1-\mu^2+\frac{3}{8}\int_0^1\frac{d\mu'}{\mu'}(1-\mu'^2)S_{ll}^{(0)}(\mu,\mu'),$$

$$\chi(\mu) = 1+\frac{3}{8}\int_0^1\frac{d\mu'}{\mu'}[\mu'^2 S_{rl}^{(0)}(\mu,\mu')+S_{rr}^{(0)}(\mu,\mu')],$$

$$\zeta(\mu) = \frac{3}{8}\int_0^1\frac{d\mu'}{\mu'}(1-\mu'^2)S_{rl}^{(0)}(\mu,\mu'),$$

$$\xi(\mu) = \mu^2 e^{-\tau_1/\mu}+\frac{3}{8}\int_0^1\frac{d\mu'}{\mu'}[\mu'^2 T_{ll}^{(0)}(\mu,\mu')+T_{lr}^{(0)}(\mu,\mu')],$$

$$\eta(\mu) = (1-\mu^2)e^{-\tau_1/\mu}+\frac{3}{8}\int_0^1\frac{d\mu'}{\mu'}(1-\mu'^2)T_{ll}^{(0)}(\mu,\mu'),$$

$$\sigma(\mu) = e^{-\tau_1/\mu}+\frac{3}{8}\int_0^1\frac{d\mu'}{\mu'}[\mu'^2 T_{rl}^{(0)}(\mu,\mu')+T_{rr}^{(0)}(\mu,\mu')],$$

and
$$\theta(\mu) = \frac{3}{8}\int_0^1\frac{d\mu'}{\mu'}(1-\mu'^2)T_{rl}^{(0)}(\mu,\mu'). \quad (176)$$

Substituting for $S_{ll}^{(0)}$, etc., according to equations (175) in equations (176), we shall obtain a simultaneous system of integral equations of order eight. In solving this system of integral equations we shall be guided, as in past instances, by the form of the solutions obtained in the direct solution of the equations of transfer in the general nth approximation. In this manner† we are led to assume the following

† Cf. S. Chandrasekhar, *Astrophys. J.* **106**, 152, 1947 (eqs. [539]–[546]).

forms for $S^{(0)}$ and $T^{(0)}$:

$$\left(\frac{1}{\mu'}+\frac{1}{\mu}\right)S_{ll}^{(0)}(\mu,\mu') = 2\{X_l(\mu)X_l(\mu')[1+\nu_4(\mu+\mu')+\mu\mu']-$$
$$-Y_l(\mu)Y_l(\mu')[1-\nu_4(\mu+\mu')+\mu\mu']-$$
$$-\nu_3(\mu+\mu')[X_l(\mu)Y_l(\mu')+Y_l(\mu)X_l(\mu')]\},$$

$$\left(\frac{1}{\mu'}+\frac{1}{\mu}\right)S_{lr}^{(0)}(\mu,\mu') = (\mu+\mu')\{\nu_1[Y_l(\mu)X_r(\mu')+X_l(\mu)Y_r(\mu')]-$$
$$-\nu_2[X_l(\mu)X_r(\mu')+Y_l(\mu)Y_r(\mu')]+$$
$$+Q(\nu_2-\nu_1)\mu'[X_l(\mu)+Y_l(\mu)][X_r(\mu')-Y_r(\mu')]\},$$

$$\left(\frac{1}{\mu'}+\frac{1}{\mu}\right)S_{rl}^{(0)}(\mu,\mu') = (\mu+\mu')\{\nu_1[X_r(\mu)Y_l(\mu')+Y_r(\mu)X_l(\mu')]-$$
$$-\nu_2[X_r(\mu)X_l(\mu')+Y_r(\mu)Y_l(\mu')]+$$
$$+Q(\nu_2-\nu_1)\mu[X_r(\mu)-Y_r(\mu)][X_l(\mu')+Y_l(\mu')]\},$$

$$\left(\frac{1}{\mu'}+\frac{1}{\mu}\right)S_{rr}^{(0)}(\mu,\mu') = X_r(\mu)X_r(\mu')[1-u_4(\mu+\mu')+u_5\mu\mu']-$$
$$-Y_r(\mu)Y_r(\mu')[1+u_4(\mu+\mu')+u_5\mu\mu']+$$
$$+u_3(\mu+\mu')[X_r(\mu)Y_r(\mu')+Y_r(\mu)X_r(\mu')]-$$
$$-Qu_5\mu\mu'(\mu+\mu')[X_r(\mu)-Y_r(\mu)][X_r(\mu')-Y_r(\mu')]+$$
$$+Q(u_4-u_3)\{\mu^2[X_r(\mu)-Y_r(\mu)][X_r(\mu')+Y_r(\mu')]+$$
$$+\mu'^2[X_r(\mu)+Y_r(\mu)][X_r(\mu')-Y_r(\mu')]\}\}; \qquad (177)$$

$$\left(\frac{1}{\mu'}-\frac{1}{\mu}\right)T_{ll}^{(0)}(\mu,\mu') = 2\{Y_l(\mu)X_l(\mu')[1-\nu_4(\mu-\mu')-\mu\mu']-$$
$$-X_l(\mu)Y_l(\mu')[1+\nu_4(\mu-\mu')-\mu\mu']+$$
$$+\nu_3(\mu-\mu')[X_l(\mu)X_l(\mu')+Y_l(\mu)Y_l(\mu')]\},$$

$$\left(\frac{1}{\mu'}-\frac{1}{\mu}\right)T_{lr}^{(0)}(\mu,\mu') = (\mu-\mu')\{\nu_2[X_l(\mu)Y_r(\mu')+Y_l(\mu)X_r(\mu')]-$$
$$-\nu_1[X_l(\mu)X_r(\mu')+Y_l(\mu)Y_r(\mu')]-$$
$$-Q(\nu_2-\nu_1)\mu'[X_l(\mu)+Y_l(\mu)][X_r(\mu')-Y_r(\mu')]\},$$

$$\left(\frac{1}{\mu'}-\frac{1}{\mu}\right)T_{rl}^{(0)}(\mu,\mu') = (\mu-\mu')\{\nu_2[X_r(\mu)Y_l(\mu')+Y_r(\mu)X_l(\mu')]-$$
$$-\nu_1[X_r(\mu)X_l(\mu')+Y_r(\mu)Y_l(\mu')]-$$
$$-Q(\nu_2-\nu_1)\mu[X_r(\mu)-Y_r(\mu)][X_l(\mu')+Y_l(\mu')]\},$$

and

$$\left(\frac{1}{\mu'}-\frac{1}{\mu}\right)T^{(0)}_{rr}(\mu,\mu') = Y_r(\mu)X_r(\mu')[1+u_4(\mu-\mu')-u_5\,\mu\mu']-$$
$$-X_r(\mu)Y_r(\mu')[1-u_4(\mu-\mu')-u_5\,\mu\mu']-$$
$$-u_3(\mu-\mu')[X_r(\mu)X_r(\mu')+Y_r(\mu)Y_r(\mu')]+$$
$$+Qu_5\,\mu\mu'(\mu-\mu')[X_r(\mu)-Y_r(\mu)][X_r(\mu')-Y_r(\mu')]-$$
$$-Q(u_4-u_3)\{\mu^2[X_r(\mu)-Y_r(\mu)][X_r(\mu')+Y_r(\mu')]-$$
$$-\mu'^2[X_r(\mu)+Y_r(\mu)][X_r(\mu')-Y_r(\mu')]\}, \tag{178}$$

where ν_1, ν_2, ν_3, ν_4, u_3, u_4, and Q are constants;

$$u_5 = 1+2Q(u_4-u_3); \tag{179}$$

$X_r(\mu)$ and $Y_r(\mu)$ are defined in terms of the characteristic function

$$\Psi_r(\mu) = \tfrac{3}{8}(1-\mu^2); \tag{180}$$

and $X_l(\mu)$ and $Y_l(\mu)$ are the *standard solutions* of the X- and Y-equations for the conservative case

$$\Psi_l(\mu) = \tfrac{3}{4}(1-\mu^2), \tag{181}$$

having the property

$$\tfrac{3}{4}\int_0^1 (1-\mu^2)X_l(\mu)\,d\mu = \tfrac{3}{4}(\alpha_0-\alpha_2) = 1,$$

and

$$\int_0^1 (1-\mu^2)Y_l(\mu)\,d\mu = \beta_0-\beta_2 = 0. \tag{182}$$

The verification that the solutions for $S^{(0)}$ and $T^{(0)}$ have the forms given in equations (177) and (178) will proceed as in the other cases we have considered.

First, we must evaluate ψ, ϕ, etc., according to equations (176) for $S^{(0)}$ and $T^{(0)}$ given by equations (177) and (178); then require that when the resulting expressions for ψ, ϕ, etc., are substituted back into equations (175), we shall recover the form of the solution assumed. This procedure will lead to several conditions† among the constants ν_1, ν_2, ν_3, ν_4, u_3, u_4, and Q introduced in the solution. It is found that all these conditions can be met and that six of the constants (ν_1, ν_2, ν_3, ν_4, u_3, and u_4) can be expressed, uniquely in terms of Q and the various moments α_n, β_n, A_n, and B_n of X_l, Y_l, X_r, and Y_r. The constant Q itself is found to be left arbitrary. This is a further example of the one-parameter nature of the solutions of the integral equations incorporating the invariances of the problem, in conservative cases. However, as

† Actually twelve are found.

in the other cases of conservative scattering we have considered (Chap. IX, §§ 62.2 and 64), the arbitrariness can be removed by appealing to the K-integral which the problem admits. In this manner we find that†

$$\psi(\mu) = \mu[\nu_1 Y_l(\mu) - \nu_2 X_l(\mu)]; \qquad \xi(\mu) = \mu[\nu_2 Y_l(\mu) - \nu_1 X_l(\mu)],$$

$$\phi(\mu) = (1 + \nu_4 \mu) X_l(\mu) - \nu_3 \mu Y_l(\mu); \qquad \eta(\mu) = (1 - \nu_4 \mu) Y_l(\mu) + \nu_3 \mu X_l(\mu),$$

$$\chi(\mu) = (1 - u_4 \mu) X_r(\mu) + u_3 \mu Y_r(\mu) + Q(u_4 - u_3)\mu^2 [X_r(\mu) - Y_r(\mu)],$$

$$\sigma(\mu) = (1 + u_4 \mu) Y_r(\mu) - u_3 \mu X_r(\mu) - Q(u_4 - u_3)\mu^2 [X_r(\mu) - Y_r(\mu)],$$

$$\zeta(\mu) = \tfrac{1}{2}\mu[\nu_1 Y_r(\mu) - \nu_2 X_r(\mu)] + \tfrac{1}{2}Q(\nu_2 - \nu_1)\mu^2 [X_r(\mu) - Y_r(\mu)],$$

$$\theta(\mu) = \tfrac{1}{2}\mu[\nu_2 Y_r(\mu) - \nu_1 X_r(\mu)] - \tfrac{1}{2}Q(\nu_2 - \nu_1)\mu^2 [X_r(\mu) - Y_r(\mu)], \qquad (183)$$

where the constants ν_1, ν_2, ν_3, ν_4, u_3, u_4, and Q are to be determined by the following formulae:

$$\nu_2 + \nu_1 = 2\Delta_1(\kappa_1 \delta_1 - \kappa_2 \delta_2); \qquad \nu_2 - \nu_1 = 2\Delta_2(\kappa_1 \delta_1 - \kappa_2 \delta_2),$$

$$\nu_4 + \nu_3 = \Delta_1(d_1 \kappa_1 - d_0 \kappa_2); \qquad \nu_4 - \nu_3 = \Delta_2[c_1 \delta_1 - c_0 \delta_2 - 2Q(d_0 \delta_1 - d_1 \delta_2)],$$

$$u_4 + u_3 = \Delta_1(c_1 \delta_1 - c_0 \delta_2); \quad u_4 - u_3 = \Delta_2(d_1 \kappa_1 - d_0 \kappa_2); \quad u_5 = \Delta_2(c_0 \kappa_1 - c_1 \kappa_2),$$

$$\frac{1}{\Delta_1} = d_0 \delta_1 - d_1 \delta_2; \qquad \frac{1}{\Delta_2} = c_0 \kappa_1 - c_1 \kappa_2 - 2Q(d_1 \kappa_1 - d_0 \kappa_2), \qquad (184)$$

$$Q = \frac{c_0 - c_2}{(d_0 - d_2)\tau_1 + 2(d_1 - d_3)}; \qquad (185)$$

$$c_0 = A_0 + B_0 - \tfrac{8}{3}; \qquad d_0 = A_0 - B_0 - \tfrac{8}{3},$$

$$c_n = A_n + B_n; \qquad d_n = A_n - B_n; \qquad \kappa_n = \alpha_n + \beta_n; \qquad \delta_n = \alpha_n - \beta_n$$

$$(n = 1, 2, 3, \ldots); \qquad (186)$$

α_n, β_n, A_n, and B_n are the moments of order n of X_l, Y_l, X_r, and Y_r, respectively.

The foregoing solutions for $S^{(0)}$ and $T^{(0)}$, combined with the solutions for $S^{(1)}$, $T^{(1)}$, $S^{(2)}$, $T^{(2)}$, S_V, and T_V given in § 69 (eqs. [123] and [126]) in accordance with equations (122), provides the complete law of diffuse reflection and transmission on Rayleigh scattering.

72. The fundamental problem in the theory of scattering by planetary atmospheres and its solution in terms of the standard problem of diffuse reflection and transmission

In our discussion of the problem of diffuse reflection and transmission we have, so far, restricted ourselves to the *standard problem*, i.e. one in which a parallel beam is incident on a plane-parallel atmosphere, at $\tau = 0$, and no radiation is incident on the surface $\tau = \tau_1$ from below.

---

† The details of the analysis are, unfortunately, too long to be given here (see S. Chandrasekhar, *Astrophys. J.* **107**, 188, 1948, Section V).

The boundary condition $I(\tau_1, +\mu, \varphi) \equiv 0$ $(0 \leqslant \mu \leqslant 1)$ which we have used in the solution of the equation of transfer is, of course, equivalent to this assumption of the non-existence of any radiation incident on the atmosphere from below. However, in the important context of the illumination of planetary atmospheres by the sun, this last assumption is not realized: the atmosphere rests on solid ground, ocean, or cloud bank. The presence of the ground at $\tau = \tau_1$ will modify the law of diffuse reflection; also, the intensity as measured on the ground will differ from the law of diffuse transmission in the standard problem. The solution of this more general problem when there is a plane surface with known reflecting properties, at the bottom of the atmosphere, can be reduced to that of the standard problem. While there is no formal difficulty in carrying out this reduction for any given law of reflection of the ground, in view of practical difficulties of specifying, in given cases, the reflecting properties of the ground accurately, it should suffice, for most purposes, to idealize the 'ground' as a surface which reflects according to Lambert's law (cf. Chap. VI, § 47, p. 147) with some albedo λ_0; i.e. we shall suppose that the light reflected by the ground is unpolarized and uniform in the outward hemisphere, independently of the state of polarization and the angular distribution of the incident light, and that, further, the outward normal flux of the reflected light is *always* a certain fraction λ_0 of the inward normal flux of the radiation incident on the surface. The fundamental problem in the theory of planetary illumination and of the illumination of the sky can therefore be formulated as follows:

A parallel beam of light of net flux πF (or $\pi \mathbf{F}$ when polarization is taken into account) per unit area normal to itself is incident on a plane-parallel atmosphere in some specified direction $(-\mu_0, \varphi_0)$. The atmosphere is of optical thickness τ_1. At τ_1 there is a 'ground' which reflects according to Lambert's law with an albedo λ_0. It is required to find the state of polarization and the angular distribution of the radiation diffusely reflected from the surface $\tau = 0$ and also to specify the illumination and polarization of the 'sky' as seen by an observer at $\tau = \tau_1$.

We shall refer to the foregoing as the *planetary problem*.

72.1. *The reduction to the standard problem in case of scattering according to a phase function*

To illustrate the manner of the reduction of the planetary problem to the standard problem, we shall consider, first, the case of scattering according to a phase function. The case of scattering according to a phase-matrix will be considered in § 72.3.

Let the laws of diffuse reflection and transmission for the standard problem be given, as usual, in terms of a scattering and a transmission function in the form

$$I(0,\mu,\varphi) = \frac{F}{4\mu} S(\mu,\varphi;\mu_0,\varphi_0) \quad \text{and} \quad I(\tau_1,-\mu,\varphi) = \frac{F}{4\mu} T(\mu,\varphi;\mu_0,\varphi_0).$$

$$(187)\dagger$$

Let $S^{(0)}(\mu,\mu_0)$ and $T^{(0)}(\mu,\mu_0)$ denote the azimuth independent terms in S and T:

$$S^{(0)}(\mu,\mu_0) = \frac{1}{2\pi} \int_0^{2\pi} S(\mu,\varphi;\mu_0,\varphi_0)\, d\varphi_0,$$

and

$$T^{(0)}(\mu,\mu_0) = \frac{1}{2\pi} \int_0^{2\pi} T(\mu,\varphi;\mu_0,\varphi_0)\, d\varphi_0. \tag{188}$$

Further, let

$$s(\mu) = \tfrac{1}{2} \int_0^1 S^{(0)}(\mu,\mu')\, d\mu' = \tfrac{1}{2} \int_0^1 S^{(0)}(\mu',\mu)\, d\mu',$$

$$t(\mu) = \tfrac{1}{2} \int_0^1 T^{(0)}(\mu,\mu')\, d\mu' = \tfrac{1}{2} \int_0^1 T^{(0)}(\mu',\mu)\, d\mu', \tag{189}$$

$$\bar{s} = 2 \int_0^1 s(\mu)\, d\mu; \qquad \bar{t} = 2 \int_0^1 t(\mu)\, d\mu, \tag{190}$$

and

$$\gamma_1(\mu) = e^{-\tau_1/\mu} + \frac{t(\mu)}{\mu}. \tag{191}$$

The alternative forms for $s(\mu)$ and $t(\mu)$ in equations (189) arise from the principle of reciprocity.

To distinguish the solution of the planetary problem from that of the standard problem we shall put an asterisk to all quantities referring to the planetary problem.

Since the ground is supposed to reflect according to Lambert's law, it is clear that at $\tau = \tau_1$ the intensity will be the same in all *outward* directions. Let I_g denote this constant intensity in the outward hemisphere. The precise value of I_g will, of course, depend on circumstances, and has to be determined from the conditions of the problem; but once I_g has been found, the determination of $I^*(0,\mu,\varphi)$ and $I^*(\tau_1,-\mu,\varphi)$ is immediate: For, considering the emergent intensity $I^*(0,\mu,\varphi)$, we

† For the sake of convenience, we have omitted indicating the variable τ_1 in S and T.

observe that it must equal the intensity $I(0, \mu, \varphi)$ diffusely reflected in the direction (μ, φ), in the absence of the ground together with the intensity transmitted by the atmosphere (also, under conditions of the standard problem) of the radiation I_g $(0 \leqslant \mu' \leqslant 1, \ 0 \leqslant \varphi' \leqslant 2\pi)$ incident on the surface $\tau = \tau_1$. Thus

$$I^*(0, \mu, \varphi) = I(0, \mu, \varphi) + \frac{1}{4\pi\mu} \int_0^1 \int_0^{2\pi} T(\mu, \varphi; \mu', \varphi') I_g \, d\mu' d\varphi' + I_g e^{-\tau_1/\mu},$$

$$(192)$$

where the last term on the right-hand side represents the transmission, without scattering, of the intensity I_g, already, in the direction (μ, φ). According to equations (187)–(189), we can rewrite equation (192) in the form

$$I^*(0, \mu, \varphi) = \frac{F}{4\mu} S(\mu, \varphi; \mu_0, \varphi_0) + I_g \left\{ \frac{t(\mu)}{\mu} + e^{-\tau_1/\mu} \right\},$$

or (cf. eq. [191])

$$I^*(0, \mu, \varphi) = \frac{F}{4\mu} S(\mu, \varphi; \mu_0, \varphi_0) + I_g \gamma_1(\mu). \tag{193}$$

Now the isotropic radiation field, I_g, incident on $\tau = \tau_1$, will also be reflected by the atmosphere; this will contribute to the intensity of the 'sky' in the direction $(-\mu, \varphi)$, the additional amount

$$I_g^{(\text{ref})}(-\mu) = \frac{1}{4\pi\mu} \int_0^1 \int_0^{2\pi} S(\mu, \varphi; \mu', \varphi') I_g \, d\mu' d\varphi',$$

or (cf. eq. [189]) $$I_g^{(\text{ref})}(-\mu) = I_g \frac{s(\mu)}{\mu}; \tag{194}$$

combining this with the intensity $I(\tau_1, -\mu, \varphi)$ diffusely transmitted in the direction $(-\mu, \varphi)$ from the incident beam πF, we have

$$I^*(\tau_1, -\mu, \varphi) = \frac{F}{4\mu} T(\mu, \varphi; \mu_0, \varphi_0) + I_g \frac{s(\mu)}{\mu}. \tag{195}$$

It remains to determine I_g. For this purpose we shall make use of the condition

$$\pi I_g = \text{outward normal flux} = \lambda_0 \times \text{inward normal flux}. \tag{196}$$

The flux of radiation, directed inward at $\tau = \tau_1$, consists of three parts: first, the reduced incident flux

$$\pi\mu_0 \, F e^{-\tau_1/\mu_0};$$

second, the flux

$$\int_0^1 \int_0^{2\pi} \frac{F}{4\mu} T(\mu, \varphi; \mu_0, \varphi_0) \mu \, d\mu d\varphi = \pi F \tfrac{1}{2} \int_0^1 T^{(0)}(\mu, \mu_0) \, d\mu$$

$$= \pi F t(\mu_0), \tag{197}$$

of the diffusely transmitted intensity $I(\tau_1, -\mu, \varphi)$; and finally, the flux of the radiation $I_g^{(\mathrm{ref})}(-\mu)$ which comes from the reflection of I_g by the atmosphere; this last provides the additional flux (cf. eqs. [190] and [194])

$$\int_0^1 \int_0^{2\pi} I_g^{(\mathrm{ref})}(-\mu) \mu \, d\mu d\varphi = 2\pi I_g \int_0^1 s(\mu) \, d\mu = \pi I_g \bar{s}. \tag{198}$$

Hence the total normal inward flux at τ_1 is

$$\pi\left[\mu_0 F\left\{e^{-\tau_1/\mu_0} + \frac{t(\mu_0)}{\mu_0}\right\} + I_g \bar{s}\right] = \pi[\mu_0 F \gamma_1(\mu_0) + I_g \bar{s}]; \tag{199}$$

and λ_0 times this must equal πI_g. Therefore

$$I_g = \lambda_0[\mu_0 F \gamma_1(\mu_0) + I_g \bar{s}],$$

or
$$I_g = \frac{\lambda_0}{1 - \lambda_0 \bar{s}} \mu_0 F \gamma_1(\mu_0). \tag{200}$$

Substituting this value of I_g in equations (193) and (195), we have

$$I^*(0, \mu, \varphi) = \frac{F}{4\mu}\left\{S(\mu, \varphi; \mu_0, \varphi_0) + \frac{4\lambda_0}{1 - \lambda_0 \bar{s}} \mu\mu_0 \gamma_1(\mu_0)\gamma_1(\mu)\right\}, \tag{201}$$

and $\quad I^*(\tau_1, -\mu, \varphi) = \frac{F}{4\mu}\left\{T(\mu, \varphi; \mu_0, \varphi_0) + \frac{4\lambda_0}{1 - \lambda_0 \bar{s}} \mu\mu_0 \gamma_1(\mu_0)\frac{s(\mu)}{\mu}\right\}. \tag{202}$

It will be noticed that $I^*(0, \mu, \varphi)$ given by equation (201) has the symmetry required by the principle of reciprocity; but $I^*(\tau_1, -\mu, \varphi)$ does not have this symmetry and this is not required, either.

In conservative cases $\gamma_1(\mu)$ has another meaning which is of some importance:

In conservative cases the (standard) problem of diffuse reflection and transmission admits the flux integral (cf. Chap. IX, eqs. [29] and [95])

$$\pi F(\tau) = \pi\mu_0 F[e^{-\tau/\mu_0} - \gamma_1(\mu_0)]. \tag{203}$$

The constant γ_1 in this integral (which is a function of μ_0) is the same as the function $\gamma_1(\mu)$ we have introduced; for, according to equation (203), the *outward* normal flux in the *diffuse* radiation, at $\tau = \tau_1$, is

$$\pi F(\tau_1) = \pi\mu_0 F[e^{-\tau_1/\mu_0} - \gamma_1(\mu_0)]. \tag{204}$$

Comparing this with equation (197) and remembering that this equation (contrary to eq. [204]) refers the flux to the inward normal, we have

$$-t(\mu_0) = \mu_0[e^{-\tau_1/\mu_0} - \gamma_1(\mu_0)], \tag{205}$$

and this agrees with our earlier definition of γ_1 (cf. eq. [191]).

Since the flux of the diffusely reflected light is given by

$$\int_0^1 \int_0^{2\pi} \frac{F}{4\mu} S(\mu, \varphi; \mu_0, \varphi_0)\mu \, d\mu d\varphi = \pi F s(\mu_0), \tag{206}$$

it follows from equation (203), for $\tau = 0$, that

$$s(\mu) = \mu[1 - \gamma_1(\mu)]. \tag{207}$$

From this last equation we have

$$\bar{s} = 1 - 2 \int_0^1 \mu\gamma_1(\mu) \, d\mu. \tag{208}$$

72.2. *Illustrations of the formulae of § 72.1*

As illustrations of the formulae of § 72.1, we shall consider the various cases of scattering for which the standard problem has been solved.

(i) *Isotropic scattering with an albedo* $\varpi_0 < 1$. In this case the scattering and the transmission functions are (Chap. IX, eqs. [4] and [5])

$$S(\mu, \mu_0) = \varpi_0 \frac{\mu\mu_0}{\mu+\mu_0} [X(\mu)X(\mu_0) - Y(\mu)Y(\mu_0)]$$

and

$$T(\mu, \mu_0) = \varpi_0 \frac{\mu\mu_0}{\mu-\mu_0} [Y(\mu)X(\mu_0) - X(\mu)Y(\mu_0)], \tag{209}$$

where X and Y are defined in terms of the characteristic function $\frac{1}{2}\varpi_0$. From Chapter VIII, equations (9) and (10), we now find that

$$s(\mu) = \frac{1}{2} \int_0^1 S(\mu, \mu') \, d\mu' = \mu[1 - \{(1-x_0)X(\mu) + y_0 Y(\mu)\}]$$

and

$$t(\mu) = \frac{1}{2} \int_0^1 T(\mu, \mu') \, d\mu'$$

$$= \mu[-e^{-\tau_1/\mu} + y_0 X(\mu) + (1-x_0)Y(\mu)], \tag{210}$$

where it may be recalled that (cf. Chap. VIII, eqs. [7])

$$x_0 = \frac{1}{2}\varpi_0\alpha_0 \quad \text{and} \quad y_0 = \frac{1}{2}\varpi_0\beta_0, \tag{211}$$

α_0 and β_0 denoting the moments of order zero of $X(\mu)$ and $Y(\mu)$.

From equations (210) and (211) we have

$$s(\mu) = \mu[1 - \tfrac{1}{2}\{(2 - \varpi_0\alpha_0)X(\mu) + \varpi_0\beta_0 Y(\mu)\}],$$
$$\gamma_1(\mu) = \tfrac{1}{2}\varpi_0\beta_0 X(\mu) + \tfrac{1}{2}(2 - \varpi_0\alpha_0)Y(\mu),$$
$$\bar{s} = 1 - \{(2 - \varpi_0\alpha_0)\alpha_1 + \varpi_0\beta_0\beta_1\}. \tag{212}$$

(ii) *Scattering according to the phase function* $\varpi_0(1 + x\cos\Theta)$. According to Chapter IX, equations (109), (112), and (115), we have

$$s(\mu) = \mu - \phi(\mu) = \mu[1 - \{q_1 X(\mu) + p_1 Y(\mu)\}],$$
$$\gamma_1(\mu) = \zeta(\mu)/\mu = p_1 X(\mu) + q_1 Y(\mu),$$

and
$$\bar{s} = 1 - 2(q_1\alpha_1 + p_1\beta_1), \tag{213}$$

where the constants q_1 and p_1 are given by Chapter IX, equations (123).

(iii) *The conservative case of isotropic scattering.* In this case the identification of γ_1 with the constant in the flux integral enables us to write (Chap. IX, eqs. [39] and [43])

$$\gamma_1(\mu) = -\tfrac{1}{2}Q(\alpha_1 + \beta_1)[X(\mu) + Y(\mu)]. \tag{214}$$

Therefore, $s(\mu) = \mu[1 + \tfrac{1}{2}Q(\alpha_1 + \beta_1)\{X(\mu) + Y(\mu)\}],$

and
$$\bar{s} = 1 + Q(\alpha_1 + \beta_1)^2. \tag{215}$$

(iv) *Scattering according to Rayleigh's phase function.* Again, as we can identify γ_1 with the constant in the flux integral, we have (cf. Chap. IX, eq. [104])

$$\gamma_1(\mu) = \tfrac{3}{32}[(c_1 + c_2)(\alpha_1 + \beta_1) - (\alpha_2 - \beta_2)][X(\mu) + Y(\mu)], \tag{216}$$

where c_1 and c_2 are constants given by Chapter IX, equations (107) and (108).

We further have

$$s(\mu) = \mu[1 - \gamma_1(\mu)]; \qquad \bar{s} = 1 - \tfrac{3}{16}[(c_1 + c_2)(\alpha_1 + \beta_1) - (\alpha_2 - \beta_2)](\alpha_1 + \beta_1). \tag{217}$$

72.3. *The reduction to the standard problem in case of scattering according to a phase-matrix*

As in § 72.1, the problem again consists of two parts: the determination of the intensity I_g at the ground which is unpolarized and uniform in the outward hemisphere, and the evaluation of the effect of this ground radiation on the intensities $I(0, \mu, \varphi)$ and $I(\tau_1, -\mu, \varphi)$ at $\tau = 0$ and at $\tau = \tau_1$ which will prevail in the absence of the ground.

Since the radiation incident on $\tau = \tau_1$, from below, is unpolarized and isotropic, it is clearly sufficient to consider the azimuth independent part of the scattering and the transmission matrices, referring to I_l and

I_r. In terms of these sub-matrices $\tfrac{3}{4}\boldsymbol{S}^{(0)}$ and $\tfrac{3}{4}\boldsymbol{T}^{(0)}$ of order 2, the azimuth independent terms $I_l^{(0)}(0,\mu)$, $I_r^{(0)}(0,\mu)$, $I_l^{(0)}(\tau_1,-\mu)$, and $I_r^{(0)}(\tau_1,-\mu)$ in the reflected and transmitted intensities, in the standard problem, will be given by (cf. eqs. [173])

$$\boldsymbol{I}^{(0)}(0,\mu) = \begin{pmatrix} I_l^{(0)}(0,\mu) \\ I_r^{(0)}(0,\mu) \end{pmatrix} = \frac{3}{16\mu}\begin{pmatrix} S_{ll}^{(0)}(\mu,\mu_0) & S_{lr}^{(0)}(\mu,\mu_0) \\ S_{rl}^{(0)}(\mu,\mu_0) & S_{rr}^{(0)}(\mu,\mu_0) \end{pmatrix}\begin{pmatrix} F_l \\ F_r \end{pmatrix},$$

and

$$\boldsymbol{I}^{(0)}(\tau_1,-\mu) = \begin{pmatrix} I_l^{(0)}(\tau_1,-\mu) \\ I_r^{(0)}(\tau_1,-\mu) \end{pmatrix} = \frac{3}{16\mu}\begin{pmatrix} T_{ll}^{(0)}(\mu,\mu_0) & T_{lr}^{(0)}(\mu,\mu_0) \\ T_{rl}^{(0)}(\mu,\mu_0) & T_{rr}^{(0)}(\mu,\mu_0) \end{pmatrix}\begin{pmatrix} F_l \\ F_r \end{pmatrix}.$$

(218)

For later use we shall define the following quantities:

$$s_l(\mu) = \tfrac{3}{8}\int_0^1 [S_{ll}^{(0)}(\mu',\mu)+S_{rl}^{(0)}(\mu',\mu)]\,d\mu'$$
$$= \tfrac{3}{8}\int_0^1 [S_{ll}^{(0)}(\mu,\mu')+S_{lr}^{(0)}(\mu,\mu')]\,d\mu',$$

$$s_r(\mu) = \tfrac{3}{8}\int_0^1 [S_{lr}^{(0)}(\mu',\mu)+S_{rr}^{(0)}(\mu',\mu)]\,d\mu'$$
$$= \tfrac{3}{8}\int_0^1 [S_{rl}^{(0)}(\mu,\mu')+S_{rr}^{(0)}(\mu,\mu')]\,d\mu',$$

$$t_l(\mu) = \tfrac{3}{8}\int_0^1 [T_{ll}^{(0)}(\mu',\mu)+T_{rl}^{(0)}(\mu',\mu)]\,d\mu'$$
$$= \tfrac{3}{8}\int_0^1 [T_{ll}^{(0)}(\mu,\mu')+T_{lr}^{(0)}(\mu,\mu')]\,d\mu',$$

$$t_r(\mu) = \tfrac{3}{8}\int_0^1 [T_{lr}^{(0)}(\mu',\mu)+T_{rr}^{(0)}(\mu',\mu)]\,d\mu'$$
$$= \tfrac{3}{8}\int_0^1 [T_{rl}^{(0)}(\mu,\mu')+T_{rr}^{(0)}(\mu,\mu')]\,d\mu',$$

and $$\gamma_l^{(1)}(\mu) = \frac{t_l(\mu)}{\mu}+e^{-\tau_1/\mu}; \qquad \gamma_r^{(1)}(\mu) = \frac{t_r(\mu)}{\mu}+e^{-\tau_1/\mu},$$ (219)

where the alternative forms for $s_l(\mu)$, etc., arise from the principle of reciprocity (Chap. VII, § 52).

Considering, first, the effect of the ground radiation on the intensities $\boldsymbol{I}(0,\mu,\varphi)$ and $\boldsymbol{I}(\tau_1,-\mu,\varphi)$ in the absence of the ground, we observe that we may replace the unpolarized intensity I_g by two independent plane-polarized components of intensities, each $\tfrac{1}{2}I_g$, in the states of polarization signified by l and r. The transmission of this field $(\tfrac{1}{2}I_g,\tfrac{1}{2}I_g,0,0)$, in all outward directions at τ_1, will contribute to the emergent intensity, in the direction μ and in the state of polarization l, an amount (cf. eqs. [219])

$$\tfrac{1}{2}I_g e^{-\tau_1/\mu} + \frac{1}{4\pi\mu} \int\limits_0^1 \int\limits_0^{2\pi} T(\mu,\varphi;\mu',\varphi') \begin{pmatrix} \tfrac{1}{2}I_g \\ \tfrac{1}{2}I_g \\ 0 \end{pmatrix}_l d\mu' d\varphi'$$

$$= \tfrac{1}{2}I_g e^{-\tau_1/\mu} + \frac{I_g}{2\mu}\frac{3}{8} \int\limits_0^1 [T_{ll}^{(0)}(\mu,\mu') + T_{lr}^{(0)}(\mu,\mu')]\, d\mu'$$

$$= \tfrac{1}{2}I_g\left[e^{-\tau_1/\mu} + \frac{t_l(\mu)}{\mu}\right] = \tfrac{1}{2}I_g \gamma_l^{(1)}(\mu). \tag{220}$$

Similarly, the contribution to the emergent intensity, in the direction μ and in the state of polarization r, is

$$\tfrac{1}{2}I_g \gamma_r^{(1)}(\mu). \tag{221}$$

In view of the isotropy and the natural character of the ground radiation, it is clear that the presence of the ground will not affect the Stokes parameters U and V. We can thus write

$$I_l^*(0,\mu,\varphi) = I_l(0,\mu,\varphi) + \tfrac{1}{2}I_g \gamma_l^{(1)}(\mu),$$
$$I_r^*(0,\mu,\varphi) = I_r(0,\mu,\varphi) + \tfrac{1}{2}I_g \gamma_r^{(1)}(\mu),$$
$$U^*(0,\mu,\varphi) = U(0,\mu,\varphi) \quad \text{and} \quad V^*(0,\mu,\varphi) = V(0,\mu,\varphi). \tag{222}$$

Again, the radiation I_g ($0 \leqslant \mu' \leqslant 1$ and $0 \leqslant \phi' \leqslant 2\pi$) incident on τ_1 will be reflected by the atmosphere according to the standard law of diffuse reflection, and contribute to the intensity in the direction $-\mu$ and in the state of polarization l the additional amount

$$I_{g,l}^{(\text{ref})}(-\mu) = \frac{1}{4\pi\mu} \int\limits_0^1 \int\limits_0^{2\pi} S(\mu,\varphi;\mu',\varphi') \begin{pmatrix} \tfrac{1}{2}I_g \\ \tfrac{1}{2}I_g \\ 0 \end{pmatrix}_l d\mu' d\varphi'$$

$$= \tfrac{1}{2}I_g \frac{3}{8\mu} \int\limits_0^1 [S_{ll}^{(0)}(\mu,\mu') + S_{lr}^{(0)}(\mu,\mu')]\, d\mu' = \tfrac{1}{2}I_g \frac{s_l(\mu)}{\mu}. \tag{223}$$

Similarly, the contribution to the intensity in the direction $(-\mu)$ and in the state of polarization r, by the reflection of the ground radiation, is

$$I_{g,r}^{(\text{ref})}(-\mu) = \tfrac{1}{2}I_g \frac{s_r(\mu)}{\mu}. \tag{224}$$

Hence, combining the results of equations (223) and (224), we have

$$I_l^*(\tau_1, -\mu, \varphi) = I_l(\tau_1, -\mu, \varphi) + \tfrac{1}{2}I_g \frac{s_l(\mu)}{\mu},$$

$$I_r^*(\tau_1, -\mu, \varphi) = I_r(\tau_1, -\mu, \varphi) + \tfrac{1}{2}I_g \frac{s_r(\mu)}{\mu}, \tag{225}$$

and the other Stokes parameters will be unaffected by the presence of the ground.

Turning now to the determination of I_g, we shall again make use of the flux condition (196) which, of course, continues to be valid.

The outward normal flux is πI_g as before. However, the inward normal flux now consists of several parts: Considering, first, that part of the inward normal flux prevailing at τ_1 which is proportional to the incident flux πF_l we have the reduced incident flux (per unit area of the ground)

$$\pi \mu_0 F_l e^{-\tau_1/\mu_0},$$

the flux

$$\int_0^1 \int_0^{2\pi} \frac{1}{4\mu} T_{ll}(\mu,\varphi;\mu_0,\varphi_0) F_l \mu \, d\mu \, d\varphi = \pi F_l \frac{3}{8} \int_0^1 T_{ll}^{(0)}(\mu,\mu_0) \, d\mu$$

in the state l, the flux $\quad \pi F_l \frac{3}{8} \int_0^1 T_{rl}^{(0)}(\mu,\mu_0) \, d\mu$

in the state r, giving a total

$$\pi\mu_0 F_l \left\{ e^{-\tau_1/\mu_0} + \frac{3}{8\mu_0} \int_0^1 [T_{ll}^{(0)}(\mu,\mu_0) + T_{rl}^{(0)}(\mu,\mu_0)] \, d\mu \right\}$$

$$= \pi\mu_0 F_l \left\{ e^{-\tau_1/\mu_0} + \frac{t_l(\mu_0)}{\mu_0} \right\} = \pi\mu_0 F_l \gamma_l^{(1)}(\mu_0). \tag{226}$$

Similarly, the inward normal flux prevailing at τ_1 which is proportional to the incident flux πF_r is

$$\pi\mu_0 F_r \gamma_r^{(1)}(\mu_0). \tag{227}$$

And, finally, we have the contribution to the inward normal flux at τ_1, by the reflection of the ground radiation by the atmosphere above. This last gives (cf. eqs. [223] and [224])

$$2\pi \int_0^1 [I_{g,l}^{(\text{ref})}(-\mu) + I_{g,r}^{(\text{ref})}(-\mu)] \mu \, d\mu = \pi I_g \int_0^1 [s_l(\mu) + s_r(\mu)] \, d\mu. \tag{228}$$

Defining $\qquad \qquad \bar{s} = \int_0^1 [s_l(\mu) + s_r(\mu)] \, d\mu, \tag{229}$

we have $\qquad \qquad \qquad \pi I_g \bar{s}. \tag{230}$

The total inward flux at $\tau = \tau_1$ is therefore

$$\pi[\mu_0 F_l \gamma_l^{(1)}(\mu_0) + \mu_0 F_r \gamma_r^{(1)}(\mu_0) + I_g \bar{s}]; \tag{231}$$

and λ_0 times this must equal πI_g. We thus obtain

$$I_g = \lambda_0[\mu_0 F_l \gamma_l^{(1)}(\mu_0) + \mu_0 F_r \gamma_r^{(1)}(\mu_0) + I_g \bar{s}],$$

or,
$$I_g = \frac{\lambda_0}{1 - \lambda_0 \bar{s}} \mu_0[F_l \gamma_l^{(1)}(\mu_0) + F_r \gamma_r^{(1)}(\mu_0)]. \tag{232}$$

Inserting this last expression for I_g in equations (222) and (225), we observe that the effect of the ground can be described by the matrices

$$\boldsymbol{\Gamma}(\mu, \mu_0) = \begin{pmatrix} \gamma_l^{(1)}(\mu)\gamma_l^{(1)}(\mu_0) & \gamma_l^{(1)}(\mu)\gamma_r^{(1)}(\mu_0) & 0 \\ \gamma_r^{(1)}(\mu)\gamma_l^{(1)}(\mu_0) & \gamma_r^{(1)}(\mu)\gamma_r^{(1)}(\mu_0) & 0 \\ 0 & 0 & 0 \end{pmatrix}, \tag{233}$$

and
$$\boldsymbol{\Lambda}(\mu, \mu_0) = \begin{bmatrix} \dfrac{s_l(\mu)}{\mu}\gamma_l^{(1)}(\mu_0) & \dfrac{s_l(\mu)}{\mu}\gamma_r^{(1)}(\mu_0) & 0 \\ \dfrac{s_r(\mu)}{\mu}\gamma_l^{(1)}(\mu_0) & \dfrac{s_r(\mu)}{\mu}\gamma_r^{(1)}(\mu_0) & 0 \\ 0 & 0 & 0 \end{bmatrix}. \tag{234}$$

Thus,
$$\boldsymbol{I}^*(0, \mu, \varphi) = \frac{1}{4\mu}\left[\boldsymbol{S}(\mu, \varphi; \mu_0, \varphi_0) + \frac{2\lambda_0}{1 - \lambda_0 \bar{s}}\mu\mu_0 \boldsymbol{\Gamma}(\mu, \mu_0) \right] \boldsymbol{F} \tag{235}$$

and
$$\boldsymbol{I}^*(\tau_1, -\mu, \varphi) = \frac{1}{4\mu}\left[\boldsymbol{T}(\mu, \varphi; \mu_0, \varphi_0) + \frac{2\lambda_0}{1 - \lambda_0 \bar{s}}\mu\mu_0 \boldsymbol{\Lambda}(\mu, \mu_0) \right] \boldsymbol{F}. \tag{236}$$

In conservative cases, the γ's are the same as the constants (which are functions of μ_0) which appear in the flux integral. In the present instance, there are (formally) two flux integrals corresponding to the fact that F_l and F_r can be specified independently of each other. Indeed, starting from the equation of transfer (129), for example, we can show that the problem admits the integrals[†]

$$F_l(\tau) = 2\int_{-1}^{+1} [I_{ll}^{(0)}(\tau, \mu) + I_{rl}^{(0)}(\tau, \mu)]\mu \, d\mu = \mu_0 F_l[e^{-\tau/\mu_0} - \gamma_l^{(1)}(\mu_0)], \tag{237}$$

and
$$F_r(\tau) = 2\int_{-1}^{+1} [I_{lr}^{(0)}(\tau, \mu) + I_{rr}^{(0)}(\tau, \mu)]\mu \, d\mu = \mu_0 F_r[e^{-\tau/\mu_0} - \gamma_r^{(1)}(\mu_0)], \tag{238}$$

where $(I_{ll} + I_{rl})$ and $(I_{lr} + I_{rr})$ are the *total* intensities in the diffuse radiation field which are proportional, respectively, to F_l and F_r. Comparing equations (237) and (238) with the expressions (226) and

[†] The problem also admits the additional K-integrals
$$K_l(\tau) = \tfrac{1}{4}\mu_0 F_l(-\mu_0 e^{-\tau/\mu_0} - \gamma_l^{(1)}\tau + \gamma_l^{(2)})$$
and
$$K_r(\tau) = \tfrac{1}{4}\mu_0 F_r(-\mu_0 e^{-\tau/\mu_0} - \gamma_r^{(1)}\tau + \gamma_r^{(2)}).$$

(227), and remembering that equations (237) and (238) (contrary to [226] and [227]) refer to fluxes measured in the direction of the *outward* normal and do not include the reduced incident fluxes, we identify the γ's which occur in the flux integrals with the γ's as originally defined (eq. [219]).

In conservative cases we have the further relations (which also follow from the flux integrals):

$$s_l(\mu) = \mu[1-\gamma_l^{(1)}(\mu)], \qquad s_r(\mu) = \mu[1-\gamma_r^{(1)}(\mu)]$$

and
$$\bar{s} = 1 - \int_0^1 [\gamma_l^{(1)}(\mu)+\gamma_r^{(1)}(\mu)]\mu\, d\mu. \tag{239}$$

72.4. *Expressions for $\gamma_l^{(1)}(\mu)$, $\gamma_r^{(1)}(\mu)$, $s_l(\mu)$, $s_r(\mu)$, and $\bar{s}$ in the case of Rayleigh scattering*

In terms of the solutions for $S^{(0)}$ and $T^{(0)}$ given in § 71 (eqs. [177] and [178]), the various integrals defining $s_l(\mu)$, $s_r(\mu)$, etc., can be evaluated. We find†

$$\gamma_l^{(1)}(\mu) = \tfrac{3}{8}Q(\nu_2-\nu_1)(d_0-d_2)[X_l(\mu)+Y_l(\mu)], \tag{240}$$
and
$$\gamma_r^{(1)}(\mu) = \tfrac{3}{8}Q(d_0-d_2)[(u_4-u_3)\{X_r(\mu)+Y_r(\mu)\}-u_5\mu\{X_r(\mu)-Y_r(\mu)\}], \tag{241}$$

where the various constants have the same meanings as in § 71.

Since this is a case of conservative scattering

$$s_l(\mu) = \mu[1-\gamma_l^{(1)}(\mu)] \quad \text{and} \quad s_r(\mu) = \mu[1-\gamma_r^{(1)}(\mu)]. \tag{242}$$

From equations (240)–(242) we now obtain

$$\bar{s} = 1 - \tfrac{3}{8}Q(d_0-d_2)[(\nu_2-\nu_1)\kappa_1+(u_4-u_3)c_1-u_5 d_2]. \tag{243}$$

With this we have completed the exact solution of a problem which will serve as a basis for interpreting the illumination and the polarization of the sky.

73. The intensity and polarization of the sky radiation

The most important application of the theory developed in the two preceding sections is to the illumination and polarization of the sunlit sky. We shall accordingly illustrate the theory in this context.

As is well known, Lord Rayleigh accounted for the principal features of the brightness and polarization of the sky radiation in terms of the laws of molecular scattering now associated with his name. On Rayleigh's theory, the deep blue colour of the sky at the zenith is related to the λ^{-4} dependence of the scattering coefficient on wave-length (cf.

† S. Chandrasekhar, *Astrophys. J.* **107**, 188, 1948 (eqs. [467] and [475]).

Chap. I, eq. [205]) and to the fact that for small optical thicknesses, τ_1, the intensity of the transmitted light is proportional to τ_1. Similarly, the strongly varying polarization of the sky radiation is related to the fact that, on Rayleigh's law, the light scattered at right angles to the direction of incidence is completely polarized, while the light scattered in the forward or the backward directions has the same polarization as the incident light.

However, not all features of the sky radiation can be accounted for, so simply, in terms of the laws of single scattering. Thus, it is known that at right angles to the sun the polarization is not complete ($\delta \sim 87$ per cent.). Similarly, the polarization in the direction of the sun is not zero: instead, it is weakly, *negatively*, polarized; and for general angles of incidence there are, in the meridian plane, two points where the light is unpolarized. These points of zero polarization, called the *neutral points*, occur in the following positions: For angles of incidence not exceeding 70° the neutral points occur about 10–20° above and below the sun; these are the *Babinet* and the *Brewster* points, respectively. And when the sun is low, then near the horizon, opposite the sun, and about 20° above the anti-solar point a neutral point occurs: this is the *Arago* point.†

It is evident (and as was, indeed, emphasized by Lord Rayleigh in his earliest publications) that to account for these further facts concerning the sky radiation we must allow for the effects of scattering of orders higher than the first.

Several attempts to include the effects of higher-order scattering have been made, notably by L. V. King, J. J. Tichanowsky, and S. Chapman and A. Hammad. However, in almost all of these investigations, 'Rayleigh scattering' is taken to mean scattering according to Rayleigh's phase function; and, as we have seen, this is incorrect. Moreover, in these investigations, the correct equations of transfer in terms of a phase-matrix are not formulated and, consequently, the effects of scattering of orders higher than the second are not included; also, the effects of the reflection of the ground are not included.

We shall now show how the theory developed in §§ 71 and 72 provides an exact and a satisfactory basis for interpreting the features of the sky radiation in a detailed manner.

Considering $\tau_1 = 0.2$ as typical, we have illustrated in Figs. 27 and 28 the angular distribution and the state of polarization of the transmitted

† Sometimes all these neutral points occur simultaneously and continue to persist even after the sun has set. It is clear that these latter effects must be due to the curvature of the earth's atmosphere and are, therefore, beyond the scope of the theory based on plane-parallel atmospheres.

light for various angles of incidence according to the formulae of §§ 71 and 72. In Fig. 27 the laws of diffuse transmission for the case when there is no ground (i.e. for the standard problem) are illustrated. Fig. 27 clearly exhibits the sharp rise in intensity as we approach the horizon; this rise and the subsequent fall are steeper the lower the sun. Also, it will be observed that the theory predicts neutral points in the positions known for the Babinet, Brewster, and Arago points. Thus, for an angle of incidence of 60° ($\mu_0 = 0\cdot5$) the neutral points occur at angular distances 15° and 20° above and below the sun, respectively; i.e. exactly in the positions appropriate for the Babinet and the Brewster points. Again it is seen that when the angle of incidence slightly exceeds 66° ($\mu_0 < 0\cdot4$) the Brewster point sets and the Arago point rises in the opposite sky. For larger angles of incidence we have only the Babinet and the Arago points.

The effects of ground reflection are illustrated in **Fig. 28**. Various angles of incidence and $\tau_1 = 0\cdot1$ and $0\cdot2$ are considered. It will be noticed from these figures that the effect of a ground surface with an albedo even as high as $0\cdot20$ does not make any essential difference to the predicted positions of the neutral points. This independence of the neutral points on ground reflection is not difficult to understand. As is well known, the laws of Rayleigh scattering give the maximum polarization for the scattered light; all other laws give much less polarization. A ground reflecting according to Lambert's law can, therefore, hardly compete with Rayleigh scattering for producing polarization. The effect of reflection by the ground is therefore essentially one of adding a component of natural light to the polarized light already present.

74. Resonance line scattering and scattering by anisotropic particles

The law of scattering by anisotropic particles given by the classical theory (Chap. I, § 17) is of the same general type as the law of resonance line scattering given by the quantum theory (Chap. I, § 18). On these laws (as on Rayleigh's law) the Stokes parameter V is scattered independently of the others and according to a phase function $\frac{3}{2}q_3\cos\Theta$. The constant q_3 has the value $1-3\gamma$ ($0 < \gamma < \frac{1}{3}$) on the classical theory (Chap. I, eqs. [250] and [251]); on the quantum theory of line scattering it has a value E_3 depending on the initial state (j) and the Δj involved in the transition (Chap. I, § 18, Table II). The solutions for S_V and T_V have, therefore, the same forms as on Rayleigh scattering; but the right-hand sides of equations (126)–(128) (giving the solutions for Rayleigh scattering) must now be multiplied by q_3.

Considering, next, the part of the phase-matrix referring to I_l, I_r,

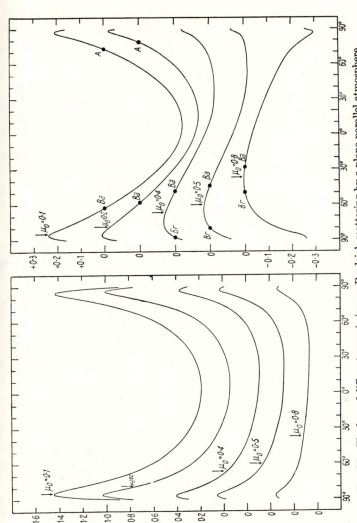

Fig. 27. The law of diffuse transmission on Rayleigh scattering by a plane-parallel atmosphere.

The ordinates represent the intensities in the unit $\mu_0 F$ and the abscissae the angle in degrees. Angles of incidence corresponding to $\mu_0 = 0.1, 0.2, 0.4, 0.5,$ and 0.8 are considered; also $\tau_1 = 0.2$. (The arrows indicate the directions of incidence.)

The curves on the left illustrate the variation of the net diffusely transmitted light $(I_l + I_r)$, in the meridian plane. For the sake of clarity, the curves for the different angles of incidence have been displaced with respect to one another: the scale of ordinates running from 0 to 1·6 refers to the topmost curve (i.e. for $\mu_0 = 0.1$) and the zero ordinates for the successive ones are indicated.

The curves on the right illustrate the variation of $I_l - I_r$ (also in the meridian plane) for various angles of incidence. The scale of ordinates has been staggered, as on the left-hand side. It will be noticed that for $\mu_0 = 0.1$ and 0.2, the neutral points occur in positions appropriate for the Babinet (Ba) and the Arago (A) points. For larger angles of incidence, the neutral points occur on either side of the 'sun' in the positions of the Babinet and Brewster (Br) points. The 'setting' of the Arago point therefore coincides with the 'rising' of the Brewster point and conversely.

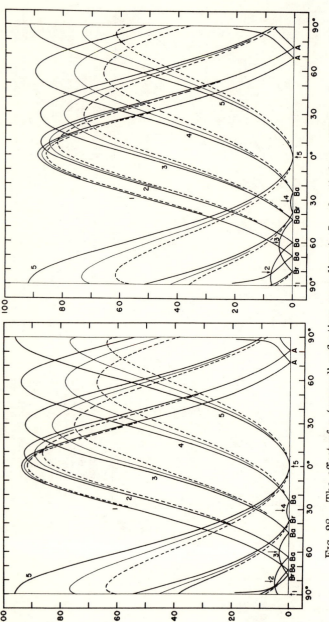

Fig. 28. The effect of a 'ground' reflecting according to Lambert's law with an albedo λ_o, on the diffuse transmission on Rayleigh scattering, by a plane-parallel atmosphere of optical thickness $\tau_1 = 0.1$ (the figure on the left) and 0.2 (the figure on the right).

Variation of the degree of polarization in the principal meridian for various angles of incidence. The abscissa gives the zenith distance and the ordinate gives the degree of polarization in per cent. The curves marked 1, 2, 3, 4, and 5 represent the variation for the angles of incidence $\theta_o = 90°$, $80.8°$, $60°$, $30.7°$, and $0°$, respectively. The thick solid curves are obtained before any ground corrections have been applied. The dashed curves are obtained if we allow for a ground reflecting according to Lambert's law with an albedo $\lambda_o = 0.20$. The thin intermediate curves are obtained if $\lambda_o = 0.10$. The positions of the neutral points are also indicated.

Rayleigh scattering) must now be multiplied by q_3.

Considering, next, the part of the phase-matrix referring to I_l, I_r, and U, we observe that it can be expressed as the sum of the phase-matrix for Rayleigh scattering and the matrix

$$E = \tfrac{1}{2}\begin{pmatrix} 1 & 1 & 0 \\ 1 & 1 & 0 \\ 0 & 0 & 0 \end{pmatrix}, \tag{244}$$

with certain weights q_1 and q_2 (cf. Chap. I, eqs. [257] and [259]). On the classical theory q_1 and q_2 have the values $(1-\gamma)/(1+2\gamma)$ and $3\gamma/(1+2\gamma)$ (Chap. I, eq. [257]); for resonance line scattering they have the values $E_1(j;\Delta j)$ and $E_2(j;\Delta j)$ given in Table II (p. 52). Consequently, the solutions for the azimuth *dependent* terms $S^{(1)}$, $S^{(2)}$, $T^{(1)}$, and $T^{(2)}$ in the scattering and the transmission matrices (eq. [122]) have the same forms as on Rayleigh scattering; but the right-hand sides of equations (123)–(125) (giving the solutions for Rayleigh scattering) must now be multiplied by q_1.

The solution for the azimuth independent terms $I_l^{(0)}(\tau,\mu)$ and $I_r^{(0)}(\tau,\mu)$ cannot be disposed of so easily. The equation of transfer we have to consider is (cf. eqs. [129]–[131])

$$\mu\frac{dI^{(0)}(\tau,\mu)}{d\tau} = I^{(0)}(\tau,\mu) - \tfrac{3}{8}q_1\int_{-1}^{+1} J(\mu,\mu')I^{(0)}(\tau,\mu')\,d\mu' -$$

$$-\tfrac{1}{2}q_2 E\int_{-1}^{+1} I^{(0)}(\tau,\mu')\,d\mu' - \left(\frac{3q_1}{16}J(\mu,\mu_0)+\frac{q_2}{4}E\right)Fe^{-\tau/\mu_0}. \tag{245}$$

The solution of this equation in the general nth approximation (along the lines of § 68) presents no difficulty. However, an essential feature which simplifies the treatment of Rayleigh scattering is absent from this case: The characteristic equation does not factorize; instead, we have (cf. eq. [23])

$$q_1(D_0-D_2-\tfrac{8}{3})(D_0-D_2-\tfrac{4}{3}) = \tfrac{16}{9}q_2(D_0-2),$$

or, alternatively,

$$(k^2-1)^2 D_2^2 - \frac{16}{9}\frac{q_2}{q_1}D_2 - \frac{4}{9} = 0. \tag{246}$$

On account of this, it does not seem possible to carry out the elimination of the constants and express the angular distributions of the emergent radiations in closed forms. Related difficulties arise in the reduction of the integral equations for $S^{(0)}$ and $T^{(0)}$. Thus, while it is not difficult to solve the equations of transfer for the case of scattering by anisotropic particles (or, of resonance line scattering) in any finite approximation, the determination of the exact solutions for the azimuth independent terms is likely to present some difficulty.

BIBLIOGRAPHICAL NOTES

§ 68. The analysis in this section is derived from

1. S. CHANDRASEKHAR, *Astrophys. J.* **103**, 351, 1946.

2. —— ibid. **105**, 164, 1947 (Section II of this paper).

The H-functions and the exact laws of darkening tabulated in this section (Tables XXII–XXIV) are taken from

3. S. CHANDRASEKHAR and FRANCES H. BREEN, ibid., p. 435, 1947 (Section II of this paper).

§§ 69–71. The problem of diffuse reflection by a semi-infinite atmosphere on Rayleigh scattering, in the nth approximation, is treated in

4. S. CHANDRASEKHAR, ibid. **104**, 110, 1946.

The exact solution derived from the integral equation for S is given in reference 2. The numerical data given in this section are taken from reference 3.

The problem of diffuse reflection and transmission on Rayleigh scattering is considered in

5. S. CHANDRASEKHAR, ibid. **106**, 152, 1947 (Section V of this paper).

6. —— ibid. **107**, 188, 1948 (Section V of this paper).

In reference 5 the problem is treated in the nth approximation and the elimination of the constants is achieved. The exact solutions are obtained in reference 6.

§ 72. The reduction of the planetary problem to the standard problem was first carried out by van de Hulst in the context of isotropic scattering:

7. H. VAN DE HULST, ibid. **107**, 220, 1948.

However, the general treatment given in this section and, in particular, the reduction for the case of scattering according to a phase-matrix, are new.

§ 73. The classical papers on the subject of sky radiation are those of:

8. LORD RAYLEIGH, *Phil. Mag.* **41**, 107, 274, 447 (1871;) also *Scientific Papers* (Cambridge, England, 1899), i, pp. 87, 104, 518.

Among the later theoretical investigations, reference may be made to:

9. L. V. KING, *Phil. Trans. Roy. Soc. London*, A. **212**, 375 (1913).

10. J. J. TICHANOWSKY, *Physik. Zeits.* **28**, 252, 680 (1927).

11. A. HAMMAD and S. CHAPMAN, *Phil. Mag.* Ser. 7, **28**, 99 (1939).

12. A. HAMMAD, ibid. **36**, 434 (1945); **38**, 515 (1947); and *Astrophys. J.* **108**, 338 (1948).

13. A. HAMMAD and S. A. M. HASSANEIN, *Phil. Mag.* Ser. 7, **39**, 956 (1948).

It is of interest to recall that the importance of correctly formulating the equation of transfer for Rayleigh scattering was recognized, already, by King. Thus in the paper quoted (ref. 9) he writes: 'The complete solution of the problem from this aspect would require us to split up the incident radiation into two components, one of which is polarized in the principal plane, the other at right angles to it: the effect of self-illumination would lead to two simultaneous integral equations in three variables, the solution of which would be much too complicated to be useful.' However, it should be noted that for a complete description of a partially plane polarized radiation field it is not sufficient to consider only the intensities I_l and I_r in two directions at right angles to each other; a third parameter U is necessary to allow for the variation in the plane of polarization. Even so, we have found that it is not too difficult a matter to formulate the correct equations and solve them exactly for the basic problems.

General accounts of the subject will be found in:

14. Chr. Jensen, *Handbuch der Physik* (Berlin, Springer, 1928), xix, Kap. iv, pp. 70–171.

15. M. Minnaert, *Light and Colour in the Open Air* (G. Bell, London, 1940).

16. H. C. van de Hulst, *The Atmospheres of the Earth and the Planets* (editor G. P. Kuiper, University of Chicago Press, 1949), chap. iii, pp. 49–112.

The application of the theory of diffuse reflection and transmission, developed in this chapter, to the problem of sky radiation, is presented here for the first time.

THE RADIATIVE EQUILIBRIUM OF A STELLAR ATMOSPHERE

75. The concept of local thermodynamic equilibrium

AN atmosphere is said to be in local thermodynamic equilibrium when it is possible to define at each point in the atmosphere a temperature T such that the coefficients of absorption (κ_ν) and emission (j_ν) are related according to the Kirchhoff–Planck equation

$$j_\nu = \kappa_\nu B_\nu(T) \quad \text{and} \quad B_\nu(T) = \frac{2h\nu^3}{c^2} \frac{1}{e^{h\nu/kT}-1}. \tag{1}$$

The usefulness of the concept of local thermodynamic equilibrium, in astrophysics, arises from the fact that it provides a convenient abstraction for situations in which a multiplicity of processes participate in the absorption and emission of radiation of any given wave-length and no unique correlation exists between particular acts of absorption and emission. Such situations are encountered when one considers the origin of the continuous spectrum of the stars. Thus in the solar atmosphere the principal source of continuous absorption is the negative hydrogen ion. Since the electron affinity of hydrogen is 0·75 e.v., the photo-ionization of the ion by radiation of wave-length, say, 5,000 A, will result in the ejection of an electron with a kinetic energy of about 3·6 e.v.; and as the mean kinetic energy of the electrons in the solar atmosphere is of the order of 1 e.v., the ejected electron will rapidly lose its energy by elastic and inelastic collisions. Moreover, the electrons in the solar atmosphere are, largely, supplied by the ionization of elements like Ca, Na, Mg, Fe, Si, etc., by radiation beyond their respective series limits. Consequently a strict evaluation of the coefficients of continuous absorption and emission in the solar atmosphere must include the consideration of the ionization-recombination equilibria of H^- and of all the elements which contribute to the concentration of the free electrons; of the elastic and inelastic collisions of the electrons with neutral hydrogen atoms; and generally of such other processes as are relevant to the establishing of a velocity distribution among the electrons. It is evident that a microscopic analysis of all these and other contributory effects will be excessively complicated. Moreover, a theory which starts out on such detailed premises will, by its very nature, obscure the essential factors which are operative. Indeed, a theory

along such lines ten years ago would not have included what has since
been established as the most important source of continuous absorption
in the solar atmosphere, namely, H^-. The importance of H^- as a major
factor in the analysis of stellar atmospheres actually became clear by
the application of methods derived from a less specific theory concerning
the mode of origin of the continuous spectrum of stars. These remarks
underline the importance, for astrophysical theory, of developing
methods of analysis on a basis broad enough for disentangling the
various factors which are operative towards particular ends. Once the
principal factors have been identified by such methods of analysis, they
can later be incorporated in a less idealized, more specific, picture.
The concept of local thermodynamic equilibrium is one which has been
exceedingly useful in the development of such general methods for
understanding and interpreting the continuous spectrum of stars. In
this chapter we shall describe the nature of some of these methods.

76. The radiative equilibrium of a stellar atmosphere in local thermodynamic equilibrium

According to equation (1), the source function for radiation of frequency ν is

$$\mathfrak{J}_\nu = B_\nu(T), \tag{2}$$

where T is the temperature prevailing at the point considered. For
a plane-parallel atmosphere, in local thermodynamic equilibrium, the
equation of transfer is, therefore,

$$-\mu \frac{dI_\nu(z,\mu)}{\rho\, dz} = \kappa_\nu I_\nu(z,\mu) - \kappa_\nu B_\nu(T_z). \tag{3}$$

The *formal* simplification which the concept of local thermodynamic
equilibrium introduces in the theory is apparent: it enables us to specify
the entire radiation field in terms of the march of a single parameter,
namely, T; for, once the temperature distribution in the atmosphere has
been determined, the source function becomes known, and the radiation field follows from the equation (cf. Chap. I, eq. [90])

$$I_\nu(\tau_\nu,\mu) = \int_{\tau_\nu}^{\infty} B_\nu(t_\nu) e^{-(t_\nu - \tau_\nu)/\mu} \frac{dt_\nu}{\mu} \quad (0 < \mu \leqslant 1)$$

$$= -\int_{0}^{\tau_\nu} B_\nu(t_\nu) e^{-(t_\nu - \tau_\nu)/\mu} \frac{dt_\nu}{\mu} \quad (-1 \leqslant \mu < 0), \tag{4}$$

where
$$\tau_\nu = \int_{z}^{\infty} \kappa_\nu \rho\, dz, \tag{5}$$

is the normal optical thickness for radiation of frequency ν.

We shall now introduce the further assumption that the stellar atmosphere is in *radiative equilibrium*, i.e. there are no mechanisms, other than radiation, for transporting heat in the atmosphere. We shall also suppose that there are no sources of heat in the atmosphere. Under these circumstances, the net flux of radiation, in all wave-lengths, must be constant:

$$\pi F = \pi \int_0^\infty F_\nu(z) \, dv = 2\pi \int_0^\infty \int_{-1}^{+1} I_\nu(z, \mu)\mu \, d\mu dv = \text{constant.} \qquad (6)$$

This constant net integrated flux must be derived from the energy generated in the 'deep interior' of the star; but this is a fact not specially relevant to the study of stellar atmospheres.

The emergent flux will have the same value πF. In astrophysics, it is customary to introduce an *effective temperature* T_e related to the constant net flux πF by

$$\sigma T_e^4 = \pi F, \qquad (7)$$

where σ $(= 5\cdot75 \times 10^{-5}$ erg/sec. cm.^{2} degree^4) is Stefan's radiation constant. A similar relation holds between the integrated Planck intensity, B, and the local temperature, T:

$$B(T) = \int_0^\infty B_\nu(T) \, dv = \frac{\sigma}{\pi} T^4. \qquad (8)$$

The meaning of the flux condition (6) can be seen by integrating the equation of transfer (3), first over μ and then over ν. Thus,

$$-\frac{1}{4} \frac{dF_\nu}{\rho \, dz} = \kappa_\nu J_\nu - \kappa_\nu B_\nu$$

and

$$-\frac{1}{4} \frac{dF}{\rho \, dz} = \int_0^\infty \kappa_\nu(J_\nu - B_\nu) \, dv. \qquad (9)$$

The constancy of the net integrated flux therefore requires that

$$\int_0^\infty \kappa_\nu J_\nu \, dv = \int_0^\infty \kappa_\nu B_\nu \, dv. \qquad (10)$$

The left-hand side of this equation is proportional to the absorption of radiation in all wave-lengths by an element of mass exposed to the radiation $I_\nu(\mu)$, and the right-hand side is proportional to the total emission by the same element. Equation (10) therefore implies that every element of mass in the atmosphere emits exactly as much radiation as it absorbs. The equivalence of this last statement with the constancy of the net integrated flux is, of course, obvious on physical grounds.

We shall now formulate a fundamental problem in the theory of stellar atmospheres:

To solve the equation of transfer (3) in a semi-infinite atmosphere, under conditions of radiative equilibrium, for assigned variations of κ_ν with frequency and depth and with the boundary conditions

$$\left.\begin{array}{ll} I_\nu(\tau_\nu, -\mu) \equiv 0 & \text{for} \quad \tau_\nu = 0 \\[2mm] I_\nu(\tau_\nu, \mu) \prec e^{\tau_\nu} & \text{as} \quad \tau_\nu \to \infty. \end{array}\right\} \tag{11}$$

and

For arbitrary variations of κ_ν with frequency it would seem that this problem can be solved only by numerical methods of iteration. However, when the variations of κ_ν with frequency are not large, it is possible to obtain approximate solutions which seem adequate for many stellar applications. It is with such cases that we shall be principally concerned in this chapter.

The case when κ_ν is independent of frequency, i.e. when the atmosphere is *grey*, is of particular interest as it provides a physically significant *standard of comparison* for interpreting the continuous spectra of stars; for the departures of the observed intensity distribution in the continuous spectrum of a star from that predicted for the radiation from a grey atmosphere must be directly traceable to the *variation* of the stellar absorption coefficient with frequency. The determination of this variation of κ_ν with frequency from observations is, clearly, important for identifying and accounting for the source of continuous absorption in stellar atmospheres.

77. The method of solution

As we have already stated, we shall be principally concerned only with those cases in which the *effects* of the departures from greyness of the stellar material can be treated as small. We shall therefore write

$$\kappa_\nu = \bar{\kappa}(1+\delta_\nu), \tag{12}$$

where $\bar{\kappa}$ is a certain mean absorption coefficient. We shall leave $\bar{\kappa}$ undefined for the present, so that we may later have the choice of defining it in a manner which is most advantageous. Further, we shall let

$$\tau = \int_z^\infty \bar{\kappa}\rho \, dz \tag{13}$$

denote the optical thickness in $\bar{\kappa}$.

With the foregoing definitions, the equation of transfer (3) can be written in the form

$$\mu \frac{dI_\nu}{d\tau} = I_\nu - B_\nu + \delta_\nu(I_\nu - B_\nu). \tag{14}$$

We shall suppose that this equation can be solved, in successively higher approximations, by the following procedure:

First, we neglect the term in δ_ν altogether and write

$$\mu \frac{dI_\nu^{(1)}}{d\tau} = I_\nu^{(1)} - B_\nu^{(1)}. \tag{15}$$

We solve this equation appropriate to the problem on hand and use this solution in the term which occurs as the factor of δ_ν in equation (14). Thus the next approximation will be given by the solution of

$$\mu \frac{dI_\nu^{(2)}}{d\tau} = I_\nu^{(2)} - B_\nu^{(2)} + \delta_\nu \mu \frac{dI_\nu^{(1)}}{d\tau}. \tag{16}$$

The higher approximations can be found by extending this method of iteration. Thus in the nth approximation we consider:

$$\mu \frac{dI_\nu^{(n)}}{d\tau} = I_\nu^{(n)} - B_\nu^{(n)} + \delta_\nu(I_\nu^{(n-1)} - B_\nu^{(n-1)}). \tag{17}$$

This method of iteration can be expected to converge if δ_ν is sufficiently small. However, it appears that the practical utility of the method is not impaired even when δ_ν takes moderately large values of the order of 2 or 3.

Integrating equations (15) and (16) over all frequencies, we have

$$\mu \frac{dI^{(1)}}{d\tau} = I^{(1)} - B^{(1)} \tag{18}$$

and

$$\mu \frac{dI^{(2)}}{d\tau} = I^{(2)} - B^{(2)} + \mu \int_0^\infty \delta_\nu \frac{dI_\nu^{(1)}}{d\tau} \, d\nu. \tag{19}$$

Equations (18) and (19) have to be solved under the condition of a constant net integrated flux. In the first two approximations this condition requires that

$$B^{(1)} = J^{(1)} \tag{20}$$

and

$$B^{(2)} = J^{(2)} + \frac{1}{2} \int_0^\infty \int_{-1}^{+1} \mu \delta_\nu \frac{dI_\nu^{(1)}}{d\tau} \, d\nu \, d\mu, \tag{21}$$

respectively. Equation (21) can be written alternatively in the form

$$B^{(2)} = J^{(2)} + \frac{1}{4} \int_0^\infty \delta_\nu \frac{dF_\nu^{(1)}}{d\tau} d\nu. \tag{22}$$

In the second approximation, the integrated Planck intensity, B, therefore differs from the mean intensity, J, by an amount which depends on the non-constancy of the monochromatic fluxes $F_\nu^{(1)}$ in a grey atmosphere.

78. The temperature distribution in a grey atmosphere

According to equations (18) and (20), the equation of transfer for the integrated intensity in a grey atmosphere is

$$\mu \frac{dI^{(1)}(\tau, \mu)}{d\tau} = I^{(1)}(\tau, \mu) - \tfrac{1}{2} \int_{-1}^{+1} I^{(1)}(\tau, \mu') \, d\mu'. \tag{23}$$

It is seen that this equation is the same as the equation of transfer for the case of conservative isotropic scattering considered in Chapters III (§ 25), IV (§ 33.2), and V (§ 42). Accordingly, the exact solution for the angular distribution of the emergent radiation is given by (Chap. IV, eq. [52])

$$I^{(1)}(0, \mu) = \frac{\sqrt{3}}{4} FH(\mu), \tag{24}$$

where $H(\mu)$ is defined in terms of the characteristic function $\Psi(\mu) = \tfrac{1}{2}$. The law of darkening given by equation (24) has already been tabulated in a different context (Chap. VI, Table XV, p. 135).

From the property (Chap. V, eq. [25])

$$\int_0^1 H(\mu) \, d\mu = 2, \tag{25}$$

of the H-function which occurs in equation (24), we conclude that

$$J^{(1)}(0) = \frac{\sqrt{3}}{4} F. \tag{26}$$

Expressing $J^{(1)}(0)$ and F in terms of the *boundary temperature*, $T_0^{(1)}$, and the effective temperature T_e, in accordance with equations (7) and (8), we have the relation

$$(T_0^{(1)})^4 = \frac{\sqrt{3}}{4} T_e^4 \quad \text{or} \quad T_0^{(1)} = 0.8112 \, T_e. \tag{27}$$

Turning next to the solution of equation (23) in the nth approximation, we have (cf. Chap. III, § 25, eq. [40])

$$I_i^{(1)} = \tfrac{3}{4}F\left\{\sum_{\alpha=1}^{n-1}\frac{L_\alpha^{(1)}e^{-k_\alpha\tau}}{1+\mu_i k_\alpha}+Q^{(1)}+\tau+\mu_i\right\} \quad (i = \pm 1,...,\pm n), \quad (28)$$

where the n constants $L_\alpha^{(1)}$ ($\alpha = 1,...,n-1$) and $Q^{(1)}$ are to be determined from the equations (Chap. III, eq. [17])

$$\sum_{\alpha=1}^{n-1}\frac{L_\alpha^{(1)}}{1-\mu_i k_\alpha}+Q^{(1)}-\mu_i = 0 \quad (i = 1,...,n), \quad (29)$$

and the k_α's ($\alpha = 1,...,n-1$) are the positive (non-zero) characteristic roots of Chapter III, equation (7). On this approximation, the solution for $J^{(1)}$ is (Chap. III, eqs. [44]–[46])

$$J^{(1)} = \tfrac{3}{4}F\{\tau+q(\tau)\} = \tfrac{3}{4}F\left\{\tau+Q^{(1)}+\sum_{\alpha=1}^{n-1}L_\alpha^{(1)}e^{-k_\alpha\tau}\right\}. \quad (30)$$

The corresponding law of temperature distribution is (cf. eq. [20])

$$(T^{(1)})^4 = \tfrac{3}{4}T_e^4\{\tau+q(\tau)\}. \quad (31)$$

The function $q(\tau)$, in the second, third, and fourth approximations, has been tabulated in Chapter III (Table X, p. 80).

In terms of the temperature distribution (31), the radiation field in the atmosphere can be determined according to

$$I_\nu^{(1)}(\tau\ \mu) = \int_\tau^\infty B_\nu(T_t^{(1)})e^{-(t-\tau)/\mu}\frac{dt}{\mu} \quad (0 < \mu \leqslant 1)$$

$$= -\int_0^\tau B_\nu(T_t^{(1)})e^{-(t-\tau)/\mu}\frac{dt}{\mu} \quad (-1 \leqslant \mu < 0), \quad (32)$$

where $B_\nu(T_t^{(1)})$ is the Planck function for the temperature at optical depth t as given by equation (31).

From the solution (32) for the radiation field, we derive (cf. Chap. I, eqs. [91] and [96])

$$\int_{-1}^{+1} I_\nu^{(1)}(\tau,\mu)\mu^j\,d\mu = \int_\tau^\infty B_\nu(T_t^{(1)})E_{j+1}(t-\tau)\,dt+$$

$$+(-1)^j\int_0^\tau B_\nu(T_t^{(1)})E_{j+1}(\tau-t)\,dt. \quad (33)$$

In particular, we have

$$F_\nu^{(1)}(\tau) = 2\int_\tau^\infty B_\nu(T_t^{(1)})E_2(t-\tau)\,dt- 2\int_0^\tau B_\nu(T_t^{(1)})E_2(\tau-t)\,dt. \quad (34)$$

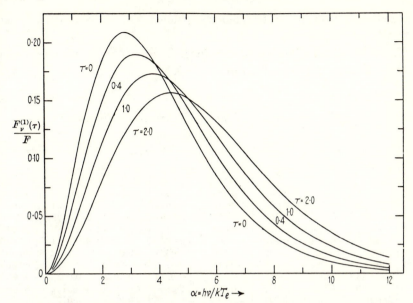

FIG. 29. The frequency distribution of the net flux of radiation at various depths in a grey atmosphere. The abscissa measures the frequency (in units of kT_e/h) and the ordinate measures the flux (in units of the constant net integrated flux). The curves refer to the depths $\tau = 0$, $0\cdot4$, $1\cdot0$, and $2\cdot0$.

TABLE XXVI

The Monochromatic Fluxes $F_\nu^{(1)}(\tau)$ in Units of F

						α			
τ		*1*	*2*	*3*	*4*	*6*	*8*	*10*	*12*
0.	.	0·0864	0·1837	0·2074	0·1772	0·0856	0·0314	0·01013	0·00305
0·1.	.	0·0784	0·1732	0·2027	0·1791	0·0911	0·0346	0·01135	0·00345
0·2.	.	0·0719	0·1640	0·1981	0·1801	0·0961	0·0376	0·01261	0·00388
0·3.	.	0·0663	0·1556	0·1933	0·1804	0·1005	0·0406	0·01390	0·00434
0·4.	.	0·0615	0·1480	0·1886	0·1802	0·1044	0·0435	0·01521	0·00482
0·5.	.	0·0571	0·1407	0·1835	0·1793	0·1078	0·0462	0·01652	0·00532
0·6.	.	0·0532	0·1342	0·1789	0·1783	0·1110	0·0489	0·01785	0·00584
0·7.	.	0·0498	0·1282	0·1745	0·1772	0·1138	0·0515	0·01918	0·00638
0·8.	.	0·0466	0·1225	0·1699	0·1755	0·1163	0·0539	0·02050	0·00693
0·9.	.	0·0439	0·1174	0·1656	0·1740	0·1186	0·0563	0·02182	0·00749
1·0.	.	0·0413	0·1128	0·1616	0·1724	0·1207	0·0586	0·02314	0·00807
1·2.	.	0·0370	0·1042	0·1540	0·1688	0·1240	0·0628	0·02572	0·00925
1·4.	.	0·0335	0·0968	0·1470	0·1651	0·1268	0·0667	0·02824	0·01046
1·6.	.	0·0305	0·0906	0·1406	0·1614	0·1289	0·0702	0·03069	0·01169
1·8.	.	0·0280	0·0850	0·1349	0·1578	0·1306	0·0735	0·03306	0·01293
2·0.	.	0·0259	0·0803	0·1295	0·1543	0·1319	0·0764	0·03534	0·01417

The monochromatic fluxes $F_\nu^{(1)}(\tau)$ computed according to this formula for various values of $\alpha = h\nu/kT_e$ and τ are given in Table XXVI; they are further illustrated in Fig. 29. It is of particular interest to observe the redistribution in the frequencies which takes place as we descend into the atmosphere.

The derivatives of the monochromatic fluxes, $dF_\nu^{(1)}/d\tau$, which are also useful (cf. eq. [22]), are given in Table XXVII.

TABLE XXVII

The Derivatives of the Monochromatic Fluxes:
$dF_\nu^{(1)}/d\tau$ *in Units of F*

τ	α							
	1	2	3	4	6	8	10	12
0	−0·086	−0·11	−0·044	+0·024	+0·0585	+0·0323	+0·0120	+0·0039
0·1	−0·072	−0·098	−0·046	+0·015	+0·0523	+0·0311	+0·0124	+0·0044
0·2	−0·060	−0·088	−0·047	+0·007	+0·0467	+0·0302	+0·0128	+0·0048
0·3	−0·052	−0·080	−0·048	+0·001	+0·0417	+0·0293	+0·0130	+0·00465
0·4	−0·046	−0·074	−0·0485	−0·005	+0·0367	+0·0281	+0·0131	+0·00490
0·5	−0·041	−0·068	−0·048	−0·008	+0·0327	+0·0270	+0·0132	+0·00510
0·6	−0·036	−0·063	−0·046	−0·010	+0·0295	+0·0260	+0·0133	+0·00530
0·7	−0·033	−0·058	−0·045	−0·013	+0·0264	+0·0250	+0·0133	+0·00544
0·8	−0·030	−0·054	−0·043	−0·015	+0·0240	+0·0242	+0·0132	+0·00558
0·9	−0·027	−0·050	−0·041	−0·016	+0·0215	+0·0233	+0·0132	+0·00573
1·0	−0·024	−0·046	−0·040	−0·017	+0·0193	+0·0222	+0·0131	+0·00583
1·2	−0·020	−0·040	−0·036	−0·018	+0·0152	+0·0202	+0·0128	+0·00598
1·4	−0·016	−0·034	−0·033	−0·0185	+0·0121	+0·0186	+0·0124	+0·00610
1·6	−0·014	−0·029	−0·030	−0·0185	+0·0095	+0·0170	+0·0120	+0·00617
1·8	−0·011	−0·026	−0·028	−0·018	+0·0075	+0·0156	+0·0116	+0·0062
2·0	−0·010	−0·022	−0·025	−0·017	+0·0060	+0·0142	+0·0112	+0·0062

79. The temperature distribution in a slightly non-grey atmosphere

For an atmosphere which departs only slightly from greyness the approximate form of the equation of transfer is (cf. eqs. [19]–[22])

$$\mu\frac{dI^{(2)}}{d\tau} = I^{(2)}-\frac{1}{2}\int_{-1}^{+1} I^{(2)}\,d\mu+\mu\int_{0}^{\infty}\delta_\nu\frac{dI_\nu^{(1)}}{d\tau}\,d\nu-\frac{1}{2}\int_{0}^{\infty}\int_{-1}^{+1}\delta_\nu\,\mu\frac{dI_\nu^{(1)}}{d\tau}\,d\mu d\nu.$$

(35)

In solving this equation, in turn, in the nth approximation, we replace it by the system of $2n$ linear equations ,

$$\mu_i\frac{dI_i^{(2)}}{d\tau} = I_i^{(2)}-\tfrac{1}{2}\sum a_j I_j^{(2)}+\mu_i\int_{0}^{\infty}\delta_\nu\frac{dI_{\nu,i}^{(1)}}{d\tau}\,d\nu-\frac{1}{2}\int_{0}^{\infty}\delta_\nu\sum a_j\mu_j\frac{dI_{\nu,j}^{(1)}}{d\tau}\,d\nu$$

$$(i = \pm 1,...,\pm n), \quad (36)$$

where the various symbols have their usual meanings.

The system of equations represented by (36) is most conveniently solved by the method of the variation of the parameters. Thus, since the homogeneous part of the system (36) is the same as that considered in Chapter III, § 25, we seek a solution of the form (cf. Chap. III, eq. [14])

$$I_i^{(2)} = \tfrac{3}{4}F\left\{ \sum_{\alpha=-n+1}^{n-1} \frac{L_\alpha^{(2)} e^{-k_\alpha \tau}}{1+\mu_i k_\alpha} + Q^{(2)}(\tau)+\tau+\mu_i\right\} \quad (i = \pm 1,\dots, \pm n),$$

(37)

where $L_\alpha^{(2)}$ ($\alpha = \pm 1,\dots, \pm n \mp 1$) and $Q^{(2)}$ are all to be considered as functions of τ. It will be noticed that in writing the solution in the form (37) we have treated as variable only $2n-1$ of the $2n$ constants of integration which the general solution of the homogeneous system associated with (36) involves. This is, however, permissible in view of the fact that (36) admits the flux integral

$$2 \sum a_i \mu_i I_i^{(2)} = \text{constant},$$

(38)

and this must have the assigned value F.

The mean intensity $J^{(2)}$ $(= \tfrac{1}{2} \sum a_j I_j^{(2)})$ corresponding to the solution (37) for the $I_i^{(2)}$'s, is (cf. Chap. III, eqs. [26] and [43])

$$J^{(2)} = \tfrac{3}{4}F\left\{ \tau+Q^{(2)}(\tau)+ \sum_{\alpha=-n+1}^{n-1} L_\alpha^{(2)}(\tau)e^{-k_\alpha \tau}\right\}.$$

(39)

Now substituting for $I_i^{(2)}$ according to equation (37) in (36) we obtain the variational equations

$$\tfrac{3}{4}F\mu_i\left\{ \sum_{\alpha=-n+1}^{n-1} \frac{e^{-k_\alpha \tau}}{1+\mu_i k_\alpha} \frac{dL_\alpha^{(2)}}{d\tau}+ \frac{dQ^{(2)}}{d\tau}\right\}$$

$$= \mu_i \int_0^\infty \delta_\nu \frac{dI_{\nu,i}^{(1)}}{d\tau} d\nu-\frac{1}{2} \int_0^\infty \delta_\nu \sum a_j \mu_j \frac{dI_{\nu,j}^{(1)}}{d\tau} d\nu \quad (i = \pm 1,\dots, \pm n). \quad (40)$$

Of the $2n$ equations represented by equation (40), only $2n-1$ are linearly independent, since the equation derived from (40) by multiplying by a_i and summing over all i is identically satisfied in virtue of Chapter III, equation (27).

The order of the system (40) can be, further, reduced to $2n-2$ by a proper choice of the mean absorption coefficient $\bar{\kappa}$. Thus, multiplying equation (40) by $a_i \mu_i$ and summing over all i, we obtain (cf. Chap. II, eq. [35] and Chap. III, eq. [27])

$$\tfrac{1}{2}F\frac{dQ^{(2)}}{d\tau} = \int_0^\infty \delta_\nu \sum a_i \mu_i^2 \frac{dI_{\nu,i}^{(1)}}{d\tau} d\nu.$$

(41)

Accordingly, if we arrange that

$$\int_0^\infty \delta_\nu \sum a_i \mu_i^2 \frac{dI_{\nu,i}^{(1)}}{d\tau} d\nu = 2 \int_0^\infty \delta_\nu \frac{dK_\nu^{(1)}}{d\tau} d\nu = 0, \tag{42}$$

we shall have the integral

$$Q^{(2)} = \text{constant} = Q^{(1)} + \Delta Q \text{ (say)}, \tag{43}$$

where $Q^{(1)}$ is the value of the constant in the solution for the grey atmosphere (eq. [28]).

Since (cf. eq. [15])

$$\frac{dK_\nu^{(1)}}{d\tau} = \tfrac{1}{4} F_\nu^{(1)}, \tag{44}$$

we can rewrite the condition (42) as

$$\int_0^\infty \delta_\nu F_\nu^{(1)} d\nu = 0. \tag{45}$$

To satisfy this condition on δ_ν, we must define the mean absorption coefficient $\bar{\kappa}$ (which we have so far left undefined) by

$$\bar{\kappa} \int_0^\infty F_\nu^{(1)} d\nu = \int_0^\infty \kappa_\nu F_\nu^{(1)} d\nu,$$

or

$$\bar{\kappa} = \frac{1}{F} \int_0^\infty \kappa_\nu F_\nu^{(1)} d\nu; \tag{46}$$

for with this choice of $\bar{\kappa}$, $\delta_\nu = (\kappa_\nu - \bar{\kappa})/\bar{\kappa}$ (cf. eq. [12]), when similarly averaged, will be zero, as required.

With $\bar{\kappa}$ chosen in this manner, $Q^{(2)}$ is a constant, and the variational equations become

$$\tfrac{3}{4} F \mu_i \left\{ \sum_{\alpha=-n+1}^{n-1} \frac{e^{-k_\alpha \tau}}{1 + \mu_i k_\alpha} \frac{dL_\alpha^{(2)}}{d\tau} \right\}$$

$$= \mu_i \int_0^\infty \delta_\nu \frac{dI_{\nu,i}^{(1)}}{d\tau} d\nu - \frac{1}{2} \int_0^\infty \delta_\nu \sum a_j \mu_j \frac{dI_{\nu,j}^{(1)}}{d\tau} d\nu \quad (i = \pm 1, ..., \pm n). \tag{47}$$

In view of the fact that the rank of the system (47) is less than the number of equations, it appears that the most symmetrical way of treating the equations is the following:

Multiplying equation (47) by $a_i \mu_i^{m-1}$ $(m = 1, ..., 2n)$ and summing over all i, we obtain

$$\tfrac{3}{4} F \sum_{\alpha=-n+1}^{n-1} D_m(k_\alpha) e^{-k_\alpha \tau} \frac{dL_\alpha^{(2)}}{d\tau} = \delta_m - \frac{1}{m} \epsilon_{m,\text{odd}} \delta_1 \quad (m = 1, ..., 2n), \tag{48}$$

where
$$\delta_m = \int_0^\infty \delta_\nu \sum a_i \mu_i^m \frac{dI_{\nu,i}^{(1)}}{d\tau} d\nu, \tag{49}$$

and D_m and $\epsilon_{m,\text{odd}}$ have the same meanings as in Chapter III, equations (18) and (20).

Since D_1, D_2, and δ_2 are all zero, of the $2n$ equations represented by (48) those for $m = 1$ and 2 are identically satisfied. The remaining $2n-2$ equations will suffice to determine the $L_\alpha^{(2)}$'s apart from $(2n-2)$ constants of integration. These constants of integration must be determined from the boundary conditions (11).

The boundary conditions at infinity require that
$$L_{-\alpha}^{(2)}(\tau) \to 0 \quad \text{as} \quad \tau \to \infty \quad \text{for} \quad \alpha = 1,...,n-1, \tag{50}$$

since $L_{-\alpha}^{(2)}$ occurs with $e^{+k_\alpha\tau}$ as a factor in the solution (37). The conditions (50) will determine the constants of integration in the solutions for $L_{-\alpha}^{(2)}(\tau)$ ($\alpha = 1,...,n-1$). However, the solutions for $L_\alpha^{(2)}(\tau)$ ($\alpha = 1,..., n-1$) will have undetermined constants in them. Therefore, writing
$$L_\alpha^{(2)}(\tau) = L_\alpha^{(1)} + \Delta L_\alpha(\tau) \quad (\alpha = 1,...,n-1), \tag{51}$$

where $L_\alpha^{(1)}$ is the value of the constant in the solution for a grey atmosphere (eq. [28]), we determine $\Delta L_\alpha(0)$ from the conditions at $\tau = 0$. Thus, we must have (cf. eq. [37])
$$\sum_{\alpha=1}^{n-1} \frac{L_\alpha^{(1)} + \Delta L_\alpha(0)}{1 - \mu_i k_\alpha} + Q^{(1)} + \Delta Q - \mu_i = -\sum_{\alpha=1}^{n-1} \frac{L_{-\alpha}^{(2)}(0)}{1 + \mu_i k_\alpha}$$
$$(i = 1,...,n). \tag{52}$$

On the other hand, $L_\alpha^{(1)}$ ($\alpha = 1,...,n-1$) and $Q^{(1)}$, being the constants in the solution for a grey atmosphere, satisfy equation (29). Equation (52), therefore, reduces to
$$\sum_{\alpha=1}^{n-1} \frac{\Delta L_\alpha(0)}{1 - \mu_i k_\alpha} + \Delta Q = -\sum_{\alpha=1}^{n-1} \frac{L_{-\alpha}^{(2)}(0)}{1 + \mu_i k_\alpha} \quad (i = 1,...,n). \tag{53}$$

These equations make the solution determinate.

With the $L_\alpha^{(2)}$'s expressed in the manner (51), the expression (39) for the mean intensity becomes
$$J^{(2)} = J^{(1)} + \tfrac{3}{4} F \Big\{ \Delta Q + \sum_{\alpha=1}^{n-1} \Delta L_\alpha(\tau) e^{-k_\alpha\tau} + \sum_{\alpha=1}^{n-1} L_{-\alpha}^{(2)} e^{+k_\alpha\tau} \Big\}. \tag{54}$$

79.1. *The solution in the* (2, 1) *approximation*

When the equation of transfer (35) is solved in the first approximation there are no L's to determine and (cf. eq. [53])
$$\Delta Q = 0. \tag{55}$$

Hence in this approximation (cf. eqs. [30] and [54])

$$J^{(2)} = J^{(1)} = \tfrac{3}{4}F(\tau + Q^{(1)}) \qquad (56)$$

and
$$B^{(2)} = \tfrac{3}{4}F(\tau + Q^{(1)}) + \frac{1}{4}\int_0^\infty \delta_\nu \frac{dF_\nu^{(1)}}{d\tau}\,d\nu. \qquad (57)$$

With the boundary conditions we have adopted, $Q^{(1)} = 1/\sqrt{3}$ (cf. Chap. III, Table VIII); however, in the first approximation it is more satisfactory to take $Q^{(1)} = \tfrac{2}{3}$, so that†

$$B^{(2)}(\tau) = \tfrac{3}{4}F(\tau + \tfrac{2}{3}) + \frac{1}{4}\int_0^\infty \delta_\nu \frac{dF_\nu^{(1)}}{d\tau}\,d\nu. \qquad (58)$$

Now if δ_ν is independent of τ

$$\int_0^\infty \delta_\nu \frac{dF_\nu^{(1)}}{d\tau}\,d\nu = \frac{d}{d\tau}\int_0^\infty \delta_\nu F_\nu^{(1)}\,d\nu = 0, \qquad (59)$$

and the temperature distributions for the grey and the non-grey atmospheres agree on this approximation. This agreement is directly a result of the manner in which we have defined $\bar{\kappa}$; its particular advantage is therefore apparent.

79.2. *The solution in the $(2,2)$ approximation*

When the equation of transfer (35) is solved in the second approximation, there is only one characteristic root,

$$k_1 = \sqrt{35}/3 = 1\cdot 97203, \qquad (60)$$

and the solution for $J^{(1)}$ is (cf. Chap. III, Table VIII)

$$J^{(1)} = \tfrac{3}{4}F(\tau + 0\cdot 6940 - 0\cdot 1167e^{-k_1\tau}). \qquad (61)$$

The variational equations for $L_1^{(2)}$ and $L_{-1}^{(2)}$ follow from equation (48) by setting $m = 3$ and 4, and, further, using the relations (cf. Chap. III, eqs. [28] and [29])

$$D_3(k_1) = -D_3(-k_1) = -k_1 D_4(k_1) = -k_1 D_4(-k_1) = \frac{2}{3k_1}. \qquad (62)$$

Thus
$$\frac{F}{2k_1}\left(e^{-k_1\tau}\frac{dL_1^{(2)}}{d\tau} - e^{+k_1\tau}\frac{dL_{-1}^{(2)}}{d\tau}\right) = \delta_3 - \tfrac{1}{3}\delta_1$$

and
$$\frac{F}{2k_1}\left(e^{-k_1\tau}\frac{dL_1^{(2)}}{d\tau} + e^{+k_1\tau}\frac{dL_{-1}^{(2)}}{d\tau}\right) = -k_1\delta_4. \qquad (63)$$

Solving these equations, we obtain:

$$F\frac{dL_1^{(2)}}{d\tau} = +e^{+k_1\tau}(\delta_3 - \tfrac{1}{3}\delta_1 - k_1\delta_4)k_1$$

† Though the value $Q^{(1)} = 2/3$ does not predict the correct boundary temperature, it is nevertheless more satisfactory on account of the fact that it predicts a law of darkening [namely, $I(0, \mu) = \tfrac{3}{4}F(\mu + \tfrac{2}{3})$] which satisfies the flux condition at $\tau = 0$. This is Milne's original argument for this choice of $Q^{(1)}$.

and
$$F\frac{dL_{-1}^{(2)}}{d\tau} = -e^{-k_1\tau}(\delta_3-\tfrac{1}{3}\delta_1+k_1\delta_4)k_1. \tag{64}$$

Integrating these equations, we have (cf. eq. [51])

$$FL_1^{(2)} = F[L_1^{(1)}+\Delta L_1(0)]+\int_0^\tau e^{+k_1\tau}(\delta_3-\tfrac{1}{3}\delta_1-k_1\delta_4)\,d(k_1\tau)$$

and
$$FL_{-1}^{(2)} = \int_\tau^\infty e^{-k_1\tau}(\delta_3-\tfrac{1}{3}\delta_1+k_1\delta_4)\,d(k_1\tau), \tag{65}$$

where, in integrating the equation for $L_{-1}^{(2)}$, we have made use of the boundary condition (50).

The remaining constants $\Delta L_1(0)$ and ΔQ now follow from equation (53). We find

$$\Delta L_1(0) = \frac{(1-\mu_1 k_1)(1-\mu_2 k_1)}{(1+\mu_1 k_1)(1+\mu_2 k_1)}L_{-1}^{(2)}(0)$$

and
$$\Delta Q = -\frac{2}{(1+\mu_1 k_1)(1+\mu_2 k_1)}L_{-1}^{(2)}(0). \tag{66}$$

With $L_1^{(2)}$ and $L_{-1}^{(2)}$ given by equations (65), the solution (54) for $J^{(2)}$ becomes

$$J^{(2)} = J^{(1)}+\tfrac{3}{4}F[\Delta L_1(0)e^{-k_1\tau}+\Delta Q]+$$

$$+\tfrac{3}{4}e^{-k_1\tau}\int_0^\tau e^{+k_1\tau}(\delta_3-\tfrac{1}{3}\delta_1-k_1\delta_4)\,d(k_1\tau)+$$

$$+\tfrac{3}{4}e^{+k_1\tau}\int_\tau^\infty e^{-k_1\tau}(\delta_3-\tfrac{1}{3}\delta_1+k_1\delta_4)\,d(k_1\tau). \tag{67}$$

Now substituting for $J^{(1)}$, $\Delta L_1(0)$, and ΔQ according to equations (61) and (66) and inserting the numerical values for the various constants, we finally obtain

$$B^{(2)} = J^{(2)}+\tfrac{1}{2}\delta_1 = \tfrac{3}{4}F(\tau+0.6940-0.1167e^{-k_1\tau})+\tfrac{1}{2}\delta_1-$$

$$-0.1664(2+0.2301e^{-k_1\tau})\int_0^\infty e^{-k_1\tau}(\delta_3-\tfrac{1}{3}\delta_4+k_1\delta_4)\,d(k_1\tau)+$$

$$+\tfrac{3}{4}e^{-k_1\tau}\int_0^\tau e^{+k_1\tau}(\delta_3-\tfrac{1}{3}\delta_1-k_1\delta_4)\,d(k_1\tau)+$$

$$+\tfrac{3}{4}e^{+k_1\tau}\int_\tau^\infty e^{-k_1\tau}(\delta_3-\tfrac{1}{3}\delta_1+k_1\delta_4)\,d(k_1\tau). \tag{68}$$

The practical use of the foregoing solution in the (2, 2) approximation requires, in addition to the monochromatic fluxes $F_\nu^{(1)}$ and their derivatives which have been tabulated (Tables XXVI and XXVII), a knowledge also of the quantities

$$W_{j,\nu} = \sum a_i \mu_i^j \frac{dI_{\nu,i}^{(1)}}{d\tau} \simeq \int_{-1}^{+1} \frac{dI_\nu^{(1)}}{d\tau} \mu^j \, d\mu \quad (j = 3, 4), \tag{69}$$

which are needed in the evaluation of δ_3 and δ_4. These 'weight functions' have been evaluated by numerical methods by Mrs. Frances H. Breen and the writer, and are given in Table XXVIII.

TABLE XXVIII
The Weight Functions $W_{3,\alpha}(\tau)$ and $W_{4,\alpha}(\tau)$

τ	$\alpha = 1$		$\alpha = 2$		$\alpha = 3$		$\alpha = 4$	
	$W_3(\tau)$	$W_4(\tau)$	$W_3(\tau)$	$W_4(\tau)$	$W_3(\tau)$	$W_4(\tau)$	$W_3(\tau)$	$W_4(\tau)$
0	-0.0127	0.0224	-0.0126	0.0487	$+0.00244$	0.0561	$+0.0153$	0.0490
0·1	-0.0127	0.0212	-0.0152	0.0473	-0.00259	0.0561	$+0.0102$	0.0503
0·2	-0.0121	0.0199	-0.0162	0.0457	-0.00573	0.0557	$+0.00657$	0.0511
0·3	-0.0114	0.0187	-0.0165	0.0441	-0.00771	0.0550	$+0.00387$	0.0516
0·4	-0.0107	0.0176	-0.0162	0.0425	-0.00892	0.0542	$+0.00186$	0.0519
0·5	-0.00986	0.0166	-0.0157	0.0409	-0.00962	0.0532	$+0.000351$	0.0520
0·6	-0.00910	0.0157	-0.0150	0.0393	-0.00998	0.0522	-0.00080	0.0520
0·7	-0.00838	0.0148	-0.0143	0.0379	-0.01015	0.0512	-0.00172	0.0519
0·8	-0.00771	0.0140	-0.0135	0.0365	-0.01016	0.0502	-0.00243	0.0516
0·9	-0.00709	0.0133	-0.0128	0.0351	-0.01008	0.0492	-0.00301	0.0514
1·0	-0.00653	0.0126	-0.0121	0.0339	-0.00991	0.0482	-0.00346	0.0511
1·2	-0.00554	0.0114	-0.0107	0.0316	-0.00948	0.0463	-0.00412	0.0503
1·4	-0.00470	0.0103	-0.00945	0.0296	-0.00889	0.0444	-0.00446	0.0494
1·6	-0.00400	0.00960	-0.00837	0.0278	-0.00831	0.0427	-0.00467	0.0485
1·8	-0.00343	0.00897	-0.00742	0.0263	-0.00774	0.0411	-0.00476	0.0476
2·0	-0.00294	0.00833	-0.00657	0.0249	-0.00716	0.0396	-0.00474	0.0466

TABLE XXVIII (continued)

τ	$\alpha = 6$		$\alpha = 8$		$\alpha = 10$		$\alpha = 12$	
	$W_3(\tau)$	$W_4(\tau)$	$W_3(\tau)$	$W_4(\tau)$	$W_3(\tau)$	$W_4(\tau)$	$W_3(\tau)$	$W_4(\tau)$
0	0.0177	0.0247	0.00906	0.00947	0.00344	0.00317	0.00113	0.000535
0·1	0.0158	0.0264	0.00889	0.0104	0.00357	0.00352	0.00122	0.000652
0·2	0.0141	0.0279	0.00866	0.0112	0.00367	0.00389	0.00130	0.000778
0·3	0.0127	0.0292	0.00841	0.0121	0.00374	0.00426	0.00137	0.000911
0·4	0.0114	0.0304	0.00815	0.0129	0.00379	0.00463	0.00143	0.00105
0·5	0.0103	0.0315	0.00789	0.0137	0.00383	0.00501	0.00149	0.00120
0·6	0.00932	0.0325	0.00764	0.0145	0.00385	0.00540	0.00154	0.00135
0·7	0.00843	0.0334	0.00737	0.0153	0.00386	0.00578	0.00159	0.00151
0·8	0.00763	0.0342	0.00711	0.0160	0.00386	0.00617	0.00164	0.00167
0·9	0.00690	0.0349	0.00685	0.0167	0.00385	0.00656	0.00167	0.00183
1·0	0.00623	0.0356	0.00659	0.0174	0.00382	0.00694	0.00171	0.00200
1·2	0.00503	0.0367	0.00607	0.0186	0.00375	0.00770	0.00176	0.00235
1·4	0.00406	0.0376	0.00559	0.0198	0.00367	0.00844	0.00180	0.00270
1·6	0.00321	0.0383	0.00511	0.0209	0.00356	0.00916	0.00182	0.00307
1·8	0.00248	0.0389	0.00466	0.0218	0.00343	0.00986	0.00182	0.00343
2·0	0.00187	0.0393	0.00424	0.0227	0.00330	0.01054	0.00182	0.00380

80. The nature and the origin of the stellar continuous absorption coefficient as inferred from the theory of radiative equilibrium

The intensity distribution in the continuous spectrum of the sun at the centre of the solar disk and the law of darkening in the different wave-lengths have been the subject of careful measurements. As a

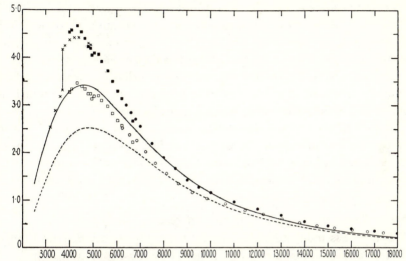

FIG. 30. The comparison of the intensity distribution in the continuous spectrum of the sun with that in a grey atmosphere with the same net integrated flux. The abscissae denote the wave-lengths in angstroms and the ordinates the intensities in the unit 10^{14} erg/cm.^{2} sec.

The crosses, solid circles, and squares represent the observations made at the centre of the solar disk: the crosses denote the measures of D. Chalonge (*Annales d'Astrophysique*, **9**, 143, 1946), the dots the measures of C. G. Abbott as reduced to the absolute scale by M. Minnaert (*Bull. Astron. Netherlands*, **2**, 75, 1924) and the squares the measures of H. H. Plaskett as reduced by G. Mulders (*Dissertation*, Utrecht, 1934). The open circles and squares represent the observations in the emergent flux, the circles again representing the measures of Abbott as reduced by Minnaert and the squares the measures of Plaskett as reduced by Mulders.

The full-line curve is the intensity distribution at the centre of the solar disk that would be expected were the solar atmosphere grey; the dashed curve, similarly, represents the distribution to be expected in the emergent flux.

result, our knowledge of the continuous spectrum of the sun is exceptionally complete: in no other case do we know, in as much detail, the character of the emergent radiation in its dependence on the two variables, angle of emergence and wave-length. It is therefore natural that all discussions relating to the continuous spectrum of the stars should begin with the sun.

In Fig. 30 the measures on the frequency distribution in the continuous spectrum of the sun at the centre of the disk and in the emergent flux are plotted and compared with the distributions to be expected from a grey atmosphere having the same effective temperature

$(T_e = 5{,}740°\,\mathrm{K}.)$ as the sun. The departures of the observed distributions from those predicted for a grey atmosphere must clearly be attributed to the variation of the solar absorption coefficient with wave-length. The question now arises as to what can be inferred concerning this variation of κ_ν with ν from the observations. It is evident that there is some ambiguity in formulating the question in this way, for the dependence of κ_ν on frequency is not the only factor which is involved: its dependence on the depth must also have an influence on the emergent radiation. Nevertheless, in a first exploratory attempt it may not be misleading to make the assumption that $\kappa_\nu/\bar\kappa$ is independent of depth, in which case the departures of the observed distributions from those predicted for a grey atmosphere must be attributed, entirely, to the dependence of the absorption coefficient on wave-length. And if it be also assumed that the departures from greyness of the stellar material are not large, the solution of the transfer problem in the $(2,1)$ approximation (§ 79.1) would seem to provide the most convenient starting-point for such discussions.

80.1. *The method of analysis and inference*

When $\kappa_\nu/\bar\kappa$ is independent of depth, the temperature distribution in the $(2,1)$ approximation of § 79.1 is given by (eqs. [58] and [59])

$$T^4 = \tfrac{1}{2}T_e^4(1+\tfrac{3}{2}\tau), \tag{70}$$

where τ is measured in the mean absorption coefficient

$$\bar\kappa = \frac{1}{F}\int\limits_0^\infty \kappa_\nu\,F_\nu^{(1)}\,d\nu. \tag{71}$$

On these assumptions, the intensity of the radiation emergent in the direction μ and in the frequency ν is given by (cf. eq. [4])

$$I_\nu(0,\mu) = \int\limits_0^\infty B_\nu(T_\tau)\exp\!\left(-\frac{\kappa_\nu}{\bar\kappa}\frac{\tau}{\mu}\right)d\!\left(\frac{\kappa_\nu}{\bar\kappa}\frac{\tau}{\mu}\right). \tag{72}$$

Similarly, the distribution in frequency of the emergent flux will be given by (cf. eq. [34])

$$F_\nu(0) = 2\int\limits_0^\infty B_\nu(T_\tau)E_2\!\left(\frac{\kappa_\nu}{\bar\kappa}\tau\right)d\!\left(\frac{\kappa_\nu}{\bar\kappa}\tau\right). \tag{73}$$

Equations (72) and (73) suffice to predict the entire character of the emergent radiation in its dependence on the two variables μ and ν in terms of a function of one variable only, namely, $\kappa_\nu/\bar\kappa$. Consequently, while we have $\kappa_\nu/\bar\kappa$ at our disposal to bring about an agreement between the observed intensities and those predicted by (72) for a particular value of μ (or between the observed fluxes and those predicted by [73]), a comparison of the predictions of (72) with all the observations relating to the darkening in the different wave-lengths and the absolute intensities at the centre will already provide a valuable guide as to the adequacy or otherwise of a theory based on the concept of radiative equilibrium under conditions of local thermodynamic equilibrium.

For carrying out the comparisons between the predictions of (72) and (73) and observations, it is convenient to rewrite equations (72) and (73) in the forms

$$I_\nu(0,\mu) = B_\nu(T_0) \int_0^\infty b_\nu(\tau)\exp\left(-\frac{\kappa_\nu}{\kappa}\frac{\tau}{\mu}\right)d\left(\frac{\kappa_\nu}{\kappa}\frac{\tau}{\mu}\right),$$

and

$$F_\nu(0) = 2B_\nu(T_0) \int_0^\infty b_\nu(\tau)E_2\left(\frac{\kappa_\nu}{\kappa}\tau\right)d\left(\frac{\kappa_\nu}{\kappa}\tau\right), \tag{74}$$

where

$$b_\nu(\tau) = \frac{B_\nu(T_\tau)}{B_\nu(T_0)} = \frac{e^{h\nu/kT_0}-1}{e^{h\nu/kT_\tau}-1}, \tag{75}$$

and $B_\nu(T_0)$ is the Planck function for the boundary temperature T_0.

Since the dependence of the emergent intensity and flux, on the effective temperature, is determined only by the value of $h\nu/kT_e$, we shall write

$$\alpha = h\nu/kT_e \tag{76}$$

and express $I_\nu(0,\mu)$ and $F_\nu(0)$ in terms of the two integrals

$$\mathscr{I}(\alpha,\beta) = \int_0^\infty e^{-\beta\tau}b_\alpha(\tau)\,d(\beta\tau)$$

and

$$\mathscr{F}(\alpha,\beta) = 2\int_0^\infty E_2(\beta\tau)b_\alpha(\tau)\,d(\beta\tau). \tag{77}$$

Thus

$$I_\nu(0,\mu) = B_\nu(T_0)\mathscr{I}\left(\frac{h\nu}{kT_e},\frac{\kappa_\nu}{\kappa\mu}\right)$$

and

$$F_\nu(0) = B_\nu(T_0)\mathscr{F}\left(\frac{h\nu}{kT_e},\frac{\kappa_\nu}{\kappa}\right). \tag{78}$$

From the expressions for $I_\nu(0,\mu)$ and $F_\nu(0)$ in terms of the two integrals $\mathscr{I}(\alpha,\beta)$ and $\mathscr{F}(\alpha,\beta)$, it is clear that both for deriving the theoretical consequences of a known source of continuous absorption and for inferring the variation of the continuous absorption coefficient with wave-length required by the results of spectrophotometry, we must have adequate tables of the basic integrals.

Integrals equivalent to $\mathscr{I}(\alpha,\beta)$ and $\mathscr{F}(\alpha,\beta)$ have been evaluated for certain ranges of α and β by E. A. Milne, B. Lindblad, and G. Burkhardt;

but the most complete tabulations are those of S. Chandrasekhar and Frances H. Breen. The computations of these last authors are given in Tables XXIX and XXX.†

TABLE XXIX

$\log \mathscr{I}(\alpha, \beta)$

$(0 \leqslant \alpha \leqslant 12; \ 0.2 \leqslant \beta \leqslant 2)$

β	$\alpha = 0$	$\alpha = 1$	$\alpha = 2$	$\alpha = 3$	$\alpha = 4$	$\alpha = 6$	$\alpha = 8$	$\alpha = 10$	$\alpha = 12$
0·2 . .	0·2223	0·3480	0·5085	0·6984	0·9108	1·3815	1·8917	2·4273	2·9810
0·3 . .	0·1880	0·2988	0·4416	0·6114	0·8021	1·2270	1·6905	2·1796	2·6879
0·4 . .	0·1653	0·2655	0·3954	0·5504	0·7250	1·1156	1·5435	1·9974	2·4712
0·5 . .	0·1487	0·2408	0·3606	0·5040	0·6659	1·0289	1·4283	1·8537	2·2996
0·6 . .	0·1359	0·2214	0·3330	0·4669	0·6182	0·9583	1·3337	1·7352	2·1574
0·7 . .	0·1256	0·2056	0·3104	0·4361	0·5785	0·8991	1·2540	1·6347	2·0365
0·8 . .	0·1170	0·1924	0·2912	0·4101	0·5447	0·8483	1·1853	1·5478	1·9315
0·9 . .	0·1097	0·1811	0·2749	0·3877	0·5156	0·8042	1·1252	1·4712	1·8389
1·0 . .	0·1035	0·1713	0·2606	0·3681	0·4899	0·7652	1·0719	1·4035	1·7564
1·1 . .	0·0981	0·1628	0·2480	0·3507	0·4672	0·7305	1·0243	1·3426	1·6821
1·2 . .	0·0933	0·1552	0·2368	0·3352	0·4469	0·6993	0·9813	1·2876	1·6148
1·3 . .	0·0890	0·1484	0·2268	0·3213	0·4285	0·6710	0·9423	1·2373	1·5532
1·4 . .	0·0852	0·1423	0·2177	0·3086	0·4118	0·6453	0·9067	1·1914	1·4968
1·5 . .	0·0817	0·1368	0·2095	0·2971	0·3966	0·6218	0·8740	1·1492	1·4447
1·6 . .	0·0785	0·1317	0·2019	0·2866	0·3827	0·6001	0·8439	1·1101	1·3965
1·7 . .	0·0757	0·1270	0·1950	0·2769	0·3698	0·5800	0·8159	1·0737	1·3516
1·8 . .	0·0730	0·1228	0·1885	0·2679	0·3579	0·5615	0·7899	1·0400	1·3097
1·9 . .	0·0706	0·1188	0·1826	0·2595	0·3468	0·5442	0·7657	0·0086	1·2704
2·0 . .	0·0683	0·1152	0·1771	0·2518	0·3365	0·5280	0·7432	0·9788	1·2338

TABLE XXIX (continued)

$\log \mathscr{I}(\alpha, \beta)$

$(0 \leqslant \alpha \leqslant 12; \ 0 \leqslant \beta^{-1} \leqslant 0.5)$

$1/\beta$	$\alpha = 0$	$\alpha = 1$	$\alpha = 2$	$\alpha = 3$	$\alpha = 4$	$\alpha = 6$	$\alpha = 8$	$\alpha = 10$	$\alpha = 12$
0 . .	0·0000	0·0000	0·0000	0·0000	0·0000	0·0000	0·0000	0·0000	0·0000
0·05 .	0·0118	0·0204	0·0317	0·0449	0·0595	0·0908	0·1242	0·1595	0·1974
0·10 .	0·0212	0·0366	0·0569	0·0809	0·1076	0·1662	0·2302	0·2994	0·3741
0·15 .	0·0293	0·0503	0·0781	0·1113	0·1483	0·2306	0·3217	0·4211	0·5290
0·20 .	0·0364	0·0623	0·0966	0·1377	0·1838	0·2870	0·4018	0·5279	0·6653
0·25 .	0·0428	0·0731	0·1131	0·1612	0·2154	0·3371	0·4731	0·6228	0·7862
0·30 .	0·0487	0·0828	0·1281	0·1824	0·2439	0·3821	0·5372	0·7078	0·8941
0·40 .	0·0591	0·1001	0·1543	0·2197	0·2937	0·4609	0·6486	0·8548	1·0788
0·50 .	0·0683	0·1152	0·1771	0·2518	0·3365	0·5280	0·7432	0·9788	1·2338

† It should be pointed out in this connexion that in evaluating the integrals $\mathscr{I}$ and $\mathscr{F}$, Chandrasekhar and Breen used the temperature distribution given by the fourth approximation of Chapter III (Table X). While it is not strictly justifiable to use the solution in this 'high approximation' in the framework of the (2, 1) approximation which is the basis of equations (78), the use of the higher approximation for the temperature distribution in a grey atmosphere corrects, semi-empirically, for certain of the inaccuracies involved in the (2, 1) approximation (cf. § 81 below).

TABLE XXX

$\log \mathscr{F}(\alpha, \beta)$

$(0 \leqslant \alpha \leqslant 12;\ 0.2 \leqslant \beta \leqslant 2)$

β	$\alpha = 0$	$\alpha = 1$	$\alpha = 2$	$\alpha = 3$	$\alpha = 4$	$\alpha = 6$	$\alpha = 8$	$\alpha = 10$	$\alpha = 12$
0·2 . .	0·183	0·294	0·438	0·607	0·798	1·224	1·693	2·193	2·716
0·3 . .	0·151	0·247	0·373	0·522	0·690	1·069	1·489	1·940	2·415
0·4 . .	0·130	0·216	0·328	0·463	0·615	0·958	1·341	1·756	2·195
0·5 . .	0·115	0·194	0·296	0·419	0·558	0·873	1·227	1·612	2·021
0·6 . .	0·104	0·176	0·270	0·384	0·512	0·805	1·135	1·495	1·879
0·7 . .	0·095	0·162	0·249	0·355	0·475	0·749	1·057	1·395	1·759
0·8 . .	0·087	0·150	0·232	0·331	0·444	0·701	0·991	1·311	1·655
0·9 . .	0·081	0·140	0·217	0·311	0·417	0·659	0·934	1·237	1·564
1·0 . .	0·075	0·131	0·204	0·293	0·393	0·623	0·883	1·171	1·484
1·1 . .	0·071	0·124	0·193	0·277	0·373	0·591	0·839	1·113	1·412
1·2 . .	0·067	0·117	0·183	0·263	0·354	0·562	0·799	1·061	1·347
1·3 . .	0·063	0·111	0·174	0·251	0·338	0·537	0·763	1·014	1·289
1·4 . .	0·060	0·106	0·166	0·240	0·323	0·513	0·730	0·971	1·235
1·5 . .	0·057	0·101	0·159	0·230	0·310	0·492	0·700	0·932	1·186
1·6 . .	0·054	0·097	0·153	0·220	0·297	0·473	0·673	0·896	1·141
1·7 . .	0·052	0·093	0·147	0·212	0·286	0·455	0·647	0·862	1·099
1·8 . .	0·050	0·089	0·141	0·204	0·276	0·439	0·624	0·831	1·060
1·9 . .	0·047	0·086	0·136	0·197	0·266	0·423	0·603	0·803	1·024
2·0 . .	0·046	0·082	0·131	0·190	0·257	0·409	0·582	0·776	0·990

TABLE XXX (continued)

$\log \mathscr{F}(\alpha, \beta)$

$(0 \leqslant \alpha \leqslant 12;\ 0 \leqslant \beta^{-1} \leqslant 0.5)$

$1/\beta$	$\alpha = 0$	$\alpha = 1$	$\alpha = 2$	$\alpha = 3$	$\alpha = 4$	$\alpha = 6$	$\alpha = 8$	$\alpha = 10$	$\alpha = 12$
0 . .	0	0	0	0	0	0	0	0	0
0·05 .	0·001	0·007	0·015	0·025	0·035	0·057	0·080	0·104	0·130
0·10 .	0·008	0·019	0·034	0·051	0·071	0·113	0·159	0·208	0·261
0·15 .	0·014	0·030	0·050	0·075	0·102	0·162	0·229	0·302	0·382
0·20 .	0·020	0·039	0·065	0·096	0·130	0·207	0·293	0·388	0·493
0·25 .	0·025	0·048	0·078	0·114	0·155	0·247	0·351	0·466	0·594
0·30 .	0·030	0·056	0·090	0·132	0·179	0·285	0·404	0·538	0·687
0·40 .	0·038	0·070	0·112	0·163	0·221	0·351	0·500	0·666	0·849
0·50 .	0·046	0·082	0·131	0·190	0·257	0·409	0·582	0·776	0·990

80.2. The continuous absorption coefficient of the solar atmosphere

Using the method of analysis described in § 80.1, G. Münch has analysed the observations relating to the continuous spectrum of the sun. In particular, he has derived the values of $\kappa_\nu/\bar{\kappa}$ from the solar observations on the frequency distribution of the intensities at the centre of the disk ($\mu = 1$) and of the emergent flux. The values of $\kappa_\nu/\bar{\kappa}$ derived by Münch are shown in Fig. 31. It is seen that the two sets of values agree very satisfactorily over the entire spectrum, within the limits of the observational errors. This agreement clearly implies that the same variation of κ_ν with wave-length is adequate (within limits) to account for both the laws of darkening and the intensity distribution in the spectrum. (A more detailed comparison by Münch between the theory and the observations confirms this view.)

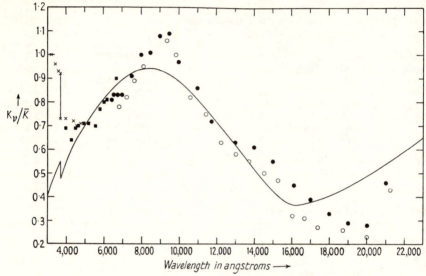

FIG. 31. The variation of the continuous absorption coefficient with wave-length in the solar atmosphere as derived from the fluxes (open circles) and the intensities at the centre (crosses and solid circles and squares). The observations used are the same as those shown in Fig. 30: the crosses, circles, and squares in the two figures correspond.

The full line curve is the theoretical curve for H⁻ at the solar effective temperature ($T_e = 5{,}740°$ K.).

It will also be noticed that the departure of κ_ν from constancy is not so large as to invalidate the use of the solution of the transfer problem in the approximations of § 79.

80.3. *The negative hydrogen ion as the source of continuous absorption in the atmospheres of the sun and the stars*

The analysis of the continuous spectrum of the sun described in § 80.2 shows that the coefficient of continuous absorption in the solar atmosphere increases by a factor of the order of 2 as the wave-length increases from 4,000 A to 9,000 A; beyond λ 9,000 A the absorption coefficient decreases until at about λ 16,000 A it has a pronounced minimum; and beyond λ 16,000 A the absorption coefficient increases again. This behaviour of the solar continuous absorption coefficient has been known for many years; indeed, it was disclosed already in

Milne's classic discussion of this problem in 1922. Further, the analysis of the results of spectrophotometry and of the data on the colour and effective temperatures of stars have all given evidence that in stellar atmospheres with effective temperatures less than 10,000° K. the continuous absorption coefficient has a dependence on wave-length similar to that in the solar atmosphere. To identify the source of continuous absorption in stellar atmospheres which will account for these results was one of the principal problems of astrophysics.

For many years (1928–38) it was believed that the principal contributions to the continuous opacity in stellar atmospheres must arise from the absorption beyond the various series limits of hydrogen and the commoner elements like Na, Mg, Ca, Fe, and Si. However, this assumption concerning the source of continuous absorption leads to a dependence on wave-length which is contrary to all astrophysical evidence; moreover, it leads to other serious discrepancies, the nature of which we cannot go into here (cf. the references in the Bibliographical Notes at the end of the chapter). The key to the solution of these difficulties was provided by R. Wildt, who pointed out that we should perhaps look for the source of continuous absorption in the solar atmosphere, in the presence of the negative ions of hydrogen. The ground for this expectation was simply that in the solar atmosphere there is an abundance of neutral hydrogen atoms and there is also a supply of free electrons from the easily ionized atoms such as sodium, calcium, magnesium, iron, silicon, and the rest; and in view of the positive electron affinity of hydrogen, a certain determinable proportion of the electrons must attach themselves to the neutral hydrogen atoms. Since the electron affinity of hydrogen is 0·75 e.v., the photo-ionization of the negative ions will provide a source of continuous absorption for wave-lengths less than 16,500 A. To this absorption arising from the 'bound-free' transitions we must add the contribution from the free-free transitions which begins to be important beyond λ 12,000 A. However, before definite conclusions could be drawn, the continuous absorption coefficient of H^- had to be derived from physical theory. The first determinations of this by Massey and Bates and others were disappointing and were at variance with the astrophysical demands; but it was soon realized that a reliable determination of the absorption coefficient of H^- from physical theory is a difficult matter on account of its relatively large size and weak binding. These difficulties have been overcome, and the last determinations by Chandrasekhar and Breen (which are illustrated in Fig. 32) are found to be entirely in accord with the requirements of astrophysics. Thus, using the determination of the absorption coefficient of H⁻ by the last-mentioned authors, the formulae of § 80.1 have been used to *predict* the emergent flux on the assumption that H⁻ provides the absorption in all wavelengths to the red of λ 4,000 A and that the total mean absorption

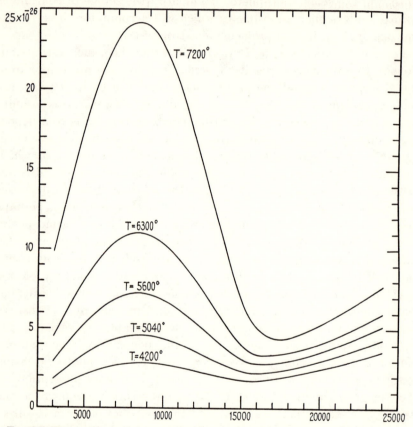

FIG. 32. The continuous absorption of H⁻ as derived by S. Chandrasekhar and Frances H. Breen (*Astrophys. J.* **104**, 430, 1946). The abscissae denote the wavelength in angstroms and the ordinates represent the absorption coefficient per negative hydrogen ion and per unit electron pressure at the various temperatures indicated.

coefficient is 1·4 times the mean absorption coefficient due to H⁻ alone at $T = T_e$.† The corresponding values of $\kappa_\nu(H^-)/1\cdot4\bar{\kappa}(H^-)$ are shown by the curve in Fig. 31. It is seen that the agreement is very satisfactory; and it may be added that equally satisfactory agreements have

† The value of $\bar{\kappa}$ was found at $\tau = 0\cdot7$, where $T \simeq T_e$; the factor 1·4 is to allow for the absorption in wave-lengths less than 4,000 A.

been found between the predictions and observations relating to the colour temperatures and 'gradients' of stars.

81. Model stellar atmospheres

If the conclusion drawn in § 80 that the negative hydrogen ion is the main source of continuous absorption in the solar atmosphere be accepted, then the determination of the entire structure of the atmosphere is a relatively straightforward matter. The principal requirements for the construction of such *model stellar atmospheres* are (i) a law governing the distribution of the temperature in the atmosphere, (ii) a physical theory giving the continuous absorption coefficient κ_ν as a function of the local temperature and electron pressure, and (iii) a knowledge of the relative abundances of the different elements, particularly the so-called 'hydrogen-metal' ratio, A, which is essentially the ratio of the numbers of atoms of those elements (like H, He, C, N, O, Fl, and Ne) which are practically un-ionized and those (like Fe, Si, Mg, Ca, Al, and Na) which are nearly once ionized. In addition, we require to know also the effective temperature T_e and the surface gravity g.

As regards the hydrogen-metal ratio, it may be observed that in the solar atmosphere the separation of the common elements into those which are mostly once ionized and those which are practically not ionized at all is so sharp that the structure of the derived atmosphere is hardly sensitive to the distribution of the abundances among the elements of the two groups. Indeed, in a first approximation, A may be equated to the ratio of the total pressure, p, to the electron pressure, p_e:

$$A \simeq p/p_e. \tag{79}$$

The departure of A from the ratio p/p_e depends only very insensitively on the particular choice we may make concerning the relative abundances of the 'metals'. In practice the distribution which is assumed is $\mathrm{Mg:Si:Fe:Ca:Al:Na} = 30:33:30:2:3:2$; these are the relative abundances with which the elements occur in the meteorites, and there is evidence that they occur with approximately the same relative abundances in the solar (and, generally, also in stellar) atmospheres.

The construction of model stellar atmospheres proceeds, then, on first assuming a set of values for the three basic parameters T_e, g, and A; then deriving the march of the variables p, p_e, T, $\bar{\kappa}$, etc., from a physical theory of the continuous absorption coefficient, the temperature distribution in the atmosphere, and the equation of hydrostatic equilibrium.

The theory of model stellar atmospheres along these lines has been developed extensively in recent years, particularly by B. Strömgren. However, in our brief discussion of this topic we shall restrict ourselves to certain examples of model atmospheres worked out by G. Münch, as these are more directly related to the theory as outlined in the earlier sections of this chapter.

81.1. *A model solar atmosphere in the* $(2, 1)$ *approximation*

When $\kappa_\nu/\bar{\kappa}$ is independent of depth, the temperature distribution on the $(2, 1)$ approximation is given by equation (70); this equation relating T and τ must be combined with the equation of hydrostatic equilibrium

$$dp = -g\rho \, dz. \tag{80}$$

Since

$$d\tau = -\bar{\kappa}\rho \, dz, \tag{81}$$

we have the differential equation

$$\frac{dp}{d\tau} = \frac{g}{\bar{\kappa}}. \tag{82}$$

To integrate this equation, we require to know $\bar{\kappa}$ as a function of p and τ. This latter function can be derived in the following manner:

Assuming that the source of continuous absorption is provided by H^- and H, we write

$$\bar{\kappa}(\tau, p_e) = \frac{1-x_H}{m_H} \int\limits_0^\infty \{p_e \kappa_\nu(H^-) + \kappa_\nu(H)\}(1 - e^{-h\nu/kT}) \frac{F_\nu^{(1)}(\tau)}{F} \, d\nu, \tag{83}$$

where m_H is the mass of the hydrogen atom; x_H is the degree of ionization of hydrogen at the temperature prevailing at τ and for an electron pressure p_e; and $\kappa_\nu(H^-)$ and $\kappa_\nu(H)$ are the continuous absorption coefficients, per neutral hydrogen atom, of H^- per unit electron pressure and of hydrogen, respectively. The factor $(1 - e^{-h\nu/kT})$ under the integral sign in (83) is to allow for stimulated emission. The coefficients $\kappa_\nu(H^-)$ have been evaluated by Chandrasekhar and Breen† and the coefficients $\kappa_\nu(H)$ have been tabulated in a convenient manner by Strömgren.‡

After $\bar{\kappa}$ has been determined according to equation (83), as a function of τ and p_e, we must next eliminate the electron pressure from $\bar{\kappa}(\tau, p_e)$ by expressing p_e in terms of the total pressure p and the temperature at τ.

As we have already remarked, in a first approximation, we may set $p_e = p/A$ (eq. [79]). However, in an accurate calculation, the elimination of p_e can be carried out by considering the ionization equilibria of

† *Astrophys. J.* **104**, 430, 1946.
‡ *Pub. mind. Med. Köbenhavns Obs.*, No. 138, 1944.

hydrogen and the metals. Thus, if x_M denotes the mean degree of ionization of the metals at the temperature at τ and for an electron pressure p_e, then

$$\frac{p_e}{p} = \frac{x_H}{1+x_H} + \frac{1}{A}\frac{x_M}{1+x_H}. \tag{84}$$

(It is seen that eq. [84] reduces to (79) in the approximation $x_H = 0$ and $x_M = 1$.)

Tables giving x_M and x_H as functions of p_e and T have been provided by Strömgren (loc. cit.). Using these tables we can eliminate p_e from equation (83) and obtain $\bar{\kappa}$ as a function of p and τ. Once $\bar{\kappa}$ has been determined, in this fashion, as a function of p and τ, equation (82) can be integrated numerically. In this manner Münch has constructed a series of model atmospheres appropriate for the sun (i.e. for $T_e = 5{,}740°$ K. and $g = 2{\cdot}740 \times 10^4$ cm./sec.^{2}) and for various values of A. We reproduce in Table XXXI his results for the case $\log A = 3{\cdot}8$.

TABLE XXXI

A Model Solar Atmosphere

($T_e = 5{,}740°$ K.; $g = 2{\cdot}74 \times 10^4$ cm./sec.^{2}; $\log A = 3{\cdot}8$)

τ	$\log p$	$\log p_e$	$\log \bar{\kappa}$	τ	$\log p$	$\log p_e$	$\log \bar{\kappa}$
0·01	3·74	9·85	8·69	0·28	4·68	0·82	9·46
0·02	4·01	0·09	8·92	0·32	4·71	0·87	9·49
0·04	4·19	0·27	9·06	0·36	4·74	0·92	9·51
0·06	4·30	0·36	9·15	0·40	4·77	0·96	9·54
0·08	4·38	0·44	9·21	0·50	4·82	1·06	9·60
0·10	4·43	0·50	9·25	0·60	4·86	1·15	9·65
0·12	4·46	0·56	9·29	0·70	4·89	1·24	9·70
0·14	4·51	0·61	9·32	0·80	4·91	1·32	9·75
0·16	4·54	0·64	9·35	0·90	4·93	1·40	9·81
0·18	4·57	0·68	9·38	1·0	4·94	1·48	9·86
0·20	4·60	0·71	9·40	1·2	4·97	1·64	9·96
0·24	4·64	0·77	9·43	1·4	4·99	1·78	0·06

81.2. *A model solar atmosphere in the* (2, 2) *approximation*

When a model atmosphere has been constructed in the (2, 1) approximation and on the assumption that $\kappa_\nu/\bar{\kappa}$ is independent of depth, it can be improved by allowing for the variation of δ_ν with depth and by using the solution, in a higher approximation, of the basic transfer problem. Thus, in the (2, 2) approximation, the temperature distribution can be corrected in accordance with equation (68) by evaluating δ_1, etc., for the pressures and temperatures derived on the lower approximation. Münch has carried out this revision for the model atmosphere given in

Table XXXI. The results of his calculations, including the revised temperature distribution, are given in Table XXXII.

From Table XXXII it is seen that, of the two corrections which have to be made to the temperature distribution $J^{(1)}$ derived on the grey atmosphere approximation, that due to the difference between $B^{(2)}$ and $J^{(2)}$ is much the larger: the difference between the solutions $J^{(2)}$ and $J^{(1)}$ (denoted by ΔJ in Table XXXII) is much smaller than $\tfrac{1}{2}\delta_1$ which represents the difference between $B^{(2)}$ and $J^{(2)}$. It is this last circumstance which justifies the use of the solution, for the temperature distribution in a grey atmosphere, in a higher approximation when the term $\tfrac{1}{2}\delta_1$, arising from the non-constancy of $\kappa_\nu/\bar\kappa$ with depth, is ignored (cf. § 80.1).

TABLE XXXII

The Temperature Distribution $T^{(2)}(\tau)$ in a Non-grey Model Solar Atmosphere

τ	$\bar\delta_1$	$\bar\delta_3$	$\bar\delta_4$	ΔJ	$B^{(1)} = J^{(1)}$	$J^{(2)}$	$B^{(2)}$	$T^{(1)}$	$T^{(2)}$
0	−0·0357	−0·0156	+0·0011	−0·0025	0·4330	0·4305	0·4127	4,634° K.	4,579° K.
0·05	−0·0397	..	..	−0·0030	0·4787	0·4757	0·4559	4,752	4,694
0·1	−0·0416	−0·0191	+0·0005	−0·0035	0·5236	0·5201	0·4993	4,860	4,802
0·2	−0·0440	−0·0216	+0·0000	−0·0045	0·6116	0·6071	0·5851	5,052	4,996
0·3	−0·0461	..	..	−0·0054	0·6971	0·6917	0·6687	5,220	5,166
0·4	−0·0477	−0·0242	−0·0002	−0·0060	0·7808	0·7748	0·7510	5,370	5,318
0·5	−0·0482	..	..	−0·0065	0·8629	0·8564	0·8323	5,506	5,457
0·6	−0·0482	..	..	−0·0068	0·9437	0·9369	0·9128	5,631	5,584
0·7	−0·0481	−0·0262	+0·0007	−0·0069	1·0235	1·0166	0·9926	5,746	5,702
0·8	−0·0479	..	..	−0·0069	1·1024	1·0955	1·0716	5,854	5,813
1·0	−0·0471	−0·0261	+0·0021	−0·0065	1·2584	1·2519	1·2283	6,051	6,014
1·2	−0·0457	..	..	−0·0058	1·4123	1·4065	1·3837	6,228	6,196
1·4	−0·0441	−0·0258	+0·0031	−0·0048	1·5650	1·5602	1·5381	6,390	6,362
2·0	−0·0388	..	..	−0·0018	2·0188	2·0170	1·9976	6,810	6,792

81.3. Model atmospheres in higher approximations

If we wish to obtain solutions in approximations higher than those considered in §§ 81.1 and 81.2, it would appear that numerical methods of iteration are to be preferred to extending the analytical methods of solution to higher approximations. Thus we can try to correct, by trial and error, the temperature distribution derived on the (2, 2) approximation, for example, by arranging that the flux condition (6) is strictly satisfied at each level. For this purpose we can determine the monochromatic fluxes, at each level, with the aid of the formula

$$F_\nu(\tau_\nu) = 2 \int_{\tau_\nu}^{\infty} B_\nu(T_{l_\nu})E_2(t_\nu-\tau_\nu)\,dt_\nu - 2 \int_0^{\tau_\nu} B_\nu(T_{l_\nu})E_2(\tau_\nu-t_\nu)\,dt_\nu, \quad (85)$$

and then evaluate F by integrating F_ν over ν. The condition is that F determined in this fashion should be a constant. Using Reiz's quadrature formula of Chapter II, § 23 (Tables VI and VII), Strömgren has used this method successfully for correcting the temperature distributions in model atmospheres, derived on the basis of solutions of the transfer problem in lower approximations.

BIBLIOGRAPHICAL NOTES

The following general references relating to the subject matter of this chapter may be noted:

1. A. S. EDDINGTON, *The Internal Constitution of the Stars*, chap. xii, Cambridge, 1926.

2. E. A. MILNE, *Handbuch der Astrophysik*, vol. **3**, part i, pp. 109–26 and 131–55, Springer, Berlin, 1930.

3. A. PANNEKOEK, ibid. vol. **3**, part i, pp. 291–7, Springer, Berlin, 1930.

4. A. UNSÖLD, *Physik der Sternatmosphären*, chaps. v, vi, and vii, Springer, Berlin, 1938.

5. B. STRÖMGREN, *Handbuch der Astrophysik*, vol. **7**, pp. 203–21, Springer, 1936.

6. S. ROSSELAND, *Theoretical Astrophysics*, chap. ix, Oxford, 1936.

§§ 75 and 76. The first introduction of the concept of radiative equilibrium is due to R. A. Sampson:

7. R. A. SAMPSON, *Memoirs of the Royal Astron. Soc. London*, **51**, 123 (1895).

However, the concept of radiative equilibrium under conditions of local thermodynamic equilibrium, as we now understand it, is due to K. Schwarzschild:

8. K. SCHWARZSCHILD, *Göttinger Nachrichten Math.-Phys. Klasse*, p. 41 (1906).

The basic concepts are further analysed in—

9. E. A. MILNE, *Monthly Notices Roy. Astron. Soc. London*, **88**, 493 (1927).

§§ 76–79. The radiative equilibrium of a grey atmosphere was treated by Schwarzschild in his first paper (ref. 8).

Further developments along these lines are due to—

10. A. S. EDDINGTON, *Monthly Notices Roy. Astron. Soc. London*, **77**, 16 (1916); *Zeits. f. Physik*, **7**, 351 (1921).

11. J. H. JEANS, *Monthly Notices Roy. Astron. Soc. London*, **78**, 28 (1917).

The method of analysis of the continuous spectrum of stars described in these sections derive from:

12. E. A. MILNE, ibid. **81**, 361, 375 (1921) and **82**, 368 (1922).

13. —— *Philos. Trans. Roy. Soc. (London)*, A, **223**, 201 (1922).

14. B. LINDBLAD, *Uppsala Univ. Årsskrift*, **1** (1920).

15. —— *Acta Reg. Soc. Sci. Uppsala*, iv. **6**, 1 (1923).

16. R. LUNDBLAD, *Astrophys. J.* **58**, 113 (1923).

17. G. BURKHARDT, *Zeits. f. Astrophysik*, **13**, 56 (1936).

The presentation of the subject in the text follows:

18. S. CHANDRASEKHAR, *Astrophys. J.* **101**, 328 (1945).

19. —— and FRANCES H. BREEN, ibid. **105**, 461 (1947).

The numerical data on the grey atmosphere given in Tables XXVI–XXVIII are taken from references 18 and 19.

§ 80. The first discussion of the dependence on wave-length of the continuous absorption coefficient in the solar atmosphere is due to Milne (ref. 13). Later discussions are those of—

20. A. Unsöld and A. W. Maue, *Zeits. f. Astrophysik*, **5**, 1 (1932).
21. G. Mülders, ibid. **11**, 132 (1935).
22. G. Münch, *Astrophys. J.* **102**, 385 (1945); also S. Chandrasekhar and G. Münch, ibid. **104**, 446 (1946).
23. D. Chalonge and V. Kourganoff, *Annales d'Astrophysique*, **9**, 69 (1946).
24. D. Chalonge, *Physica*, **12**, 721 (1947).
25. B. Strömgren, ibid. **12**, 701 (1947).

In the text we have followed the discussion of Münch (ref. 22). The writer is indebted to Dr. Münch for revising his calculations for the purposes of this book.

The fact that H⁻ must be present in the solar atmosphere and contribute to the opacity is the discovery of—

26. R. Wildt, *Astrophys. J.* **89**, 295 (1939); **90**, 611 (1939), and **93**, 47 (1941).

The early determinations of the absorption coefficient of H⁻ are due to—

27. C. K. Jen, *Phys. Rev.* **43**, 540 (1933).
28. H. S. W. Massey and D. R. Bates, *Astrophys. J.* **91**, 202 (1940).
29. R. E. Williamson, ibid. **96**, 438 (1942).
30. L. R. Henrich, ibid. **99**, 59 (1943).
31. J. A. Wheeler and R. Wildt, ibid. **95**, 281 (1942).

These early determinations are unsatisfactory for reasons pointed out in—

32. S. Chandrasekhar, ibid. **100**, 176 (1944) and *Reviews of Modern Physics*, **16**, 301 (1944).

The coefficients currently adopted are those of—

33. S. Chandrasekhar, *Astrophys. J.* **102**, 223, 395 (1945).
34. —— and Frances H. Breen, ibid. **104**, 430 (1946).

For a comparison similar to that described in § 80.2 but in the context of a star see—

35. A. B. Underhill, *Astrophys. J.* **108**, 83 (1948).

§ 81. The early work on model stellar atmospheres are those of:

36. E. A. Milne, *Philos. Trans. Roy. Soc.* (*London*), A, **228**, 421 (1929).
37. W. H. McCrea, *Monthly Notices Roy. Astron. Soc. London*, 91, 836 (1931).
38. H. N. Russell, *Astrophys. J.* **78**, 239 (1933).
39. A. Unsöld, *Zeits. f. Astrophysik*, **8**, 225 (1934).
40. A. Pannekoek, *Publ. Astr. Inst. Univ.* Amsterdam No. 4 (1935).

The recent advances by Strömgren are contained in—

41. B. Strömgren, *Festschrift für Elis Strömgren*, pp. 218–57, Einar Munksgaard, Copenhagen, 1940.
42. —— *Publ. mind. Med. Københavns Obs.*, No. 138 (1944).
43. —— *Physica*, **12**, 701 (1947).

Further developments along the same lines are due to—

44. M. Rudkjöbing, *Zeits. f. Astrophysik*, **21**, 254 (1942).
45. —— *Publ. mind. Med. Københavns Obs.*, No. 145 (1947).

46. A. B. Underhill, *Astrophys. J.* **107**, 349 (1948).

47. L. H. Aller, ibid. **109**, 244 (1948).

The model atmospheres described in §§ 81.1 and 81.2 are taken from—

48. G. Münch, ibid. **106**, 217 (1947) and **107**, 265 (1948).

Related matters not treated in the book are contained in—

49. S. Chandrasekhar, *Monthly Notices Roy. Astron. Soc. London*, **96**, 21 (1936).

50. V. Ambarzumian, *Pub. Astr. Obs. U. Leningrad*, new ser., **1**, 7 (1936).

51. G. Münch, *Astrophys. J.* **104**, 87 (1946).

FURTHER ASTROPHYSICAL PROBLEMS

82. Introduction

In Chapter XI we have considered a transfer problem which is fundamental in the interpretation of the continuous spectra of stars. In this
chapter we shall consider certain other problems which occur in the
study of the formation of absorption lines in stellar atmospheres. These
problems provide interesting examples of the application of the methods
of the preceding chapters to different problems.

83. Schuster's problem in the theory of line formation

In his pioneering paper on the interpretation of absorption lines in
stellar spectra Schuster introduced an idealization of a stellar atmosphere which has proved extremely useful in the preliminary investigations of many astrophysical problems.

On Schuster's idealization, we distinguish between the *reversing layers*
in which the lines are formed and the deeper *photospheric layers* in which
the continuous spectrum is formed. More particularly, we suppose that
there is a definite *photospheric surface* which radiates in the outward
directions in a known manner: this is the emergent continuous radiation
of the star. Overlying this photospheric surface is a scattering atmosphere with a scattering coefficient, σ_ν, which is different from zero only
in the immediate neighbourhood of certain frequencies ν_0 where σ_ν has
pronounced maxima. The emergent radiation in the vicinities of the
frequencies ν_0 will, therefore, be less than in the continuous spectrum
which forms the background. On Schuster's picture, this is the interpretation of the formation of absorption lines in stellar atmospheres.

We shall now formulate Schuster's problem in the following manner:

An infinite plane surface radiates in the outward directions with a
known angular distribution. Above this radiating surface is a plane-
parallel, perfectly scattering, atmosphere. The optical thickness of the
atmosphere is τ_1 and the phase function is that for isotropic scattering.
It is required to find the angular distribution of the emergent radiation.

In terms of the laws of diffuse reflection and transmission for an
isotropically scattering atmosphere (Chap. IX, § 62), the solution of
Schuster's problem can be readily written down: For, if $I^{(s)}(\tau_1, +\mu')$
$(0 \leqslant \mu' \leqslant 1)$ is the intensity of the light emitted by the radiating
surface at τ_1, then the emergent light can be regarded as arising from the

transmission of the radiation $I^{(s)}(\tau_1, \mu')$, by the overlying atmosphere in accordance with the standard laws of diffuse reflection and transmission. We can, therefore, write

$$I(0, \mu) = I^{(s)}(\tau_1, +\mu)e^{-\tau_1/\mu} + \frac{1}{2\mu} \int_0^1 T(\mu, \mu') I^{(s)}(\tau_1, \mu') \, d\mu'. \tag{1}$$

The first term on the right-hand side of (1) represents the contribution to the emergent intensity from the direct transmission of the light emitted by the radiating surface in the direction μ and the second term represents the contribution from the diffuse transmission of the radiation incident on the atmosphere from below.

For the case under discussion the transmission function is given by (cf. Chap. IX, eqs. [27] and [43])

$$\left(\frac{1}{\mu'} - \frac{1}{\mu}\right) T(\mu, \mu') = Y(\mu)X(\mu') - X(\mu)Y(\mu') -$$
$$- Q(\mu - \mu')[X(\mu) + Y(\mu)][X(\mu') + Y(\mu')], \tag{2}$$

where $X(\mu)$ and $Y(\mu)$ are the standard solutions for the conservative case $\Psi(\mu) = \frac{1}{2}$ and

$$Q = -\frac{\alpha_1 - \beta_1}{(\alpha_1 + \beta_1)\tau_1 + 2(\alpha_2 + \beta_2)}. \tag{3}$$

In equation (3) α_n and β_n are the moments of order n of $X(\mu)$ and $Y(\mu)$, respectively.

Substituting for T according to equation (2) in equation (1), we have

$$I(0, \mu) = I^{(s)}(\tau_1, +\mu)e^{-\tau_1/\mu} +$$
$$+ \frac{1}{2} \int_{-1}^{+1} \frac{\mu'}{\mu - \mu'} [Y(\mu)X(\mu') - X(\mu)Y(\mu')] I^{(s)}(\tau_1, \mu') \, d\mu' -$$
$$- \frac{1}{2} Q[X(\mu) + Y(\mu)] \int_0^1 [X(\mu') + Y(\mu')] I^{(s)}(\tau_1, \mu')\mu' \, d\mu'. \tag{4}$$

Equation (4) represents the complete solution of Schuster's problem.

83.1. *The case* $I^{(s)}(\tau_1, +\mu) = I^{(0)} + I^{(1)}\mu$

In astrophysical contexts, greatest interest is attached to the case when the photospheric surface radiates in the outward directions with an angular distribution which is linear in μ.

Let $\qquad\qquad I^{(s)}(\tau_1, +\mu') = I^{(0)} + I^{(1)}\mu' \qquad (0 \leqslant \mu' \leqslant 1), \tag{5}$

where $I^{(0)}$ and $I^{(1)}$ are certain constants.

When $I^{(s)}(\tau_1, +\mu')$ has the form (5), the integrals over μ' in equation (4) can be evaluated and expressed in terms of the moments of X and Y.

Thus by using the relations of Chapter VIII, Theorem 8, equations (70) and (72), we find

$$I(0,\mu) = \tfrac{1}{2}I^{(1)}[\beta_1 X(\mu) - \alpha_1 Y(\mu)] - \\ - \tfrac{1}{2}Q[X(\mu)+Y(\mu)][I^{(0)}(\alpha_1+\beta_1)+I^{(1)}(\alpha_2+\beta_2)]. \tag{6}$$

Now substituting for Q according to equation (3) and making use of the relation (Chap. IX, eq. [46])

$$\alpha_1^2 - \beta_1^2 = \tfrac{4}{3}, \tag{7}$$

we have

$$I(0,\mu) = \frac{X(\mu)+Y(\mu)}{(\alpha_1+\beta_1)\tau_1+2(\alpha_2+\beta_2)}[\tfrac{2}{3}I^{(0)}+\tfrac{1}{2}I^{(1)}(\alpha_2+\beta_2)(\alpha_1-\beta_1)] + \\ + \tfrac{1}{2}I^{(1)}[\beta_1 X(\mu)-\alpha_1 Y(\mu)]. \tag{8}$$

A quantity of particular interest in the application of Schuster's model to the interpretation of the contours of absorption lines in stellar spectra is the ratio of the emergent flux in the lines to the outward flux of $I^{(s)}(\tau_1, +\mu)$ which represents the background continuum. This ratio defines the *residual intensity* in the line.

Evaluating

$$F(0) = 2\int_0^1 I(0,\mu)\mu\,d\mu \tag{9}$$

according to equation (8) and making use of equation (7), we obtain

$$F(0) = \frac{4}{3}\frac{I^{(0)}(\alpha_1+\beta_1)+I^{(1)}(\alpha_2+\beta_2)}{(\alpha_1+\beta_1)\tau_1+2(\alpha_2+\beta_2)}, \tag{10}$$

or, alternatively,

$$F(0) = \frac{1}{\tfrac{3}{4}\tau_1+\dfrac{3}{2}\dfrac{\alpha_2+\beta_2}{\alpha_1+\beta_1}}\left[I^{(0)}+I^{(1)}\frac{\alpha_2+\beta_2}{\alpha_1+\beta_1}\right]. \tag{11}$$

For obtaining the residual intensity we must compare (11) with

$$F^{(\text{cont})} = 2\int_0^1 I^{(s)}(\tau_1,\mu)\mu\,d\mu = I^{(0)}+\tfrac{2}{3}I^{(1)}. \tag{12}$$

Thus,

$$\frac{F(0)}{F^{(\text{cont})}} = \frac{I^{(0)}+I^{(1)}(\alpha_2+\beta_2)/(\alpha_1+\beta_1)}{(I^{(0)}+\tfrac{2}{3}I^{(1)})[\tfrac{3}{4}\tau_1+\tfrac{3}{2}(\alpha_2+\beta_2)/(\alpha_1+\beta_1)]}. \tag{13}$$

For $\tau_1 \to \infty$, $(\alpha_2+\beta_2)/(\alpha_1+\beta_1)$ will tend to the ratio of the second and the first moments of the corresponding H-function (i.e. the entry under $\varpi_0 = 1$ in Chap. V, Table XI). Hence (cf. Table XXXIII)

$$\left(\frac{\alpha_2+\beta_2}{\alpha_1+\beta_1}\right)_{\tau_1\to\infty} = \frac{\int_0^1 H(\mu)\mu^2\,d\mu}{\int_0^1 H(\mu)\mu\,d\mu} = \frac{0\cdot820352}{1\cdot154701} = 0\cdot710447. \tag{14}$$

On the other hand, for $\tau_1 \to 0$ (cf. Chap. VIII, § 60)

$$X(\mu)+Y(\mu) \to 1+e^{-\tau_1/\mu} \qquad (15)$$

and $\qquad (\alpha_2+\beta_2)/(\alpha_1+\beta_1) \to \tfrac{2}{3}(1+\tfrac{1}{4}\tau_1) \quad (\tau_1 \to 0). \qquad (16)$

84. The theory of line formation, including the effects of scattering and absorption

In the preceding section we have described the theory of line formation in stellar atmospheres under the conditions of Schuster's problem. Now Schuster's idealization of a stellar atmosphere as consisting of a scattering atmosphere overlying a photospheric surface, while it is suitable for a first analysis of a novel situation, is inadequate as a basis for a satisfactory theory. Thus in a stellar atmosphere we cannot, strictly, distinguish between the 'reversing layers' and the 'photospheric layers' except in so far as the layers in which the lines may be said to be formed are higher than those in which the continuum is principally formed. Moreover, the continuous absorption coefficient, κ_ν, is never really negligible compared to the line-scattering coefficient σ_ν. A correct theory of stellar absorption lines must therefore include both the effects of line scattering and continuous absorption. A theory along these more general lines was initiated by Eddington and Milne, who included the effects of scattering and absorption by writing the equation of transfer in the form

$$-\mu \frac{dI_\nu}{\rho\, dz} = (\kappa_\nu+\sigma_\nu)I_\nu - \tfrac{1}{2}\sigma_\nu \int_{-1}^{+1} I_\nu\, d\mu' - \kappa_\nu B_\nu, \qquad (17)$$

where B_ν is the Planck intensity for the prevailing temperature.

In writing the equation of transfer in the form (17), the assumption is made that the emission consists of two parts: a part arising from the conservative isotropic scattering with a scattering coefficient σ_ν, and a part arising from thermal radiation, according to Kirchhoff's law, for an absorption coefficient κ_ν.

Sometimes it is convenient to generalize equation (17) to allow for the possibility that a certain amount of thermal (Kirchhoff) emission may also be associated with the coefficient σ_ν.† We then write

$$-\mu \frac{dI_\nu}{\rho\, dz} = (\kappa_\nu+\sigma_\nu)I_\nu - \tfrac{1}{2}\sigma_\nu(1-\epsilon_\nu) \int_{-1}^{+1} I_\nu\, d\mu' - (\kappa_\nu+\epsilon_\nu\sigma_\nu)B_\nu \quad (\epsilon_\nu \leqslant 1). \qquad (18)$$

---

† The origin of this may lie either in collisions of the second kind (Milne, Eddington, Pannekoek) or 'fluorescence' resulting from ionization and recombination (Woolley, Strömgren).

In terms of the optical thickness

$$t_\nu = \int_z^\infty (\kappa_\nu + \sigma_\nu)\rho \, dz \qquad (19)$$

of the combined line-scattering and continuous absorption coefficients, the equation of transfer becomes

$$\mu \frac{dI_\nu}{dt_\nu} = I_\nu - \tfrac{1}{2}(1 - \lambda_\nu) \int_{-1}^{+1} I_\nu(t_\nu, \mu') \, d\mu' - \lambda_\nu B_\nu(t_\nu), \qquad (20)$$

where

$$\lambda_\nu = \frac{\kappa_\nu + \epsilon_\nu \sigma_\nu}{\kappa_\nu + \sigma_\nu} = \frac{1 + \epsilon_\nu \eta_\nu}{1 + \eta_\nu} \quad \left(\eta_\nu = \frac{\sigma_\nu}{\kappa_\nu}\right). \qquad (21)$$

An assumption which appears not unreasonable in the context of the theory of line formation is that B_ν increases linearly with the optical thickness τ_ν in the *continuum*. In other words, we may assume that

$$B_\nu = B_\nu^{(0)} + B_\nu^{(1)} \tau_\nu, \qquad (22)$$

where $B_\nu^{(0)}$ and $B_\nu^{(1)}$ are certain appropriately chosen constants. It may be remarked here that in $B_\nu^{(0)}$ and $B_\nu^{(1)}$ we may replace ν by the frequency, ν_0, at the centre of the line without introducing any sensible error. (This cannot, of course, be done for I_ν and λ_ν.)

We shall now, further, suppose (and this is not always as justifiable an assumption as [22]) that $\eta_\nu = \sigma_\nu/\kappa_\nu$ is constant with depth. Then

$$B_\nu = B_\nu^{(0)} + B_\nu^{(1)} \frac{\kappa_\nu}{\kappa_\nu + \sigma_\nu} t_\nu = B_\nu^{(0)} + \frac{\lambda_\nu B_\nu^{(1)}}{1 + \epsilon_\nu \eta_\nu} t_\nu. \qquad (23)$$

With B_ν given by equation (23) the equation of transfer takes the form

$$\mu \frac{dI}{dt} = I - \tfrac{1}{2}(1 - \lambda) \int_{-1}^{+1} I(t, \mu') \, d\mu' - \lambda \left[B^{(0)} + \frac{\lambda B^{(1)}}{1 + \epsilon \eta} t \right], \qquad (24)$$

where, for convenience, we have suppressed the subscript ν to the various quantities.

84.1. *The solution of the equation of transfer* (24) *in the n-th approximation*

In the nth approximation, we replace equation (24) by the system of $2n$ linear equations

$$\mu_i \frac{dI_i}{dt} = I_i - \tfrac{1}{2}(1 - \lambda) \sum a_j I_j - \lambda \left[B^{(0)} + \frac{\lambda B^{(1)}}{1 + \epsilon \eta} t \right] \quad (i = \pm 1, ..., \pm n), \qquad (25)$$

where the various symbols have their usual meanings.

The homogeneous system associated with equation (25) has already been considered in Chapter III, § 26.1. The solution can now be completed by observing that

$$I_i = B^{(0)} + \frac{\lambda B^{(1)}}{1+\epsilon\eta}(t+\mu_i) \quad (i = \pm1,...,\pm n) \tag{26}$$

satisfies equation (25). The solution appropriate to the problem on hand can therefore be written in the form

$$I_i = \frac{\lambda B^{(1)}}{1+\epsilon\eta}\left\{\sum_{\alpha=1}^{n}\frac{L_\alpha e^{-k_\alpha t}}{1+\mu_i k_\alpha}+t+\mu_i+\frac{1+\epsilon\eta}{\lambda}\frac{B^{(0)}}{B^{(1)}}\right\} \quad (i = \pm1,...,\pm n), \tag{27}$$

where the k_α's $(\alpha = 1,...,n)$ are the positive roots of the characteristic equation

$$1 = (1-\lambda)\sum_{j=1}^{n}\frac{a_j}{1-\mu_j^2 k^2} \tag{28}$$

and the L_α's $(\alpha = 1,...,n)$ are n constants of integration to be determined from the boundary conditions

$$I_{-i} = 0 \quad \text{at} \quad \tau = 0 \quad \text{for} \quad i = 1,...,n. \tag{29}$$

84.2. *The elimination of the constants and the expression of the solution in closed form*

Quite generally, in transfer problems in semi-infinite plane-parallel atmospheres, the boundary conditions which determine the constants of integration and the equation which governs the angular distribution of the emergent radiation can both be expressed in terms of the same function. Thus, in the present instance, the boundary conditions and the emergent intensity $I(0,\mu)$ can be expressed in the forms:

$$S(\mu_i) = 0 \quad (i = 1,...,n) \tag{30}$$

and

$$I(0,\mu) = \frac{\lambda B^{(1)}}{1+\epsilon\eta}S(-\mu), \tag{31}$$

where

$$S(\mu) = \sum_{\alpha=1}^{n}\frac{L_\alpha}{1-k_\alpha\mu}-\mu+\frac{1+\epsilon\eta}{\lambda}\frac{B^{(0)}}{B^{(1)}}. \tag{32}$$

In view of the boundary conditions (30), we can write

$$S(\mu) = (-1)^{n+1}k_1...k_n\frac{P(\mu)}{R(\mu)}(\mu-c), \tag{33}$$

where c is a constant and $P(\mu)$ and $R(\mu)$ have their standard meanings (cf. Chap. III, § 25, eqs. [54] and [55] and Chap. X, eq. [55]).

Moreover, according to equations (32) and (33)

$$L_\alpha = (-1)^{n+1}k_1...k_n\frac{P(1/k_\alpha)}{R_\alpha(1/k_\alpha)}\left(\frac{1}{k_\alpha}-c\right) \quad (\alpha = 1,...,n). \tag{34}$$

For the roots of the characteristic equation (28), it can be readily shown that

$$k_1...k_n \mu_1...\mu_n = \sqrt{\lambda}. \tag{35}\dagger$$

Equation (33) therefore becomes

$$S(\mu) = -\lambda^{\frac{1}{2}} H(-\mu)(\mu-c). \tag{36}$$

To determine the constant c, we proceed as follows:
First, putting $\mu = 0$ in equations (32) and (36), we have

$$\sum_{\alpha=1}^{n} L_\alpha + \frac{1+\epsilon\eta}{\lambda} \frac{B^{(0)}}{B^{(1)}} = c\sqrt{\lambda}. \tag{37}$$

We next evaluate $\sum L_\alpha$ in accordance with equation (34). Thus

$$\sum_{\alpha=1}^{n} L_\alpha = (-1)^{n+1} k_1...k_n \sum_{\alpha=1}^{n} \frac{P(1/k_\alpha)}{R_\alpha(1/k_\alpha)} \left(\frac{1}{k_\alpha} - c\right)$$

$$= (-1)^{n+1} k_1...k_n f(0), \tag{38}$$

where

$$f(x) = \sum_{\alpha=1}^{n} \frac{P(1/k_\alpha)}{R_\alpha(1/k_\alpha)} \left(\frac{1}{k_\alpha} - c\right) R_\alpha(x). \tag{39}$$

Defined in this manner, $f(x)$ is a polynomial of degree $(n-1)$ in x which takes the values

$$\left(\frac{1}{k_\alpha} - c\right) P(1/k_\alpha) \tag{40}$$

for $x = 1/k_\alpha$ $(\alpha = 1,..., n)$. There must, accordingly, exist a relation of the form

$$f(x) = (x-c)P(x) + R(x)(Ax+C), \tag{41}$$

where A and C are certain constants; and the constants can be determined from the condition that the coefficients x^{n+1} and x^n must vanish on the right-hand side. We thus find that

$$A = \frac{(-1)^{n+1}}{k_1...k_n} \quad \text{and} \quad C = \frac{(-1)^n}{k_1...k_n} \left(\sum_{j=1}^{n} \mu_j - \sum_{\alpha=1}^{n} \frac{1}{k_\alpha} + c\right). \tag{42}$$

Hence,

$$f(0) = (-1)^{n+1} \mu_1...\mu_n c + \frac{(-1)^n}{k_1...k_n} \left(\sum_{j=1}^{n} \mu_j - \sum_{\alpha=1}^{n} \frac{1}{k_\alpha} + c\right) \tag{43}$$

and (cf. eqs. [35] and [38])

$$\sum_{\alpha=1}^{n} L_\alpha = c\sqrt{\lambda} + \sum_{\alpha=1}^{n} \frac{1}{k_\alpha} - \sum_{j=1}^{n} \mu_j - c. \tag{44}$$

---

† This relation can be established in a manner analogous to Chap. III, eq. (36), and Chap. X, eq. (58).

Inserting this value of $\sum L_\alpha$ in equation (37), we obtain

$$c = \frac{1+\epsilon\eta}{\lambda}\frac{B^{(0)}}{B^{(1)}} + \sum_{\alpha=1}^{n}\frac{1}{k_\alpha} - \sum_{j=1}^{n}\mu_j. \qquad (45)$$

Finally, substituting for c from equation (45) in (36), we have

$$S(\mu) = -\lambda^{\frac{3}{2}}H(-\mu)\left[\mu - \left(\frac{1+\epsilon\eta}{\lambda}\frac{B^{(0)}}{B^{(1)}} + \sum_{\alpha=1}^{n}\frac{1}{k_\alpha} - \sum_{j=1}^{n}\mu_j\right)\right]. \qquad (46)$$

Equation (31) governing the angular distribution of the emergent radiation now becomes

$$I(0,\mu) = \frac{\lambda^{\frac{3}{2}}B^{(1)}}{1+\epsilon\eta}H(\mu)\left(\mu + \frac{1+\epsilon\eta}{\lambda}\frac{B^{(0)}}{B^{(1)}} + \sum_{\alpha=1}^{n}\frac{1}{k_\alpha} - \sum_{j=1}^{n}\mu_j\right). \qquad (47)$$

This is the required solution in closed form.

84.3. *Passage to the limit of infinite approximation*

From the theory of the H-functions given in Chapter V it follows that the *exact* solution for the emergent intensity for the problem under consideration must be of the form

$$I(0,\mu) = \frac{\lambda^{\frac{3}{2}}B^{(1)}}{1+\epsilon\eta}H(\mu)\left(\mu + \frac{1+\epsilon\eta}{\lambda}\frac{B^{(0)}}{B^{(1)}} + q\right), \qquad (48)$$

where $H(\mu)$ is the solution of the equation

$$H(\mu) = 1 + \tfrac{1}{2}(1-\lambda)\mu H(\mu)\int_0^1\frac{H(\mu')}{\mu+\mu'}d\mu', \qquad (49)$$

which is bounded in the entire half-plane $R(z) \geqslant 0$ and

$$q = \lim_{n\to\infty}\left(\sum_{\alpha=1}^{n}\frac{1}{k_\alpha} - \sum_{j=1}^{n}\mu_j\right). \qquad (50)$$

The exact H-functions characterizing the problem of line formation are therefore the same as those which occur in the problem of diffuse reflection by a semi-infinite plane-parallel atmosphere with an albedo

$$\varpi_0 = 1-\lambda. \qquad (51)$$

The H-functions tabulated in Chapter V (Table XI) are therefore equally applicable to this problem. However, to complete the solution we must determine q as defined in equation (50).

84.4. *The evaluation of* $\lim\limits_{n \to \infty} \left(\sum\limits_{\alpha=1}^{n} k_\alpha^{-1} - \sum\limits_{j=1}^{n} \mu_j \right)$. *The exact solution*

Consider the function

$$s(\mu) = \sum_{\alpha=1}^{n} \frac{l_\alpha}{1 - k_\alpha \mu} + 1, \tag{52}$$

where the k_α's ($\alpha = 1,...,n$) are the positive roots of the characteristic equation (28) and the l_α's ($\alpha = 1,...,n$) are n constants determined by the conditions

$$s(\mu_i) = 0 \quad (i = 1,...,n). \tag{53}$$

From these definitions it follows that (cf. eq. [35])

$$s(\mu) = (-1)^n k_1 ... k_n \frac{P(\mu)}{R(\mu)} = \lambda^{\ddagger} H(-\mu). \tag{54}$$

Moreover, according to equations (52) and (54),

$$l_\alpha = (-1)^n k_1 ... k_n \frac{P(1/k_\alpha)}{R_\alpha(1/k_\alpha)} \quad (\alpha = 1,...,n). \tag{55}$$

Now consider

$$\sum_{\alpha=1}^{n} \frac{l_\alpha}{k_\alpha} = (-1)^n k_1 ... k_n \sum_{\alpha=1}^{n} \frac{P(1/k_\alpha)}{R_\alpha(1/k_\alpha)} \frac{1}{k_\alpha}. \tag{56}$$

It is seen that the summation on the right-hand side is a special case ($c = 0$) of the one considered in § 84.2 (eq. [38]). We therefore have (cf. eq. [43])

$$\sum_{\alpha=1}^{n} \frac{l_\alpha}{k_\alpha} = \sum_{j=1}^{n} \mu_j - \sum_{\alpha=1}^{n} \frac{1}{k_\alpha}. \tag{57}$$

But l_α/k_α can also be expressed in terms of $H(\mu)$. Thus, consider

$$\sum_{i=1}^{n} a_i \mu_i s(-\mu_i). \tag{58}$$

Since $s(\mu_i) = 0$ ($i = 1,...,n$) we can extend the summation in (58) for negative values of i also. Hence

$$\sum_{i=1}^{n} a_i \mu_i s(-\mu_i) = \sum_{i=-n}^{+n} a_i \mu_i s(-\mu_i). \tag{59}$$

Now substitute for $s(-\mu_i)$ according to equation (52). We obtain

$$\sum_{i=1}^{n} a_i \mu_i s(-\mu_i) = \sum_{i=-n}^{+n} a_i \mu_i \left(\sum_{\alpha=1}^{n} \frac{l_\alpha}{1 + k_\alpha \mu_i} + 1 \right) = \sum_{i=-n}^{+n} a_i \mu_i \sum_{\alpha=1}^{n} \frac{l_\alpha}{1 + k_\alpha \mu_i}, \tag{60}$$

or, inverting the order of the summation, we have

$$\sum_{i=1}^{n} a_i \mu_i s(-\mu_i) = \sum_{\alpha=1}^{n} l_\alpha \sum_{i=-n}^{+n} \frac{a_i \mu_i}{1 + k_\alpha \mu_i} = \sum_{\alpha=1}^{n} \frac{l_\alpha}{k_\alpha} \sum_{i=-n}^{+n} a_i \left(1 - \frac{1}{1 + k_\alpha \mu_i} \right). \tag{61}$$

Using the equation defining the characteristic roots (eq. [28]) and remembering that $\sum a_i = 2$, we have

$$\sum_{i=1}^{n} a_i \mu_i s(-\mu_i) = \left(2 - \frac{2}{1-\lambda}\right) \sum_{\alpha=1}^{n} \frac{l_\alpha}{k_\alpha} = -\frac{2\lambda}{1-\lambda} \sum_{\alpha=1}^{n} \frac{l_\alpha}{k_\alpha}. \tag{62}$$

Hence (cf. eqs. [54] and [57])

$$\sum_{i=1}^{n} a_i \mu_i H(\mu_i) = \frac{2\sqrt{\lambda}}{1-\lambda}\left(\sum_{\alpha=1}^{n} \frac{1}{k_\alpha} - \sum_{j=1}^{n} \mu_j\right), \tag{63}$$

or

$$\sum_{\alpha=1}^{n} \frac{1}{k_\alpha} - \sum_{j=1}^{n} \mu_j = \frac{1-\lambda}{2\sqrt{\lambda}} \sum_{i=1}^{n} a_i \mu_i H(\mu_i). \tag{64}$$

Now pass to the limit $n \to \infty$. Then $H(\mu)$ becomes the solution of equation (49) which is bounded in the half-plane $R(z) \geqslant 0$ and

$$\lim_{n \to \infty}\left(\sum_{\alpha=1}^{n} \frac{1}{k_\alpha} - \sum_{j=1}^{n} \mu_j\right) = \frac{1-\lambda}{2\sqrt{\lambda}} \int_0^1 H(\mu)\mu \, d\mu = \frac{1-\lambda}{2\sqrt{\lambda}} \alpha_1. \tag{65}$$

The exact solution for $I(0,\mu)$ is therefore given by

$$I(0,\mu) = \frac{\lambda^{\frac{3}{2}} B^{(1)}}{1+\epsilon\eta} H(\mu)\left(\mu + \frac{1+\epsilon\eta}{\lambda} \frac{B^{(0)}}{B^{(1)}} + \frac{1-\lambda}{2\sqrt{\lambda}} \alpha_1\right). \tag{66}$$

84.5. *Exact formulae for the residual intensity. A table of the moments of $H(\mu)$*

The intensity, $I^{(\mathrm{cont})}(0,\mu)$, in the continuum is obtained by letting $\lambda \to 1$ and $\eta \to 0$. In this limit

$$H(\mu) \equiv 1 \quad (\lambda = 1) \tag{67}$$

and

$$I^{(\mathrm{cont})}(0,\mu) = B^{(1)}\left(\mu + \frac{B^{(0)}}{B^{(1)}}\right). \tag{68}$$

(This solution for the case $\lambda = 1$ can, of course, be obtained directly from [24].) The residual intensity, r, in the line is therefore given by

$$r(\mu) = \frac{\lambda^{\frac{3}{2}}}{1+\epsilon\eta} \frac{H(\mu)}{\mu + B^{(0)}/B^{(1)}}\left(\mu + \frac{1+\epsilon\eta}{\lambda} \frac{B^{(0)}}{B^{(1)}} + \frac{1-\lambda}{2\sqrt{\lambda}} \alpha_1\right). \tag{69}$$

Again, according to equation (66) the emergent flux is given by

$$F(0) = \frac{2\lambda^{\frac{3}{2}}}{1+\epsilon\eta} B^{(1)}\left(\alpha_2 + \frac{1+\epsilon\eta}{\lambda} \frac{B^{(0)}}{B^{(1)}} \alpha_1 + \frac{1-\lambda}{2\sqrt{\lambda}} \alpha_1^2\right), \tag{70}$$

where α_2 is the second moment of $H(\mu)$. The residual intensity R in the emergent flux is therefore given by

$$R = \frac{\lambda^{\frac{3}{2}}}{(1+\epsilon\eta)[\frac{1}{3}+\frac{1}{2}B^{(0)}/B^{(1)}]}\left(\alpha_2 + \frac{1+\epsilon\eta}{\lambda} \frac{B^{(0)}}{B^{(1)}} \alpha_1 + \frac{1-\lambda}{2\sqrt{\lambda}} \alpha_1^2\right). \tag{71}$$

To facilitate the use of the solutions (66), (69), and (71) we provide below a table of the first and second moments of $H(\mu)$ for various values of λ.

The predicted variation of the residual intensity $r(\mu)$ as a function of λ is illustrated in Table XXXIV, in which the values of r at the centre of the disk ($\mu = 1$) for the case $\epsilon = 0$ and $B^{(0)}/B^{(1)} = \frac{2}{3}$ are given. For comparison, the residual intensities in the emergent flux (again, for $\epsilon = 0$ and $B^{(0)}/B^{(1)} = \frac{2}{3}$) are also given.

TABLE XXXIII

The First and the Second Moments of $H(\mu)$

ϖ_0	λ	First moment	Second moment	ϖ_0	λ	First moment	Second moment
0	1·0	0·500000	0·333333	0·7	0·3	0·678674	0·461423
0·1	0·9	0·515609	0·344357	0·8	0·2	0·735808	0·503218
0·2	0·8	0·533154	0·356787	0·85	0·15	0·774376	0·531645
0·3	0·7	0·553123	0·370985	0·90	0·10	0·825318	0·569449
0·4	0·6	0·576210	0·387466	0·925	0·075	0·858734	0·594404
0·5	0·5	0·603495	0·407030	0·950	0·050	0·901864	0·626785
0·6	0·4	0·636636	0·430922	0·975	0·025	0·964471	0·674134
				1·000	0	1·154701	0·820352

TABLE XXXIV

The Residual Intensities at the Centre of the Disk ($\mu = 1$) and in the Emergent Flux for the Case $\epsilon = 0$ and $B^{(0)}/B^{(1)} = \frac{2}{3}$

λ	$r(1)$	R	λ	$r(1)$	R
1·0	1·0000	1·0000	0·3	0·5208	0·5580
0·9	0·9390	0·9481	0·2	0·4281	0·4616
0·8	0·8766	0·8939	0·15	0·3746	0·4036
0·7	0·8124	0·8369	0·10	0·3104	0·3340
0·6	0·7458	0·7762	0·075	0·2722	0·2919
0·5	0·6760	0·7109	0·050	0·2264	0·2412
0·4	0·6018	0·6391	0·025	0·1650	0·1735
			0	0	0

85. The softening of radiation by multiple Compton scattering

As is well known, when a light quantum is scattered by a free electron (at rest) its wave-length is increased by the amount

$$\delta\lambda = \frac{h}{mc}(1-\cos\Theta), \tag{72}$$

where Θ is the angle of scattering, h is Planck's constant, m is the mass of the electron, and c is the velocity of light. A problem in the theory of radiative transfer which arises in this context is the manner in which radiation of a particular wave-length gets modified by repeated *Compton scatterings* in transmission through an atmosphere of free electrons. In considering this problem, we shall suppose that an infinite plane surface radiates uniformly in the outward directions with a known spectral distribution and that above such a surface there is an atmosphere of free electrons. It is required to find the modified distribution in wave-length of the emergent radiation.

In formulating the equation of transfer for this problem of the softening of radiation by multiple Compton scattering, we shall suppose that for the wave-lengths which come under discussion it is an adequate approximation to consider the coefficient of scattering as independent of wave-length and having the classical (Thomson) value

$$\sigma_e = \frac{8\pi}{3} \frac{e^4}{m^2 c^4} N_e, \tag{73}$$

where e denotes the charge on the electron and N_e the number of electrons per unit mass.

85.1. *The equation of transfer and its approximate form*
Under the assumptions of the preceding paragraphs, the equation of transfer governing the radiation field is

$$\mu \frac{\partial I(\tau, \mu, \lambda)}{\partial \tau} = I(\tau, \mu, \lambda) - \frac{1}{4\pi} \int_{-1}^{+1} \int_{0}^{2\pi} I(\tau, \mu', \lambda - \gamma[1 - \cos\Theta]) \, d\mu' d\varphi', \tag{74}$$

where

$$\gamma = h/mc = 0\cdot 024 \text{ A} \tag{75}$$

$$\cos\Theta = \mu\mu' + (1 - \mu^2)^{\frac{1}{2}} (1 - \mu'^2)^{\frac{1}{2}} \cos\varphi' \tag{76}$$

and τ is the optical thickness in σ_e measured, as usual, from the boundary inward.

The form of the source function represented by the second term on the right-hand side of equation (74) arises from the fact that, in accordance with equation (72), we must consider radiation of wave-length

$$\lambda - \gamma[1 - \cos\Theta]$$

in the direction (μ', φ') in order that when scattered in the direction $(\mu, 0)$ it may have the wave-length λ considered.

We shall now suppose that $I(\tau, \mu', \lambda - \gamma[1 - \cos\Theta])$ can be expanded as a Taylor series in the form

$$I(\tau, \mu', \lambda - \gamma[1 - \cos\Theta]) = I(\tau, \mu', \lambda) - \gamma(1 - \cos\Theta) \frac{\partial I(\tau, \mu', \lambda)}{\partial \lambda} + \dots \tag{77}$$

and that it is sufficient to retain only the first two terms in the expansion. (The limitations on the solution implied by this assumption will become apparent later.) We therefore replace equation (74) by

$$\mu \frac{\partial I(\tau,\mu,\lambda)}{\partial \tau} = I(\tau,\mu,\lambda) - \frac{1}{2} \int_{-1}^{+1} \left[I(\tau,\mu',\lambda) - \gamma(1-\mu\mu') \frac{\partial I(\tau,\mu',\lambda)}{\partial \lambda} \right] d\mu'.$$

(78)

In the nth approximation we replace equation (78), in turn, by the system of $2n$ equations

$$\mu_i \frac{\partial I_i(\tau,\lambda)}{\partial \tau} = I_i(\tau,\lambda) - \frac{1}{2} \sum_j a_j I_j(\tau,\lambda) + \frac{1}{2}\gamma \sum_j a_j(1-\mu_i\mu_j) \frac{\partial I_j(\tau,\lambda)}{\partial \lambda}$$

$$(i = \pm 1,...,\pm n), \quad (79)$$

where the various symbols have their usual meanings.

In the first approximation

$$a_{+1} = a_{-1} = 1 \quad \text{and} \quad \mu_{+1} = -\mu_{-1} = \frac{1}{\sqrt{3}},$$

(80)

and equation (79) leads to the pair of equations

$$\frac{1}{\sqrt{3}} \frac{\partial I_{+1}}{\partial \tau} - \frac{1}{3}\gamma \frac{\partial I_{+1}}{\partial \lambda} - \frac{2}{3}\gamma \frac{\partial I_{-1}}{\partial \lambda} = \frac{1}{2}(I_{+1}-I_{-1})$$

and

$$\frac{1}{\sqrt{3}} \frac{\partial I_{-1}}{\partial \tau} + \frac{2}{3}\gamma \frac{\partial I_{+1}}{\partial \lambda} + \frac{1}{3}\gamma \frac{\partial I_{-1}}{\partial \lambda} = \frac{1}{2}(I_{+1}-I_{-1}).$$

(81)

Introducing the variables

$$x = \frac{3}{2}\tau \quad \text{and} \quad y = \frac{3}{2\gamma}(\lambda-\lambda_0),$$

(82)

where λ_0 is some suitably chosen constant wave-length, we find that equations (81) become

$$\frac{\sqrt{3}}{2} \frac{\partial I_{+1}}{\partial x} - \frac{1}{2} \frac{\partial I_{+1}}{\partial y} - \frac{\partial I_{-1}}{\partial y} = \frac{1}{2}(I_{+1}-I_{-1})$$

(83)

and

$$\frac{\sqrt{3}}{2} \frac{\partial I_{-1}}{\partial x} + \frac{\partial I_{+1}}{\partial y} + \frac{1}{2} \frac{\partial I_{-1}}{\partial y} = \frac{1}{2}(I_{+1}-I_{-1}).$$

(84)

We require to solve equations (83) with the boundary conditions

$$I_{+1}(x_1,y) = \text{a known function of } y = \psi(y) \quad \text{(say)}$$

and

$$I_{-1}(0,y) \equiv 0,$$

(85)

specifying, respectively, the known spectral distribution of the outward radiation at the base of the atmosphere at $x = x_1$ and the absence of any radiation incident on the atmosphere at $\tau = 0$.

85.2. *The reduction to a boundary-value problem*

Subtracting equation (84) from (83) we have

$$\frac{\sqrt{3}}{2}\frac{\partial}{\partial x}(I_{+1}-I_{-1}) = \frac{3}{2}\frac{\partial}{\partial y}(I_{+1}+I_{-1}). \tag{86}$$

We can therefore write

$$I_{+1}-I_{-1} = \sqrt{3}\frac{\partial S(x,y)}{\partial y} \quad \text{and} \quad I_{+1}+I_{-1} = \frac{\partial S(x,y)}{\partial x}. \tag{87}$$

The conservation of the net integrated flux of radiation readily follows from these equations.

Next adding equations (83) and (84) we have

$$\frac{\sqrt{3}}{2}\frac{\partial}{\partial x}(I_{+1}+I_{-1}) + \frac{1}{2}\frac{\partial}{\partial y}(I_{+1}-I_{-1}) = I_{+1}-I_{-1}, \tag{88}$$

or, substituting for $I_{+1}+I_{-1}$ and $I_{+1}-I_{-1}$ according to equations (87), we have

$$\frac{\partial^2 S}{\partial x^2} + \frac{\partial^2 S}{\partial y^2} = 2\frac{\partial S}{\partial y}. \tag{89}$$

Now put

$$S = e^y f(x,y). \tag{90}$$

Equation (89) reduces to the standard elliptic equation in two variables. We have

$$\frac{\partial^2 f}{\partial x^2} + \frac{\partial^2 f}{\partial y^2} - f = 0. \tag{91}$$

Returning to equations (87) and (90), we find that

$$I_{+1}(x,y) = \tfrac{1}{2}e^y\left[\frac{\partial f}{\partial x} + \sqrt{3}\left(f+\frac{\partial f}{\partial y}\right)\right]$$

and

$$I_{-1}(x,y) = \tfrac{1}{2}e^y\left[\frac{\partial f}{\partial x} - \sqrt{3}\left(f+\frac{\partial f}{\partial y}\right)\right]. \tag{92}$$

The boundary conditions (85) are therefore equivalent to

$$\left[\frac{\partial f}{\partial x} + \sqrt{3}\left(f+\frac{\partial f}{\partial y}\right)\right]_{x=x_1} = 2e^{-y}\psi(y),$$

and

$$\left[\frac{\partial f}{\partial x} - \sqrt{3}\left(f+\frac{\partial f}{\partial y}\right)\right]_{x=0} \equiv 0. \tag{93}$$

Our problem then is to solve the elliptic equation (91) with the boundary conditions (93).

85.3. *The solution of the boundary-value problem*

The boundary-value problem formulated in § 85.2 can be solved by a special method based on Green's theorem. We find†

$$f(0, y) = \frac{1}{\sqrt{(2\pi)}} \int_{-\infty}^{+\infty} \frac{e^{-i\beta y}\Psi(\beta)[(p-\beta^2 q)+i\beta(p+q)]}{(p^2+\beta^2 q^2)\sqrt{(1+\beta^2)}} d\beta, \tag{94}$$

and

$$f(x_1, y) = \frac{1}{\sqrt{(6\pi)}} \int_{-\infty}^{+\infty} \frac{e^{-i\beta y}\Psi(\beta)[p+(1-3i\beta)q]}{(1-i\beta)(p-i\beta q)} d\beta, \tag{95}$$

where

$$\Psi(\beta) = \frac{1}{\sqrt{(2\pi)}} \int_{-\infty}^{+\infty} e^{i\beta y} e^{-y} \psi(y) \, dy, \tag{96}$$

and

$$p = \sqrt{\{3(1+\beta^2)\}}\cosh\{x_1\sqrt{(1+\beta^2)}\}+2\sinh\{x_1\sqrt{(1+\beta^2)}\},$$
$$q = \sinh\{x_1\sqrt{(1+\beta^2)}\}. \tag{97}$$

85.4. *The spectral distribution of the emergent radiation*

From the linearity of the equation of transfer and more particularly of the reduced equation (91) it follows that the solution for an arbitrary continuous function $\psi(y)$ can be derived simply from the solution for the case

$$\psi(y) = \delta(y), \tag{98}$$

where $\delta(y)$ is the δ-function of Dirac. Thus, if $I_{+1}(0, y; \delta)$ denotes the solution for the emergent radiation for the case when the surface at $x = x_1$ radiates *monochromatically* at a wave-length λ (cf. eq. [82]), the solution when the surface radiates with a spectral distribution corresponding to an assigned $\psi(y)$, is given by

$$I_{+1}(0, y; \psi) = \int_{-\infty}^{+\infty} I_{+1}(0, y-\eta; \delta)\psi(\eta) \, d\eta. \tag{99}$$

It might be thought that the assumption of monochromatic emission by the radiating surface is incompatible with our earlier expansion of $I(\tau, \mu, \lambda)$ in a Taylor series (cf. eq. [77]). But this is not the case. For the manner of deriving the solution for an arbitrary ψ in terms of the solution for the case (98) is a mathematical statement strictly concerning the solution of the boundary-value problem formulated in § 85.2 which has no bearing on what has gone before. However, our earlier assumption concerning $I(\tau, \mu, \lambda)$ means, now, that physically significant solutions are obtained only after 'smearing' $I_{+1}(0, y; \delta)$ by a relatively

† For the details of the derivation see S. Chandrasekhar, *Proc. Roy. Soc.* (*London*), A, **192**, 508 (1948).

smooth function $\psi(y)$ according to equation (99). With this understanding we proceed to consider the basic solution $I_{+1}(0, y; \delta)$ of the problem.

For $\psi(y) = \delta(y)$ it is apparent that (cf. eq. [96])

$$\Psi(\beta) = 1/\sqrt{(2\pi)},$$

and equation (94) reduces to

$$f(0, y) = \frac{1}{2\pi} \int_{-\infty}^{+\infty} \frac{e^{-i\beta y}[(p - \beta^2 q) + i\beta(p + q)]}{(p^2 + \beta^2 q^2)\sqrt{(1 + \beta^2)}} d\beta, \tag{100}$$

or, equivalently,

$$f(0, y) = \frac{1}{\pi} \int_0^\infty \frac{(p - \beta^2 q)\cos\beta y + (p + q)\beta\sin\beta y}{(p^2 + \beta^2 q^2)\sqrt{(1 + \beta^2)}} d\beta. \tag{101}$$

We are, of course, particularly interested in the distribution with wave-length of the emergent radiation. According to equation (92)

$$I_{+1}(0, y) = \tfrac{1}{2}e^y \left[\frac{\partial f}{\partial x} + \sqrt{3}\left(f + \frac{\partial f}{\partial y}\right) \right]_{x=0}, \tag{102}$$

or, using equation (93), we have

$$I_{+1}(0, y) = \sqrt{3} \cdot e^y \left(f + \frac{\partial f}{\partial y} \right)_{x=0}. \tag{103}$$

Now substituting for $f(0, y)$ according to equation (101) we find, after some minor rearranging of the terms, that

$$I_{+1}(0, y; \delta) = \frac{\sqrt{3}}{\pi} e^y \int_0^\infty \frac{(p\cos\beta y + \beta q\sin\beta y)\sqrt{(1 + \beta^2)}}{(p^2 + \beta^2 q^2)} d\beta, \tag{104}$$

where it may be recalled that p and q are given by equations (97).

For any given x_1 and y, the emergent intensity can be found by evaluating the integral (104). The results of such calculations for $x_1 = 1, 2,$ and 3 are illustrated in Fig. 33. In examining these results it should be remembered that y measures the wave-length shifts in units of $\tfrac{2}{3}$ the Compton wave-length. And the values of x_1 for which the calculations have been made correspond to optical thicknesses equal to $\tfrac{2}{3}, \tfrac{4}{3},$ and 2, respectively.

From Fig. 33 we observe that the calculated distributions predict finite intensities to the violet as well. An exact solution of the equation of transfer (74) will not, of course, predict this. The error in the treatment was clearly introduced in passing from equation (74) to (78).

Indeed, the area to the left of $y = 0$ may be regarded as a measure of the inaccuracy introduced by our approximative treatment of the equation of transfer (74). It is therefore gratifying to note that even for $\tau_1 = \frac{2}{3}$ the contribution to the violet in the total intensity is less than 15 per cent. A further point to which attention may be drawn in this connexion is that the derived spectral distributions are probably

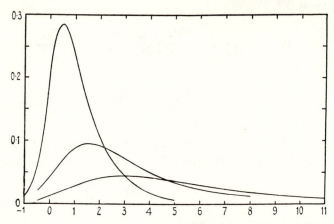

Fig. 33. The softening of radiation by multiple Compton scattering: the spectral distribution, $I_{+1}(0, y; \delta)$ of the radiation emergent after transmission through optical thicknesses $\frac{2}{3}$, $\frac{4}{3}$ and 2 of an incident monochromatic flux of radiation. The abscissae denote the wave-length shifts in units of $\frac{2}{3}$ Compton wave-length ($= 0.016$ A) and the ordinates the emergent intensities in units of the integrated outward intensity at the base of the atmosphere.

less affected by the passage from the exact equation (74) to the approximate equation (78), in the region of large wave-length shifts, than in the region of small shifts; for as τ becomes different from τ_1, the circumstances become more favourable for the use of the Taylor expansion (77) when $\delta\lambda/\gamma$ is large than when $\delta\lambda/\gamma$ is small.

One somewhat unexpected result which emerges from the calculations is the relatively high probability of quite large shifts after transmission through optical thicknesses of order even unity. This result has some interesting astrophysical applications; but it will take us too far outside the scope of this book to go into them here.

86. The broadening of lines by electron scattering

In § 85 we have considered the softening of radiation by multiple Compton scattering when each scattering results in an increase of wave-length of amount given by equation (72). However, when a light quantum is scattered by an electron in motion, the change in wave-length is not given by equation (72); instead, we have the formula†

$$(mc-p_x)\frac{v}{v'}-(mc-p_x\cos\Theta-p_y\sin\Theta) = \frac{hv}{c}(1-\cos\Theta), \qquad (105)$$

which relates the frequencies v and v' of the incident and the scattered quanta with the angle of scattering Θ and the momentum of the scattering electron, $\mathbf{p} = (p_x, p_y, p_z)$. In writing equation (105), we have assumed that the direction of the incident quantum is the x-axis and that the plane of scattering is the xy-plane.

In equation (105) the term in hv/c on the right-hand side is peculiar to the quantum theory and is characteristic of the Compton effect. The terms on the left-hand side, which do not involve h, represent the classical Doppler effect. In a classical approximation, therefore, the equation relating v, v', Θ, and $\mathbf{p}$ is

$$\left(\frac{v}{v'}-\cos\Theta\right)p_x-p_y\sin\Theta = mc\left(\frac{v}{v'}-1\right), \qquad (106)$$

or, to a sufficient accuracy ($|\mathbf{p}| \ll mc$),

$$(1-\cos\Theta)p_x-p_y\sin\Theta = mc\left(\frac{v}{v'}-1\right). \qquad (107)$$

Let radiation of intensity I_v in the frequency v, confined to an element of solid angle $d\omega$, be incident on an element of mass dm containing $N_e\,dm$ free electrons with a Maxwellian distribution of velocities corresponding to a temperature T. The scattering of the incident radiation will result in a re-distribution over the frequencies, the electrons with momentum $\mathbf{p} = (p_x, p_y, p_z)$ scattering, in the direction Θ, radiation of frequency v' given by equation (107).

In our present classical approximation the scattering of radiation by the free electrons may be assumed to take place according to Thomson's laws.

Now in the element of mass dm there will be

$$dm\,\frac{N_e}{(2\pi mkT)^{\frac{3}{2}}}\,e^{-(p_x^2+p_y^2+p_z^2)/2mkT}\,dp_x\,dp_y\,dp_z \qquad (108)$$

electrons with momenta in the range (p_x, p_y, p_z) and $(p_x+dp_x, p_y+dp_y, p_z+dp_z)$. Let the scattering by these electrons result in radiation, in the direction Θ, having frequencies in the interval $(v', v'+dv')$. If the incident radiation be assumed to have frequencies in the interval $(v, v+dv)$, it is evident that not all dv, dv', dp_x, and dp_y can be prescribed arbitrarily: they must be related in conformity with equation (107).

† Cf. P. A. M. Dirac, *Monthly Notices Roy. Astron. Soc.*, London, **85**, 825 (1925) (eq. [6]).

On the assumptions made, the *total* amount of radiant energy scattered per unit time, in the direction Θ, confined to an element of solid angle $d\omega'$ and in the frequency interval $(\nu',\ \nu'+d\nu')$, will be given by

$$I_\nu\, d\nu d\omega \times \sigma_e\, dm \times \tfrac{3}{4}(1+\cos^2\Theta)\frac{d\omega'}{4\pi} \times \psi(\nu;\nu')\, d\nu', \qquad (109)$$

where σ_e denotes Thomson's scattering coefficient (eq. [73]) and

$$\psi(\nu;\nu') = \frac{1}{(2\pi mkT)^{\frac{3}{2}}} \int\limits_{-\infty}^{+\infty}\int\limits_{0}^{+\infty} e^{-(p_x^2+p_y^2+p_z^2)/2mkT}\frac{\partial p_x}{\partial \nu}dp_y\,dp_z. \qquad (110)$$

In (110), $\partial p_x/\partial \nu$ must be evaluated according to equation (107) and expressed in terms of ν, ν', and p_y; similarly, p_x must also be expressed in terms of ν, ν', and p_y.

It will be noticed that in writing (109) we have allowed for an anisotropy of the scattered radiation according to Rayleigh's phase function; (109) is therefore valid only for incident unpolarized light; its use for light scattered more than once is, strictly, incorrect. However, in the first approximation, in which we shall solve the equation of transfer, the more refined considerations do not make any difference.

Returning to (110), we easily verify that according to equation (107)

$$p_x^2+p_y^2 = \frac{2}{1-\cos\Theta}\left[p_y+\frac{mc(\nu-\nu')\sin\Theta}{2\nu'(1-\cos\Theta)}\right]^2 + \frac{m^2c^2(\nu-\nu')^2}{2\nu'^2(1-\cos\Theta)}. \qquad (111)$$

Also,

$$\frac{\partial p_x}{\partial \nu} = \frac{mc}{\nu'(1-\cos\Theta)}. \qquad (112)$$

Hence,

$$\psi(\nu;\nu') = \frac{mce^{-mc^2(\nu-\nu')^2/4kT\nu'^2(1-\cos\Theta)}}{\nu'(1-\cos\Theta)(2\pi mkT)^{\frac{3}{2}}} \times$$

$$\times \int\limits_{-\infty}^{+\infty}\int\limits_{0}^{+\infty} \exp\left\{-\left[\frac{2}{1-\cos\Theta}\left(p_y+\frac{mc(\nu-\nu')\sin\Theta}{2\nu'(1-\cos\Theta)}\right)^2+p_z^2\right]\bigg/2mkT\right\}dp_y\,dp_z. \qquad (113)$$

The integral on the right-hand side is readily evaluated and we are left with

$$\psi(\nu;\nu') = \left[\frac{mc^2}{4\pi kT\nu'^2(1-\cos\Theta)}\right]^{\frac{1}{2}} e^{-mc^2(\nu-\nu')^2/4kT\nu'^2(1-\cos\Theta)}. \qquad (114)$$

According to this last equation, for incident monochromatic light, the 'line' scattered in the direction Θ has the 'width'

$$\left[\frac{4kT}{mc^2}\lambda^2(1-\cos\Theta)\right]^{\frac{1}{2}}; \qquad (115)$$

for $T = 10{,}000°$, $\lambda = 4{,}000$ A, and $\Theta = \tfrac{1}{2}\pi$, this amounts to 10·4 A.

Combining (109) and (114), we can write for the source function for radiation of frequency ν,† in the direction (μ, φ), the expression

$$\Im(\mu, \varphi, \nu) = \frac{3}{16\pi} \int_0^{2\pi} d\varphi' \int_{-1}^{+1} d\mu'(1+\cos^2\Theta) \times$$

$$\times \int_0^\infty d\nu' I(\mu', \varphi', \nu') \left[\frac{mc^2}{4\pi k T \nu^2(1-\cos\Theta)} \right]^{\frac{1}{2}} e^{-mc^2(\nu'-\nu)^2/4kT\nu^2(1-\cos\Theta)}, \quad (116)$$

where $\cos\Theta$ has now the value given by equation (76). The corresponding equation of transfer for plane-parallel atmospheres is

$$\mu \frac{\partial I(\tau, \mu, \nu)}{\partial \tau} = I(\tau, \mu, \nu) - \Im(\tau, \mu, \nu). \quad (117)$$

In the context of equations (116) and (117) we shall consider the following problem:

An infinite plane surface radiates uniformly in the outward directions with a known spectral distribution. Overlying this radiating surface is an atmosphere of free electrons at a temperature T. It is required to find the modified spectral distribution of the emergent radiation.

In most practical applications it is a sufficient approximation to replace ν and ν' by a constant frequency ν_0 (of the 'centre' of the line) except when the difference $(\nu-\nu')$ is involved; moreover, we may let the range of $(\nu-\nu')$ extend from $-\infty$ to $+\infty$. With this understanding, we may write the equation of transfer (117), explicitly, in the form

$$\mu \frac{\partial I(\tau, \mu, \alpha)}{\partial \tau} = I(\tau, \mu, \alpha) -$$

$$- \frac{3}{16\pi} \int_0^{2\pi} d\varphi' \int_{-1}^{+1} d\mu'(1+\cos^2\Theta) \int_{-\infty}^{+\infty} \frac{d\alpha'\, I(\tau, \mu', \alpha')}{\sqrt{\{\pi(1-\cos\Theta)\}}} e^{-(\alpha'-\alpha)^2/(1-\cos\Theta)}, \quad (118)$$

where, instead of ν, we have introduced the variable

$$\alpha = \left(\frac{mc^2}{4kT} \right)^{\frac{1}{2}} \frac{\nu-\nu_0}{\nu_0}, \quad (119)$$

ν_0 being some suitably chosen constant frequency. (When dealing with absorption or emission lines, ν_0 may be taken to be the centre of the line.)

We require to solve equation (118) with the boundary conditions

$$I(\tau_1, \mu, \alpha) = I^{(0)}(\alpha) \quad \text{and} \quad I(0, -\mu, \alpha) \equiv 0 \quad (0 \leqslant \mu \leqslant 1), \quad (120)$$

† Notice that this requires the interchange of the role of ν and ν'.

corresponding, respectively, to the known spectral distribution of the outward radiation at the base of the atmosphere at $\tau = \tau_1$ and the absence of any radiation incident on the atmosphere at $\tau = 0$.

The solution of the problem, we shall now present, is due to G. Münch.

86.1. *The Fourier transform of equation* (118)

Applying to equation (118) a Fourier transformation with respect to α, i.e. multiplying the equation by $e^{i\alpha\xi}/\sqrt{(2\pi)}$ and integrating over the entire range $(-\infty, +\infty)$ of α, we obtain

$$\mu \frac{\partial \mathscr{I}(\tau, \mu; \xi)}{\partial \tau} = \mathscr{I}(\tau, \mu; \xi) -$$

$$-\frac{3}{16\pi} \int_0^{2\pi} d\varphi' \int_{-1}^{+1} d\mu' \, (1+\cos^2\Theta) \int_{-\infty}^{+\infty} \frac{d\alpha \, e^{i\alpha\xi}}{\sqrt{(2\pi)}} \int_{-\infty}^{+\infty} \frac{d\alpha' \, I(\tau, \mu, \alpha')}{\sqrt{\{\pi(1-\cos\Theta)\}}} \, e^{-(\alpha'-\alpha)^2/(1-\cos\Theta)}, \tag{121}$$

where
$$\mathscr{I}(\tau, \mu; \xi) = \frac{1}{\sqrt{(2\pi)}} \int_{-\infty}^{+\infty} e^{i\alpha\xi} I(\tau, \mu, \alpha) \, d\alpha, \tag{122}$$

is the Fourier transform of $I(\tau, \mu, \alpha)$. The integrations over α and α' in the multiple integral on the right-hand side of equation (121) can be performed and expressed in terms of $\mathscr{I}(\tau, \mu', \xi)$. We have

$$\int_{-\infty}^{+\infty} \frac{d\alpha \, e^{i\alpha\xi}}{\sqrt{(2\pi)}} \int_{-\infty}^{+\infty} \frac{d\alpha' \, I(\tau, \mu', \alpha')}{\sqrt{\{\pi(1-\cos\Theta)\}}} \, e^{-(\alpha'-\alpha)^2/(1-\cos\Theta)}$$

$$= \int_{-\infty}^{+\infty} \frac{d\alpha'}{\sqrt{(2\pi)}} e^{i\alpha'\xi} I(\tau, \mu', \alpha') \int_{-\infty}^{+\infty} \frac{d\alpha}{\sqrt{\{\pi(1-\cos\Theta)\}}} e^{-i(\alpha'-\alpha)\xi - (\alpha'-\alpha)^2/(1-\cos\Theta)}$$

$$= e^{-\xi^2(1-\cos\Theta)/4} \mathscr{I}(\tau, \mu'; \xi). \tag{123}$$

The application of the Fourier transformation, therefore, reduces the equation of transfer to the form

$$\mu \frac{\partial \mathscr{I}(\tau, \mu; \xi)}{\partial \tau} = \mathscr{I}(\tau, \mu; \xi) - \tfrac{3}{8} \int_{-1}^{+1} p(\mu, \mu'; \xi) \mathscr{I}(\tau, \mu'; \xi) \, d\mu', \tag{124}$$

where
$$p(\mu, \mu'; \xi) = \frac{1}{2\pi} \int_0^{2\pi} (1+\cos^2\Theta) e^{-\xi^2(1-\cos\Theta)/4} \, d\varphi'. \tag{125}$$

Using the integral representation

$$I_n(x) = \frac{1}{\pi} \int_0^{\pi} e^{x\cos\varphi'} \cos n\varphi' \, d\varphi' \tag{126}$$

of the Bessel function of order n for a purely imaginary argument, we can easily establish the formula

$$p(\mu,\mu';\xi) = \{[1+\mu^2\mu'^2+\tfrac{1}{2}(1-\mu^2)(1-\mu'^2)]I_0(\tfrac{1}{4}\xi^2\sqrt{(1-\mu^2)}\cdot\sqrt{(1-\mu'^2)})+$$
$$+2\mu\mu'(1-\mu^2)^{\frac{1}{2}}(1-\mu'^2)^{\frac{1}{2}}I_1(\tfrac{1}{4}\xi^2\sqrt{(1-\mu^2)}\cdot\sqrt{(1-\mu'^2)})+$$
$$+\tfrac{1}{2}(1-\mu^2)(1-\mu'^2)I_2(\tfrac{1}{4}\xi^2\sqrt{(1-\mu^2)}\cdot\sqrt{(1-\mu'^2)})\}e^{-\xi^2(1-\mu\mu')/4}. \quad (127)$$

The boundary conditions with respect to which equation (124) must be solved are the Fourier transforms of the boundary conditions (120). Thus,

$$\mathscr{I}(\tau_1,\mu;\xi) = \mathscr{I}^{(0)}(\xi) = \frac{1}{\sqrt{(2\pi)}}\int_{-\infty}^{+\infty}e^{i\alpha\xi}I^{(0)}(\alpha)\,d\alpha$$

and

$$\mathscr{I}(0,-\mu;\xi) \equiv 0 \quad (0 \leqslant \mu \leqslant 1). \quad (128)$$

86.2. *The solution of equation* (124) *in the first approximation*

It is seen that equation (124) is of the form to which our standard methods of solution apply. Restricting ourselves to the first approximation (eq. [80]), we have the pair of equations

$$\frac{1}{\sqrt 3}\frac{\partial\mathscr{I}_{+1}}{\partial\tau} = \mathscr{I}_{+1}-[M(\xi)\mathscr{I}_{+1}+N(\xi)\mathscr{I}_{-1}]$$

and

$$-\frac{1}{\sqrt 3}\frac{\partial\mathscr{I}_{-1}}{\partial\tau} = \mathscr{I}_{-1}-[N(\xi)\mathscr{I}_{+1}+M(\xi)\mathscr{I}_{-1}], \quad (129)$$

where (cf. eq. [127])

$$M(\xi) = \tfrac{1}{2}[I_0(\tfrac{1}{6}\xi^2)+\tfrac{1}{3}I_1(\tfrac{1}{6}\xi^2)+\tfrac{1}{6}I_2(\tfrac{1}{6}\xi^2)]e^{-\xi^2/6}$$

and

$$N(\xi) = \tfrac{1}{2}[I_0(\tfrac{1}{6}\xi^2)-\tfrac{1}{3}I_1(\tfrac{1}{6}\xi^2)+\tfrac{1}{6}I_2(\tfrac{1}{6}\xi^2)]e^{-\xi^2/3}. \quad (130)$$

It is seen that the general solution of equation (129) is of the form

$$\mathscr{I}_{+1}(\tau;\xi) = A(\xi)e^{+\chi\tau}+B(\xi)e^{-\chi\tau}$$

and

$$\mathscr{I}_{-1}(\tau;\xi) = \frac{A(\xi)}{N}\left(1-M-\frac{\chi}{\sqrt 3}\right)e^{+\chi\tau}+\frac{B(\xi)}{N}\left(1-M+\frac{\chi}{\sqrt 3}\right)e^{-\chi\tau}, \quad (131)$$

where $A(\xi)$ and $B(\xi)$ are arbitrary functions of the argument and

$$\chi \equiv \chi(\xi) = \{[1-M(\xi)]^2-N^2(\xi)\}^{\frac{1}{2}}\sqrt 3. \quad (132)$$

The boundary conditions (128) determine $A(\xi)$ and $B(\xi)$. We find

$$A(\xi) = \frac{\tfrac{1}{2}(1-M+\chi/\sqrt 3)}{(1-M)\sinh\chi\tau_1+(\chi/\sqrt 3)\cosh\chi\tau_1}\mathscr{I}^{(0)}(\xi)$$

and

$$B(\xi) = \frac{-\tfrac{1}{2}(1-M-\chi/\sqrt 3)}{(1-M)\sinh\chi\tau_1+(\chi/\sqrt 3)\cosh\chi\tau_1}\mathscr{I}^{(0)}(\xi). \quad (133)$$

The solutions for $\mathscr{I}_{+1}(\tau;\xi)$ and $\mathscr{I}_{-1}(\tau;\xi)$ now follow from equations (131) and (133). In particular we have

$$\mathscr{I}_{+1}(0;\xi) = \frac{1}{\sqrt{(2\pi)}} \int_{-\infty}^{+\infty} e^{i\alpha\xi} I_{+1}(0,\alpha)\, d\alpha = G(\xi;\tau_1)\mathscr{I}^{(0)}(\xi), \qquad (134)$$

where
$$G(\xi;\tau_1) = \frac{\chi/\sqrt{3}}{(1-M)\sinh\chi\tau_1 + (\chi/\sqrt{3})\cosh\chi\tau_1} \qquad (135)$$

From the inverse property of the Fourier transforms, it follows from equation (134) that

$$I_{+1}(0,\alpha) = \frac{1}{\sqrt{(2\pi)}} \int_{-\infty}^{+\infty} e^{-i\alpha\xi} G(\xi;\tau_1)\mathscr{I}^{(0)}(\xi)\, d\xi. \qquad (136)$$

This represents the solution for the emergent radiation.

Now it can be shown that

$$G(\xi;\tau_1) \to e^{-\tau_1\sqrt{3}} \quad \text{as} \quad \xi \to \pm\infty. \qquad (137)$$

We therefore rewrite equation (136) in the form

$$I_{+1}(0,\alpha) = \frac{e^{-\tau_1\sqrt{3}}}{\sqrt{(2\pi)}} \int_{-\infty}^{+\infty} e^{-i\alpha\xi}\mathscr{I}^{(0)}(\xi)\, d\xi +$$

$$+ \frac{1}{\sqrt{(2\pi)}} \int_{-\infty}^{+\infty} e^{-i\alpha\xi}[G(\xi;\tau_1) - e^{-\tau_1\sqrt{3}}]\mathscr{I}^{(0)}(\xi)\, d\xi,$$

or,
$$I_{+1}(0,\alpha) = e^{-\tau_1\sqrt{3}} I^{(0)}(\alpha) + \frac{1}{\sqrt{(2\pi)}} \int_{-\infty}^{+\infty} e^{-i\alpha\xi}[G(\xi;\tau_1) - e^{-\tau_1\sqrt{3}}]\mathscr{I}^{(0)}(\xi)\, d\xi. \qquad (138)$$

The second term on the right-hand side of equation (138) can be reduced in the following manner:

$$\frac{1}{\sqrt{(2\pi)}} \int_{-\infty}^{+\infty} e^{-i\alpha\xi}[G(\xi;\tau_1) - e^{-\tau_1\sqrt{3}}]\mathscr{I}^{(0)}(\xi)\, d\xi$$

$$= \frac{1}{2\pi} \int_{-\infty}^{+\infty} d\xi\, [G(\xi;\tau_1) - e^{-\tau_1\sqrt{3}}]e^{-i\alpha\xi} \int_{-\infty}^{+\infty} d\beta\, e^{i\beta\xi} I^{(0)}(\beta)$$

$$= \frac{1}{2\pi} \int_{-\infty}^{+\infty} d\beta\, I^{(0)}(\beta) \int_{-\infty}^{+\infty} d\xi\, [G(\xi;\tau_1) - e^{-\tau_1\sqrt{3}}]e^{-i\xi(\alpha-\beta)}$$

$$= \int_{-\infty}^{+\infty} I^{(0)}(\beta)\Delta(\alpha-\beta;\tau_1)\, d\beta, \qquad (139)\dagger$$

† We are effectively using here one of Parseval's formulae in the theory of Fourier integrals.

where $\qquad \Delta(\alpha; \tau_1) = \dfrac{1}{2\pi} \displaystyle\int_{-\infty}^{+\infty} e^{-i\alpha\xi}[G(\xi; \tau_1) - e^{-\tau_1\sqrt{3}}]\, d\xi.$ $\qquad$ (140)

The solution for $I_{+1}(0, \alpha)$ therefore takes the form

$$I_{+1}(0, \alpha) = e^{-\tau_1\sqrt{3}} I^{(0)}(\alpha) + \int_{-\infty}^{+\infty} I^{(0)}(\beta)\Delta(\alpha - \beta; \tau_1)\, d\beta. \qquad (141)$$

The solution for the emergent radiation in the form (141) allows the following simple interpretation: The first term represents the part of

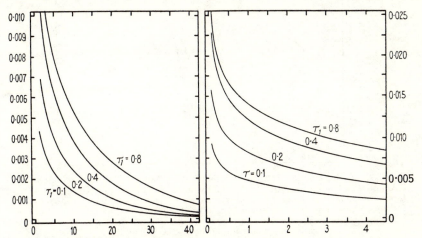

FIG. 34. The broadening of lines by electron scattering: the spectral distribution $\Delta(\alpha; \tau_1)$ of the diffuse emergent radiation for an incident monochromatic flux of radiation. The abscissae denote the wave-length shifts in angstroms and the ordinates the intensities in the unit of the integrated outward intensity at the base of the atmosphere. The particular case $\lambda_0 = 5,000$ A and $T = 30,000°$ K. is illustrated (cf. eq. [119]). (Since the curves are symmetrical about $\lambda = \lambda_0$, only halves of the curves are illustrated.)

the radiation incident on the atmosphere at $\tau = \tau_1$, from below, which emerges without having suffered any scattering process in the atmosphere (cf. eq. [1]). The second term, therefore, represents the diffuse radiation which has suffered one or more scattering processes in the atmosphere. It is also clear that $\Delta(\alpha; \tau_1)$ represents the spectral distribution of the emergent diffuse radiation for incident *monochromatic* light, i.e. for $I^{(0)}(\alpha) = \delta(\alpha)$ where δ denotes Dirac's δ-function (cf. eq. [99]). It can be shown that $\Delta(\alpha; \tau_1)$ has a logarithmic singularity as $\alpha \to 0$:

$$\Delta(\alpha; \tau_1) \to -\frac{9\tau_1 e^{-\tau_1\sqrt{3}}}{4\pi^{\frac{3}{2}}}\log\alpha \quad (\alpha \to 0). \qquad (142)$$

However, when a continuous distribution $I^{(0)}(\alpha)$ is 'smeared' according to equation (141), the resulting distribution has no singularity.

The function $\Delta(\alpha; \tau_1)$ has been evaluated by Münch for various values of τ_1. His results are shown in Fig. 34. The effect of electron scattering in broadening stellar absorption lines (in accordance with eq. [141]) is further illustrated in Fig. 35. It is seen that the effect consists, essentially, in producing very extended wings and in 'shallowing' the lines,

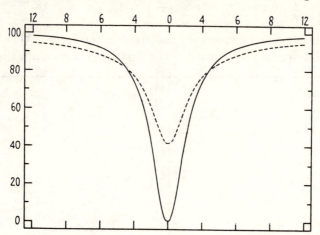

Fig. 35. The broadening of an absorption line at 5,000 A, initially of half-width 3 A and total equivalent width 9·4 A, by an atmosphere of free electrons at temperature 30,000° K. and optical thickness 0·8. The initial line is shown by the full-line curve and the broadened line by the dashed curve. The abscissae measure the wave-length shifts in angstroms from the centre of the line and the ordinates the residual intensities in the line.

generally. It appears that in these features of the effects of electron scattering we have a means of interpreting certain peculiarities in the absorption lines in early type stars.

BIBLIOGRAPHICAL NOTES

§ 83. The problem considered in this section derives from:

1. A. SCHUSTER, *Astrophys. J.* **21**, 1 (1905).

Later discussions of the same problem are those of—

2. K. SCHWARZSCHILD, *Sitzungsberichte d. Preuss. Akad. Berlin*, 1183 (1914).
3. E. A. MILNE, *Philos. Trans. Roy. Soc. (London)*, A, **223**, 201 (1922).
4. —— *Monthly Notices Roy. Astron. Soc. London*, **89**, 3 (1928).
5. —— *Philos. Trans. Roy. Soc. (London)*, A, **228**, 421 (1929).
6. E. HOPF, *Mathematical Problems of Radiative Equilibrium*, chap. iii (Cambridge Mathematical Tracts, No. 31), Cambridge, 1934.
7. A. UNSÖLD, *Physik der Sternatmosphären*, § 64, Berlin, Springer, 1938.

The solution of the problem in terms of X- and Y-functions given in the text, is however, new.

§ 84. The premisses on which the problem of line formation is considered in this section is sometimes referred to as the 'Milne–Eddington model'. The first discussion of the problem along these general lines was given by—

8. A. S. Eddington, *Internal Constitution of the Stars*, §§ 234–8, pp. 337–43, Cambridge, 1926.

9. —— *Monthly Notices Roy. Astron. Soc. London*, **89**, 620 (1929).

See also Milne (refs. 4 and 5).

The exact solution given in the text is due to—

10. S. Chandrasekhar, *Astrophys. J.* **106**, 145 (1947).

The H-functions and their moments which occur in the exact solution were computed by—

11. S. Chandrasekhar and Frances H. Breen, ibid. **106**, 143 (1947).

Methods of analysis suitable for cases in which the ratio $\eta_\nu = \sigma_\nu/\kappa_\nu$, of the line to the continuous absorption coefficient, is not constant, are described by—

12. B. Strömgren, ibid. **86**, 1 (1937).

13. Merle Tuberg, ibid. **103**, 145 (1946).

Applications of methods derived from those described in the text will be found in—

14. J. L. Greenstein, ibid. **107**, 151 (1948).

15. A. B. Underhill, ibid. **107**, 349 (1948).

16. L. H. Aller, ibid. **109**, 244 (1948).

17. D. Harris, ibid. **109**, 53 (1949); M. Wrubel, ibid. **109**, 66 (1949).

§ 85. The analysis given in this section is based on—

18. S. Chandrasekhar, *Proc. Roy. Soc. (London)*, A, **192**, 508 (1948).

§ 86. The physical theory for considering the effects of electron scattering on line formation in stellar atmospheres, is due to—

19. P. A. M. Dirac, *Monthly Notices Roy. Astron. Soc. London*, **86**, 825 (1925).

The related problems in transfer theory have been considered by Chandrasekhar (ref. 18) and G. Münch.

20. G. Münch, *Astrophys. J.* **108**, 116 (1948).

The analysis in this section follows Münch's paper.

A problem in transfer theory not discussed in this chapter is the influence of Doppler effect consequent to macroscopic differential motions. The discussion of this problem leads to some rather technical matters relating to the solution of certain novel types of boundary-value problems in hyperbolic equations; it is omitted on that account. Reference may, however, be made to the following papers in which the problem is treated:

21. W. H. McCrea and K. K. Mitra, *Zeits. f. Astrophysik*, **11**, 359 (1936).

22. S. Chandrasekhar, *Rev. Mod. Phys.* **17**, 138 (1945).

23. —— *Astrophys. J.* **102**, 402 (1945).

24. —— *Proc. Camb. Philos. Soc.* **42**, 250 (1946).

XIII

MISCELLANEOUS PROBLEMS

87. Introduction

IN this last chapter we shall treat a number of miscellaneous problems which have bearings on several points on which we have had no occasion to touch. The problems have been selected with a view to indicating the directions of further advance of the theory of radiative transfer. The topics discussed include the relation of the H-functions to solutions of integral equations of the Schwarzschild–Milne type; a time-dependent problem in the theory of the escape of imprisoned radiation; and an elementary problem in the theory of transfer in atmospheres with spherical symmetry.

88. An example of a problem in semi-infinite atmospheres with no incident radiation and in non-conservative cases

In the context of the equation of transfer

$$\mu \frac{dI(\tau,\mu)}{d\tau} = I(\tau,\mu) - \tfrac{1}{2}\varpi_0 \int\limits_{-1}^{+1} I(\tau,\mu')\,d\mu', \tag{1}$$

the following problem was formulated in Chapter I, § 12:

To solve equation (1) with the boundary conditions

$$I(0,-\mu) = 0 \quad (0 \leqslant \mu \leqslant 1) \tag{2}$$

and
$$I(\tau,\mu) \to L_0 \frac{e^{k\tau}}{1-k\mu} \quad (\tau \to \infty), \tag{3}$$

where L_0 is a constant and k is the positive root, less than one, of the transcendental equation (Chap. I, eq. [113], Table I)

$$2k = \varpi_0 \log\left(\frac{1+k}{1-k}\right). \tag{4}$$

We shall show how, for this problem, the angular distribution of the emergent radiation can be found without explicitly solving the equation of transfer by appealing to a principle of invariance of the type considered in Chapter IV, § 29.3.

The equation of transfer (1) admits the integral

$$I(\tau,\mu) = L_0 \frac{e^{k\tau}}{1-k\mu} \quad (L_0 = \text{a constant}). \tag{5}$$

This solution does not, of course, satisfy the boundary condition (2). We shall therefore let

$$I(\tau, \mu) = L_0 \frac{e^{k\tau}}{1-k\mu} + I^*(\tau, \mu), \tag{6}$$

represent the solution of the problem. With this expression of the solution as the sum of two terms, one representing the solution for an infinite unbounded atmosphere and the other representing the departure of the asymptotic solution (5) as we approach $\tau = 0$, it is evident that at any level τ, the intensity $I^*(\tau, +\mu)$ in the outward direction $(0 < \mu \leqslant 1)$ must result from the reflection of the inward directed radiation $I^*(\tau, -\mu')$ $(0 \leqslant \mu' \leqslant 1)$ by the semi-infinite atmosphere below τ. Accordingly,

$$I(\tau, +\mu) = L_0 \frac{e^{k\tau}}{1-k\mu} + \frac{1}{2\mu} \int_0^1 S(\mu, \mu') I^*(\tau, -\mu') \, d\mu', \tag{7}$$

where $S(\mu, \mu')$ is the scattering function defined in the usual manner.

Applying equation (7) to the boundary of the atmosphere at $\tau = 0$, we have

$$I(0, \mu) = \frac{L_0}{1-k\mu} + \frac{1}{2\mu} \int_0^1 S(\mu, \mu') I^*(0, -\mu') \, d\mu'; \tag{8}$$

but according to equations (2) and (6)

$$I^*(0, -\mu') = -\frac{L_0}{1+k\mu'}. \tag{9}$$

Hence
$$I(0, \mu) = L_0 \left\{ \frac{1}{1-k\mu} - \frac{1}{2\mu} \int_0^1 \frac{S(\mu, \mu')}{1+k\mu'} \, d\mu' \right\}. \tag{10}$$

Now for the problem on hand (Chap. IV, § 33.1, eqs. [41] and [42])

$$S(\mu, \mu') = \varpi_0 \frac{\mu\mu'}{\mu+\mu'} H(\mu) H(\mu'), \tag{11}$$

where $H(\mu)$ is defined in terms of the characteristic function $\frac{1}{2}\varpi_0$. With this value of S, equation (10) becomes

$$I(0, \mu) = L_0 \left\{ \frac{1}{1-k\mu} - \frac{1}{2}\varpi_0 H(\mu) \int_0^1 \frac{\mu' H(\mu')}{(\mu+\mu')(1+k\mu')} \, d\mu' \right\}. \tag{12}$$

Expressing $\mu'/(\mu+\mu')\cdot(1+k\mu')$ in partial fractions and rearranging the terms, we obtain

$$I(0,\mu) = \frac{L_0}{1-k\mu}\left\{1+\tfrac{1}{2}\varpi_0\,\mu H(\mu)\int_0^1 \frac{H(\mu')}{\mu+\mu'}\,d\mu'-\right.$$
$$\left.-\tfrac{1}{2}\varpi_0\,H(\mu)\int_0^1 \frac{H(\mu')}{1+k\mu'}\,d\mu'\right\}. \quad (13)$$

On using the integral equation satisfied by $H(\mu)$, the foregoing equation reduces to

$$I(0,\mu) = L_0\frac{H(\mu)}{1-k\mu}\left\{1-\tfrac{1}{2}\varpi_0\int_0^1 \frac{H(\mu')}{1+k\mu'}\,d\mu'\right\}. \quad (14)$$

Again, from the integral equation satisfied by $H(\mu)$ (cf. Chap. V, eq. [68]) it follows that the quantity in braces in equation (14) is $[H(1/k)]^{-1}$.

Hence
$$I(0,\mu) = \frac{L_0}{H(1/k)}\frac{H(\mu)}{1-k\mu}. \quad (15)$$

This is the required solution.

The density of radiation in the atmosphere is directly related to the inverse Laplace transform of $I(0,\mu)$ (cf. § 89.2 below).†

There is, of course, no difficulty in adapting the method of solution described here to other laws of scattering. An example is given in Appendix III.

89. The relation of H-functions to solutions of integral equations of the Schwarzschild–Milne type. The 'pseudo-problems' in transfer theory

In our discussion of transfer problems we have laid the principal emphasis on the angular distributions of the emergent radiations. For these we have obtained exact solutions under a variety of conditions; but for the radiation field in the interior we have obtained only solutions in finite approximations. However, we have shown already in Chapter I (§§ 11.1 and 11.2) that linear integral equations can be obtained for the source functions of the problems as well. Thus, for the simplest problem of constant net flux and isotropic scattering we have the Schwarzschild–Milne integral equation (Chap. I, eq. [97])

$$J(\tau) = \tfrac{1}{2}\int_0^\infty J(t)E_1(|t-\tau|)\,dt. \quad (16)$$

† This form for the solution can also be obtained, directly, by solving equation (1) in a finite approximation and eliminating the constants in the usual fashion.

In terms of $J(\tau)$ the emergent radiation is given by

$$I(0,\mu) = \int_0^\infty J(\tau)e^{-\tau/\mu}\frac{d\tau}{\mu}. \tag{17}$$

On the other hand, from the integral equations derived from the principles of invariance, we have shown that

$$I(0,\mu) = \frac{\sqrt{3}}{4}FH(\mu), \tag{18}$$

where $H(\mu)$ is defined in terms of the characteristic function $\frac{1}{2}$. It is therefore apparent that the non-linear integral equation governing $H(\mu)$ must be related in some way with the Laplace transform of equation (16). We shall now analyse the nature of the relationship suggested.

89.1. *The H-function in conservative cases as the Laplace transform of the solution of an integral equation of the Schwarzschild–Milne type*

The relation between H-functions in conservative cases and solutions of integral equations of the Schwarzschild–Milne type is expressed by the following theorem:

THEOREM 1. Let

$$\Psi(\mu) = \sum a_{2j}\mu^{2j}, \tag{19}$$

be an even polynomial in μ satisfying the condition

$$\int_0^1 \Psi(\mu)\,d\mu = \tfrac{1}{2}. \tag{20}$$

Further, let

$$I(0,\mu) = \int_0^\infty \mathfrak{J}(\tau)e^{-\tau/\mu}\frac{d\tau}{\mu}, \tag{21}$$

where $\mathfrak{J}(\tau)$ satisfies the linear integral equation

$$\mathfrak{J}(\tau) = \int_0^\infty \mathfrak{J}(t)\sum a_{2j}E_{2j+1}(|t-\tau|)\,dt. \tag{22}$$

Then

$$I(0,\mu) = \text{constant } H(\mu), \tag{23}$$

where $H(\mu)$ is the unique solution of the non-linear integral equation

$$H(\mu) = 1+\mu H(\mu)\int_0^1 \frac{H(\mu')}{\mu+\mu'}\Psi(\mu')\,d\mu', \tag{24}$$

which is bounded in the interval $(0,1)$.

PROOF. Multiplying equation (22) by $e^{-\tau/\mu}\,d\tau/\mu$ and integrating over μ from 0 to ∞, we obtain

$$I(0,\mu) = \int_0^\infty \int_0^\infty \mathfrak{J}(t) \sum a_{2j} E_{2j+1}(|t-\tau|)e^{-\tau/\mu}\,dt\,\frac{d\tau}{\mu}. \tag{25}$$

Breaking the range of integration at $t = \tau$, we write

$$I(0,\mu) = \int_0^\tau dt \int_0^\infty \frac{d\tau}{\mu}\mathfrak{J}(t) \sum a_{2j} E_{2j+1}(\tau-t)e^{-\tau/\mu} +$$

$$+ \int_\tau^\infty dt \int_0^\infty \frac{d\tau}{\mu}\mathfrak{J}(t) \sum a_{2j} E_{2j+1}(t-\tau)e^{-\tau/\mu}. \tag{26}$$

Inverting the order of the integrations over t and τ in (26), we obtain

$$I(0,\mu) = \int_0^\infty dt\,\mathfrak{J}(t) \int_t^\infty \frac{d\tau}{\mu}e^{-\tau/\mu} \sum a_{2j} E_{2j+1}(\tau-t) +$$

$$+ \int_0^\infty dt\,\mathfrak{J}(t) \int_0^t \frac{d\tau}{\mu}e^{-\tau/\mu} \sum a_{2j} E_{2j+1}(t-\tau). \tag{27}$$

Now

$$\int_t^\infty \frac{d\tau}{\mu}e^{-\tau/\mu}E_{2j+1}(\tau-t) = \int_t^\infty \frac{d\tau}{\mu}e^{-\tau/\mu} \int_0^1 \frac{d\mu'}{\mu'}\mu'^{2j}e^{-(\tau-t)/\mu'}$$

$$= \int_0^1 \frac{d\mu'}{\mu'}\mu'^{2j}e^{t/\mu'} \int_t^\infty \frac{d\tau}{\mu}\exp\left\{-\tau\left(\frac{1}{\mu}+\frac{1}{\mu'}\right)\right\} = e^{-t/\mu}\int_0^1 \frac{\mu'^{2j}}{\mu+\mu'}\,d\mu'. \tag{28}$$

Similarly,

$$\int_0^t \frac{d\tau}{\mu}e^{-\tau/\mu}E_{2j+1}(t-\tau) = \int_0^1 \frac{\mu'^{2j}}{\mu-\mu'}\left[e^{-t/\mu}-e^{-t/\mu'}\right]d\mu'. \tag{29}$$

Inserting the foregoing relations in (27) and recalling the definitions (19) and (21), we have

$$I(0,\mu) = I(0,\mu)\int_0^1 \frac{\mu\Psi(\mu')}{\mu+\mu'}d\mu' + \int_0^1 \frac{\Psi(\mu')}{\mu-\mu'}[\mu I(0,\mu)-\mu'I(0,\mu')]\,d\mu'. \tag{30}$$

Equation (30) can be written, somewhat differently, in the form

$$I(0,\mu)\left[1-\mu\int_0^1\frac{\Psi'(\mu')}{\mu+\mu'}\,d\mu'\right]$$

$$=\int_0^1 I(0,\mu')\Psi'(\mu')\,d\mu'+\mu\int_0^1\frac{\Psi'(\mu')}{\mu-\mu'}[I(0,\mu)-I(0,\mu')]\,d\mu'. \quad (31)$$

On the other hand, $H(\mu)$ satisfies the equation (cf. Chap. V, eq. [75])

$$H(\mu)\left[1-\mu\int_0^1\frac{\Psi'(\mu')}{\mu+\mu'}\,d\mu'\right]=1+\mu\int_0^1\frac{\Psi'(\mu')}{\mu-\mu'}[H(\mu)-H(\mu')]\,d\mu'. \quad (32)$$

Remembering that in conservative cases (Chap. V, eq. [16])

$$\int_0^1 H(\mu)\Psi'(\mu)\,d\mu = 1, \quad (33)$$

we observe the equivalence of equations (31) and (32). The truth of the theorem is now apparent.

89.2. *The relation between H-functions and the Laplace transforms of solutions of integral equations of the Schwarzschild–Milne type in non-conservative cases*

In non-conservative cases the relationship between H-equations and equations of the Schwarzschild–Milne type is expressed by the following theorem:

THEOREM 2. Let $\Psi'(\mu)=\sum a_{2j}\mu^{2j}$ be an even polynomial in μ satisfying the condition

$$\int_0^1\Psi'(\mu)\,d\mu<\tfrac{1}{2}. \quad (34)$$

Further, let the transcendental equation

$$1=2\int_0^1\frac{\Psi'(\mu)}{1-k^2\mu^2}d\mu \quad (35)$$

admit a root $0<k^2<1$.

Consider the solution of the integral equation

$$\mathfrak{I}(\tau)=\int_0^\infty\mathfrak{I}(t)\sum a_{2j}E_{2j+1}(|t-\tau|)\,dt, \quad (36)$$

which behaves like $e^{k\tau}$ as $\tau \to \infty$. [It can be shown that such solutions exist (cf. Chap. I, p. 19).] Then

$$I(0,\mu) = \int_0^\infty \Im(\tau)e^{-\tau/\mu}\frac{d\tau}{\mu}, \tag{37}$$

is given by

$$I(0,\mu) = \text{constant} \frac{H(\mu)}{1-k\mu}, \tag{38}$$

where $H(\mu)$ is the unique solution of the corresponding H-equation which is bounded in the entire half-plane $R(z) > 0$.

PROOF. As in the proof of Theorem 1 we first establish relation (30); then substituting for $I(0,\mu)$ according to equation (38), we obtain

$$\frac{H(\mu)}{1-k\mu}\left[1-\mu\int_0^1\frac{\Psi(\mu')}{\mu+\mu'}d\mu'\right] = \int_0^1\frac{\Psi(\mu')}{\mu-\mu'}\left[\frac{\mu H(\mu)}{1-k\mu}-\frac{\mu'H(\mu')}{1-k\mu'}\right]d\mu'. \tag{39}$$

This equation can be readily transformed to

$$H(\mu)\left[1-\mu\int_0^1\frac{\Psi(\mu')}{\mu+\mu'}d\mu'\right]$$

$$= \int_0^1\frac{\Psi(\mu')}{1-k\mu'}H(\mu')\,d\mu'+\mu\int_0^1\frac{\Psi(\mu')}{\mu-\mu'}[H(\mu)-H(\mu')]\,d\mu'. \tag{40}$$

On the other hand, since $-1/k$ is a pole of $H(\mu)$ (cf. the remarks preceding eq. [106] in Chap. V)

$$0 = \frac{1}{H(-1/k)} = 1+\frac{1}{k}\int_0^1\frac{\Psi(\mu')}{-1/k+\mu'}H(\mu')\,d\mu'. \tag{41}$$

Equation (40), therefore, reduces to equation (32) which $H(\mu)$ satisfies quite generally. The truth of the theorem is now apparent.

From Theorem 2 it follows, in particular, that when J satisfies the integral equation

$$J(\tau) = \tfrac{1}{2}\varpi_0\int_0^1 J(t)E_1(|t-\tau|)\,dt, \tag{42}$$

the 'emergent intensity' defined in the manner (37) is of the form (38) where k is now the positive root of equation (4). Equation (42) represents the integral equation for the problem considered in § 88 and the result just stated is in agreement with the solution obtained in § 88 (eq. [15]).

89.3. *The 'pseudo-problems' in transfer theory*

In § 89.2 we saw that in conservative cases the H-function is related in a direct way with the Laplace transform of a function $\mathfrak{J}(\tau)$ which satisfies a linear integral equation with a symmetric kernel of the form (22). We now ask whether there exists an equation of transfer the source function of which satisfies the integral equation (22) in the same way as equation (16) represents the integral equation for the problem of isotropic scattering and constant net flux. We shall see that such an equation exists, though in general it cannot be regarded as the equation of transfer of a real physical problem.

Consider the 'equation of transfer'

$$\mu \frac{dI(\tau, \mu)}{d\tau} = I(\tau, \mu) - \int\limits_{-1}^{+1} I(\tau, \mu')\Psi(\mu')\, d\mu', \qquad (43)$$

for the case of no incident radiation.

Since

$$\Phi = \int\limits_{-1}^{+1} I(\tau, \mu)\Psi(\mu)\mu\, d\mu = \text{constant}, \qquad (44)\dagger$$

the problem with a 'constant net Φ' has a meaning.

The source function appropriate to equation (43) is

$$\mathfrak{J}(\tau) = \int\limits_{-1}^{+1} I(\tau, \mu)\Psi(\mu)\, d\mu = \sum a_{2j} \int\limits_{-1}^{+1} I(\tau, \mu)\mu^{2j}\, d\mu. \qquad (45)$$

Since (cf. Chap. I, eqs. [91] and [96])

$$\int\limits_{-1}^{+1} I(\tau, \mu)\mu^{2j}\, d\mu = \int\limits_{0}^{\infty} \mathfrak{J}(t)E_{2j+1}(|t-\tau|)\, dt, \qquad (46)$$

it follows that the integral equation governing $\mathfrak{J}(\tau)$ is

$$\mathfrak{J}(\tau) = \int\limits_{0}^{\infty} \mathfrak{J}(t) \sum a_{2j} E_{2j+1}(|t-\tau|)\, dt, \qquad (47)$$

in agreement with (22).

From Theorem 1 we now conclude that the law of darkening for the problem with a constant net Φ is given by $H(\mu)$.‡

---

† This relation readily follows from equation (43) by multiplying by $\Psi(\mu)$, integrating over the range of μ, and using the evenness of Ψ and the condition of conservativeness (20).

‡ This result can also be established directly by solving equation (43) in a finite approximation in the usual manner, eliminating the constants of integration in the solution for $I(0, \mu)$ and, finally, passing to the limit of infinite approximation.

In non-conservative cases, when the characteristic equation (35) admits a root $0 < k^2 < 1$, equation (43) allows an integral of the form

$$I(\tau, \mu) = \text{constant} \, \frac{e^{\pm k\tau}}{1 \mp k\mu}, \qquad (48)$$

and if we consider (as in § 88) solutions for the case of no incident radiation which have the behaviour

$$I(\tau, \mu) \to \text{constant} \, \frac{e^{+k\tau}}{1 - k\mu} \quad \text{as} \quad \tau \to \infty, \qquad (49)$$

the corresponding integral equation for $\mathfrak{J}(\tau)$ is again of the form (47). The angular distribution of the emergent radiation will therefore be governed (as in § 88) by $H(\mu)/(1 - k\mu)$.

As we have already stated, equations of the form (43) cannot, in general, be regarded as equations of transfer of genuine physical problems. Nevertheless, the solutions of these 'pseudo'-equations of transfer are related to the real physical problems. Thus, for the problem with a constant net flux and Rayleigh's phase function we have shown (Chap. VI, § 45) that the angular distribution is directly governed by an H-function defined in terms of the characteristic function

$$\Psi(\mu) = \tfrac{3}{16}(3 - \mu^2). \qquad (50)$$

The law of darkening for this physical problem is therefore the same as for the 'pseudo-problem' in the context of the equation

$$\mu \frac{dI(\tau, \mu)}{d\tau} = I(\tau, \mu) - \tfrac{3}{16} \int_{-1}^{+1} I(\tau, \mu')(3 - \mu'^2) \, d\mu', \qquad (51)$$

and for a constant net

$$\Phi = \tfrac{3}{16} \int_{-1}^{+1} I(\tau, \mu)(3 - \mu^2)\mu \, d\mu. \qquad (52)$$

This identity in the laws of darkening for the solutions of the 'pseudo'-equation of transfer (51) and the true equation of transfer (Chap. I, eq. [102]) has the following consequence:

The equation of transfer Chapter I, equation (102), for scattering according to Rayleigh's phase-function, leads to a pair of simultaneous linear integral equations for $J(\tau)$ and $K(\tau)$ (Chap. I, eqs. [106] and [107]). With J and K determined as solutions of these equations, the emergent intensity is given by

$$I(0, \mu) = \tfrac{3}{8} \int_{0}^{\infty} e^{-\tau/\mu}[(3 - \mu^2)J(\tau) + (3\mu^2 - 1)K(\tau)] \frac{d\tau}{\mu}; \qquad (53)$$

and this is, apart from a determinable constant of proportionality, the same as

$$I(0,\mu) = \int\limits_0^\infty e^{-\tau/\mu}\mathfrak{J}(\tau)\frac{d\tau}{\mu},\qquad(54)$$

where $\mathfrak{J}(\tau)$ is the solution of the equation

$$\mathfrak{J}(\tau) = \tfrac{3}{16}\int\limits_0^\infty \mathfrak{J}(t)[3E_1(|t-\tau|)-E_3(|t-\tau|)]\,dt.\qquad(55)$$

The pair of equations (Chap. I, eqs. [106] and [107]) governing J and K is therefore, in some sense, equivalent to the single equation (55). The nature of this equivalence is, however, not quite clear.

One of the chief reasons for the tractableness of the integral equations derived from the principles of invariance now becomes apparent: The angular distributions of the emergent radiations are reducible to H-functions; and, as we have seen, these are related, through a Laplace transformation, to simpler source functions than occur in the equations of transfer. The characteristic functions $\Psi(\mu)$ define 'pseudo-problems' of transfer which are much simpler than the original ones; nevertheless they provide the basic functions in terms of which the solutions of the physical problems are expressed. We shall further illustrate this aspect of the matter:

In the treatment of the problem of diffuse reflection and transmission according to the phase-function $\varpi_0(1+x\cos\Theta)$ (Chap. VI, § 46, and Chap. IX, § 65) it was found that the solution for the azimuth independent term involves the H- or the X- and Y-functions for

$$\Psi(\mu) = \tfrac{1}{2}\varpi_0[1+x(1-\varpi_0)\mu^2].\qquad(56)$$

Consequently, the 'pseudo-problem' of reflection and transmission in the context of the equation (cf. Chap. X, § 69, eq. [118] and the remarks following)

$$\mu\frac{dI(\tau,\mu)}{d\tau} = I(\tau,\mu)-\tfrac{1}{2}\varpi_0\int\limits_{-1}^{+1} I(\tau,\mu')[1+x(1-\varpi_0)\mu'^2]\,d\mu'-e^{-\tau/\mu_0},\quad(57)$$

provides the basic functions which occur in the solution of the original problem. Equation (57) is, of course, much simpler than the one governing the azimuth independent intensity in the physical problem.

The foregoing observations suggest that a full discussion of the relation of the 'pseudo-problems' to the real ones, and particularly the relation of the systems of integral equations derived from the formal solution of the equation of transfer to the simpler equations of the type (22), will throw some light on the basic structure of the theory of radiative transfer.

90. The diffusion of imprisoned radiation through a gas

In our discussion of transfer problems we have so far restricted ourselves to time-independent situations in which the radiation field is stationary. In this section we shall consider a first example of a time-dependent problem which arises in the following context:

To be concrete, suppose that a slab of mercury vapour having been illuminated from one side by light from a mercury arc for a length of time sufficient to establish a steady state is suddenly cut off from the source of illumination. Then on account of the fact that the excited atoms of mercury will decay with a certain mean life emitting radiation, the radiation field in the gas will not cease instantaneously with the source of illumination. The radiation field will die down gradually and the problem is to specify the manner of this decay.

In considering the foregoing problem, we shall suppose that the steady state existing before the illumination is cut off is one of radiative equilibrium. The situation then is one which has been idealized in Schuster's problem (Chap. XII, § 83), and there will be a determinate march of the density of radiation through the gas. The concentration of excited atoms, previous to the cutting off of the illumination, will follow the density of the radiation (cf. eqs. [69], [70], and [87] below). After the illumination has been cut off, the excited atoms will decay in a manner calculable from the Einstein coefficients of absorption and emission. The radiation field will have a non-stationary character, and it is this we wish to describe.

90.1. *The equations of the problem*

Let the suffixes 1 and 2 denote, respectively, the normal and the excited states of the atom in question. We shall let the Einstein coefficients B_{12}, A_{21}, and B_{21} have the following meanings: $B_{12} I_\nu$ is the probability, per unit time, that an atom exposed to isotropic radiation of intensity $I_\nu \, d\nu$ will absorb the quantum $h\nu$ and pass to the state 2; A_{21} is the probability, per unit time, that an atom in the state 2 will spontaneously emit a quantum $h\nu$ and pass to the state 1, and $B_{21} I_\nu$ is the additional probability that the same atom will be induced to undergo the same transition when exposed to isotropic radiation I_ν. The Einstein coefficients defined in this manner are related by

$$\frac{A_{21}}{B_{12}} = \frac{2h\nu^3}{c^2} \frac{q_1}{q_2} \quad \text{and} \quad \frac{B_{21}}{B_{12}} = \frac{q_1}{q_2}, \tag{58}$$

where q_1 and q_2 are the statistical weights of the states 1 and 2 and c is the velocity of light.

If $\sigma(\nu)$ is the atomic absorption coefficient for the frequency ν, then

$$\int \sigma(\nu')\, d\nu' = B_{12} \frac{h\nu}{4\pi}, \tag{59}$$

where the integral is extended over the absorption line corresponding to the transition 1→2. We shall approximate the relation (59) by

$$\sigma(\nu)\, \Delta\nu = B_{12} \frac{h\nu}{4\pi}; \tag{60}$$

this is equivalent to supposing that the absorption through the line is uniform and has a breadth $\Delta\nu$.

Let n_1 and n_2 denote the number of atoms per unit volume in the states 1 and 2; n_1 and n_2 must, of course, be considered as varying through the gas and dependent on the time.

Counting the gains and losses of a pencil of radiation through a length of path ds in the gas, we find (cf. Chap. I, § 6)

$$\Delta\nu \frac{dI_\nu}{ds} = [n_2(A_{21} + B_{21} I_\nu) - n_1 B_{12} I_\nu] \frac{h\nu}{4\pi}, \tag{61}$$

where the quantities proportional to n_2 and n_1 represent the number of emissions and absorptions (per unit time) of the quantum $h\nu$. Dividing equation (61) by $B_{12} h\nu/4\pi$ and making use of the relations (58) and (60), we have

$$\frac{dI_\nu}{\sigma\, ds} = -\left(n_1 - n_2 \frac{q_1}{q_2}\right) I_\nu + \frac{2h\nu^3}{c^2} \frac{q_1}{q_2} n_2. \tag{62}$$

This is the equation of transfer.

Turning next to the condition of radiative equilibrium, we note that the number of emissions per unit volume and per unit time, in an element of the gas, is

$$n_2 \int (A_{21} + B_{21} I_\nu) \frac{d\omega}{4\pi}, \tag{63}$$

where the integration is extended over the entire solid angle. The corresponding number of absorptions is

$$n_1 \int B_{12} I_\nu \frac{d\omega}{4\pi}. \tag{64}$$

The excess of the number of absorptions over the number of emissions must equal the rate of increase of the number of atoms in the excited state. Hence

$$n_1 \int B_{12} I_\nu \frac{d\omega}{4\pi} - n_2 \int (A_{21} + B_{21} I_\nu) \frac{d\omega}{4\pi} = \frac{\partial n_2}{\partial t}. \tag{65}$$

Dividing this equation by B_{12} and rearranging the terms, we have

$$\left(n_1-n_2\frac{q_1}{q_2}\right)J_\nu-\frac{2h\nu^3}{c^2}\frac{q_1}{q_2}n_2 = \frac{2h\nu^3}{c^2}\frac{q_1}{q_2}\frac{1}{A_{21}}\frac{\partial n_2}{\partial t}, \tag{66}$$

where J_ν, as usual, denotes the mean intensity of the radiation.

Equations (62) and (66) were first derived by Milne.

In most problems of practical interest n_2 may be neglected in comparison with n_1; also, n_1 may be treated as a constant and independent of time. On this approximation, equations (62) and (66) become

$$\frac{dI_\nu}{n_1\sigma\,ds} = -I_\nu + \frac{2h\nu^3}{c^2}\frac{q_1}{q_2}\frac{n_2}{n_1}$$

and

$$J_\nu = \frac{2h\nu^3}{c^2}\frac{q_1}{q_2}\frac{1}{n_1}\left(n_2+\frac{\partial n_2}{A_{21}\,\partial t}\right). \tag{67}$$

Writing

$$N = \frac{2h\nu^3}{c^2}\frac{q_1}{q_2}\frac{n_2}{n_1}, \tag{68}$$

and measuring time in the unit $1/A_{21}$ (i.e. the unit is the mean life of the excited state), we obtain the equations

$$\mu\frac{\partial I(t,\tau,\mu)}{\partial\tau} = I(t,\tau,\mu)-N(t,\tau) \tag{69}$$

and

$$J(t,\tau) = N(t,\tau)+\frac{\partial N(t,\tau)}{\partial t}, \tag{70}$$

where it will be noted that we have suppressed the subscript ν and further specialized the equations for plane-parallel slabs and introduced the normal optical thickness, τ, through the slab.

For the particular problem of the diffusion of imprisoned radiation through a plane-parallel slab of optical thickness $2\tau_1$ ($-\tau_1 \leqslant \tau \leqslant +\tau_1$) when the surface $\tau = +\tau_1$ is illuminated by radiation of intensity $I^{(0)}$ up to a certain instant $t = 0$ and then suddenly cut off from the source of illumination, the boundary conditions for solving equations (69) and (70) are

$$\left.\begin{array}{l}I(t,+\tau_1,+\mu) = I^{(0)}\\ I(t,-\tau_1,-\mu) = 0\end{array}\right\} \quad \text{for} \quad t \leqslant 0 \quad \text{and} \quad 0 \leqslant \mu \leqslant 1, \tag{71}$$

and

$$\left.\begin{array}{l}I(t,+\tau_1,+\mu) = 0\\ I(t,-\tau_1,-\mu) = 0\end{array}\right\} \quad \text{for all} \quad t > 0 \quad \text{and} \quad 0 \leqslant \mu \leqslant 1. \tag{72}$$

This is the boundary-value problem we have to solve.

90.2. *The general method of solution*

The boundary-value problem formulated at the end of § 90.1 can be solved in a manner similar to analogous boundary-value problems in the theory of heat conduction:

First we seek a set of fundamental solutions of the equations (69) and (70), which are separable in the variables t and τ, and which will satisfy the boundary conditions (72) for $t > 0$. These solutions provide a complete set of orthogonal functions $\psi^{(m)}(\tau)$ in the interval $(-\tau_1, +\tau_1)$ in terms of which any arbitrary (continuous) function defined in the interval $(-\tau_1, +\tau_1)$ can be expanded.

Next, considering the solution for $t \leqslant 0$, we observe that all the quantities are independent of time and that the conditions (71) suffice to determine the solution uniquely. The solution $J(t, \tau) \equiv J(\tau)$ for $t \leqslant 0$ is then expanded as a series in $\psi^{(m)}(\tau)$. In terms of such an expansion the evolution of J for $t > 0$ can be written down at once.

Following the procedure outlined above, we first seek solutions of equations (69) and (70) which are of the form

$$I(t, \tau, \mu) = e^{-(\varpi-1)t/\varpi}\phi(\tau, \mu),$$

$$J(t, \tau) = \tfrac{1}{2}e^{-(\varpi-1)t/\varpi} \int_{-1}^{+1} \phi(\tau, \mu)\, d\mu = e^{-(\varpi-1)t/\varpi}\psi(\tau),$$

and $N(t, \tau) = e^{-(\varpi-1)t/\varpi}\nu(\tau),$ (73)

and which satisfy the boundary conditions (cf. eq. [72])

$$\phi(+\tau_1, +\mu) = \phi(-\tau_1, -\mu) = 0 \quad (0 \leqslant \mu \leqslant 1). \tag{74}$$

In equations (73) $\varpi > 1$ is a certain constant unspecified for the present.

It will be noticed that in writing the solution (73) we have implicitly assumed that
$$\varpi > 1. \tag{75}$$

Strictly, we must expect this to be disclosed by the subsequent analysis; but we have already assumed it as it is apparent on physical grounds that under no circumstances can the radiation field decay faster than the atoms in the excited state; and in the unit of time we have adopted the mean life of the excited state is 1.

For solutions of the form (73), equations (69) and (70) become

$$\mu \frac{d\phi(\tau, \mu)}{d\tau} = \phi(\tau, \mu) - \nu(\tau); \quad \psi(\tau) = \frac{1}{\varpi}\nu(\tau). \tag{76}$$

Accordingly, we have to find if and when an equation of the form

$$\mu \frac{d\phi(\tau, \mu)}{d\tau} = \phi(\tau, \mu) - \varpi\,\psi(\tau) \quad (\varpi > 1) \tag{77}$$

will admit solutions which will satisfy the boundary conditions (74). It is readily seen that for general values of ϖ, the problem admits only the trivial solution $\phi(\tau, \mu) \equiv 0$. However, it can be shown that there exists an infinity of values $\varpi^{(m)}$ $(m = 1, 2, ..., \infty)$ for which non-trivial solutions $\phi^{(m)}(\tau, \mu)$ and $\psi^{(m)}(\tau)$ exist. An important, and in the present context, a decisive property of the functions $\psi^{(m)}(\tau)$ is that they form an orthogonal system in the interval $(-\tau_1, +\tau_1)$:

$$\int_{-\tau_1}^{+\tau_1} \psi^{(m_1)}(\tau)\psi^{(m_2)}(\tau) \, d\tau = 0 \quad (m_1 \neq m_2). \tag{78}$$

The orthogonality of the functions $\psi^{(m)}(\tau)$ can be proved in the following manner: Consider the equations satisfied by $\phi^{(1)}(\tau, \mu)$ and $\phi^{(2)}(\tau, \mu)$ belonging to two different eigenvalues $\varpi^{(1)}$ and $\varpi^{(2)}$. We have

$$\mu \frac{d\phi^{(1)}(\tau, \mu)}{d\tau} = \phi^{(1)}(\tau, \mu) - \varpi^{(1)}\psi^{(1)}(\tau)$$

and

$$-\mu \frac{d\phi^{(2)}(\tau, -\mu)}{d\tau} = \phi^{(2)}(\tau, -\mu) - \varpi^{(2)}\psi^{(2)}(\tau). \tag{79}$$

It will be noticed that we have written the second of the equations for $-\mu$; but we are not restricting the range of μ in any way.

Multiplying the first of the equations (79) by $\phi^{(2)}(\tau, -\mu)$ and the second by $\phi^{(1)}(\tau, \mu)$ and subtracting, we have

$$\mu \frac{d}{d\tau}[\phi^{(1)}(\tau, \mu)\phi^{(2)}(\tau, -\mu)] = \varpi^{(2)}\psi^{(2)}(\tau)\phi^{(1)}(\tau, \mu) - \varpi^{(1)}\psi^{(1)}(\tau)\phi^{(2)}(\tau, -\mu). \tag{80}$$

Integrating this equation over τ from $-\tau_1$ to $+\tau_1$, we observe that the quantity

$$[\phi^{(1)}(\tau, \mu)\phi^{(2)}(\tau, -\mu)]_{-\tau_1}^{+\tau_1}$$

on the left-hand side vanishes identically (i.e. for all $-1 \leqslant \mu \leqslant +1$) in virtue of the boundary conditions (74). Consequently

$$\int_{-\tau_1}^{+\tau_1} [\varpi^{(2)}\psi^{(2)}(\tau)\phi^{(1)}(\tau, \mu) - \varpi^{(1)}\psi^{(1)}(\tau)\phi^{(2)}(\tau, -\mu)] \, d\tau = 0$$

$$(-1 \leqslant \mu \leqslant +1). \tag{81}$$

Now integrating this equation over μ from -1 to $+1$ and recalling the definition of ψ, we have

$$[\varpi^{(2)} - \varpi^{(1)}] \int_{-\tau_1}^{+\tau_1} \psi^{(2)}(\tau)\psi^{(1)}(\tau) \, d\tau = 0. \tag{82}$$

This establishes the orthogonality of the functions belonging to different eigenvalues.

The proof that the functions $\psi^{(m)}(\tau)$ form a *complete* set of orthogonal functions is not so simple and we shall not attempt it here. However, if the completeness is assumed, it follows that any arbitrary continuous function, $f(\tau)$, defined in the interval $(-\tau_1, +\tau_1)$ can be expanded as a series in $\psi^{(m)}(\tau)$ in the form

$$f(\tau) = \sum_{m=1}^{\infty} a^{(m)}\psi^{(m)}(\tau). \tag{83}$$

The coefficients, $a^{(m)}$, in the expansion are given by

$$a^{(m)} = \int_{-\tau_1}^{+\tau_1} f(\tau)\psi^{(m)}(\tau)\, d\tau \bigg/ \int_{-\tau_1}^{+\tau_1} [\psi^{(m)}(\tau)]^2\, d\tau. \tag{84}$$

In terms of the fundamental solutions $\phi^{(m)}(\tau,\mu)$, $\psi^{(m)}(\tau)$, and $\nu^{(m)}(\tau)$, the general solution of equations (69) and (70) which will satisfy the boundary conditions (72), can be written in the form

$$I(t,\tau,\mu) = \sum_{m=1}^{\infty} a^{(m)} e^{-(\varpi^{(m)}-1)t/\varpi^{(m)}} \phi^{(m)}(\tau,\mu),$$

$$J(t,\tau) = \sum_{m=1}^{\infty} a^{(m)} e^{-(\varpi^{(m)}-1)t/\varpi^{(m)}} \psi^{(m)}(\tau),$$

and
$$N(t,\tau) = \sum_{m=1}^{\infty} a^{(m)}\varpi^{(m)} e^{-(\varpi^{(m)}-1)t/\varpi^{(m)}} \psi^{(m)}(\tau), \tag{85}$$

where the $a^{(m)}$'s are arbitrary constants.

To complete the solution we must determine the particular set of constants $a^{(m)}$ in (85) which will lead to a distribution of excited atoms through the slab at time $t = 0$, that which prevailed at the instant at which the illumination was cut off. The coefficients $a^{(m)}$ must therefore be so chosen that

$$J(0,\tau) = \sum_{m=1}^{\infty} a^{(m)}\psi^{(m)}(\tau) \tag{86}$$

represents the actual expansion of the mean intensity $J(\tau)$ in the solution of the equation of transfer

$$\mu \frac{dI(\tau,\mu)}{d\tau} = I(\tau,\mu) - J(\tau) \quad (J = N,\ t \leqslant 0), \tag{87}$$

which satisfies the boundary conditions

$$I(+\tau_1, +\mu) = I^{(0)} \quad \text{and} \quad I(-\tau_1, -\mu) = 0 \quad (0 \leqslant \mu \leqslant 1). \tag{88}$$

With the coefficients $a^{(m)}$ determined in this fashion, the solution (85) satisfies all the conditions of the problem and therefore represents the required solution.

90.3. *The form of the solution in finite approximations*

We shall now illustrate how the general theory outlined in § 90.2 can be adapted to the practical solution of the problem in a finite approximation.

Considering first the problem of determining the basic set of orthogonal functions, we have to find the condition on ϖ which will lead to non-trivial solutions of equation (77) satisfying the boundary conditions (74).

In the nth approximation we replace equation (77) by the system of $2n$ linear equations

$$\mu_j \frac{d\phi_j}{d\tau} = \phi_j - \tfrac{1}{2}\varpi \sum a_j \phi_j \quad (j = \pm 1,..., \pm n), \tag{89}$$

where the various symbols have their usual meanings. Since $\varpi > 1$, we seek solutions of (89) of the form

$$\phi_j = \gamma_j e^{-ik\tau} \quad (\gamma_j = \text{constant}; \ j = \pm 1,..., \pm n), \tag{90}$$

and find that

$$\gamma_j = \frac{\text{constant}}{1 + ik\mu_j} \quad (j = \pm 1,..., \pm n), \tag{91}$$

and, further, that the characteristic equation for k is

$$1 = \varpi \sum_{j=1}^{n} \frac{a_j}{1 + k^2 \mu_j^2}. \tag{92}$$

Equation (92) admits $2n$ roots which occur in pairs as[†]

$$k_\alpha = -k_{-\alpha} \quad (\alpha = \pm 1,..., \pm n). \tag{93}$$

The general solution for ϕ_j is therefore of the form

$$\phi_j = \sum_{\alpha=-n}^{+n} \frac{\lambda_\alpha e^{-ik_\alpha \tau}}{1 + ik_\alpha \mu_j} \quad (j = \pm 1,..., \pm n), \tag{94}[‡]$$

where λ_α's ($\alpha = \pm 1,..., \pm n$) are constants.

It is convenient to rewrite the solution (94) in terms of sines and cosines instead of the imaginary exponentials. We have

$$\phi_j = \sum_{\alpha=1}^{n} \frac{1}{1 + k_\alpha^2 \mu_j^2} [A_\alpha(\cos k_\alpha \tau - k_\alpha \mu_j \sin k_\alpha \tau) +$$
$$+ B_\alpha(\sin k_\alpha \tau + k_\alpha \mu_j \cos k_\alpha \tau)] \quad (j = \pm 1,..., \pm n), \tag{95}$$

where A_α and B_α ($\alpha = 1,..., n$) are $2n$ arbitrary constants.

[†] Actually only one pair of these roots is real; the others are purely imaginary. This does not, however, effect the discussion in any way.

[‡] In the summation on the right-hand there is no term with $\alpha = 0$ (cf. eq. [93]).

Using the solution (95) we readily find that

$$\psi(\tau) = \tfrac{1}{2} \sum a_j \phi_j = \frac{1}{\varpi} \sum_{\alpha=1}^{n} (A_\alpha \cos k_\alpha \tau + B_\alpha \sin k_\alpha \tau) \qquad (96)$$

and $\qquad \sum a_j \mu_j \phi_j = \dfrac{2(\varpi-1)}{\varpi} \displaystyle\sum_{\alpha=1}^{n} \dfrac{1}{k_\alpha}(-A_\alpha \sin k_\alpha \tau + B_\alpha \cos k_\alpha \tau). \qquad (97)$

Returning to the solution (95), we find that with the definitions

$$\tan \theta_{\alpha j} = k_\alpha \mu_j; \qquad \theta_{\alpha j} = -\theta_{\alpha,-j}$$
$$(\alpha = 1,...,n; \; j = \pm 1,..., \pm n), \quad (98)$$

the solution can be expressed conveniently in the form

$$\phi_j = \sum_{\alpha=1}^{n} A_\alpha \cos \theta_{\alpha j} \cos(\theta_{\alpha j} + k_\alpha \tau) +$$
$$+ \sum_{\alpha=1}^{n} B_\alpha \cos \theta_{\alpha j} \sin(\theta_{\alpha j} + k_\alpha \tau) \quad (j = \pm 1,..., \pm n). \quad (99)$$

The boundary conditions (cf. eq. [74])

$$\phi_{+j}(+\tau_1) = \phi_{-j}(-\tau_1) = 0 \quad (j = 1,...,n) \qquad (100)$$

now lead to

$$\sum_{\alpha=1}^{n} A_\alpha \cos \theta_{\alpha j} \cos(\theta_{\alpha j} + k_\alpha \tau_1) + \sum_{\alpha=1}^{n} B_\alpha \cos \theta_{\alpha j} \sin(\theta_{\alpha j} + k_\alpha \tau_1) = 0,$$
$$\sum_{\alpha=1}^{n} A_\alpha \cos \theta_{\alpha j} \cos(\theta_{\alpha j} + k_\alpha \tau_1) - \sum_{\alpha=1}^{n} B_\alpha \cos \theta_{\alpha j} \sin(\theta_{\alpha j} + k_\alpha \tau_1) = 0$$
$$(j = 1,...,n). \quad (101)$$

The pair of equations (101) is equivalent to

$$\sum_{\alpha=1}^{n} A_\alpha \cos \theta_{\alpha j} \cos(\theta_{\alpha j} + k_\alpha \tau_1) = 0 \quad (j = 1,...,n),$$

and $\qquad \displaystyle\sum_{\alpha=1}^{n} B_\alpha \cos \theta_{\alpha j} \sin(\theta_{\alpha j} + k_\alpha \tau_1) = 0 \quad (j = 1,...,n). \qquad (102)$

Hence, *either*

$$A_\alpha \neq 0, \quad B_\alpha = 0, \quad \text{and} \quad \|\cos \theta_{\alpha j} \cos(\theta_{\alpha j} + k_\alpha \tau_1)\| = 0,$$

or $\qquad A_\alpha = 0, \quad B_\alpha \neq 0, \quad \text{and} \quad \|\cos \theta_{\alpha j} \sin(\theta_{\alpha j} + k_\alpha \tau_1)\| = 0. \qquad (103)$

Let the values of ϖ which lead to the vanishing of the determinants

$$\|\cos \theta_{\alpha j} \cos(\theta_{\alpha j} + k_\alpha \tau_1)\| \quad \text{and} \quad \|\cos \theta_{\alpha j} \sin(\theta_{\alpha j} + k_\alpha \tau_1)\|$$

be denoted by

$$\varpi^{(e,m)} \quad \text{and} \quad \varpi^{(o,m)} \qquad (m = 1, 2,..., \infty), \qquad (104)$$

respectively. Let the corresponding roots of equation (92) be

$$k_\alpha^{(e,m)} \quad \text{and} \quad k_\alpha^{(o,m)} \qquad (\alpha = 1,...,n; \; m = 1, 2,..., \infty). \qquad (105)$$

Also, let the sequences $A_\alpha^{(m)}$ and $B_\alpha^{(m)}$ ($\alpha = 1,...,n$ and $m = 1, 2,...,\infty$), which are determined apart from a constant of proportionality by the equations

$$\sum_{\alpha=1}^{n} A_\alpha^{(m)} \cos \theta_{\alpha j}^{(e,m)} \cos[\theta_{\alpha j}^{(e,m)} + k_\alpha^{(e,m)} \tau_1] = 0 \quad (j = 1,...,n)$$

and $$\sum_{\alpha=1}^{n} B_\alpha^{(m)} \cos \theta_{\alpha j}^{(o,m)} \sin[\theta_{\alpha j}^{(o,m)} + k_\alpha^{(o,m)} \tau_1] = 0 \quad (j = 1,...,n), \qquad (106)$$

be made unique by the condition that the functions†

$$\psi^{(e,m)}(\tau) = \sum_{\alpha=1}^{n} A_\alpha^{(m)} \cos k_\alpha^{(e,m)} \tau$$

and $$\psi^{(o,m)}(\tau) = \sum_{\alpha=1}^{n} B_\alpha^{(m)} \sin k_\alpha^{(o,m)} \tau \qquad (107)$$

are normalized.

The general solution for the mean intensity and the number of excited atoms for $t > 0$ can now be expressed in the forms

$$J(t,\tau) = \sum_{m=1}^{\infty} a^{(e,m)}\psi^{(e,m)}(\tau)\exp\{-[\varpi^{(e,m)}-1]t/\varpi^{(e,m)}\} +$$
$$+ \sum_{m=1}^{\infty} a^{(o,m)}\psi^{(o,m)}(\tau)\exp\{-[\varpi^{(o,m)}-1]t/\varpi^{(o,m)}\}$$

and

$$N(t,\tau) = \sum_{m=1}^{\infty} a^{(e,m)}\varpi^{(e,m)}\psi^{(e,m)}(\tau)\exp\{-[\varpi^{(e,m)}-1]t/\varpi^{(e,m)}\} +$$
$$+ \sum_{m=1}^{\infty} a^{(o,m)}\varpi^{(o,m)}\psi^{(o,m)}(\tau)\exp\{-[\varpi^{(o,m)}-1]t/\varpi^{(o,m)}\}. \qquad (108)$$

To complete the solution we must determine the coefficients $a^{(e,m)}$ and $a^{(o,m)}$ in (108) by the condition that

$$J(0,\tau) = \sum_{m=1}^{\infty} [a^{(e,m)}\psi^{(e,m)}(\tau) + a^{(o,m)}\psi^{(o,m)}(\tau)], \qquad (109)$$

at $t = 0$ represents the expansion in $\psi^{(e,m)}$ and $\psi^{(o,m)}$ of the mean intensity at the instant of the cutting off of the illumination. For this purpose we require to solve the stationary problem, i.e. in our present approximation, the system of equations

$$\mu_j \frac{dI_j}{d\tau} = I_j - \tfrac{1}{2} \sum a_j I_j \quad (j = \pm 1,..., \pm n) \qquad (110)$$

with the boundary conditions

$$I_{+j}(+\tau_1) = I^{(0)} \quad \text{and} \quad I_{-j}(-\tau_1) = 0 \quad (j = 1,...,n). \qquad (111)$$

† The orthogonality of these functions can be established as in § 90.2.

By standard methods, we find that

$$I_j = \sum_{\alpha=-n}^{+n} \frac{L_\alpha e^{-k_\alpha^{(0)}\tau}}{1+k_\alpha^{(0)}\mu_j} + L_0(\tau+\mu_j) + L_n \quad (j = \pm 1,...,\pm n), \quad (112)$$

where the $k_\alpha^{(0)}$'s ($\alpha = \pm 1,..., \pm n \mp 1$ and $k_0^{(0)} = -k_{-\alpha}^{(0)}$) are the roots of the characteristic equation (cf. Chap. III, eq. [7])

$$1 = \sum_{j=1}^{n} \frac{a_j}{1-\mu_j^2 k^2}, \quad (113)$$

and the constants L_α ($\alpha = \pm 1,..., \pm n \mp 1$), L_0 and L_n are to be determined from the equations (cf. eq. [111])

$$\sum_{\alpha=-n}^{+n} \frac{L_\alpha e^{-k_\alpha^{(0)}\tau_1}}{1+\mu_j k_\alpha^{(0)}} + L_0(\tau_1+\mu_j) + L_n = I^{(0)}$$

and
$$\sum_{\alpha=-n}^{+n} \frac{L_\alpha e^{+k_\alpha^{(0)}\tau_1}}{1-\mu_j k_\alpha^{(0)}} - L_0(\tau_1+\mu_j) + L_n = 0 \quad (j = 1,...,n). \quad (114)$$

From (112) we obtain

$$J(\tau) = \tfrac{1}{2} \sum a_j I_j = L_0\tau + L_n + \sum_{\alpha=-n}^{+n} L_\alpha e^{-k_\alpha^{(0)}\tau}. \quad (115)$$

It is the expansion of the function $J(\tau)$ in $\psi^{(e,m)}(\tau)$ and $\psi^{(o,m)}(\tau)$ that determines the coefficients $a^{(e,m)}$ and $a^{(o,m)}$ in the solution (108). And with the determination of these coefficients the solution of the problem also becomes determinate.

90.4. *The solution in the first approximation*

In the first approximation ($a_{+1} = a_{-1} = 1$ and $\mu_{+1} = -\mu_{-1} = 1/\sqrt{3}$) the equations given in § 90.3 for the characteristic roots and eigenvalues become

$$\varpi = 1+\tfrac{1}{3}k^2, \quad \tan\theta = k/\sqrt{3}, \quad (116)$$

$$\cos(\theta+k\tau_1) = 0 \quad \text{and} \quad \sin(\theta+k\tau_1) = 0. \quad (117)$$

Equations (117) can be expressed, alternatively, in the forms

$$\tan k\tau_1 = \cot\theta = \sqrt{3}/k$$

and
$$\tan k\tau_1 = -\tan\theta = -k/\sqrt{3}. \quad (118)$$

Each of these equations admits a single infinity of roots. Denoting them by $k^{(e,m)}$ and $k^{(o,m)}$ ($m = 1,2,...,\infty$) respectively, we have the normalized eigenfunctions

$$\frac{\cos k^{(e,m)}\tau}{[\tau_1+\sqrt{3}/(k^2+3)]^{\frac{1}{2}}} \quad \text{and} \quad \frac{\sin k^{(o,m)}\tau}{[\tau_1+\sqrt{3}/(k^2+3)]^{\frac{1}{2}}}. \quad (119)$$

The solution of the stationary problem in the first approximation is

$$I_{\pm 1} = \tfrac{1}{2}I^{(0)}\left(1 + \frac{\tau \pm 1/\sqrt{3}}{\tau_1 + 1/\sqrt{3}}\right)$$

and

$$J(\tau) = \tfrac{1}{2}I^{(0)}\left(1 + \frac{\tau}{\tau_1 + 1/\sqrt{3}}\right). \tag{120}$$

The coefficients in the expansion of $J(\tau)$ in the eigenfunctions (119) are readily found. They are:

$$a^{(e,m)} = \frac{I^{(0)}\sin k^{(e,m)}\tau_1}{k^{(e,m)}[\tau_1 + \sqrt{3}/(k^2+3)]^{\frac{1}{2}}}$$

and

$$a^{(o,m)} = \frac{I^{(0)}(\sin k^{(o,m)}\tau_1 - k^{(o,m)}\tau_1\cos k^{(o,m)}\tau_1)}{(\tau_1 + 1/\sqrt{3})[k^{(o,m)}]^2[\tau_1 + \sqrt{3}/(k^2+3)]^{\frac{1}{2}}}. \tag{121}$$

The solution of the problem can now be completed.

91. The transfer of radiation in atmospheres with spherical symmetry

In this book we have restricted ourselves to problems of radiative transfer in plane-parallel atmospheres, principally, with the object of illustrating and analysing, as fully as possible, the role of general principles (such as the principles of invariance) in the theory of radiative transfer; and plane-parallel geometry is the most suitable for this purpose. The extension of the approximate methods of solution to other geometries is straightforward; but it is not equally apparent what the nature of the general principles are which will play the same unique role as the principles of invariance in plane-parallel atmospheres. Nevertheless, it is of interest to provide at least one example of the manner in which the method of solution of replacing integrals by sums in finite approximations can be adapted to transfer problems in geometries other than plane-parallel.

The example we shall consider is the spherically symmetric problem under conditions of conservative isotropic scattering. The equation of transfer appropriate to this problem is (Chap. I, eq. [136])

$$\mu\frac{\partial I(r,\mu)}{\partial r} + \frac{1-\mu^2}{r}\frac{\partial I(r,\mu)}{\partial\mu} = -\kappa\rho I(r,\mu) + \tfrac{1}{2}\kappa\rho\int_{-1}^{+1} I(r,\mu')\,d\mu'. \tag{122}$$

On the method of replacing integrals by sums, equation (122) would lead, in the nth approximation, to the system of $2n$ equations,

$$\mu_i\frac{dI_i}{dr} + \frac{1-\mu_i^2}{r}\left(\frac{\partial I}{\partial\mu}\right)_{\mu=\mu} = -\kappa\rho I_i + \tfrac{1}{2}\kappa\rho\sum a_j I_j \quad (i = \pm 1,...,\pm n), \tag{123}$$

where the various symbols have their usual meanings; but the question is immediately raised as to the values we are to assign to the derivatives $(\partial I/\partial\mu)_{\mu=\mu_i}$ at the points of the Gaussian division. The question which is raised here is of general significance: The Gaussian and other quadrature formulae are devised with the intent of evaluating integrals 'most accurately' for a given number of divisions; the question we now ask is, essentially, the converse one: what are the analogous formulae for differentiation? Without going into these more general implications of the question, we shall indicate how the explicit appearance of the derivatives in equation (123) can be eliminated.

Let
$$Q_l(\mu) = \frac{1}{2l+1}[P_{l-1}(\mu)-P_{l+1}(\mu)] = \frac{1-\mu^2}{l(l+1)}P'_l(\mu). \tag{124}$$

From this definition it follows that

$$\frac{dQ_l}{d\mu} = -P_l(\mu) \quad \text{and} \quad Q_l(\pm 1) = 0. \tag{125}$$

Now consider
$$\int_{-1}^{+1} Q_l(\mu)\frac{\partial I}{\partial\mu}d\mu \quad (l = 1,...,2n). \tag{126}$$

Integrating (126) by parts and making use of the relations (125), we arrive at the result

$$\int_{-1}^{+1} Q_l(\mu)\frac{\partial I}{\partial\mu}d\mu = \int_{-1}^{+1} P_l(\mu)I\,d\mu. \tag{127}$$

Replacing the integrals on either side of this equation by the corresponding Gauss sums, we have

$$\sum_{i=-n}^{+n} a_i\,Q_l(\mu_i)\left(\frac{\partial I}{\partial\mu}\right)_{\mu=\mu_i} = \sum_{i=-n}^{+n} a_i\,P_l(\mu_i)I_i \quad (l = 1,...,2n). \tag{128}$$

Equation (128) provides $2n$ equations for determining the derivatives at the points of the Gaussian division in terms of the values of the function at the same points.†

Equations (123) and (128) together provide the required reduction of the equation of transfer to an equivalent system of linear equations in a finite approximation. For purposes of practical solution it appears advantageous to combine these equations in the following manner:

---

† Notice that since $P_{2n}(\mu_i) = 0$ $(i = \pm 1,..., \pm n)$, relation (128) cannot be inverted into one giving the values the function in terms of the derivatives.

Multiplying equation (123) by $a_i P'_l(\mu_i)$ and summing over all i, we obtain

$$\frac{d}{dr} \sum a_i \mu_i P'_l(\mu_i) I_i + \frac{l(l+1)}{r} \sum a_i P_l(\mu_i) I_i = -\kappa\rho \sum a_i P'_l(\mu_i) I_i +$$

$$+ \tfrac{1}{2}\kappa\rho \sum a_j I_j \sum a_i P'_l(\mu_i) \quad (l = 1,...,2n) \quad (129)\dagger$$

on using equations (124) and (128).

Using the known expressions for the Legendre polynomials, we find that equation (129) for $l = 1, 2, 3,$ and 4 takes the form

$$\frac{d}{dr} \sum a_i \mu_i I_i + \frac{2}{r} \sum a_i \mu_i I_i = 0,$$

$$\frac{d}{dr} \sum a_i \mu_i^2 I_i + \frac{1}{r} \sum a_i(3\mu_i^2-1)I_i = -\kappa\rho \sum a_i \mu_i I_i,$$

$$\frac{d}{dr} \sum a_i \mu_i(5\mu_i^2-1)I_i + \frac{4}{r} \sum a_i \mu_i(5\mu_i^2-3)I_i = -\tfrac{5}{3}\kappa\rho \sum a_i(3\mu_i^2-1)I_i,$$

$$\frac{d}{dr} \sum a_i \mu_i^2(7\mu_i^2-3)I_i + \frac{1}{r} \sum a_i(35\mu_i^4-30\mu_i^2+3)I_i = -\kappa\rho \sum a_i \mu_i(7\mu_i^2-3)I_i,$$

etc. (130)

The first of the equations (130) integrates at once and gives

$$\tfrac{1}{2}F = \sum a_i \mu_i I_i = \tfrac{1}{2}F_0 r^{-2}, \qquad (131)$$

where F_0 is a constant; this is the flux integral.

It may also be noted that in the nth approximation

$$\sum a_i P_{2n}(\mu_i)I_i = 0, \qquad (132)$$

as, by definition, the μ_i's are the zeros of $P_{2n}(\mu)$.

91.1. *The solution in the first approximation*

In the first approximation, we consider the first two of the equations (130); and remembering that $a_{+1} = a_{-1} = 1$ and $\mu_{+1} = -\mu_{-1} = 1/\sqrt{3}$, we have

$$I_{+1}-I_{-1} = \frac{\sqrt{3}}{2}\frac{F_0}{r^2} \qquad (133)$$

and

$$\frac{1}{3}\frac{d}{dr}(I_{+1}+I_{-1}) = -\frac{\kappa\rho}{\sqrt{3}}(I_{+1}-I_{-1}) = -\frac{\kappa\rho}{2}\frac{F_0}{r^2}. \qquad (134)$$

† A system of equations entirely equivalent to this set can be obtained by expanding $I(r, \mu)$ in the form

$$I(r,\mu) = \sum_{l=0}^{2n-1} \tfrac{1}{2}(2l+1)P_l(\mu)\psi_l(\mu).$$

On examination, it is seen that the two methods are equivalent in *all* details including the particular recursion formulae which are used.

Hence,
$$I_{+1}+I_{-1} = \tfrac{3}{2}F_0 \int_r^R \frac{\kappa\rho\,dr}{r^2} + \text{constant}, \tag{135}$$

where $r = R$ defines the extent of the atmosphere.

The constant of integration in equation (135) can be determined by the condition $I_{-1} = 0$ at $r = R$.† In this manner we find that the solution for the source function is

$$J = \tfrac{1}{2}(I_{+1}+I_{-1}) = \frac{\sqrt{3}}{4}\frac{F_0}{R^2} + \tfrac{3}{4}F_0 \int_r^R \frac{\kappa\rho\,dr}{r^2}. \tag{136}$$

For an atmosphere which extends to infinity, we should require that both I_{+1} and I_{-1} tend to zero as $r \to \infty$. The solution (136) for J then reduces to

$$J = \tfrac{3}{4}F_0 \int_r^\infty \frac{\kappa\rho\,dr}{r^2} = \tfrac{3}{4}F_0 \int_0^\tau \frac{d\tau}{r^2}, \tag{137}$$

where τ denotes the radial optical thickness measured from $r = \infty$ inward.

91.2. *The equations for the second approximation*

In the second approximation we consider all four of the equations (130); however, in the last of these equations the term in $1/r$ on the left-hand side vanishes: in the second approximation the μ_i's are the zeros of $P_4(\mu)$.

Writing

$$J = \tfrac{1}{2}\sum a_i I_i, \quad H = \tfrac{1}{2}\sum a_i \mu_i I_i, \quad K = \tfrac{1}{2}\sum a_i \mu_i^2 I_i$$
$$L = \tfrac{1}{2}\sum a_i \mu_i^3 I_i \quad \text{and} \quad M = \tfrac{1}{2}\sum a_i \mu_i^4 I_i, \tag{138}$$

we have the equations:

$$H = \tfrac{1}{4}F_0 r^{-2}; \quad 35M - 30K + 3J = 0,$$

$$\frac{dK}{dr} + \frac{1}{r}(3K - J) = -\kappa\rho H,$$

$$\frac{d}{dr}(5L - H) + \frac{4}{r}(5L - 3H) = -\tfrac{5}{3}\kappa\rho(3K - J),$$

and
$$\frac{d}{dr}(7M - 3K) = -\kappa\rho(7L - 3H). \tag{139}$$

In terms of the quantities

$$X = 3K - J \quad \text{and} \quad Y = 5L - 3H, \tag{140}$$

---

† In the first approximation it is more advantageous to let $J = \tfrac{1}{2}F$ at $r = R$; then $\tfrac{1}{2}$ replaces $\sqrt{3}/4$ as the factor of F_0/R^2 in (136).

equations (139) take more convenient forms. After some elementary reductions we find:

$$K = -\int^r \frac{X}{r}\,dr - \tfrac14 F_0 \int^r \frac{\kappa\rho}{r^2}\,dr, \qquad \frac{dX}{dr} - \frac{2}{r}X = -\tfrac73\kappa\rho Y,$$

and

$$\frac{dY}{dr} + \frac{4}{r}Y = -\tfrac53\kappa\rho X + \frac{F_0}{r^3}. \tag{141}$$

These equations, together with the flux integral, $H = F_0/4r^2$, provide the basic equations of the problem in the second approximation.

91.3. *The solution of the equations in the second approximation for the case* $\kappa\rho \propto r^{-n}$ $(n > 1)$

Let

$$\kappa\rho = cr^{-n} \quad (n > 1), \tag{142}$$

where c is a constant. The optical thickness, τ, measured from $r = \infty$ inward, has then the value

$$\tau = \int_r^\infty \kappa\rho\,dr = cr^{-n+1}/(n-1). \tag{143}$$

From equations (142) and (143) we obtain the relations

$$\kappa\rho r = (n-1)\tau \quad \text{and} \quad \tau = (R/r)^{n-1}, \tag{144}$$

where R denotes the radius at which $\tau = 1$.

Using the relations (142)–(144) and measuring intensity in the unit F_0/R^2, we find that equations (141) become

$$K = \frac{1}{n-1}\int_0^\tau X\frac{d\tau}{\tau} + \frac{n-1}{4(n+1)}\tau^{(n+1)/(n-1)}, \tag{145}$$

$$\frac{dX}{d\tau} + \frac{2}{(n-1)\tau}X = \tfrac73 Y \tag{146}$$

and

$$\frac{dY}{d\tau} - \frac{4}{(n-1)\tau}Y = \tfrac53 X - \frac{\tau^{(3-n)/(n-1)}}{n-1}. \tag{147}$$

It will be noticed that in equation (145) we have adjusted the limits of integration to be in accordance with the condition that under the conditions of the atmosphere extending to infinity, all quantities must vanish at $r = \infty$, $\tau = 0$.

Eliminating Y between equations (146) and (147), we obtain the differential equation

$$\frac{d^2X}{d\tau^2} - \frac{2}{(n-1)\tau}\frac{dX}{d\tau} - \frac{2(n+3)}{(n-1)^2\tau^2}X = \frac{35}{9}X - \frac{7\tau^{(3-n)/(n-1)}}{3(n-1)}. \tag{148}$$

With the substitutions

$$z = k\tau, \qquad k = \sqrt{35/3} = 1{\cdot}9720$$

and
$$X = \frac{7}{3(n-1)} k^{-(n+1)/(n-1)} z^{(n+1)/2(n-1)} \phi(z), \tag{149}$$

equation (148) becomes

$$z^2 \frac{d^2\phi}{dz^2} + z\frac{d\phi}{dz} - (z^2+\nu^2)\phi = -z^{\mu+1}, \tag{150}$$

where
$$\nu = \frac{n+5}{2(n-1)} \quad \text{and} \quad \mu = \frac{3-n}{2(n-1)}. \tag{151}$$

Equation (150) is Lommel's equation for a purely imaginary argument.† Accordingly, the solution of this equation can be written in the form

$$\phi = I_\nu(z) \int_z^{c_1} z^\mu K_\nu(z)\, dz + K_\nu(z) \int_{c_2}^z z^\mu I_\nu(z)\, dz, \tag{152}$$

where $I_\nu(z)$ and $K_\nu(z)$ are (in Watson's notation) the fundamental solutions of Bessel's equation for a purely imaginary argument and c_1 and c_2 are certain arbitrary limits.

The limits of integration in (152) can be determined by the following considerations:

First, since none of the quantities must tend to infinity exponentially as $z \to \infty$, we must require that $c_1 = \infty$. This follows from the known asymptotic behaviours of $I_\nu(z)$ and $K_\nu(z)$ as $z \to \infty$. Second, the vanishing of all the quantities at $z = 0$ requires that (cf. eq. [149])

$$z^{(n+1)/2(n-1)} \phi(z) \to 0 \quad (z \to 0). \tag{153}$$

Since $K_\nu(z)$ diverges at the origin, the condition (153) can be met only by letting $c_2 = 0$. The solution for ϕ appropriate to the problem is, therefore,

$$\phi = I_\nu(z) \int_z^\infty z^\mu K_\nu(z)\, dz + K_\nu(z) \int_0^z z^\mu I_\nu(z)\, dz. \tag{154}$$

With this expression for ϕ the solution for X given by equation (149) becomes determinate.

The solutions for X and $J = 3K - X$ now follow from equation (145). They are:

$$K = k^{-(n+1)/(n-1)} \left[\frac{7}{3(n-1)^2} \int_0^z z^\mu \phi(z)\, dz + \frac{n-1}{4(n+1)} z^{(n+1)/(n-1)} \right].$$

† Cf. G. N. Watson, *Treatise on the Theory of Bessel Functions*, § 10.7, p. 345, Cambridge, 1922.

and

$$J = k^{-(n+1)/(n-1)}\left[\frac{7}{(n-1)^2}\int_0^z z^\mu \phi(z)\,dz -\right.$$

$$\left. - \frac{7}{3(n-1)}z^{(n+1)/2(n-1)}\phi(z) + \frac{3(n-1)}{4(n+1)}z^{(n+1)/(n-1)}\right]. \quad (155)$$

For the case $n = 2$ ($\nu = 3.5$ and $\mu = 0.5$) the solutions as given by the foregoing equations have been evaluated numerically. The results are given in Table XXXV. For comparison, the solution for J given by the first approximation (eq. [137]) is also included. It is seen that the second approximation introduces corrections to the extent of 10 per cent.

TABLE XXXV

*The Solution for a Spherical Atmosphere
in which $\kappa\rho \propto r^{-2}$*

z	k^3X	k^3K	k^3J	$\frac{1}{4}z^3$	z	k^3X	k^3K	k^3J	$\frac{1}{4}z^3$
0	0	0	0	0	2·8	2·5081	2·8719	6·1075	5·4880
0·1	0·0002322	0·0001605	0·0002500	0·00025	2·9	2·7048	3·1664	6·7944	6·0973
0·2	0·0018377	0·0012825	0·0020097	0·00200	3·0	2·9068	3·4791	7·5304	6·7500
0·3	0·0061122	0·0043092	0·0068154	0·00675	3·1	3·1137	3·8103	8·3172	7·4478
0·4	0·014237	0·010158	0·016238	0·01600	3·2	3·3254	4·1606	9·1564	8·1920
0·5	0·027265	0·019718	0·031890	0·03125	3·3	3·5414	4·5303	10·0494	8·9843
0·6	0·046111	0·033844	0·055420	0·05400	3·4	3·7617	4·9199	10·9979	9·8260
0·7	0·071559	0·053357	0·088512	0·08575	3·5	3·9858	5·3297	12·0033	10·7188
0·8	0·10426	0·079047	0·13288	0·12800	3·6	4·2136	5·7603	13·0671	11·6640
0·9	0·14475	0·11167	0·19026	0·18225	3·7	4·4448	6·2119	14·1910	12·6633
1·0	0·19345	0·15195	0·26241	0·25000	3·8	4·6792	6·6852	15·3763	13·7180
1·1	0·25067	0·20059	0·35109	0·33275	3·9	4·9165	7·1804	16·6246	14·8298
1·2	0·31661	0·25824	0·45813	0·43200	4·0	5·1565	7·6979	17·9373	16·0000
1·3	0·39146	0·32557	0·58523	0·54925	4·1	5·3991	8·2383	19·3159	17·2303
1·4	0·47522	0·40317	0·73429	0·68600	4·2	5·6440	8·8019	20·7618	18·5220
1·5	0·56792	0·49165	0·90703	0·84375	4·3	5·8911	9·3892	22·2766	19·8768
1·6	0·66950	0·59158	1·10524	1·02400	4·4	6·1401	10·0006	23·8617	21·2960
1·7	0·77985	0·70352	1·33072	1·22825	4·5	6·3909	10·6365	25·5184	22·7813
1·8	0·89876	0·82800	1·58525	1·45800	4·6	6·6434	11·2973	27·2484	24·3340
1·9	1·0261	0·96556	1·8706	1·7148	4·7	6·8975	11·9835	29·0529	25·9558
2·0	1·1616	1·1167	2·1884	2·0000	4·8	7·1528	12·6954	30·9334	27·6480
2·1	1·3051	1·2819	2·5405	2·3153	4·9	7·4095	13·4336	32·8914	29·4123
2·2	1·4563	1·4616	2·9286	2·6620	5·0	7·6672	14·1985	34·9283	31·2500
2·3	1·6148	1·6564	3·3545	3·0418	5·1	7·9260	14·9905	37·0454	33·1628
2·4	1·7805	1·8667	3·8197	3·4560	5·2	8·1856	15·8100	39·2443	35·1520
2·5	1·9529	2·0930	4·3259	3·9063	5·3	8·4461	16·6574	41·5263	37·2193
2·6	2·1319	2·3356	4·8749	4·3940	5·4	8·7072	17·5333	43·8928	39·3660
2·7	2·3171	2·5951	5·4682	4·9208	5·5	8·9690	18·4381	46·3453	41·5938

BIBLIOGRAPHICAL NOTES

§ 88. The method of solution presented in this section is new.

The solution of the same problem for scattering according to the phase function $\varpi_0(1 + x\cos\Theta)$ will be found in Appendix III.

§ 89. This section provides only a first discussion of a problem which requires a much more careful and thorough investigation.

Matters pertaining to the same circle of ideas are dealt with in—

1. G. PLACZEK and W. SEIDEL, *Phys. Rev.* **72**, 550 (1947).
2. G. PLACZEK, ibid. **72**, 556 (1947).
3. J. C. MARK, ibid. **72**, 558 (1947).
4. R. E. MARSHAK, ibid. **71**, 688 (1947).
5. B. DAVISON, ibid. **71**, 694 (1947).
6. J. LeCAINE, ibid. **72**, 564 (1947).

Reference should also be made to the earlier classic investigations of—

7. N. WIENER and E. HOPF, *Sitzungsberichte der Preussischen Akad. Phys.- Math. Klasse*, 696 (1931).
8. E. HOPF, *Mathematical Problems of Radiative Equilibrium* (Cambridge Mathematical Tract, No. 31), Cambridge, 1934.

§ 90. The escape of imprisoned resonance radiation, as a problem in radiative transfer, was first considered by—

9. E. A. MILNE, *Jour. London Math. Soc.* **1**, 40 (1926).

In this paper Milne derives the basic equations (eqs. [69] and [70]), but proceeds, at once, to their discussion in the approximation of Schuster and Schwarzschild (cf. Chap. II, § 20). The general consideration of the mathematical problem, given in the text, is new. However, the solution in the first approximation given in § 90.4 is, essentially, equivalent to what was obtained by Milne.

Discussions of the same physical problem, but from slightly different premisses, are contained in—

10. C. KENTY, *Phys. Rev.* **42**, 823 (1932).
11. T. HOLSTEIN, ibid. **72**, 1213 (1947).

§ 91. The analysis in this section follows:

12. S. CHANDRASEKHAR, *Astrophys. J.* **101**, 95 (1945).

See also—

13. A. B. UNDERHILL, ibid. **107**, 247 (1948).

Earlier considerations of the problem of transfer in atmospheres with spherical symmetry, are those of:

14. W. H. McCREA, *Monthly Notices Roy. Astron. Soc. London*, **88**, 729 (1928).
15. N. A. KOSIREV, ibid. **94**, 430 (1934)
16. S. CHANDRASEKHAR, ibid. **94**, 444 (1934).
17. L. GRATTON, *Soc. Astron. Italiana*, 10, 309 (1937).

Later discussions are:

18. W. BOTHE, *Zeits. f. Physik*, **119**, 493 (1942).
19. R. E. MARSHAK, *Phys. Rev.* **71**, 443 (1947).

20. G. PLACZEK and G. M. VOLKOFF, *Notes on Diffusion of Neutrons without Change of Energy*, National Research Council of Canada, Division of Atomic Energy, MT-4, Chalk River, Ontario (1943).

21. —— —— *Canadian Journal of Research*, A, **25,** 276 (1947).

22. V. A. AMBARZUMIAN, *Bulletin of the Erevan Astronomical Observatory*, No. 6, 3 (1945).

In references 18–22 the problem of transfer in homogeneous spheres is considered and the emphasis is on *infinite* homogeneous spheres.

Particular attention may be drawn to Ambarzumian's paper (ref. 22) in which by a most ingenious set of transformations the problem of an infinite homogeneous sphere with a source at the centre is reduced to one in a plane-parallel medium.

APPENDIX I

92. The exponential integrals

THE nth exponential integral $E_n(x)$ for positive real arguments is defined by

$$E_n(x) = \int_1^\infty e^{-xt}\frac{dt}{t^n} = \int_0^1 e^{-x/\mu}\mu^{n-1}\frac{d\mu}{\mu}. \tag{1}$$

(We shall be concerned with only positive integral values for n.)

The functions $E_n(x)$ defined in this manner satisfy the recursion formulae:

$$nE_{n+1}(x) = e^{-x} - xE_n(x) \quad (n \geqslant 1), \tag{2}$$

$$E'_{n+1}(x) = -E_n(x) \quad (n \geqslant 1) \quad \text{and} \quad E'_1(x) = -e^{-x}/x. \tag{3}$$

The first of these relations follows by writing

$$nE_{n+1}(x) = -\int_1^\infty e^{-xt}\frac{d}{dt}t^{-n}\,dt, \tag{4}$$

and integrating by parts. The second follows by direct differentiation of (1).

A further relation which is useful is

$$E_n(0) = \int_0^1 \frac{dt}{t^n} = \frac{1}{n-1} \quad (n > 1). \tag{5}$$

An integral which occurs frequently is

$$\int_0^x x^l E_n(x)\,dx.$$

By repeated integration by parts and using (3), we obtain

$$\int_0^x x^l E_n(x)\,dx = \frac{x^{l+1}}{l+1}E_n(x) + \frac{x^{l+2}}{(l+1)(l+2)}E_{n-1}(x) + \ldots +$$

$$+ \frac{x^{l+n}}{(l+1)(l+2)\ldots(l+n)}E_1(x) + \frac{1}{(l+1)(l+2)\ldots(l+n)}\int_0^x x^{l+n-1}e^{-x}\,dx, \tag{6}$$

in which the last remaining integral is an elementary one.

According to (2) all exponential integrals can be reduced to the first exponential integral

$$E_1(x) = \int_1^\infty e^{-xt}\frac{dt}{t} = \int_x^\infty e^{-t}\frac{dt}{t}. \tag{7}$$

For large values of x an asymptotic expansion for $E_1(x)$ can be readil found. Thus, by continuing the process

$$E_1(x) = -\int\limits_x^\infty \frac{d}{dt}(e^{-t})\frac{dt}{t} = \frac{e^{-x}}{x} + \int\limits_x^\infty \frac{d}{dt}(e^{-t})\frac{dt}{t^2},$$

we obtain
$$E_1(x) = \frac{e^{-x}}{x}\left[1 - \frac{1}{x} + \frac{2}{x^2} - \frac{6}{x^3} + \cdots\right]. \tag{8}$$

For $x \to 0$, $E_1(x)$ has a logarithmic singularity, and a series expansio appropriate to this case can be found in the following manner:

$$E_1(x) = \int\limits_1^\infty e^{-t}\frac{dt}{t} + \int\limits_x^1 e^{-t}\frac{dt}{t}$$

$$= \left\{\int\limits_1^\infty e^{-t}\frac{dt}{t} - \int\limits_0^1 (1-e^{-t})\frac{dt}{t}\right\} + \int\limits_x^1 \frac{dt}{t} + \int\limits_0^x (1-e^{-t})\frac{dt}{t}. \tag{9}$$

Hence
$$E_1(x) = -\gamma - \log x + \int\limits_0^x (1-e^{-t})\frac{dt}{t}, \tag{10}$$

where $\gamma = 0{\cdot}5772156\ldots$ is the Euler–Mascheroni constant. The integral in (9) has a convergent series expansion for $x \to 0$. We find,

$$E_1(x) = -\gamma - \log x + \sum_{n=1}^\infty (-1)^{n-1}\frac{x^n}{n.n!}. \tag{11}$$

In addition to $E_1(x)$ it is sometimes convenient to introduce, also, the function

$$\text{Ei}(x) = \int\limits_{-\infty}^x e^t\frac{dt}{t} \quad (-\infty < x < +\infty). \tag{12}$$

For negative values of the argument, $\text{Ei}(x)$ is, apart from sign, the same as $E_1(x)$:
$$\text{Ei}(-x) = -E_1(x) \quad (x > 0). \tag{13}$$

On the other hand, for positive values of the argument, the integral (12) must be understood in the sense of the Cauchy principal value

$$\text{Ei}(x) = \lim_{\delta\to 0}\left[\int\limits_\delta^x e^t\frac{dt}{t} + \int\limits_{-\infty}^{-\delta} e^t\frac{dt}{t}\right] \quad (x > 0). \tag{14}$$

3. The functions $F_j(\tau, \mu)$

In Chapter VIII, § 60 (eq. [118]), we introduced the function

$$F_j(\tau, \mu) = \mu \int_0^1 \frac{\mu'^{j-1}}{\mu - \mu'} \left[1 - \exp\left\{-\tau\left(\frac{1}{\mu'} - \frac{1}{\mu}\right)\right\}\right] d\mu', \tag{15}$$

here j is an integer $\geqslant 1$. This function is related to the exponential integral E_j:

First writing $\qquad s = 1/\mu \quad \text{and} \quad s' = 1/\mu', \tag{16}$

ve have the alternative expression

$$F_j(\tau, s) = \int_1^\infty \frac{ds'}{s'^j(s'-s)} [1 - e^{-\tau(s'-s)}]. \tag{17}$$

ince $\qquad \displaystyle\int_0^\tau e^{-t(s'-s)} \, dt = \frac{1}{s'-s}[1 - e^{-\tau(s'-s)}], \tag{18}$

ve can rewrite (17) in the form

$$F_j(\tau, s) = \int_1^\infty \frac{ds'}{s'^j} \int_0^\tau dt \, e^{-t(s'-s)} = \int_0^\tau dt \, e^{ts} \int_1^\infty \frac{ds'}{s'^j} e^{-ts'}. \tag{19}$$

Hence, $\qquad \displaystyle F_j(\tau, \mu) = \int_0^\tau e^{t/\mu} E_j(t) \, dt. \tag{20}$

A recursion formula which enables the reduction of $F_j(\tau, \mu)$ to $F_1(\tau, \mu)$ can be derived from equation (20). Thus, from

$$F_j(\tau, \mu) = \mu \int_0^\tau E_j(t) \frac{d}{dt} e^{t/\mu} \, dt, \tag{21}$$

we obtain after an integration by parts and making use of (3),(5), and (20)

$$F_j(\tau, \mu) = \mu\left[F_{j-1}(\tau, \mu) + e^{\tau/\mu} E_j(\tau) - \frac{1}{j-1}\right]. \tag{22}$$

According to equation (22), it is sufficient to consider only

$$F_1(\tau, \mu) = \int_0^1 \frac{\mu}{\mu - \mu'}\left[1 - \exp\left\{-\tau\left(\frac{1}{\mu'} - \frac{1}{\mu}\right)\right\}\right] d\mu'; \tag{23}$$

however, the cases $\mu \leqslant 0$, $0 < \mu < 1$, and $\mu = 1$ must be distinguished. For $\mu \leqslant 0$ there is no difficulty in reducing (23) to $E_1(\tau)$. We have

$$F_1(\tau, -\mu) = \mu\left[\log\left(1 + \frac{1}{\mu}\right) - e^{-\tau/\mu} E_1(\tau) + E_1\left(\frac{\tau}{\mu} + \tau\right)\right] \quad (\mu > 0). \tag{24}$$

The evaluation of the integral (23) for $0 < \mu < 1$ and $\mu = 1$ requires some care. Thus, considering the case $0 < \mu < 1$, we rewrite the integral (cf. eq. [17])

$$F_1(\tau, s) = \frac{1}{s} \int_1^\infty \left(\frac{1}{s' - s} - \frac{1}{s'} \right) (1 - e^{\tau s - \tau s'}) \, ds' \quad (s > 1), \tag{25}$$

in the form

$$F_1(\tau, s) = \frac{1}{s} \lim_{\delta \to 0} \left[\int_{s+\delta}^\infty + \int_1^{s-\delta} \right] \left(\frac{1}{s' - s} - \frac{1}{s'} \right) (1 - e^{\tau s - \tau s'}) \, ds', \tag{26}$$

and find after some elementary reductions that

$$F_1(\tau, s) = \frac{1}{s} \left[-\log(s-1) + e^{\tau s} E_1(\tau) + \lim_{\delta \to 0} \left(\int_\delta^{s-1} + \int_{-\infty}^{-\delta} \right) e^{\tau y} \frac{dy}{y} \right]. \tag{27}$$

Hence (cf. eq. [14])

$$F_1(\tau, \mu) = \mu \left[-\log \left(\frac{1}{\mu} - 1 \right) + e^{\tau/\mu} E_1(\tau) + \mathrm{Ei} \left(\frac{\tau}{\mu} - \tau \right) \right] \quad (0 < \mu < 1). \tag{28}$$

In like manner, we find that

$$F_1(\tau, 1) = \gamma + \log \tau + e^\tau E_1(\tau). \tag{29}$$

94. The integrals $G_{n,m}(\tau)$ and $G'_{n,m}(\tau)$

In Chapter VIII, § 60.1, we introduced the integrals

$$G_{n,m}(\tau) = \int_1^\infty F_n(\tau, -s) \frac{ds}{s^m}$$

and

$$G'_{n,m}(\tau) = \int_1^\infty e^{-\tau s} F_n(\tau, s) \frac{ds}{s^m}. \tag{30}$$

These integrals can be expressed more conveniently by introducing in them, explicitly, the expression for F_n according to (20). Thus

$$G_{n,m}(\tau) = \int_1^\infty \frac{ds}{s^m} \int_0^\tau dt \, e^{-ts} E_n(t) = \int_0^\tau dt \, E_n(t) \int_1^\infty \frac{ds}{s^m} e^{-ts},$$

or

$$G_{n,m}(\tau) = \int_0^\tau E_n(t) E_m(t) \, dt. \tag{31}$$

Similarly,

$$G'_{n,m}(\tau) = \int_0^\tau E_n(t) E_m(\tau - t) \, dt = \int_0^\tau E_n(\tau - t) E_m(t) \, dt. \tag{32}$$

From (31) and (32) it is apparent that $G_{n,m}(\tau)$ and $G'_{n,m}(\tau)$ are symmetrical in the indices n and m: no loss of generality will therefore be entailed by requiring $m \geqslant n \geqslant 1$.

The function $G_{n,m}(\tau)$ can be expressed directly in terms of the exponential integrals. Thus by (2) and (20)

$$(m-1)G_{n,m}(\tau) = \int_0^\tau E_n(t)[e^{-t}-tE_{m-1}(t)]\, dt$$

$$= F_n(\tau, -1) - \int_0^\tau tE_n(t)E_{m-1}(t)\, dt. \tag{33}$$

Similarly

$$(n-1)G_{n,m}(\tau) = F_m(\tau, -1) - \int_0^\tau tE_m(t)E_{n-1}(t)\, dt. \tag{34}$$

On the other hand, by an integration by parts we have

$$G_{n,m}(\tau) = \tau E_n(\tau)E_m(\tau) + \int_0^\tau tE_n(t)E_{m-1}(t)\, dt + \int_0^\tau tE_{n-1}(t)E_m(t)\, dt. \tag{35}$$

Combining equations (33)–(35), we obtain

$$(m+n-1)G_{n,m}(\tau) = \tau E_n(\tau)E_m(\tau) + F_n(\tau, -1) + F_m(\tau, -1). \tag{36}$$

The functions $G'_{n,m}(\tau)$ cannot be similarly reduced to known functions. However, recursion formulae can be derived which will relate all of them to $G'_{11}(\tau)$. Thus using the recursion formula (22) for the F-functions, we have

$$G'_{n,m}(\tau) = \int_1^\infty \frac{ds}{s^{m+1}}\left[e^{-\tau s}F_{n-1}(\tau, s) - \frac{1}{n-1}e^{-\tau s} + E_n(\tau)\right]$$

or

$$G'_{n,m}(\tau) = G'_{n-1,m+1}(\tau) - \frac{1}{n-1}E_{m+1}(\tau) + \frac{1}{m}E_n(\tau). \tag{37}$$

From the symmetry of $G'_{n,m}$ in the indices, we can conclude from (37) that

$$G'_{n,m}(\tau) = G'_{m-1,n+1}(\tau) - \frac{1}{m-1}E_{n+1}(\tau) + \frac{1}{n}E_m(\tau). \tag{38}$$

A more symmetrical form of these relations is

$$G'_{n,m-1}(\tau) - \frac{E_n(\tau)}{m-1} = G'_{n-1,m}(\tau) - \frac{E_m(\tau)}{n-1}. \tag{39}$$

It remains to consider $G'_{11}(\tau)$. By a somewhat elaborate reduction van de Hulst has shown that

$$G'_{11}(\tau) = 2[E_1(\tau) + (\log \tau + \gamma)E_2(\tau) - \tau E_1^{(2)}(\tau)], \tag{40}$$

where

$$E_1^{(2)}(\tau) = \int_\tau^\infty E_1(t)\frac{dt}{t}. \tag{41}$$

It does not seem that this new function $E_1^{(2)}(\tau)$, can be reduced to the ordinary exponential integrals; but van de Hulst has given the following series expansion which is convenient for the computation of the function for $\tau \leqslant 1$

$$E_1^{(2)}(\tau) = \tfrac{1}{2}(\log\tau+\gamma)^2 + \frac{\pi^2}{12} - \tau + \frac{\tau^2}{2^2 \cdot 2!} - \frac{\tau^3}{3^2 \cdot 3!} + \dots. \tag{42}$$

BIBLIOGRAPHICAL NOTES

§ 92. A complete account of the more classical parts of the theory of the exponential integrals will be found in—

1. N. NIELSEN, *Theorie des Integrallogarithmus*, Leipzig (1906). In this connexion, see also:

2. E. T. WHITTAKER and G. N. WATSON, *Modern Analysis*, 4th ed., chap. xvi, Cambridge, 1927.

3. V. KOURGANOFF, *Annales d'Astrophysique*, **10**, 282, 329 (1947).

The most complete tabulation of the exponential integrals is Placzek's:

4. G. PLACZEK, *The Functions $E_n(x)$*, National Research Council of Canada, Atomic Energy Project, Chalk River, Ontario (1947).

In Placzek's tables the first 20 exponential integrals are tabulated with sufficient accuracy and at close enough intervals for all practical purposes. Of the other tabulations, reference need be made only to—

5. A. HAMMAD, *Philos. Mag.*, Ser. 7, **38**, 515 (1947), where the first five exponential integrals are tabulated to seven decimals and at intervals of 0·01 in the range (0,1).

§ 93. The F-functions were apparently first introduced by L. V. King:

6. L. V. KING, *Philos. Trans. Roy. Soc. London*, A, **212**, 375 (1913).

They have since been used by—

7. A. HAMMAD and S. CHAPMAN, *Philos. Mag.*, Ser. 7, **28**, 99 (1939).

8. H. C. VAN DE HULST, *Astrophys. J.* **107**, 220 (1948).

9. S. CHANDRASEKHAR, ibid. **108**, 92 (1948).

§ 94. The functions $G_{n,m}(\tau)$ and $G'_{n,m}(\tau)$ were introduced and studied by van de Hulst (ref. 8). They occur also in reference 9.

The functions F_j of the odd orders, $j = 1, 3$, and 5, and the integrals $G_{n,m}(\tau)$ and $G'_{n,m}(\tau)$ for $m = 1,\dots, 6$ and $n \leqslant m$ have been tabulated by—

10. S. CHANDRASEKHAR and FRANCES H. BREEN, *Astrophys. J.* **108**, 92 (1948): the Appendix to this paper; S. CHANDRASEKHAR, ibid. **109**, 555 (1949).

The functions F_2 and F_4 were also computed by these authors; but they have not been published.

Finally reference may be made to the following valuable compilation:

11. J. LeCAINE, *A Table of Integrals involving the Functions $E_n(x)$*, National Research Council of Canada, Atomic Energy Project, Chalk River, Ontario (1947).

APPENDIX II

95. A problem in interpolation theory

In Chapter VIII, § 59.1, we encountered the following problem:

To determine two polynomials $s(x)$ and $t(x)$ such that

$$s(x_j) = \lambda_j t(-x_j) \quad \text{and} \quad t(x_j) = \lambda_j s(-x_j) \qquad (j = 1,...,n), \qquad (1)$$

where x_j $(j = 1,...,n)$ are n distinct values of the argument and λ_j $(j = 1,...,n)$ are n assigned numbers all different from one another.

If
$$F(x) = s(x) + t(x) \quad \text{and} \quad G(x) = s(x) - t(x), \qquad (2)$$
then

$$F(x_j) = \lambda_j F(-x_j) \quad \text{and} \quad G(x_j) = -\lambda_j G(-x_j) \qquad (j = 1,...,n). \quad (3)$$

Certain consequences which follow directly from (3) are:

(i) For sufficiently general (x_j) and (λ_j), n is the lowest degree of a polynomial (not identically zero) which will satisfy the conditions of the problem; further, under the conditions stated, F and G are uniquely determined apart from a constant of proportionality.

(ii) Polynomials of degree higher than n can be constructed which will satisfy the conditions (3). Thus if F and G are polynomials of degree n which satisfy the conditions (3), then $aF + b\mu G$ (where a and b are constants) is a polynomial of degree $n+1$ which also satisfies the conditions on F; similarly, $aG + b\mu F$ satisfies the conditions on G.

We shall now obtain explicit formulae for polynomials of degree n which satisfy the conditions of the problem:†

Let
$$F(x) = \sum_{j=0}^{n} a_j x^j. \qquad (4)$$

The conditions on F require

$$\sum_{j=0}^{n} a_j [1 + \lambda_i(-1)^{j+1}] x_i^j = 0 \quad (i = 1,...,n). \qquad (5)$$

The determinant of the $(n+1)$ equations represented by (4) and (5) namely,
$$F(x) = \|b_{i,j}\|$$

---

† The problem considered in this Appendix has been treated by S. Chandrasekhar, *Astrophys. J.* **106**, 152 (1947); see particularly § 4 (pp. 158–65) of this paper. Explicit formulae for F and G are obtained in this paper by a somewhat indirect method; the more direct solution given in the text is due to Dr. H. Kestelman, to whom, and to Dr. H. Davenport, the author is indebted in this context.

where

$$b_{1,j} = x^{j-1} \quad \text{and} \quad b_{i,j} = [1+\lambda_{i-1}(-1)^j]x_{i-1}^{j-1}$$

$$(1 \leqslant j \leqslant n+1; \; 2 \leqslant i \leqslant n+1), \qquad (6$$

will satisfy the required conditions.

Let P denote a permutation of the integers $1, 2,..., n+1$; and let P_i be the 'image' of i. Taking the rows (i) in the order $1, 2,..., n+1$, we shall select the elements $P_1, P_2,..., P_{n+1}$ in the various columns. Then

$$F(x) = \sum_P [P]x^{P_1-1} \prod_{i=2}^{n+1} \{1+\lambda_{i-1}(-1)^{P_i}\}x_{i-1}^{P_i-1}, \qquad (7$$

where $[P]$ is $+1$ or -1 according as the permutation is even or odd.

We shall now group the terms of $F(x)$ according to the factors $\lambda_1,..., \lambda_n$ which they contain. Let L denote any one of the 2^n possible sets of integers chosen from among $1, 2,..., n$ and L' denote the set of integers among $1, 2,..., n$ which are not in L. (L or L' may be empty.) Then, by (7),

$$F(x) = \sum_L \beta_L \prod_{s \in L} \lambda_s \quad \left(\prod_{s \in L} \lambda_s = 1 \text{ if } L = 0\right), \qquad (8)$$

where

$$\beta_L = \sum_P [P]x^{P_1-1} \prod_{i=2}^{n+1} x_{i-1}^{P_i-1} \prod_{i \in L} (-1)^{P_{i+1}}$$

$$= \sum_P [P]x^{P_1-1} \prod_{m \in L'} x_m^{P_{m+1}-1} \prod_{m \in L} x_m^{P_{m+1}-1}(-1)^{P_{m+1}}. \qquad (9)$$

Thus

$$\beta_L = (-1)^l \sum_P [P]x^{P_1-1} \prod_{m=1}^{n} y_m^{P_{m+1}-1}, \qquad (10)$$

where

$$\begin{aligned} y_m &= -x_m \quad \text{if} \quad m \in L \\ &= +x_m \quad \text{if} \quad m \in L', \end{aligned} \Bigg\} \qquad (11)$$

and l is the number of members in L.

Hence

$$\beta_L = (-1)^l \begin{vmatrix} 1 & x & x^2 & . & . & . & x^n \\ 1 & y_1 & y_1^2 & . & . & . & y_1^n \\ . & . & . & . & . & . & . \\ 1 & y_n & y_n^2 & . & . & . & y_n^n \end{vmatrix} = \gamma_L \prod_{j=1}^{n} (x-y_j), \qquad (12)$$

where

$$\gamma_L = (-1)^{n+l} \begin{vmatrix} 1 & y_1 & . & . & . & y_1^{n-1} \\ . & . & . & . & . & . \\ 1 & y_n & . & . & . & y_n^{n-1} \end{vmatrix} = (-1)^{n+l} \prod_{1 \leqslant s < r \leqslant n} (y_r - y_s). \qquad (13)$$

In the product (13) we divide the pairs (r, s) in three classes, according to whether 1, 2, or none of r, s belong to L. Let δ_1 denote the product of those factors $y_r - y_s$ for which (r, s) belong to the first class: i.e. let

$$\delta_1 = \prod_{\substack{r>s \\ r \text{ or } s \in L}} (y_r - y_s) = \prod_{s \in L, r \in L'} (x_r + x_s)\sigma(r, s)$$

$$\text{where} \quad \sigma(r, s) = (r-s)/|r-s|. \quad (14)$$

Let
$$\delta_2 = \prod_{\substack{r \in L, s \in L \\ r>s}} (x_s - x_r) \quad \text{and} \quad \delta_0 = \prod_{\substack{r \in L', s \in L' \\ r>s}} (x_r - x_s). \quad (15)$$

Then
$$\prod_{1 \leqslant s < r \leqslant n} (y_r - y_s) = \delta_1 \delta_2 \delta_0. \quad (16)$$

If L has l members, then δ_2 has $\frac{1}{2}l(l-1)$ factors and

$$\delta_2 \delta_0 = (-1)^{\frac{1}{2}l(l-1)} \prod_{\substack{r>s \\ r \in L \text{ and } s \in L \\ \text{or } r \in L' \text{ and } s \in L'}} (x_r - x_s) = (-1)^{\frac{1}{2}l(l-1)} \frac{\prod_{r>s} (x_r - x_s)}{\prod_{r \in L', s \in L} (x_r - x_s)\sigma(r, s)}. \quad (17)$$

Hence,
$$\delta_1 \delta_2 \delta_0 = (-1)^{\frac{1}{2}l(l-1)} \prod_{i>j} (x_i - x_j) \prod_{r \in L', s \in L} \left(\frac{x_r + x_s}{x_r - x_s}\right). \quad (18)$$

Combining equations (12), (13), (16), and (18) we have

$$\beta_L = (-1)^{\frac{1}{2}l(l-1)} \prod_{i>j} (x_i - x_j) \prod_{s \in L} (x_s + x) \prod_{r \in L'} (x_r - x) \prod_{r \in L', s \in L} \left(\frac{x_r + x_s}{x_r - x_s}\right). \quad (19)$$

Since the function $F(x)$ is determined apart from a constant of proportionality, we shall omit the factor $\prod_{i>j} (x_i - x_j)$ in (19) and choose as the solution

$$F(x) = \sum_L \epsilon_l^{(e)} \prod_{r \in L', s \in L} \left(\frac{x_r + x_s}{x_r - x_s}\right) \prod_{s \in L} \lambda_s(x_s + x) \prod_{r \in L'} (x_r - x), \quad (20)$$

where $\epsilon_l^{(e)}$ denotes the sequence of integers

$$\epsilon_l^{(e)} = +1, (-1)^{n-1}, -1, (-1)^n, +1, (-1)^{n-1}, -1, (-1)^n, \ldots. \quad (21)$$

The analogous solution for G is

$$G(x) = \sum_L \epsilon_l^{(o)} \prod_{r \in L', s \in L} \left(\frac{x_r + x_s}{x_r - x_s}\right) \prod_{s \in L} \lambda_s(x_s + x) \prod_{r \in L'} (x_r - x), \quad (22)$$

where $\epsilon_l^{(o)}$ denotes the sequence

$$\epsilon_l^{(o)} = +1, (-1)^n, -1, (-1)^{n-1}, +1, (-1)^n, -1, (-1)^{n-1}, \ldots. \quad (23)$$

On examining the sequences (21) and (23), we observe that the terms n, $n-2$, etc., in F and G agree, while the terms $n-1$, $n-3$, etc., are of opposite signs. We can, therefore, express $F(x)$ and $G(x)$ in the forms

$$F(x) = C_0(x) + C_1(x) \quad \text{and} \quad G(x) = C_0(x) - C_1(x), \quad (24)$$

where

$$C_0(x) = \sum_{\substack{l=n,n-2,\ldots \\ 2^{n-1} \text{ terms}}} \epsilon_l^{(0)} \prod_{r \in L', s \in L} \left(\frac{x_r + x_s}{x_r - x_s}\right) \prod_{s \in L} \lambda_s(x_s + x) \prod_{r \in L'} (x_r - x) \quad (25$$

and

$$C_1(x) = (-1)^{n-1} \sum_{\substack{l=n-1,n-3,\ldots \\ 2^{n-1} \text{ terms}}} \epsilon_l^{(1)} \prod_{r \in L', s \in L} \left(\frac{x_r + x_s}{x_r - x_s}\right) \prod_{s \in L} \lambda_s(x_s + x) \prod_{r \in L'} (x_r - x),$$

$$(26$$

where

$$\epsilon_l^{(0)} = +1 \quad \text{for integers of the form } n - 4m$$
$$= -1 \quad \text{for integers of the form } n - 4m - 2$$
$$= 0 \quad \text{otherwise,}$$

and

$$\epsilon_l^{(1)} = +1 \quad \text{for integers of the form } n - 4m - 1$$
$$= -1 \quad \text{for integers of the form } n - 4m - 3$$
$$= 0 \quad \text{otherwise.}$$

From equations (3) and (24) it readily follows that

$$C_0(x_j) = \lambda_j C_1(-x_j) \quad \text{and} \quad C_1(x_j) = \lambda_j C_0(-x_j) \quad (j = 1,\ldots,n). \quad (27)$$

Accordingly, returning to equation (1), we can express the general solution for $s(x)$ and $t(x)$ in the form

$$s(x) = q_0 C_0(x) + q_1 C_1(x) \quad \text{and} \quad t(x) = q_1 C_0(x) + q_0 C_1(x), \quad (28)$$

where q_0 and q_1 are arbitrary constants.

APPENDIX III

6. The problem in semi-infinite atmospheres with no incident radiation and for scattering according to the phase function $\varpi_0(1+x\cos\Theta)$

PARTIAL solutions of this problem using very complicated methods have been published.† It is, therefore, of some interest to see how very simply the angular distribution of the emergent radiation for this problem can be found by appealing to a principle of invariance and using the known law of diffuse reflection (Chap. VI, § 46). The principle of the method has already been described in Chapter XIII, § 88, in the context of isotropic scattering.

The equation of transfer appropriate to the problem is

$$\mu\frac{dI(\tau,\mu)}{d\tau} = I(\tau,\mu)-\tfrac{1}{2}\varpi_0\int\limits_{-1}^{+1} I(\tau,\mu')(1+x\mu\mu')\,d\mu'. \tag{1}$$

It is readily verified that this equation admits the integral

$$I(\tau,\mu) = \text{constant}\,\frac{1\pm x(1-\varpi_0)\mu/k}{1\mp k\mu}e^{\pm k\tau}, \tag{2}$$

where k is a root of the transcendental equation‡

$$1 = \varpi_0\int\limits_0^1 \frac{1+x(1-\varpi_0)\mu^2}{1-k^2\mu^2}d\mu$$

$$= \tfrac{1}{2}\varpi_0\left[1+\frac{x(1-\varpi_0)}{k^2}\right]\log\left(\frac{1+k}{1-k}\right)-\frac{1}{k^2}x\varpi_0(1-\varpi_0). \tag{3}$$

We therefore seek a solution of equation (1) which satisfies the boundary condition,
$$I(0,-\mu) = 0 \quad (0\leqslant\mu\leqslant 1), \tag{4}$$

for no incident radiation, and which has the behaviour

$$I(\tau,\mu) \to L_0\frac{1+x(1-\varpi_0)\mu/k}{1-k\mu}e^{k\tau} \quad \text{as} \quad \tau\to\infty. \tag{5}$$

(L_0 is some assigned constant.)

† B. Davison, *Milne Problem in a Multiplying Medium with a Linearly Anisotropic Scattering*, National Research Council of Canada, Atomic Energy Project, Chalk River, Ontario (1946). See also R. E. Marshak, *Phys. Rev.* **72**, 47 (1947).

‡ Cf. S. Chandrasekhar, *Astrophys. J.* **103**, 165 (1946); see particularly § 3 of this paper.

As in Chapter XIII, § 88, we can obtain the solution for the emergent radiation by writing

$$I(\tau,\mu) = L_0 \frac{1+x(1-\varpi_0)\mu/k}{1-k\mu} e^{k\tau} + I^*(\tau,\mu) \qquad (6)$$

and noting that $I^*(\tau,+\mu)$ $(0 \leqslant \mu \leqslant 1)$ must result from the reflection of $I^*(\tau,-\mu)$ $(0 < \mu \leqslant 1)$ by the semi-infinite atmosphere below τ. Thus,

$$I(\tau,+\mu) = L_0 \frac{1+x(1-\varpi_0)\mu/k}{1-k\mu} e^{k\tau} + \frac{\varpi_0}{2\mu} \int_0^1 S^{(0)}(\mu,\mu') I^*(\tau,-\mu')\,d\mu', \qquad (7)$$

where $\varpi_0 S^{(0)}$ is the azimuth independent term in the scattering function (cf. Chap. VI, eq. [43]).

At $\tau = 0$, $\qquad I^*(0,-\mu') = -L_0 \frac{1-x(1-\varpi_0)\mu'/k}{1+k\mu'}$, $\qquad (8)$

and equation (7) becomes

$$I(0,\mu) = L_0 \left\{ \frac{1+x(1-\varpi_0)\mu/k}{1-k\mu} - \frac{\varpi_0}{2\mu} \int_0^1 \frac{S^{(0)}(\mu,\mu')}{1+k\mu'}\left[1 - \frac{x(1-\varpi_0)}{k}\mu'\right] d\mu' \right\}. \qquad (9)$$

The integral over $S^{(0)}$ which occurs in this expression can be evaluated in terms of the known law of diffuse reflection (Chap. VI, § 46). We shall indicate the principal steps in this evaluation.

First rewriting equation (9) in the form

$$I(0,\mu) = L_0 \left\{ \frac{1+x(1-\varpi_0)\mu/k}{1-k\mu} - \left[1 + \frac{x(1-\varpi_0)}{k^2}\right] \frac{\varpi_0}{2\mu} \int_0^1 \frac{S^{(0)}(\mu,\mu')}{1+k\mu'}\,d\mu' + \right.$$

$$\left. + \frac{x(1-\varpi_0)}{k^2} \frac{\varpi_0}{2\mu} \int_0^1 S^{(0)}(\mu,\mu')\,d\mu' \right\}, \qquad (10)$$

we observe that the (second) integral over $S^{(0)}(\mu,\mu')$ can be evaluated directly in terms of known functions (Chap. VI, eqs. [47] and [58]). Next, substituting for $S^{(0)}(\mu,\mu')$ according to Chapter VI, equation (49), we obtain after some elementary reductions that

$$I(0,\mu) = \frac{L_0}{1-k\mu} \left\{ 1 + \frac{x(1-\varpi_0)}{k}\mu + \frac{x(1-\varpi_0)}{k^2}(1-k\mu)[1-qH(\mu)] + \right.$$

$$+ \left[1 + \frac{x(1-\varpi_0)}{k^2}\right]\tfrac{1}{2}\varpi_0\mu H(\mu) \int_0^1 \frac{H(\mu')}{\mu+\mu'}[1-c(\mu+\mu')-x(1-\varpi_0)\mu\mu']\,d\mu' -$$

$$\left. - \left[1 + \frac{x(1-\varpi_0)}{k^2}\right]\tfrac{1}{2}\varpi_0 H(\mu) \int_0^1 \frac{H(\mu')}{1+k\mu'}[1-c(\mu+\mu')-x(1-\varpi_0)\mu\mu']\,d\mu' \right\},$$

$$(11)$$

where it will be recalled that $H(\mu)$ is defined in terms of the characteristic function

$$\Psi(\mu) = \tfrac{1}{2}\varpi_0[1+x(1-\varpi_0)\mu^2]. \tag{12}$$

Now according to Chapter VI, equations (46), (49), and (69),

$$\tfrac{1}{2}\varpi_0\,\mu H(\mu) \int_0^1 \frac{H(\mu')}{\mu+\mu'}[1-c(\mu+\mu')-x(1-\varpi_0)\mu\mu']\,d\mu'$$

$$= \tfrac{1}{2}\varpi_0 \int_0^1 S^{(0)}(\mu,\mu')\frac{d\mu'}{\mu'} = \psi(\mu)-1 = H(\mu)(1-c\mu)-1. \tag{13}$$

Also, rewriting the second integral in (11) in the form

$$\frac{1}{k}\int_0^1 H(\mu')\left\{\frac{c+x(1-\varpi_0)\mu+k(1-c\mu)}{1+k\mu'}-[c+x(1-\varpi_0)\mu]\right\}d\mu'$$

$$= \frac{1}{k}[c+x(1-\varpi_0)\mu+k(1-c\mu)]\int_0^1 \frac{H(\mu')}{1+k\mu'}\,d\mu'-\frac{1}{k}[c+x(1-\varpi_0)\mu]\alpha_0, \tag{14}$$

we can evaluate the integral

$$\int_0^1 \frac{H(\mu')}{1+k\mu'}\,d\mu',$$

according to Chapter VI, equation (53). Using the results of these evaluations in (11), we obtain

$$I(0,\mu) = L_0\frac{H(\mu)}{1-k\mu}\left\{\left[1+\frac{x(1-\varpi_0)}{k^2}\right](1-c\mu)-\frac{x(1-\varpi_0)}{k^2}q(1-k\mu)-\right.$$

$$-\frac{1}{k}\left[c+\frac{x(1-\varpi_0)}{k}\mu\right]\left[1-\frac{1}{H(1/k)}-\frac{x\varpi_0(1-\varpi_0)}{2k}\alpha_1-\tfrac{1}{2}\varpi_0\,\alpha_0\right]-$$

$$\left.-(1-c\mu)\left[1-\frac{1}{H(1/k)}-\frac{x\varpi_0(1-\varpi_0)}{2k}\left(\alpha_1-\frac{\alpha_0}{k}\right)\right]\right\}. \tag{15}$$

The quantity in braces on the right-hand side of this equation simplifies considerably when use is made of Chapter VI, equations (59) and (61). We find:

$$I(0,\mu) = \frac{L_0}{H(1/k)}\frac{H(\mu)}{1-k\mu}\left\{1+\frac{c}{k}+\mu\left[\frac{x(1-\varpi_0)}{k}-c\right]\right\}. \tag{16}$$

An alternative form of this equation is

$$I(0,\mu) = \frac{L_0}{H(1/k)}H(\mu)\left\{\frac{1+x(1-\varpi_0)\mu/k}{1-k\mu}+\frac{c}{k}\right\}. \tag{17}$$

All the functions and constants which are necessary to make this solution determinate are known (Chap. VI, Tables XVI and XVII).

INDEX OF DEFINITIONS

**A CATALOGUE OF SELECTED DOVER BOOKS
IN ALL FIELDS OF INTEREST**

A CATALOGUE OF SELECTED DOVER BOOKS
IN ALL FIELDS OF INTEREST

AMERICA'S OLD MASTERS, James T. Flexner. Four men emerged unexpectedly from provincial 18th century America to leadership in European art: Benjamin West, J. S. Copley, C. R. Peale, Gilbert Stuart. Brilliant coverage of lives and contributions. Revised, 1967 edition. 69 plates. 365pp. of text.

21806-6 Paperbound $3.00

FIRST FLOWERS OF OUR WILDERNESS: AMERICAN PAINTING, THE COLONIAL PERIOD, James T. Flexner. Painters, and regional painting traditions from earliest Colonial times up to the emergence of Copley, West and Peale Sr., Foster, Gustavus Hesselius, Feke, John Smibert and many anonymous painters in the primitive manner. Engaging presentation, with 162 illustrations. xxii + 368pp.

22180-6 Paperbound $3.50

THE LIGHT OF DISTANT SKIES: AMERICAN PAINTING, 1760-1835, James T. Flexner. The great generation of early American painters goes to Europe to learn and to teach: West, Copley, Gilbert Stuart and others. Allston, Trumbull, Morse; also contemporary American painters—primitives, derivatives, academics—who remained in America. 102 illustrations. xiii + 306pp.

22179-2 Paperbound $3.00

A HISTORY OF THE RISE AND PROGRESS OF THE ARTS OF DESIGN IN THE UNITED STATES, William Dunlap. Much the richest mine of information on early American painters, sculptors, architects, engravers, miniaturists, etc. The only source of information for scores of artists, the major primary source for many others. Unabridged reprint of rare original 1834 edition, with new introduction by James T. Flexner, and 394 new illustrations. Edited by Rita Weiss. 6⅝ x 9⅝.

21695-0, 21696-9, 21697-7 Three volumes, Paperbound $13.50

EPOCHS OF CHINESE AND JAPANESE ART, Ernest F. Fenollosa. From primitive Chinese art to the 20th century, thorough history, explanation of every important art period and form, including Japanese woodcuts; main stress on China and Japan, but Tibet, Korea also included. Still unexcelled for its detailed, rich coverage of cultural background, aesthetic elements, diffusion studies, particularly of the historical period. 2nd, 1913 edition. 242 illustrations. lii + 439pp. of text.

20364-6, 20365-4 Two volumes, Paperbound $6.00

THE GENTLE ART OF MAKING ENEMIES, James A. M. Whistler. Greatest wit of his day deflates Oscar Wilde, Ruskin, Swinburne; strikes back at inane critics, exhibitions, art journalism; aesthetics of impressionist revolution in most striking form. Highly readable classic by great painter. Reproduction of edition designed by Whistler. Introduction by Alfred Werner. xxxvi + 334pp.

21875-9 Paperbound $2.50

PLANETS, STARS AND GALAXIES: DESCRIPTIVE ASTRONOMY FOR BEGINNERS, A. E. Fanning. Comprehensive introductory survey of astronomy: the sun, solar system, stars, galaxies, universe, cosmology; up-to-date, including quasars, radio stars, etc. Preface by Prof. Donald Menzel. 24pp. of photographs. 189pp. 5¼ x 8¼.
21680-2 Paperbound $1.50

TEACH YOURSELF CALCULUS, P. Abbott. With a good background in algebra and trig, you can teach yourself calculus with this book. Simple, straightforward introduction to functions of all kinds, integration, differentiation, series, etc. "Students who are beginning to study calculus method will derive great help from this book." Faraday House Journal. 308pp.
20683-1 Clothbound $2.00

TEACH YOURSELF TRIGONOMETRY, P. Abbott. Geometrical foundations, indices and logarithms, ratios, angles, circular measure, etc. are presented in this sound, easy-to-use text. Excellent for the beginner or as a brush up, this text carries the student through the solution of triangles. 204pp.
20682-3 Clothbound $2.00

TEACH YOURSELF ANATOMY, David LeVay. Accurate, inclusive, profusely illustrated account of structure, skeleton, abdomen, muscles, nervous system, glands, brain, reproductive organs, evolution. "Quite the best and most readable account,' Medical Officer. 12 color plates. 164 figures. 311pp. 4¾ x 7.
21651-9 Clothbound $2.50

TEACH YOURSELF PHYSIOLOGY, David LeVay. Anatomical, biochemical bases; digestive, nervous, endocrine systems; metabolism; respiration; muscle; excretion; temperature control; reproduction. "Good elementary exposition," The Lancet. 6 color plates. 44 illustrations. 208pp. 4¼ x 7.
21658-6 Clothbound $2.50

THE FRIENDLY STARS, Martha Evans Martin. Classic has taught naked-eye observation of stars, planets to hundreds of thousands, still not surpassed for charm, lucidity, adequacy. Completely updated by Professor Donald H. Menzel, Harvard Observatory. 25 illustrations. 16 x 30 chart. x + 147pp.
21099-5 Paperbound $1.25

MUSIC OF THE SPHERES: THE MATERIAL UNIVERSE FROM ATOM TO QUASAR, SIMPLY EXPLAINED, Guy Murchie. Extremely broad, brilliantly written popular account begins with the solar system and reaches to dividing line between matter and nonmatter; latest understandings presented with exceptional clarity. Volume One: Planets, stars, galaxies, cosmology, geology, celestial mechanics, latest astronomical discoveries; Volume Two: Matter, atoms, waves, radiation, relativity, chemical action, heat, nuclear energy, quantum theory, music, light, color, probability, antimatter, antigravity, and similar topics. 319 figures. 1967 (second) edition. Total of xx + 644pp.
21809-0, 21810-4 Two volumes, Paperbound $5.00

OLD-TIME SCHOOLS AND SCHOOL BOOKS, Clifton Johnson. Illustrations and rhymes from early primers, abundant quotations from early textbooks, many anecdotes of school life enliven this study of elementary schools from Puritans to middle 19th century. Introduction by Carl Withers. 234 illustrations. xxxiii + 381pp.
21031-6 Paperbound $2.50

VOLUME III, PART I: Variation of solutions, partial differential equations of the second order. Poincaré's theorem, periodic solutions, asymptotic series, wave propagation, Dirichlet's problem in space, Newtonian potential, etc. Translated by Howard G. Bergmann. 15 figures. x + 329pp. 61176-0 Paperbound $3.50

VOLUME III, PART II: Integral equations and calculus of variations: Fredholm's equation, Hilbert-Schmidt theorem, symmetric kernels, Euler's equation, transversals, extreme fields, Weierstrass's theory, etc. Translated by Howard G. Bergmann. Note on Conformal Representation by Paul Montel. 13 figures. xi + 389pp. 61177-9 Paperbound $3.00

ELEMENTARY STATISTICS: WITH APPLICATIONS IN MEDICINE AND THE BIOLOGICAL SCIENCES, Frederick E. Croxton. Presentation of all fundamental techniques and methods of elementary statistics assuming average knowledge of mathematics only. Useful to readers in all fields, but many examples drawn from characteristic data in medicine and biological sciences. vii + 376pp. 60506-X Paperbound $2.50

ELEMENTS OF THE THEORY OF FUNCTIONS. A general background text that explores complex numbers, linear functions, sets and sequences, conformal mapping. Detailed proofs. Translated by Frederick Bagemihl. 140pp. 60154-4 Paperbound $1.50

THEORY OF FUNCTIONS, PART I. Provides full demonstrations, rigorously set forth, of the general foundations of the theory: integral theorems, series, the expansion of analytic functions. Translated by Federick Bagemihl. vii + 146pp. 60156-0 Paperbound $1.50

INTRODUCTION TO THE THEORY OF FOURIER'S SERIES AND INTEGRALS, Horatio S. Carslaw. A basic introduction to the theory of infinite series and integrals, with special reference to Fourier's series and integrals. Based on the classic Riemann integral and dealing with only ordinary functions, this is an important class text. 84 examples. xiii + 368pp. 60048-3 Paperbound $3.00

AN INTRODUCTION TO FOURIER METHODS AND THE LAPLACE TRANSFORMATION, Philip Franklin. Introductory study of theory and applications of Fourier series and Laplace transforms, for engineers, physicists, applied mathematicians, physical science teachers and students. Only a previous knowledge of elementary calculus is assumed. Methods are related to physical problems in heat flow, vibrations, eletcrical transmission, electromagnetic radiation, etc. 828 problems with answers. Formerly Fourier Methods. x + 289pp. 60452-7 Paperbound $2.75

INFINITE SEQUENCES AND SERIES, Konrad Knopp. Careful presentation of fundamentals of the theory by one of the finest modern expositors of higher mathematics. Covers functions of real and complex variables, arbitrary and null sequences, convergence and divergence. Cauchy's limit theorem, tests for infinite series, power series, numerical and closed evaluation of series. Translated by Frederick Bagemihl. v + 186pp. 60153-6 Paperbound $2.00

AN ELEMENTARY INTRODUCTION TO THE THEORY OF PROBABILITY, B. V. Gnedenko and A. Ya. Khinchin. Introduction to facts and principles of probability theory. Extremely thorough within its range. Mathematics employed held to elementary level. Excellent, highly accurate layman's introduction. Translated from the fifth Russian edition by Leo Y. Boron. xii + 130pp.
60155-2 Paperbound $2.00

SELECTED PAPERS ON NOISE AND STOCHASTIC PROCESSES, edited by Nelson Wax. Six papers which serve as an introduction to advanced noise theory and fluctuation phenomena, or as a reference tool for electrical engineers whose work involves noise characteristics, Brownian motion, statistical mechanics. Papers are by Chandrasekhar, Doob, Kac, Ming, Ornstein, Rice, and Uhlenbeck. Exact facsimile of the papers as they appeared in scientific journals. 19 figures. v + 337pp. 6⅛ x 9¼.
60262-1 Paperbound $3.50

STATISTICS MANUAL, Edwin L. Crow, Frances A. Davis and Margaret W. Maxfield. Comprehensive, practical collection of classical and modern methods of making statistical inferences, prepared by U. S. Naval Ordnance Test Station. Formulae, explanations, methods of application are given, with stress on use. Basic knowledge of statistics is assumed. 21 tables, 11 charts, 95 illustrations. xvii + 288pp.
60599-X Paperbound $2.50

MATHEMATICAL FOUNDATIONS OF INFORMATION THEORY, A. I. Khinchin. Comprehensive introduction to work of Shannon, McMillan, Feinstein and Khinchin, placing these investigations on a rigorous mathematical basis. Covers entropy concept in probability theory, uniqueness theorem, Shannon's inequality, ergodic sources, the E property, martingale concept, noise, Feinstein's fundamental lemma, Shanon's first and second theorems. Translated by R. A. Silverman and M. D. Friedman. iii + 120pp.
60434-9 Paperbound $1.75

INTRODUCTION TO SYMBOLIC LOGIC AND ITS APPLICATION, Rudolf Carnap. Clear, comprehensive, rigorous introduction. Analysis of several logical languages. Investigation of applications to physics, mathematics, similar areas. Translated by Wiliam H. Meyer and John Wilkinson. xiv + 214pp.
60453-5 Paperbound $2.50

SYMBOLIC LOGIC, Clarence I. Lewis and Cooper H. Langford. Probably the most cited book in the literature, with much material not otherwise obtainable. Paradoxes, logic of extensions and intensions, converse substitution, matrix system, strict limitations, existence of terms, truth value systems, similar material. vii + 518pp.
60170-6 Paperbound $4.50

VECTOR AND TENSOR ANALYSIS, George E. Hay. Clear introduction; starts with simple definitions, finishes with mastery of oriented Cartesian vectors, Christoffel symbols, solenoidal tensors, and pplications. Many worked problems show applications. 66 figures. viii + 193pp.
60109-9 Paperbound $2.50

EAST O' THE SUN AND WEST O' THE MOON, George W. Dasent. Considered the best of all translations of these Norwegian folk tales, this collection has been enjoyed by generations of children (and folklorists too). Includes True and Untrue, Why the Sea is Salt, East O' the Sun and West O' the Moon, Why the Bear is Stumpy-Tailed, Boots and the Troll, The Cock and the Hen, Rich Peter the Pedlar, and 52 more. The only edition with all 59 tales. 77 illustrations by Erik Werenskiold and Theodor Kittelsen. xv + 418pp. 22521-6 Paperbound $3.50

GOOPS AND HOW TO BE THEM, Gelett Burgess. Classic of tongue-in-cheek humor, masquerading as etiquette book. 87 verses, twice as many cartoons, show mischievous Goops as they demonstrate to children virtues of table manners, neatness, courtesy, etc. Favorite for generations. viii + 88pp. 6½ x 9¼.
22233-0 Paperbound $1.25

ALICE'S ADVENTURES UNDER GROUND, Lewis Carroll. The first version, quite different from the final Alice in Wonderland, printed out by Carroll himself with his own illustrations. Complete facsimile of the "million dollar" manuscript Carroll gave to Alice Liddell in 1864. Introduction by Martin Gardner. viii + 96pp. Title and dedication pages in color. 21482-6 Paperbound $1.25

THE BROWNIES, THEIR BOOK, Palmer Cox. Small as mice, cunning as foxes, exuberant and full of mischief, the Brownies go to the zoo, toy shop, seashore, circus, etc., in 24 verse adventures and 266 illustrations. Long a favorite, since their first appearance in St. Nicholas Magazine. xi + 144pp. 6⅝ x 9¼.
21265-3 Paperbound $1.75

SONGS OF CHILDHOOD, Walter De La Mare. Published (under the pseudonym Walter Ramal) when De La Mare was only 29, this charming collection has long been a favorite children's book. A facsimile of the first edition in paper, the 47 poems capture the simplicity of the nursery rhyme and the ballad, including such lyrics as I Met Eve, Tartary, The Silver Penny. vii + 106pp. 21972-0 Paperbound $1.25

THE COMPLETE NONSENSE OF EDWARD LEAR, Edward Lear. The finest 19th-century humorist-cartoonist in full: all nonsense limericks, zany alphabets, Owl and Pussycat, songs, nonsense botany, and more than 500 illustrations by Lear himself. Edited by Holbrook Jackson. xxix + 287pp. (USO) 20167-8 Paperbound $2.00

BILLY WHISKERS: THE AUTOBIOGRAPHY OF A GOAT, Frances Trego Montgomery. A favorite of children since the early 20th century, here are the escapades of that rambunctious, irresistible and mischievous goat—Billy Whiskers. Much in the spirit of Peck's Bad Boy, this is a book that children never tire of reading or hearing. All the original familiar illustrations by W. H. Fry are included: 6 color plates, 18 black and white drawings. 159pp. 22345-0 Paperbound $2.00

MOTHER GOOSE MELODIES. Faithful republication of the fabulously rare Munroe and Francis "copyright 1833" Boston edition—the most important Mother Goose collection, usually referred to as the "original." Familiar rhymes plus many rare ones, with wonderful old woodcut illustrations. Edited by E. F. Bleiler. 128pp. 4½ x 6⅜. 22577-1 Paperbound $1.25

MATHEMATICAL PUZZLES FOR BEGINNERS AND ENTHUSIASTS, Geoffrey Mott-Smith. 189 puzzles from easy to difficult—involving arithmetic, logic, algebra, properties of digits, probability, etc.—for enjoyment and mental stimulus. Explanation of mathematical principles behind the puzzles. 135 illustrations. viii + 248pp.
20198-8 Paperbound $1.75

PAPER FOLDING FOR BEGINNERS, William D. Murray and Francis J. Rigney. Easiest book on the market, clearest instructions on making interesting, beautiful origami. Sail boats, cups, roosters, frogs that move legs, bonbon boxes, standing birds, etc. 40 projects; more than 275 diagrams and photographs. 94pp.
20713-7 Paperbound $1.00

TRICKS AND GAMES ON THE POOL TABLE, Fred Herrmann. 79 tricks and games— some solitaires, some for two or more players, some competitive games—to entertain you between formal games. Mystifying shots and throws, unusual catoms, tricks involving such props as cork, coins, a hat, etc. Formerly *Fun on the Pool Table*. 77 figures. 95pp.
21814-7 Paperbound $1.00

HAND SHADOWS TO BE THROWN UPON THE WALL: A SERIES OF NOVEL AND AMUSING FIGURES FORMED BY THE HAND, Henry Bursill. Delightful picturebook from great-grandfather's day shows how to make 18 different hand shadows: a bird that flies, duck that quacks, dog that wags his tail, camel, goose, deer, boy, turtle, etc. Only book of its sort. vi + 33pp. 6½ x 9¼.
21779-5 Paperbound $1.00

WHITTLING AND WOODCARVING, E. J. Tangerman. 18th printing of best book on market. "If you can cut a potato you can carve" toys and puzzles, chains, chessmen, caricatures, masks, frames, woodcut blocks, surface patterns, much more. Information on tools, woods, techniques. Also goes into serious wood sculpture from Middle Ages to present, East and West. 464 photos, figures. x + 293pp.
20965-2 Paperbound $2.00

HISTORY OF PHILOSOPHY, Julián Marias. Possibly the clearest, most easily followed, best planned, most useful one-volume history of philosophy on the market; neither skimpy nor overfull. Full details on system of every major philosopher and dozens of less important thinkers from pre-Socratics up to Existentialism and later. Strong on many European figures usually omitted. Has gone through dozens of editions in Europe. 1966 edition, translated by Stanley Appelbaum and Clarence Strowbridge. xviii + 505pp.
21739-6 Paperbound $3.00

YOGA: A SCIENTIFIC EVALUATION, Kovoor T. Behanan. Scientific but non-technical study of physiological results of yoga exercises; done under auspices of Yale U. Relations to Indian thought, to psychoanalysis, etc. 16 photos. xxiii + 270pp.
20505-3 Paperbound $2.50

Prices subject to change without notice.
Available at your book dealer or write for free catalogue to Dept. GI, Dover Publications, Inc., 180 Varick St., N. Y., N. Y. 10014. Dover publishes more than 150 books each year on science, elementary and advanced mathematics, biology, music, art, literary history, social sciences and other areas.